Neural Membranes

Experimental and Clinical Neuroscience

Neural Membranes, edited by **Grace Y. Sun, Nicolas Bazan, Jang-Yen Wu, Guiseppe Porcellati, and Albert Y. Sun,** 1983

Neural Membranes

Edited by

Grace Y. Sun, Nicolas Bazan, Jang-Yen Wu, Guiseppe Porcellati, and Albert Y. Sun

Humana Press • Clifton, New Jersey

Acknowledgments

This volume contains the proceedings of the International Symposium on Neural Membranes that was held at National Yang Ming Medical College, Taipei, Taiwan, ROC, August 2–5, 1982. The organizing committee is indebted to the following organizations for their enthusiastic support for the meeting, without which the symposium would not have been possible:

The National Science Council, Taiwan, ROC
National Yang Ming Medical College, Taipei, Taiwan, ROC
The Ching-Ling Medical Foundation, Veterans General Hospital, Taipei, Taiwan, ROC
Department of Education, Taiwan, ROC
Fidia Research Laboratories, Abano Terme, Italy
The Humana Press, Clifton, New Jersey, USA

Special thanks are also due to President Paul Han, National Yang Ming Medical College, for his enthusiasm and encouragement, and Dr. L. Foudin and Ms. D. Torres, for their faithful assistance in preparation of the manuscripts.

Library of Congress Cataloging in Publication Data

International Symposium on Neural Membranes (1982 :
Kuo li Yang-ming i hsüeh yüan)
Neural membranes.

(Experimental and clinical neuroscience)
Includes index.
1. Plasma membranes—Congresses. 2. Neurons—Congresses. I. Sun, Grace Y. II. Title. III. Series.
[DNLM: 1. Cell membrane—Congresses. 2. Neurons—Congresses. WL 102.5 I609n 1982]
QP363.I546 1982 599.01'88 83-6106
ISBN 0-89603-052-0

Crescent Manor
PO Box 2148
Clifton, NJ 07015

Printed in the United States of America

Preface

One of the most active and productive areas of biological science in the past decade has been the study of the biochemical and biophysical properties of cell membranes. There is little doubt that membranes are essential components of all cellular systems and that each type of membrane manifests specific and characteristic cellular functions. In the nervous system, important events such as neurotransmission, receptor binding, ion transport, axonal transport, and cell uptake are all known to take place within the neural cell membrane.

Phospholipids, one of the major components of membranes, not only provide the membrane with its structural integrity and physical properties, but also play an important role in regulating membrane function. Attention has recently been focused on the asymmetric localization of these molecules, the identification of discrete metabolic pools of phospholipids within the membrane matrix, and their involvement in signal transmission. Although synaptic membranes generally lack an active mechanism for the *de novo* biosynthesis of phospholipids, a number of enzymic routes are present for their interconversions and for facilitating metabolic turnover. Metabolites generated during the interconversion reactions may also exert a great influence in modulating membrane functions. The phosphogylcerides of neural membranes are especially enriched in polyunsaturated fatty acids. However, only very small amounts of these fatty acids are present in the free form, and they are maintained in dynamic equilibrium with the membrane phospholipids. In this volume, some aspects of the metabolism of arachidonoyl and docosahexaenoyl groups are described, and correlations are made with respect to the effect on brain edema and various forms of neuronal stimulation.

Studies of chemical transmission have greatly progressed to include examining enzymes involved in transmitter synthesis and degradation, modulation of transmitter binding activity, and synthesis of coated vesicles for the recycling process. Localization and identification of specific transmitter enzymes in neural membranes, e.g., L-glutamate decarboxylase (GAD), has been made possible by immunocytochemical techniques.

The horizontal cells in the vertebrate retina can provide an ideal model system for studying neurotransmitter uptake and release. New procedures for the isolation and characterization of dopamine-binding proteins from synaptosomes are described. The binding of GABA and benzodiazepine to synaptic membranes is a process easily affected by the membrane lipid environment. Therefore, treatment of the membrane with phospholipase A_2 and phospholipase C can induce an increase in the binding activity. On the other hand, post-synaptic serotonin binding activity is modulated by agents affecting the lipid peroxidation of the membrane.

A number of new biochemical and biophysical techniques are described for the study of membrane structure and function. Some valuable correlations have been made using agents such as ethanol and divalent metal ions. Understanding the properties of neural membranes may also be aided by studying specific enzymes, such as cholineacetyl transferase and amino enkephalinase, or membrane proteins, such as the myelin basic protein. A chapter is included that describes new insights into the excitation properties of the nerve membrane.

Many other membrane components are also important in regulating the structure and functions of neural membranes. This pertains especially to gangliosides, which are more abundant in neural membranes (CNS and PNS) than in nonneural membranes. Differences in the distribution of gangliosides are found among different types of neural membranes. In this volume, some new roles for gangliosides with respect to membrane functions are revealed; e.g., their involvements in neuritogenesis, in the regulation of neuronal development and regeneration, in oncogenesis, and in the internalization of choleragen are described.

It is apparent that this volume cannot possibly cover the entire spectrum of current research on neural membranes. In fact, we are only beginning to realize the complexity underlying neural membrane structure and functions. The information provided is therefore only a foretaste of the "richness" of this scientific area for future exploration.

Grace Y. Sun
Nicolas Bazan
Jang-Yen Wu
Guiseppe Porcellati
Albert Y. Sun

Contents

Part II: Receptors

Part III: Membrane Structure and Functions

List of Participants

Ata Abdel-Latif • *Medical College of Georgia, Augusta, Georgia*
Nicolas G. Bazan • *Louisiana State University Medical Center, New Orleans, Louisiana*
L. R. Chang • *National Defense Medical College, Taipei, Taiwan R.O.C.*
W. C. Chang • *National Taiwan University and Academia Sinica, Taipei, Taiwan R.O.C.*
Li-Pen Chao • *UCLA School of Medicine, Los Angeles, California*
Sylvia Fang Chen • *University of California School of Medicine, San Francisco, California*
C. Chiang • *Academia Sinica, Taipei, Taiwan R.O.C.*
Joel A. Dain • *University of Rhode Island, Kingston, Rhode Island*
Joseph Eichberg • *University of Houston, Houston, Texas*
H. C. Ha • *National Yang Ming Medical College, Taipei, Taiwan R.O.C.*
George Hauser • *Harvard Medical School, Belmont, Massachusetts*
Paul W. Han • *National Yang Ming Medical College, Taipei, Taiwan R.O.C.*
Lloyd A. Horrocks • *The Ohio State University, Columbus, Ohio*
Louise L. Hsu • *University of Texas Medical Branch, Galveston, Texas*
Koon-Sea Hui • *New York State Research Institute for Neurochemistry and Drug Addiction, Ward's Island, New York*
Ingming Jeng • *Missouri Institute of Psychiatry, St. Louis, Missouri*
T. C. Jou • *National Yang Ming Medical College, Taipei, Taiwan R.O.C.*
Firoze B. Jungalwala • *E. K. Shriver Center for Mental Retardation, Waltham, Massachusetts*
L. W. Ko • *National Yang Ming Medical College, Taipei, Taiwan R.O.C.*
Kinya Kuriyama • *Kyoto Prefectural University of Medicine, Kyoto, Japan*
James C. K. Lai • *The Burke Rehabilitation Center, White Plains, New York*
Dominic Man-Kit Lam • *Baylor College of Medicine, Houston, Texas*
Robert Ledeen • *Albert Einstein College of Medicine, Bronx, New York*
S. K. Li • *National Defense Medical College, Taipei, Taiwan, R.O.C.*
Yu-Teh Li • *Tulane University School of Medicine, New Orleans, Louisiana*
M. T. Lin • *National Defense Medical College, Taipei, Taiwan R.O.C.*
S.-Y. Lin-Shiau • *National Taiwan University, Taipei, Taiwan R.O.C.*
T. B. Lo • *National Taiwan University, Taipei, Taiwan R.O.C.*
Guiseppe Porcellati • *University of Perugia, 06110 Perugia, Italy*
Jean Shih • *University of Southern California, Los Angeles, California*
Albert Y. Sun • *University of Missouri, Columbia, Missouri*
Grace Y. Sun • *University of Missouri, Columbia, Missouri*
Gino Toffano • *Fidia Research Laboratories, 35031 Abano Terme, Italy*

KEIZO WAKU • *Teikyo University, Sagamiko, Tsukuigun, Kanagawa 199-01, Japan*
C-M WAN • *Academia Sinica, Taipei, Taiwan R.O.C.*
JANG Y. WU • *Baylor College of Medicine, Houston, Texas*
ROBERT YU • *Yale University School of Medicine, New Haven, Connecticut*

Part I: Membrane Phospholipids

PHOSPHOLIPID METABOLISM IN NEURAL MEMBRANES

Giuseppe Porcellati

Department of Biochemistry, The Medical School, Via del Giochetto, 06100 Perugia, Italy

INTRODUCTION

Phospholipids are found in high concentrations in the nervous tissue. They normally account for about 25% of the dry weight of the adult brain. Their distribution differs qualitatively and quantitatively in various membrane structures. The relative proportion of their subclasses and molecular species evidently shows that they are maintained at a steady state which is typical for the cellular and subcellular membrane components to be considered.

It is well known that membrane components, particularly phospholipids, are continuously being renewed. Various investigations on phospholipid metabolism in neural membranes have indicated that different phospholipids turn over at different rates with respect to the structure of the lipid and their localization in different cells and membranes. The renewal is obviously due to catabolic and biosynthetic reactions which underline, as a final task, the compositional maintenance of the membrane and consequently its functional integrity.

The metabolism of the phospholipids in membrane is regulated by the enzymes involved in their synthesis and catabolism, during the transport from their sites of formation to other membranes, and during the incorporation of the lipids into the membrane structure. Concurrently with the synthesis de novo of phospholipid molecules, several reactions, which lead to the "interconversion" of one molecule into another, take place. These reactions are connected with the turnover and rearrangement of membrane phospholipids, by redistributing fatty acids and bases. In addition, they are responsible for the turnover of different portions of the phospholipid molecules.

Membrane function requires a certain physical state of its structure. Membrane fluidity (including neural membrane) is a property which very often can influence processes occurring at the membrane level (enzymic reactions, receptor activity, and transport). Membrane fluidity and functions may depend on several factors including the structure and the topological distribution of various lipid species within the lipid bilayer. Probably, the membrane operates differently on the two sides it separates, and therefore it should be asymmetric; i.e., its outer surface must be chemically different from the inner one. How this asymmetry arises in neural and other membranes is not known. This may result either from an uneven distribution of the enzymes of lipid synthesis on both sides of the membrane, or a trans-bilayer movement of phospholipids must occur after synthesis.

This contribution will provide explanations on some aspects of these interrelated, yet independent subjects: a) composition of lipid molecular species of neural membranes; b) dynamics of lipid synthesis; c) interconversion reactions in membrane and relation to membrane structure; d) phospholipid and enzyme asymmetry; e) lipid modification in neural membranes and relationship to "synaptic function."

COMPOSITION OF LIPID MOLECULAR SPECIES OF NEURAL MEMBRANES

Data on neural membrane composition have often suffered in the past from the lack of assessment of their purity. However, more precise methods are now available which warrant a better purification of cell types or membranes and therefore provide more useful information (Smith and Loh, 1979; Fontaine et al., 1980; McCarthy and de Vellis, 1980; Farooq et al., 1981). Another important problem is to achieve a good ultrastructural preservation of intracellular organelles for both biochemical and cytochemical investigations. In recently developed procedures, as those in use for the isolation of neurons or various types of glia, in bulk form or in culture (see Arienti et al., 1981 for review), the amount of intact membranes as well as the presence of relationships among different cells represent features suggestive of potential viability and therefore indicates improvements over previous data.

The purity of synaptic plasma membrane (SPM), for instance, has been rather extensively examined in recent years by several groups of workers (Smith and Loh, 1977; Fontaine et al., 1980; Butler and Abood, 1982; Porcellati, unpublished results). Examples of biochemical characterization are given in Table I. The SPM fraction appears to be enriched about 4- to 5-fold with respect to homogenate. This result is satisfactory considering that about 20% of the brain homogenate is calculated to be neuronal plasma membrane (Fontaine et al., 1980). Na^+,K^+-ATPase (EC 3.6.1.4), N-acetylglucosaminidase (EC 3.2.1.30) and 5'-nucleotidase (EC 3.1.3.5) are thought to be reliable positive enzyme markers for SPM (Smith and Loh, 1977; Baker, 1979; Fontaine et al., 1980; Butler and Abood, 1982; Porcellati, unpublished results). However, recently it has been claimed that 5'-nucleotidase in brain is largely restricted to glial membranes and myelin (Kreutzberg et al., 1978; Cammer et al., 1980).

TABLE I

Characterization of Synaptic Plasma Membrane from Mouse and Rat Synaptosomes

Fraction	ATPase	5'-nucl.	Ouabain binding	SDH	NADPH-cyt c OR
Homogenate[a]	0.073		1.63	10.6	4.88
SPM[a]	0.284		5.41	4.07	0.55
Homogenate[b]	0.14	0.008		30.0	
SPM[b]	0.75	0.032		10.0	
Homogenate[c]	0.051			9.11	5.23
SPM[c]	0.189			2.62	0.41

[a]Fontaine et al. (1980), mouse; [b]Smith and Loh (1977), mouse; [c]Porcellati (unpublished results), rat. Enzyme activities measured as umol/min/mg protein, except for SDH and NADPH-cyt c OR (nmol/min/mg protein. Deviation values and number of observations have been omitted for clarity. SPM = synaptic membranes; ATPase = Na^+,K^+-ATPase; 5'nucl. = 5'-nucleotidase; SDH = succinate cytochrome c oxido-reductase; NADPH-cyt c OR = NADPH cytochrome c oxido-reductase. Oubain binding measured as pmol/mg protein.

Another interesting neural membrane recently separated from CNS myelinated axons is represented by the axolemma (DeVries et al., 1981a,b). In this respect, saxitoxin binding (a marker for sodium channel) appears to be a reliable positive marker for this type of membrane.

A large number of contributions on lipid class and subclass distribution in neural membranes has appeared. In addition, the main fatty acid residues in choline glycerophosphatides (CGP), ethanolamine glycerophosphatides (EGP) and serine glycerophosphatides (SGP) have been analyzed (Hamberger and Svennerholm, 1971; Corbin and Sun, 1978; Baker, 1979; Fontaine et al., 1980; Butler and Abood, 1982; Porcellati, unpublished results). On the whole, few distinctive differences seem to exist for each acyl group among different membrane types. This applies to the CGP 16:0 content, whose percentage over total fatty acids increases from neuronal to glial to synaptic membranes, while those of CGP 18:1 and 20:4(n-6) decrease in the same order. With regard to EGP (both diacyl- and alkenylacyl-subtypes), the percentages of 16:0, 18:0, 18:1 and 22:6(n-3) normally increase in the mentioned order, while those of 20:4(n-6) and 22:4(n-6) decrease. Finally, the main differences of SGP fatty acyl residues are represented by 16:0 and 20:4(n-6), whose percentages decrease with a concomitant increase of the 22:6(n-3) content.

More information is obtained regarding the acyl group composition of individual phosphoglycerides in a given cellular component. Results from Corbin and Sun (1978) have in fact indicated that a distinctively different acyl group profile for SGP and inositol glycerophospholipids (IGP) is present in mouse brain synaptosomes. The major polyunsaturated fatty acid in IGP is in fact 20:4(n-6), whereas that in SGP is 22:6(n-3). Moreover, 16:0 is most abundant in CGP, whereas 18:0 is predominantly present in EGP (diacyl-type), IGP and SGP, and 22:6(n-3) is also most abundant in the alkenylacyl-type of EGP (Corbin and Sun, 1978).

A comparative study of molecular species distribution of CGP, EGP and SGP in SPM and myelin from cerebral cortex of adult rats has been recently carried out (Butler and Abood, 1982). On the whole, SPM contain more polyunsaturated fatty acids than myelin, but more interesting differences are shown looking at the various lipid classes. Table II shows that EGP and SGP contain most of the 22:6(n-3) in both myelin and SPM, while CGP contain more saturated fatty acids. The 16:0 CGP content increases from myelin to SPM, while the 18:0 content of the same lipid decreases. As known, the 18:1 content of all the three lipid classes noticeably decreases from myelin to SPM, while that of 20:4(n-6) increases in EGP and decreases in SGP. Strikingly, a very large increase from myelin to SPM takes place in the EGP and SGP 22:6(n-3) content.

TABLE II

Fatty Acid Composition of Major Phospholipids in Myelin and Synaptic Membranes from Adult Rat Cerebral Cortex[a]

Lipid	Source	16:0	18:0	18:1	20:4	22:6	Other
CGP	Myelin	39.1	15.8[b]	36.1	3.3	1.9	3.8
	SPM	56.7	10.9	22.1	4.4	2.6	2.4
EGP	Myelin	7.6	23.2[b]	31.8	8.6	9.9	18.9
	SPM	9.4	29.9	10.7	12.8	23.5	13.7
SGP	Myelin	2.8	43.3[b]	31.5	5.6	10.8	6.0
	SPM	3.2	44.8	14.0	2.5	29.7	5.8

[a]From Butler and Abood (1982). Data are molar % of the total fatty acid and are expressed as mean values. Standard deviation and number of observations have been omitted for clarity. "Other" includes 16:0, 18:0 and 18:1 fatty aldehydes, and 16:1, 18:1, 20:0, 20:1, 20:3, 22:4(n-6) and 22:5(n-6) fatty acids.

[b]Includes 18:1 fatty aldehyde.

Exactly what the differences in acyl group composition might mean in terms of functional implication is hard to state. Most probably, a certain microenvironment or restricted domain of lipid molecular species in specialized neural membranes may fulfill some functional requirements. As a tentative hypothesis, the lipid environment surrounding the receptor could regulate the receptor binding activity, and/or can modulate receptor function after the interaction between the receptor and the ligand. It is worth mentioning, in this connection, that the particular richness of polyunsaturated fatty acids such as 20:4(n-6), 22:4(n-6) and 22:6(n-3), as found in SPM (Corbin and Sun, 1978; Butler and Abood, 1982), may lead to a high degree of fluidity necessary for neurotransmitter activity.

DYNAMICS OF LIPID SYNTHESIS

The bulk of CGP and EGP of neural membranes is synthesized by the cytidine pathway. The membranes of endoplasmic reticulum represent the principal site of synthesis (Porcellati et al., 1970; Ansell, 1973). The phosphorylcholine cytidylyltransferase (EC 2.7.7.15) and the phosphorylethanolamine cytidylyltransferase (EC 2.7.7.14), which convert respectively phosphorylcholine (PC) and phosphorylethanolamine (PE) into CDP-choline and CDP-ethanolamine, are rate-limiting (Porcellati, 1972). PC and PE are then transferred from CDP-choline and CDP-ethanolamine to diacylglycerol (DG) by cholinephosphotransferase (EC 2.7.8.2) and ethanolaminephosphotransferase (EC 2.7.8.1), producing respectively CGP or EGP and cytidine-5'-monophosphate (CMP). The first reaction and probably the second are reversible in brain membranes (Goracci et al., 1981). DG can be formed from glycerol phosphate or dihydroxyacetone phosphate, and phosphatidic acid (PA) is its direct precursor.

The occurrence of the cytidine pathway for CGP and EGP synthesis has also been observed in neuronal and glial homogenate and membranes (Binaglia et al., 1973). Table III indicates that a much higher activity occurs in neuronal cells than in glial cells for both syntheses. Subsequent results have appeared in this connection (Strosznajder et al., 1977).

A substantial activity of cholinephosphotransferase (CPT) and ethanolaminephosphotransferase (EPT) has been detected in rat brain synaptosomes (Strosznajder et al., 1979). Synaptosomal enzyme activities are lower but comparable with similar transferases of neuronal perikarya (Binaglia et al., 1973) and possess the same kinetic properties which have been reported for brain microsomal membranes (Porcellati et al., 1970; Ansell and Metcalfe, 1971; Radominska-Pyrek and Horrocks, 1972). The occurrence of other enzymic activities connected with phospholipid synthesis (see Strosznajder et al., 1979) would indicate that, if DG and cytidine nucleotides are available, the potential for synthesis of CGP and EGP is present in the nerve ending membranes. Therefore, part of the phospholipids in synaptosomal membranes may be synthesized in the nerve endings in addition to the glycerophospholipids supplied by axonal transport (Brunetti et al., 1981).

TABLE III

Incorporation of CDP-Choline and CDP-Ethanolamine into Corresponding Phospholipids of Neuronal and Glial Cells of the Rabbit Brain *in vitro*

Cell Type	Labelled nucleotide precursor	Activity
Neuron	CDP-choline	24.7
Glia	CDP-choline	2.7
Neuron	CDP-ethanolamine	23.6
Glia	CDP-ethanolamine	5.0

From Binaglia et al. (1973). Mean values are expressed as nmol/mg protein/30 min incubation. See Binaglia et al. (1973) for details.

Interesting data on the synthesis of CGP by CPT have been obtained by Roberti et al. (1980) by labeling microsomal DG with radioactive glycerol phosphate *in vitro* and by incubating the labeled microsomes with CDP-choline for different time intervals. In this case, labeled DG were formed from PA only, and the initial distribution of the label into CGP molecular species was very similar to that of DG before incubation. However, a certain specificity for monoenoic DG was also found in this case (Roberti et al., 1980). These data are consistent with the hypothesis that the composition of available DG is as important as the specificity of biosynthetic enzymes, at least for the last metabolic step of CGP synthesis. Relationships between molecular species of PA and DG are also important (Roberti et al., 1980; Bazan, 1982).

Probably, CPT can utilize only a certain pool of DG which differs in molecular composition from the total pool present in a particular membrane. Binaglia et al. (1982a) have approached this problem by using different experimental models and have concluded that brain microsomal membranes have two pools of DG. The first can be utilized very fast by CPT since it is probably formed in the membrane area close to the enzyme. The second can be utilized, but with a slower rate, because it needs a diffusion process which is rate-limiting, or a flip-flop from one side of the membrane to the other (Fig 1).

The pool size and composition of DG depend also upon the relative concentrations of CDP-choline, CDP-ethanolamine and CMP. CPT is competitively inhibited by CDP-ethanolamine and EPT by CDP-choline. CMP, which is the product of the phosphotransferase reactions, inhibits the incorporation of CDP-choline and CDP-ethanolamine into the corresponding phosphoglycerides (Freysz et al.,1978). The inhibition is probably connected with the reversibility of both phosphotransferases, as shown by Goracci et al. (1981) and Mozzi et al. (1982).

From this brief discussion, it seems likely that several factors, very often connected, can influence the synthesis *de novo* of glycerophospholipids.

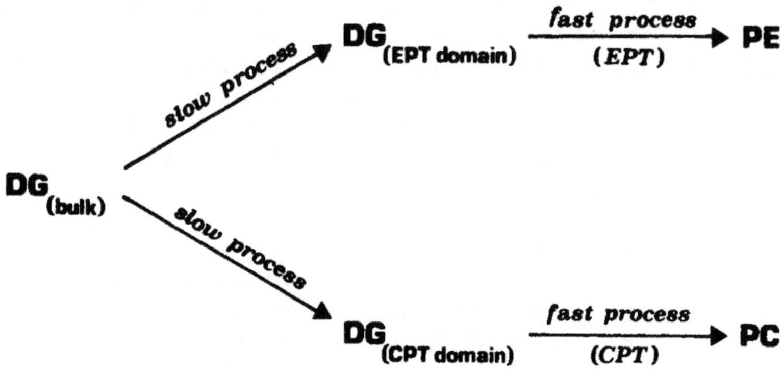

Fig 1. Pools of diglycerides (DG) at the microsomal level for choline phospholipid (PC) and ethanolamine phospholipid (PE) synthesis (Roberti et al., 1980; Binaglia et al., 1982a). EPT = ethanolaminephosphotransferase; CPT = cholinephosphotransferase.

The current knowledge of phospholipid metabolism of neural components indicates that the major mechanisms of phospholipid synthesis are located in the endoplasmic reticulum membranes of the nervous tissue. Hence, lipid molecules synthesized in the reticulum may "flow" to other cell membranes. More precisely, newly synthesized phospholipids may migrate from one cellular compartment to another, along the membranes, from one side of the membrane to the other, and from one cell to another. These migrations, which contribute to the dynamics of lipid synthesis, can take place without changes, but it is also possible that the transport of a phospholipid molecule from one place to another requires rearrangement of the membrane structure. For instance, it is known that nerve cells synthesize phospholipids mainly in the endoplasmic reticulum of the cell body. They migrate successively along the axon to the nerve ending. Studies carried out by Brunetti et al. (1981) on the transport of phospholipids in the oculomotor nerve of the chicken have shown that CGP are axonally transported. A small portion of these phospholipids translocates from the axon to the myelin sheath, probably as intact molecules. Furthermore, Schwann cells seem to reutilize the breakdown products of the transported CGP, as choline is much better reincorporated into myelin than glycerol (Brunetti et al., 1981). This model thus provides nice evidence for the complex mechanisms which are involved in the dynamics of phospholipids in tissues. The ciliary ganglion-oculomotor nerve system of the chicken is, however, not able to reutilize ethanolamine as quickly and efficiently as choline.

INTERCONVERSION REACTIONS IN NEURAL MEMBRANES AND RELATION TO MEMBRANE STRUCTURE

Several reactions, which convert one phospholipid molecule to another, are known to take place in the cell. These reactions are extremely important (see Mozzi et al., 1982 for review) since they allow the formation of new phospholipid molecules with low expenditure of energy. Furthermore, at least in some cases, this conversion occurs at the site where new molecules are required.

A phospholipid can be converted into another by reactions which produce changes in polar head groups, and these reactions are responsible for the interconversion of phospholipid classes or subclasses. Other reactions, which produce changes in the hydrocarbon chain, are involved in the interconversion of molecular species of the same phospholipid class or subclass (deacylation-reacylation reactions). The conversion of a phospholipid molecule into another is performed directly by a single reaction, or indirectly by several reactions. The most important interconversions which produce changes at the level of the polar head group are indicated in Fig 2. Only some of them will be considered in this section.

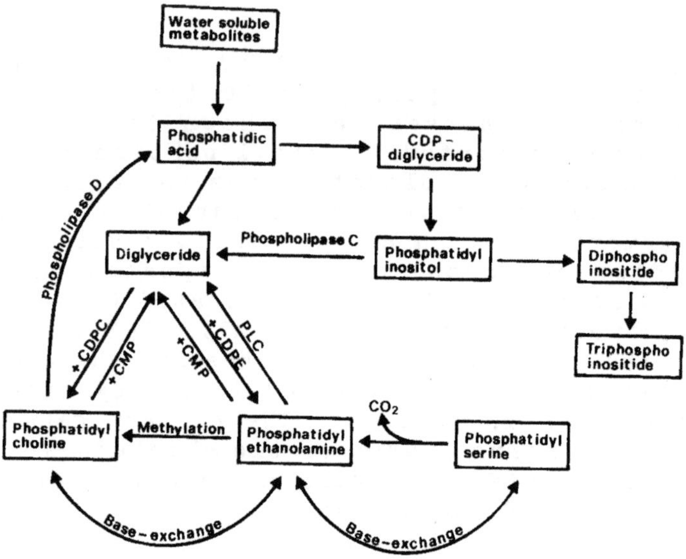

Fig 2. Interconversion reactions at the level of polar head-groups in phospholipid metabolism. Abbreviations other than those in the text: CDPC = CDP-Choline; CDPE = CDP-ethanolamine; PLC = phospholipase C.

Base-Exchange Reactions

Base-exchange reactions are interesting since they allow the rapid conversion of one phospholipid molecule to another at the membrane level without an apparent requirement for energy (see Kanfer, 1980 for review). Serine, ethanolamine and choline can be used as substrates for exchanging the free bases, but the reaction rates are different. Ethanolamine and serine are better substrates than choline. Ca^{2+} is required for the enzyme activity while other divalent cations are ineffective or strongly inhibitory. The enzyme system is present in the brain (Porcellati et al., 1971) and has been found in neuronal and glial cells (Goracci et al., 1973). With regard to subcellular localization of the relevant enzymes, they are mainly located in the microsomal fraction (Porcellati et al., 1971; Kanfer, 1972), but a high activity has also been found in the neuronal plasma membrane (Goracci et al., 1973). Table IV shows part of the results of Goracci et al. (1973). Recently, a serine exchange enzyme has been observed in purified calf brain oligodendroglia (Pleasure et al., 1981).

TABLE IV

Base-Exchange Activity for Ethanolamine, Serine and Choline in Plasma Membranes and Microsomal Membranes of Neuronal and Glial Cell-Enriched Preparations from Brain Cortex in Rabbit

Source	Specific Activity for		
	Serine	Ethanolamine	Choline
Neuronal H	3.8	6.6	0.80
Neuronal M	6.1	11.3	-
Neuronal PM	18.1	23.2	4.10
Glial H	0.6	0.8	-
Glial M	1.0	1.2	-
Glial PM	1.7	1.5	-

Incubations were carried out at 37°C (pH 8.1) with 1.5 mM labeled bases, 2.0 mM Ca^{2+} and 0.3 to 0.6 mg protein. Procedures have been reported by Goracci et al. (1973). Data, reported as nmol/mg protein/30 min, represent mean of six determinations. H = homogenate; M = microsomal membranes; PM = plasma membranes.

Apparently, a small pool of phospholipids is available for base-exchange. An estimation of the size of this pool is provided by prelabeling brain microsomal phosphoglycerides with ethanolamine, serine or choline by base-exchange and then by measuring the extent of the displacement of the labeled bases after short incubations with unlabeled or differently labeled bases (Gaiti et al., 1975; 1976). The results indicated that microsomal phosphatidylethanolamine can be the substrate for the incorporation of choline by base-exchange, but only 5-6% of the total EGP is available for the exchange reaction. On the other hand, when microsomal phosphatidylethanolamine was prelabeled *in vitro* by the cytidine pathway, choline could not displace the labeled ethanolamine. Smaller pools of exchanging phospholipid molecules are available at the membrane level for the exchange of free choline (Gaiti et al., 1976). This has been shown with microsomal membranes (Gaiti et al., 1976) and with synaptosomal membranes (De Medio et al., 1977b). Conversely, when the exchanging lipid is membrane CGP, smaller amounts of free bases can exchange with it (Gaiti et al., 1976; De Medio et al., 1977b).

TABLE V

Synthesis of Choline, Ethanolamine and Serine Phosphoglycerides in Brain Synaptosomes by the Base-exchange Reaction

Source	Specific Activity for		
	SGP	EGP	CGP
Synaptosomal H (rabbit)[a]	1.40	2.12	0.36
Synaptosomal PM (rabbit)[a]	2.71	3.83	-
Synaptosomes (rat)[b]	0.15	0.86	0.34
Synaptosomal H (rat)[b]	1.34	1.80	1.06
Synaptosomal PM (rat)[b]	0.33	1.00	0.16

[a]Incubation with adult rabbit brain synaptosomes or synaptosomal fractions (0.2-0.4 mg protein) was carried out at 37°C for 30 min (pH 8.1) with 1 mM bases and 4 mM $CaCl_2$. Data reported in nmol/mg protein/60 min. See Goracci et al. (1973) and De Medio et al. (1977b).
[b]From Buchanan and Kanfer (1980). Data in nmol/mg protein/60 min.

Synaptosomes also have base-exchange activity (Goracci et al., 1973). Table V indicates that synaptic plasma membrane from rabbit brain appears to be enriched in this system, at least for serine and ethanolamine exchange. The same does not seem to apply to rat brain (Buchanan and Kanfer, 1980).

There are indications that base-exchange can take place also *in vivo* (Arienti et al., 1976; Orlando et al., 1977), thus supporting the hypothesis that it might have a physiological role, which is not yet revealed.

Methylation Pathway

Phosphatidylcholine can be formed by the methylation of phosphatidylethanolamine. Phosphatidyl-N-monomethylethanolamine and phosphatidyl-N,N-dimethylethanolamine represent the intermediates of the pathway and S-adenosylmethionine (SAM) is the methyl donor. The mechanism is an example of a direct conversion. Although three successive methylations are required, phosphatidyl-N-monomethylethanolamine and phosphatidyl-N,N-dimethylethanolamine are, in fact, still phospholipid molecules. This methylation was first demonstrated in liver (Bremer and Greenberg, 1961). Mozzi and Porcellati (1979) and Blusztajn et al. (1979) independently reported its occurrence in brain, and a similar finding was also made by Crews et al. (1980a).

Mozzi and Porcellati (1979) have reported that the methyl groups of SAM can be incorporated into phosphatidylcholine of rat brain homogenate. The addition of N,N-dimethylphosphatidylethanolamine greatly enhanced the formation of phosphatidylcholine by this pathway. Using rat brain microsomes, the incorporation of the methyl groups showed two pH optima at 7.0 and 8.2 (Mozzi et al., 1980), thus indicating the probable existence in brain of two methyltransferases as was found in erythrocyte membrane by Hirata and Axelrod (1978).

Of particular interest is the finding that choline plasmalogens can also be synthesized by the methylation pathway from ethanolamine plasmalogens (Mozzi et al., 1980; 1981), since the metabolic mechanism leading to the synthesis of these phospholipids is still largely unknown.

Recent data of Crews et al. (1980b) have shown that rat brain synaptosomes display methyltransferase activity and can synthesize CGP by the methylation pathway. With 200 uM SAM and under proper incubation conditions (Crews et al., 1980a), about 20 pmol/mg protein of total CGP are synthesized within 60 min of incubation. Recent experimental evidence from our laboratory has indicated that rat brain neuronal homogenate under the same experimental conditions can synthesize about 200 pmol/mg protein/60 min of diacyl-glycerophosphorylcholine (diacyl-GPC) and about 350 of the alkenylacyl analogue (alkenylacyl-GPC), whereas the glial population is able only to synthesize 60-70 pmol of both subclasses (Mozzi and Porcellati, unpublished results). Therefore, a noticeable difference exists between the two cell types with regard to the activity of the methylation pathway for CGP synthesis. Moreover, neuronal plasma membrane shows some enrichment of this activity.

The stepwise methylation is a small contribution to the formation of CGP from the quantitative point of view. Nevertheless, the pathway could be of great interest because it may provide a mechanism for the formation of choline from ethanolamine. It is possible that in the nervous tissue, the mechanism might be linked to acetylcholine synthesis (Massarelli et al., 1982).

The possibility that the phosphatidylethanolamine pool, synthesized in brain microsomes by the base-exchange reaction (see the previous section), could be utilized for the methylation to phosphatidylcholine has been examined by Mozzi et al. (1982). Brain microsomes from rat were incubated for 20 min at 37°C in the presence of labeled ethanolamine and 2 mM Ca^{2+} to allow a typical base-exchange reaction for synthesis of phosphatidylethanolamine. Cold SAM was then added to the preincubated microsomes, and the methyl group transfer reactions were followed for additional time intervals. Interestingly, synthesis of phosphatidyl-N-monomethylethanolamine takes place after SAM addition, indicating that the labeled phosphatidylethanolamine coming from the base-exchange reaction is a substrate for the methylation to phosphatidyl-N-monomethylethanolamine. With longer time intervals radioactivity also appears into phosphatidyl-N,N-dimethylethanolamine and phosphatidylcholine. Further evidence for the participation of base-exchange reactions in the methylation pathway has been reported recently (Kanfer, 1982).

Back-Reaction of Phosphotransferases

As mentioned above, the reactions catalyzed by EPT and CPT are both reversible. The reversal of these reactions, also called back-reactions, utilizes membrane-bound phospholipids, which react with CMP and produce CDP-ethanolamine or CDP-choline and DG. These nucleotides or DG can be reutilized for the synthesis of new phospholipids. The interconversion reactions of phosphatidylcholine to phosphatidylethanolamine and *vice versa* are shown in Fig 3.

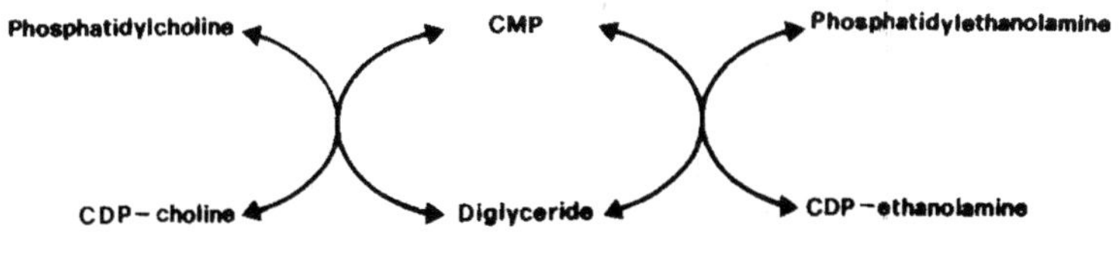

Fig 3. Interconversion of phosphatidylcholine and phosphatidylethanolamine by the back-reaction of phosphotransferase.

In brain, the reversibility of CPT has been recently demonstrated (Goracci et al., 1981), and some evidence has been reported for the reversibility of EPT (Goracci et al., 1981). In brain, DG produced by the back-reaction are rapidly hydrolyzed to fatty acids and glycerol by the combined action of DG lipase and monoglyceride lipase, both present in the microsomal membrane (Goracci et al., 1981; Mozzi et al., 1982). However, by blocking lipase activities, diglycerides accumulate and are available for the synthesis of new phospholipid molecules (Mozzi et al., 1982).

These interconversion reactions lead to the rearrangement of phospholipid molecules. Moreover, the direction of the reactions probably depends on the relative concentrations of cytidine nucleotides and on the energy state of the cell. The back-reaction may also be involved in the formation of free fatty acids during brain ischemia, a well known phenomenon which takes place in brain of rats and other animals (Bazan et al., 1971; Trovarelli et al., 1981).

Conclusions

Several reactions can lead to the synthesis of phospholipids. The bulk of phospholipid is formed by net synthesis, but this pathway alone does not account for the steady-state composition of lipids in various membranes. Therefore, interconversion reactions such as base-exchange, stepwise methylation, reversal of phosphotransferases and deacylation-reacylation reactions play a very important role.

The discovery of these pathways in most tissues has led to the concept that once a certain phospholipid molecule has been synthesized by net synthesis, it can be rapidly modified in the polar head group or in the hydrophobic tail of the molecule. These changes are probably connected with the functional properties of the membrane, thus may affect the transport of ions or other substances, modulation of membrane-bound enzymatic activities, and affinity of receptors for neurotransmitters or hormones. Unfortunately, regulation of the enzymic activities of these pathways, their pertinent localization in the cell and membranes, and their connections with specific functions are still largely unknown. This aspect will be partially covered in the last section of the present chapter.

While the enzymes for the net synthesis of phospholipids are particularly concentrated in the membranes of the endoplasmic reticulum and absent in the plasma membrane, the interconversion reactions are particularly active in plasma membranes from neurons and in synaptic membranes (Porcellati et al., 1971; Gorracci et al., 1971; Crews et al., 1980b; Mozzi and Porcellati, unpublished results; and this review). These interconversion reactions need endogenous phospholipids as substrates. Further, the net synthesis pathway provides a system for the "bulk" phospholipid structure of membranes, whereas the interconversion reactions may function to maintain a phospholipid asymmetry under varying conditions linked to pharmacological events.

PHOSPHOLIPID AND ENZYME ASYMMETRY

As previously mentioned in this chapter, the membranes of the endoplasmic reticulum play a central role in cellular phospholipid metabolism since the majority of phospholipids are synthesized by this route. Most of the enzymes are located on the cytoplasmic surface (Vance et al., 1977; Coleman and Bell, 1978). In this connection, detailed studies on the distribution, molecular arrangement, and mobility of the microsomal phospholipids have often given contradictory results (see Van Deenen, 1981 for review). This is mainly due to the fact that phospholipids, once synthesized, are subsequently transported in three different directions in neural membranes: across the membrane bilayer, along the microsomal membranes, and into different intracellular organelles. This complexity creates additional difficulties.

Generally, one of the main questions to answer regarding distribution of phospholipids and lipid-synthesizing enzymes in neural membranes is how asymmetry arises in these membranes. One explanation is that the enzymes catalyzing the last step of phospholipid synthesis, namely the phosphotransferases, are unevenly distributed on both sides of the membrane where synthesis takes place. The lipids that are formed by these enzymes would then be released on each side of the membrane bilayer, thus producing asymmetry. Another explanation might be that phospholipids are synthesized only on one side of the endoplasmic reticulum and that a rapid transbilayer movement of phospholipids would account for the membrane asymmetry. The presence of transbilayer movement of these molecules in biological membranes (see Van Deenen, 1981) makes the second mechanism an attractive hypothesis applicable also to neural membranes.

Phospholipid and enzyme asymmetry in neural membranes can become more complicated by their heterogeneity and by the fact that phospholipids are mostly synthesized in the endoplasmic reticulum (microsomal membranes), while only "interconversion reactions" (base-exchange, methylation, phosphatidylinositol stepwise phosphorylation) may be located also in plasma membrane and in synaptic membranes, particularly where no de novo synthesis occurs.

How does phospholipid asymmetry arise in plasma and synaptic membranes where no net synthesis takes place, and how is this asymmetry established in microsomal membranes, where production of lipids occurs? These questions are not easily answered, and only speculative or provisional data can be given.

Phospholipid asymmetry occurs in synaptosomal plasma membranes (SPM), as shown by various workers (Smith and Loh, 1977; 1979; Fontaine et al., 1980; Crews et al., 1980b). The results are rather controversial, but it is reasonably true that SPM possess much EGP and SGP distributed in the inner, rather than in the outer monolayer (Fontaine et al., 1980), whereas most CGP is localized on the external side. Blood lipoproteins, which are rich in CGP but not in EGP, cannot easily cross the blood-brain barrier, and therefore cannot generate a lipid asymmetry in SPM which does not synthesize phospholipids de novo. Probably, phospholipid methyltransferase asymmetry in SPM (Crews et al., 1980b) may constitute a mechanism for an enzyme-mediated flip-flop of phospholipids from the cytoplasmic side to the outer surface of SPM. Evidence has been produced that EGP of the inner side can be methylated to intermediary lipid molecules and form CGP in the outer side of SPM (Crews et al., 1980b). In addition, base-exchange may be an alternative way for producing phospholipid asymmetry. Although it is known that this enzymic mechanism exists in SPM and generally in plasma membranes (Porcellati et al., 1971; Goracci et al., 1973; and this review), the contribution it can give to lipid asymmetry in membranes which do not synthesize phospholipids, like plasma membranes, is not well established. On the other hand, these exchange reactions may well function to maintain a phospholipid asymmetry under varying pharmacological and physiological conditions (see Kanfer, 1980, for review).

Phospholipid and enzyme asymmetry has been more thoroughly examined in brain microsomal membranes. Normally, the outer leaflet of the microsomal vesicles contains about 75% of the CGP and about 25% of the aminophospholipids, whereas an opposite distribution is observed for the inner leaflet (Freysz et al., 1982). Experiments performed to establish the localization of the phosphotransferases catalyzing the synthesis of ethanolamine and choline phosphoglycerides strongly suggest that the CPT is situated on the external side of the microsomal vesicle, and the EPT is either located on the internal side of the vesicle or has a transmembrane position (Freysz et al., 1982). More recent work, however, has established that the diacyl-GPE and the corresponding plasmalogen are synthesized on the external side of the microsomal vesicle of brain tissue (Dainous et al., 1982), which corresponds to the cytoplasmic side, and that a certain amount of the synthesized molecules of EGP are discharged by the EPT into the inner leaflet.

The transfer of labeled EGP molecules from the outer to the inner microsomal leaflet has been proved also with in vivo experiments (Binaglia et al., 1982b); Butler and Morell, 1982; Roberti et al., 1982).

Interesting experiments have been carried out (Binaglia et al., 1982b) on the localization of the EGP molecules synthesized by base-exchange reaction (see previous section). The assumption made before that EGP molecules synthesized by the cytidine pathway are transferred from the outer to the inner layer by transbilayer movement cannot be made a priori for base-exchange reaction, which catalyzes the transfer of phosphatidic acid from endogenous phospholipid molecules to ethanolamine (or other bases) present in the aqueous medium. Free ethanolamine, in this case, could be easily available on both sides of the membrane, whereas cytidine-5'-diphosphoethanolamine (CDP-ethanolamine), being a much more polar molecule, probably does not cross the lipid bilayer. Binaglia et al. (1982b) have demonstrated that most of the EGP synthesis by base-exchange reaction takes place on the external side of microsomal vesicles (cytoplasmic side). It is pertinent to mention that Buchanan and Kanfer (1980) have reported that the enzyme catalyzing ethanolamine base-exchange is localized on the luminal surface of rat brain microsomes. Evidently, the phosphatidylethanolamine synthesized by the enzyme present in the inner surface of the microsomal vesicles will be immediately transferred to the outer surface.

In conclusion, most of the studies carried out on lipid and enzyme asymmetry in microsomal membranes indicate that the biosynthesis of phospholipids takes place in a compartment of the microsomes that is more accessible to surface probes and that the labeled molecules are then transferred in a time-dependent process to other compartments where the lipids are less available for enzyme action.

LIPID MODIFICATION IN NEURAL MEMBRANES AND RELATIONSHIP TO "SYNAPTIC FUNCTION"

As mentioned before, lipid modification in neural membranes has been thought to play an important role in their functional activity. However, there is still little pertinent information about the functional role of lipid modification and particularly on the enzymes connected with the interconversion reactions which modify the lipid molecule at the membrane level. Apart from the interest in the fatty acid residues and their metabolic interconversions (see specific chapters in this book) and the importance of phospholipases A_1 and A_2, little attention has been given as yet to the interconversion of polar head groups of membrane lipids or to reactions taking place at this molecular level. The most important exception is the large body of evidence concerning the importance of the phosphoinositides for membrane function (see also this Volume).

Synaptic membranes seem to be most implicated in the relationships between lipid modification and functioning. Current ideas favor the view that specific pools of phospholipids in these membranes may serve as annular lipids to maintain the right conformation of functional proteins, such as enzymes or receptors. Hence, reactions such as those represented by base-exchange, methylation or acyl interconversion may provide a reasonable mechanism for regulating the functional properties of enzymes or receptors, by modulation at the level of the polar head group or hydrophobic acyl chain.

IGP are rather rich in stearoyl and arachidonoyl groups (see chapters of Sun et al. and of Bazan et al. in this Volume). This fact has provided evidence for important metabolic functions in brain of the deacylation-reacylation reactions taking place on this lipid in neural membranes, since the release of arachidonate through various degradative events may be related to the regulation of arachidonate for prostaglandin synthesis. It is interesting to note, in this connection, that previous studies (Corbin and Sun, 1978) have indicated that newly synthesized IGP are found associated in brain synaptic membranes.

Recently, SGP in brain, together with its lyso-derivative, has been indicated to represent a lipid of noticeable pharmacological interest (Bruni and Toffano, 1982). Since the base-exchange reactions are the only known active systems for SGP biosynthesis, studies on the production of SGP may reflect with certainty base-exchange activity and therefore may point to the functional importance of this mechanism in membrane activity.

In this regard, De Medio et al. (1977a; 1980) have shown that GABA uptake is stimulated in rabbit brain synaptosomes when the synaptosomal phospholipids are enriched in SGP after a base-exchange reaction. Conversely, the uptake of GABA is inhibited when the SGP is displaced in its serine moiety by ethanolamine or choline (De Medio et al., 1977a; 1980). It has also been shown (De Medio et al., 1980) that the membrane lipid composition of the synaptosomes may be returned back to the original structure with a return to the control level of the GABA uptake. These data have been recently confirmed by Chweh and Leslie (1982), who have shown that the addition of intact SGP to a preparation of rat brain synaptosomes enhances the uptake of GABA. Base-exchange reaction has been reported to affect also the adenylate cyclase activity in the rat brain synaptosomal membranes (Francescangeli et al., 1977).

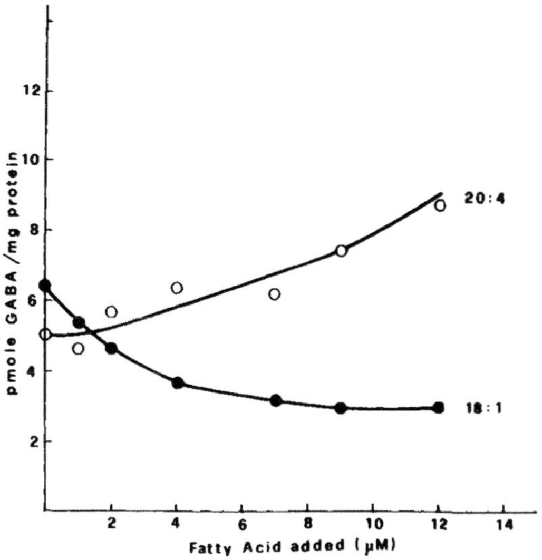

Fig 4. The effect of fatty acyl residues incorporation into rat brain synaptosomal phospholipids upon GABA uptake. Fatty acid incorporation was carried out without the addition of exogenous lipid acceptors in the incubation medium described by Corbin and Sun (1978) for 15 min. Synaptosomes were then immediately incubated in a GABA uptake medium and the extent of the neurotransmitter uptake measured as reported previously (De Medio et al., 1977a; 1980). The labeled stearate was found particularly incorporated into EGP and CGP, while arachidonate was incorporated into IGP and CGP.

Changes in acyl-group composition of synaptosomal phospholipids are also effective in modifying neurotransmitter uptake. It has been shown in our laboratory that by incubating rat brain synaptosomes with labeled stearate or arachidonate under proper experimental conditions (Corbin and Sun, 1978) and then by performing incubation in a GABA uptake medium, a noticeable gradually-increasing inhibition of uptake takes place when stearate is added, while an opposite effect is seen when arachidonate is added (Fig

4). These results indicate that changes of the hydrophobic microenvironment may lead to modulatory effects on functional activity in neural membranes. It is pertinent to note that Rhoads et al. (1982) have recently indicated that changes in the fatty acyl profile of rat synaptosomes may have effect on the amino acid uptake system.

ACKNOWLEDGEMENTS

This work has been supported by the grant No. 81.0031.804 of the Consiglio Nazionale delle Ricerche, Rome.

REFERENCES

Ansell, G.B. (1973) Phospholipids and the nervous system. In: Form and Function of Phospholipids (G.B. Ansell, J.N. Hawthorne and R.M.C. Dawson, eds.) pp. 377-422. Elsevier, Amsterdam.

Ansell, G.B. and Metcalfe, R.F. (1971) Studies on the CDP-ethanolamine-1,2-diglyceride ethanolaminephosphotransferase of rat brain. J. Neurochem. 18, 647-665.

Arienti, G., Corazzi, L., Woelk, H. and Porcellati, G. (1976) Biosynthesis of rat brain phosphatidylcholines from intracerebrally injected choline. J. Neurochem. 27, 203-210.

Arienti, G., Goracci, G. and Porcellati, G. (1981) Glycerophospholipid metabolism in neuronal and glial cell-enriched fractions. Neurochem. Res. 6, 729-742.

Arienti, G., Corazzi, L., Roberti, R., Alunni-Bistocchi, G., Binaglia, L. and Porcellati, G. (1982) Biochim. Biophys. Acta (in press)

Baker, R.R. (1979) The fatty acid composition of phosphoglycerides of nerve cell bodies isolated in bulk from rabbit cerebral cortex: Changes during development and positional distribution. Can. J. Biochem. 57, 378-384.

Bazan, N.G. (1982) Biosynthesis of phosphatidic acid and polyenoic phospholipids in the central nervous system. In: Phospholipid Metabolism in the Nervous System (L.A. Horrocks, G.B. Ansell and G. Porcellati, eds.) Vol. I, pp. 49-62. Raven Press, New York.

Bazan, N.G., Pascual de Bazan, H.E., Kennedy, W.G. and Joel, C.D. (1971) Regional distribution and rate of production of free fatty acids in rat brain. J. Neurochem. 18, 1387-1393.

Binaglia, L., Goracci, G., Porcellati, G., Roberti, R. and Woelk, H. (1973) The synthesis of choline and ethanolamine phosphoglycerides in neuronal and glial cells of rabbit in vitro. J. Neurochem. 21, 1067-1082.

Binaglia, L., Roberti, R., Vecchini, A. and Porcellati, G. (1982a) Evidence for a compartmentation of brain microsomal diacylglycerol. J. Lipid Res. 23, 955-961.

Binaglia, L., Arienti, G., Corazzi, L., Roberti, R., Freysz, L., Alunni-Bistocchi, G. and Porcellati, G. (1982b) J. Membrane Sci. (in press)

Blusztajn, J.K., Zeisel, S.H. and Wurtman, R.J. (1979) Synthesis of lecithin (phosphatidylcholine) from phosphatidylethanolamine in bovine brain. Brain Res. 179, 319-327.

Bremer, J. and Greenberg, D. (1961) Methyl transferring enzyme system of microsomes in the biosynthesis of lecithin (phosphatidylcholine). Biochim. Biophys. Acta 46, 205-216.

Brunetti, M., De Giamberardino, L., Porcellati, G. and Droz, B. (1981) Contribution of axonal transport to the renewal of myelin phospholipids in peripheral nerves. II. Biochemical study. Brain Res. 219, 73-84.

Bruni, A. and Toffano, G. (1982) Lysophosphatidylserine, a short-lived intermediate with plasma membrane regulatory properties. Pharmacol. Res. Commun. 14, 469-484.

Buchanan, A.G. and Kanfer, J.N. (1980) Topographical distribution of base exchange activities in rat brain subcellular fractions. J. Neurochem. 34, 720-725.

Butler, M. and Abood, L.G. (1982) Use of phospholipase A to compare phospholipid organization in synaptic membranes, myelin and liposomes. J. Membrane Biol. 66, 1-7.

Butler, M. and Morell, P. (1982) Sidedness of phospholipid synthesis on brain membranes. J. Neurochem. 39, 155-164.

Cammer, W., Sirota, S.R. and Norton, W.T. (1980) 5'-Nucleotidase in rat brain myelin. J. Neurochem. 35, 367-373.

Chweh, A.Y. and Leslie, S.W. (1982) Phosphatidylserine enhancement of [^{3}H] -aminobutyric acid uptake by rat whole brain synaptosomes. J. Neurochem. 38, 691-695.

Coleman, R. and Bell, R.M. (1978) Evidence that biosynthesis of phosphatidylethanolamine, phosphatidylcholine, and triacylglycerol occurs on the cytoplasmic side of microsomal vesicles. J. Cell Biol. 76, 245-253.

Corbin, D.R. and Sun, G.Y. (1978) Characterization of the enzymic transfer of arachidonoyl groups to 1-acyl-phosphoglycerides in mouse synaptosome fraction. J. Neurochem. 30, 77-82.

Crews, F.T., Hirata, F. and Axelrod, J. (1980a) Identification and properties of methyltransferases that synthesize phosphatidylcholine in rat brain synaptosomes. J. Neurochem. 34, 1491-1498.

Crews, F.T., Hirata, F. and Axelrod, J. (1980b) Phospholipid methyltransferase asymmetry in synaptosomal membranes. Neurochem. Res. 5, 983-991.

De Medio, G.E., Hamberger, A., Sellstrom, A. and Porcellati, G. (1977a) The phospholipid base-exchange system as a possible modulator of γ-aminobutyric acid transport in brain cells. Neurochem. Res. 2, 469-484.

De Medio, G.E., Trovarelli, G. and Porcellati, G. (1977b) Relationships between base-exchange reaction and synaptosomal phospholipid pool in the rabbit brain *in vitro*. Ital. J. Biochem. 26, 202-214.

De Medio, G.E., Trovarelli, G., Hamberger, A. and Porcellati, G. (1980) Synaptosomal phospholipid pool in rabbit brain and its effect on GABA uptake. Neurochem. Res. 5, 171-179.

DeVries, G.H., Payne, W. and Saul, R.G. (1981a) Composition of axolemma-enriched fractions isolated from bovine CNS myelinated axons. Neurochem. Res. 6, 521-537.

DeVries, G.H., Zetusky, W.J., Zmachinski, C. and Calabrese, V.P. (1981b) Lipid composition of axolemma-enriched fractions from human brains. J. Lipid Res. 22, 208-216.

Dainous, F., Freysz, L., Mozzi, R., Dreyfus, H., Louis, J., Porcellati, G. and Massorelli, R. (1982) Synthesis of choline phospholipids in neuronal and glial cell cultures by the methylation pathway. FEBS Lett. 146, 221-223.

Farooq, M., Cammer, W., Snyder, D.S., Raine, C.S. and Norton, W.T. (1981) Properties of bovine oligodendroglia isolated by a new procedure using physiologic conditions. J. Neurochem. 36, 431-440.

Fontaine, R.N., Harris, R.A. and Schroeder, F. (1980) Aminophospholipid asymmetry in murine synaptosomal plasma membrane. J. Neurochem. 34, 269-277.

Francescangeli, E., Brunetti, M., Gaiti, A. and Porcellati, G. (1977) Relationships between base-exchange in phospholipid metabolism and cyclic-AMP levels of synaptosomal membranes. Ital. J. Biochem. 26, 428-436.

Freysz, L., Horrocks, L.A. and Mandel, P. (1978) Ethanolamine and choline phosphotransferases of chicken brain. In: Enzymes of Lipid Metabolism (S. Gatt, L. Freysz and P. Mandel, eds.) pp. 253-268. Plenum Press, New York.

Freysz, L., Dreyfus, H., Vincendon, G., Binaglia, L., Roberti, R. and Porcellati, G. (1982) Asymmetry of brain microsomal membranes: Correlation between the asymmetric distribution of phospholipids and the enzymes involved in their synthesis. In: Phospholipid Metabolism in the Nervous System (L.A. Horrocks, G.B. Ansell and G. Porcellati, eds.) Vol. I, pp. 37-47. Raven Press, New York.

Gaiti, A., Brunetti, M. and Porcellati, G. (1975) The relationships between the phospholipid pool and the base-exchange reaction in the Ca^{++}-stimulated incorporation of ethanolamine into brain microsomal phospholipids. FEBS Lett. 49, 361-364.

Gaiti, A., Brunetti, M., Woelk, H. and Porcellati, G. (1976) Relationships between base-exchange reaction and the microsomal phospholipid pool in the rat brain in vitro. Lipids 11, 823-829.

Goracci, G., Blomstrand, C., Arienti, G., Hamberger, A. and Porcellati, G. (1973) Base-exchange enzymic system for the synthesis of phospholipids in neuronal and glial cells and their subfractions: A possible marker for neuronal membranes. J. Neurochem. 20, 1167-1180.

Goracci, G., Francescangeli, E., Horrocks, L.A. and Porcellati, G. (1981) The reverse reaction of choline phosphotransferase in rat brain microsomes. A new pathway for degradation of phosphatidylcholine. Biochim. Biophys. Acta 664, 373-379.

Hamberger, A. and Svennerholm, L. (1971) Composition of gangliosides and phospholipids of neuronal and glial cell enriched fractions. J. Neurochem. 18, 1821-1829.

Hirata, F. and Axelrod, J. (1978) Enzymatic synthesis and rapid translocation of phosphatidylcholine by two methyltransferases in erythrocyte membranes. Proc. Natl. Acad. Sci. USA 75, 2348-2352.

Kanfer, J.N. (1972) Base exchange reactions of the phospholipids in rat brain particles. J. Lipid Res. 13, 468-476.

Kanfer, J.N. (1980) The base exchange enzymes and phospholipase D of mammalian tissue. Can. J. Biochem. 58, 1370-1380.

Kanfer, J.N. (1982) Contribution of substrates for phospholipid N methylation by the base exchange enzyme in rat brain microsomes. Biochem. Biophys. Res. Comm. 106, 422-428.

Kreutzberg, G., Barron, K.D. and Schubert, P. (1978) Cytochemical localization of 5'-nucleotidase in glial plasma membranes. Brain Res. 158, 247-257.

Massarelli, R., Dainous, F., Freysz, L., Dreyfus, H., Mozzi, R., Floridi, A., Siepi, D. and Porcellati, G. (1982) In: Basic and Clinical Aspects of Molecular Neurobiology (A.M. Giuffrida, G. Gombos, G. Benzi and H.S. Bachelard, eds.) pp. 147-155. Fondazione Menarini, Milan.

McCarthy, K.D. and de Vellis, J. (1980) Preparation of separate astroglial and oligodendroglial cell cultures from rat cerebral tissue. J. Cell Biol. 85, 890-902.

Mozzi, R. and Porcellati, G. (1979) Conversion of phosphatidylethanolamine to phosphatidylcholine in rat brain by the methylation pathway. FEBS Lett. 100, 363-366.

Mozzi, R., Andreoli, V. and Porcellati, G. (1980) Phospholipid synthesis by interconversion reactions in brain tissue. In: Natural Sulphur Compounds (D. Cavallini, G.E. Gaull and V. Zappia, eds.) pp. 41-54. Plenum Press, New York.

Mozzi, R., Siepi, D., Andreoli, V. and Porcellati, G. (1981) The synthesis of choline plasmalogen by the methylation pathway in rat brain. FEBS Lett. 131, 115-118.

Mozzi, R., Goracci, G., Siepi, D., Francescangeli, E., Andreoli, V., Horrocks, L.A. and Porcellati, G. (1982) In: Phospholipid Metabolism in the Nervous System (L.A. Horrocks, G.B. Ansell and G. Porcellati, eds.) Vol. I, pp. 1-12. Raven Press, New York.

Orlando, P., Arienti, G., Cerrito, F., Massari, P. and Porcellati, G. (1977) Quantitative evaluation of two pathways for phosphatidylcholine biosynthesis in rat brain in vivo. Neurochem. Res. 2, 191-201.

Pleasure, D., Hardy, M., Johnson, G., Lisak, R. and Silberberg, D. (1981) Oligodendroglial glycerophospholipid synthesis: Incorporation of radioactive precursors into ethanolamine glycerophospholipids by calf oligodendroglia prepared by a Percoll procedure and maintained in suspension culture. J. Neurochem. 37, 452-460.

Porcellati, G. (1972) Aspects of regulatory mechanisms in phospholipid biosynthesis of nervous tissue. In: Advances in Enzyme Regulation (G. Weber, ed.) pp. 83-100. Academic Press, New York.

Porcellati, G., Biasion, M.G. and Pirotta, M.G. (1970) The labeling of brain ethanolamine phosphoglycerides from cytidine diphosphate ethanolamine *in vitro*. Lipids 5, 734-742.

Porcellati, F., Arienti, G., Pirotta, M. and Giorgini, D. (1971) Base-exchange reactions for the synthesis of phospholipids in nervous tissue: The incorporation of serine and ethanolamine into the phospholipids of isolated brain microsomes. J. Neurochem. 18, 1395-1417.

Radominska-Pyrek, A. and Horrocks, L.A. (1972) Enzymic synthesis of 1-alkyl-2-acyl-sn-glycero-3-phosphorylethanolamines by the CDP-ethanolamine: 1-radyl-2-acyl-sn-glycerol ethanolaminephosphotransferase from microsomal fraction of rat brain. J. Lipid Res. 13, 580-587.

Rhoads, D.E., Kaplan, M.A., Peterson, N.A. and Raghupathy, E. (1982) Effects of free fatty acids on synaptosomal amino acid uptake systems. J. Neurochem. 38, 1255-1260.

Roberti, R., Binaglia, L. and Porcellati, G. (1980) Synthesis of molecular species of glycerophospholipids from diglyceride-labeled brain microsomes. J. Lipid Res. 21, 449-454.

Roberti, R., Corazzi, L., Binaglia, L., Arienti, G. and Porcellati, G. (1982) Sidedness of phosphatidylethanolamine synthesis in rat brain microsomes. In: Phospholipid Metabolism in the Nervous System (L.A. Horrocks, G.B. Ansell and G. Porcellati, eds.) Vol. I, pp. 330. Raven Press, New York.

Smith, A.P. and Loh, H.H. (1977) Isolation of synaptosomal plasma membrane from mouse brain. J. Neurochem. 28, 887-891.

Smith, A.P. and Loh, H.H. (1979) Architecture of the nerve ending membrane. Life Sci. 24, 1-20.
Strosznajder, J., Radominska-Pyrek, A., Lazarewicz, J. and Horrocks, L.A. (1977) Synthesis *in vitro* of 1-alkyl-2-acyl- and 1,2-diacyl-*sn*-glycero-3-phosphorylcholines and ethanolamines by neuronal, glial and synaptosomal fractions from adult rabbit brain. Bull. Acad. Pol. Sci. Ser. Sci. Biol. 25, 363-370.
Strosznajder, J., Radominska-Pyrek, A. and Horrocks, L.A. (1979) Choline and ethanolamine glycerophospholipid synthesis in isolated synaptosomes of rat brain. Biochim. Biophys. Acta 574, 48-56.
Trovarelli, G., De Medio, G.E., Dorman, R.V., Piccinin, G.L., Horrocks, L.A. and Porcellati, G. (1981) Effect of cytidine diphosphate choline (CDP-choline) on ischemia-induced alterations of brain lipid in the gerbil. Neurochem. Res. 6, 821-833.
Vance, D.E., Choy, P.C., Farren, S.B., Lim, P.H. and Schneider, W.J. (1977) Asymmetry of phospholipid biosynthesis. Nature 270, 268-269.
Van Deenen, L.L.M. (1981) Topology and dynamics of phospholipids in membranes. FEBS Lett. 123, 3-15.

THE INVOLVEMENT OF LIPIDS IN SIGNAL TRANSMISSION ACROSS CELL MEMBRANES OF THE NERVOUS SYSTEM

George Hauser

Ralph Lowell Laboratories, McLean Hospital, Belmont, MA 02178, and Department of Biological Chemistry, Harvard Medical School, Boston, MA 02115, USA

INTRODUCTION

Plasma membranes of cells in the nervous system have, in addition to specialized tasks characteristic for each cell type or particular cell region (such as nerve endings), the same functions as those of cells in other organs. One of these functions is the transmission of information from the extracellular environment to the interior of the cell where a variety of metabolic events take place in response to the requirements of the organism.

The barrier which the signal must cross, the cell membrane, is generally accepted to be best represented by the lipid-globular protein fluid mosaic model (Singer and Nicolson, 1972), although this generalized concept fails to account for the exquisite specialization in composition of the multiplicity of domains responsible for selective functions. Although the protein constituents are represented as extrinsic or intrinsic, either penetrating the lipid bilayer or protruding from the extracellular or intracellular face of the membrane, the phospholipid portion is shown as a uniform bilayer with the hydrophilic head groups oriented towards the surfaces and the hydrophobic carbon chains facing each other in the interior of the membrane. Yet not only is there asymmetry in the distribution of different lipid classes across the membrane, but also undoubtedly non-uniformity over the surface areas, whether interstitial or cytoplasmic (Karnovsky et al., 1982). This heterogeneity applies both to lipid classes and within each to individual species, distributions which are presumably to some degree variable owing to the fluidity of the lipid matrix.

It is the nature and composition of the lipid environment of individual proteins which must to a large extent influence and regulate their functions just as the nature and activation of protein molecules can characteristically affect the metabolism of specific lipids. Among the proteins on the external cell surface are receptors for neurotransmitters and other small molecules which function to initiate a chain of events resulting in the transfer of the signal to the appropriate intracellular apparatus. In the course of this sequence messenger molecules are involved, performing whatever functions are necessary to enable the proper cellular response to occur. Two main classes of receptors and associated messenger molecules can be identified (Table 1), of which the β-adrenergic receptor-cyclic AMP system and the α_1-adrenergic or muscarinic cholinergic receptor-Ca^{2+} system are typical examples. In each of these receptor systems the initial reaction is the binding of an agonist to a receptor site followed by coupling to an effector, the generation of a second messenger and eventual

Table 1

Examples of Receptor Types

I. Second Messenger: cyclic AMP

Agonist	Receptor
Norepinephrine	α_2-adrenergic (negative)
	$\beta_{1,2}$-adrenergic
Histamine	H_2-histaminergic
5-Hydroxytryptamine	$5HT_2$-serotonergic

II. Second Messenger: calcium ions

Agonist	Receptor
Norepinephrine	α_1-adrenergic
Acetylcholine	muscarinic cholinergic
Histamine	H_1-histaminergic
5-Hydroxytryptamine	$5HT_1$-serotonergic

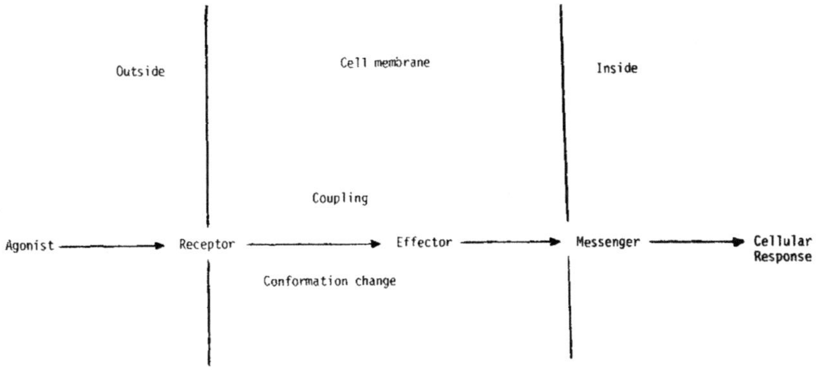

Fig. 1. Diagram of information transfer across cell membranes.

intracellular events (Fig. 1). This chapter is an overview of the participation of lipids in the functioning of the receptors with some examples from our own studies.

PHOSPHOLIPID METHYLATION AND β-ADRENERGIC RECEPTORS

In the case of β-adrenergic receptors considerable information is available on the mechanism of receptor activation and the subsequent production of cyclic AMP as second messenger (Johnson et al., 1980; Ross and Gilman, 1980; Lefkowitz et al., 1982). In brief, the sequence of events involves three proteins which are present in the membrane and upon stimulation undergo rearrangement. The components of the system are the receptor protein to which the appropriate ligand, presumably NE[1], binds, a guanine nucleotide binding protein and the enzyme protein. The guanine nucleotide binding protein apparently functions to permit regulation of the system by guanine nucleotides which are required for stimulation of adenylate cyclase by hormones. Through complex formation and dissociation after GDP- and GTP-controlled affinity changes it seems to act as a shuttle

[1]Abbreviations:

CCh = carbamylcholine	PhC = phosphatidylcholine
DAG = diacylglycerol	PhE = phosphatidylethanolamine
NE = norepinephrine	PhI = phosphatidylinositol
PhA = phosphatidic acid	PhS = phosphatidylserine

between the receptor and the catalytic enzyme protein. A model for this mechanism has been developed (Stadel et al., 1982).

Little is known about the role of lipids in these interactions which are presumed to be similar in nervous tissue and the model systems used (such as erythrocytes). It is, however, highly likely that the lipid components of the plasma membrane which the signal must traverse are involved in the mechanism and play an essential role in its regulation.

One indication along these lines comes from studies on phospholipid methylation. In many tissues, including brain, two enzymes have been shown to catalyze three methylation reactions (Blusztajn et al., 1979; Crews et al., 1980) which have the overall effect of converting PhE to PhC (Fig. 2). Since the enzymes are asymmetrically located within the membrane, this is accompanied by translocation of lipid from the cytoplasmic to the external surface of the membrane and by a decrease in membrane viscosity (Hirata and Axelrod, 1980). If one can extrapolate from the findings in a variety of systems, not only does the binding of an agonist to β-adrenergic receptors stimulate phospholipid methylation but phospholipid methylation also increases the number of β-adrenergic binding sites, apparently by uncovering cryptic receptors. Possibly as a result of the change in the fluidity of the membrane, receptor mobility may be increased and coupling of the activated receptor to adenylate cyclase and therefore cyclic AMP formation seems to be enhanced. However, considerable further work is required to establish the general applicability of these concepts and to validate the model proposed by Hirata and Axelrod (1980).

Phospholipid methyltransferase (PMT)

Intracellular | $\text{PhE} \xrightarrow{\text{I}} \text{PhE-Me} \xrightarrow{\text{II}} \text{PhE-Me}_2 \xrightarrow{\text{II}} \text{PhE-Me}_3 = \text{PhC}$ | Extracellular

Fig. 2. Enzymatic conversion of PhE to PhC in the cell membrane. Me = methyl group.

RECEPTORS FOR THE PHOSPHOINOSITIDE RESPONSE

If one considers the receptors which do not result in an elevation of cyclic AMP levels, different mechanisms are found to be operative. Actually very little is known about the exact means used by the cell to bring about the observed consequences following activation of these receptors by the appropriate agonists. However, several of the consequences are known and are thought to be interrelated in an as yet incompletely understood fashion. Thus the elevation of intracellular Ca^{2+}, enhanced turnover of acidic phospholipids, increase in cyclic GMP concentration and arachidonic acid release seem to be connected in many of the systems examined to date.

It is the change in phospholipid metabolism originally observed as an increase in the incorporation of [^{32}P]orthophosphate into PhA and PhI following the addition of acetylcholine to tissue preparations *in vitro* (Hokin and Hokin, 1955a, 1955b, 1958) which has piqued the curiosity of investigators ever since. In the context of this discussion, which emphasizes the involvement of lipids in the transfer of information across cell membranes, this phenomenon is of the greatest significance even though its exact role has not yet been elucidated. Several reviews of this topic have recently appeared (Michell, 1975, 1979; Hawthorne and Pickard, 1979; Fain and García-Sáinz, 1980; Berridge, 1981; Putney, 1981; Irvine et al., 1982; Takai et al., 1982). In nervous tissue both adrenergic and cholinergic agonists are capable of bringing about changes in phosphoinositide metabolism which have been repeatedly documented in several preparations from nervous tissue *in vitro*, including synaptosomes (Hokin and Hokin, 1958; Schacht and Agranoff, 1972; Yagihara and Hawthorne, 1972), as well as following injection of the neurotransmitters (Margolis and Heller, 1966; Friedel and Schanberg, 1972; Soukup et al., 1978).

The pertinent portions of the pathways of phospholipid metabolism are shown in Figure 3. Of particular interest in the context of the receptor-mediated changes is the closed cycle PhI - DAG - PhA - PhCMP - PhI which is normally involved in phospholipid turnover but is apparently stimulated as the result of ligand binding to an appropriate receptor. There is good evidence that this enhanced turnover is initiated through the cleavage of PhI by a specific, primarily cytoplasmic, Ca^{2+}-dependent phospholipase C (Irvine and

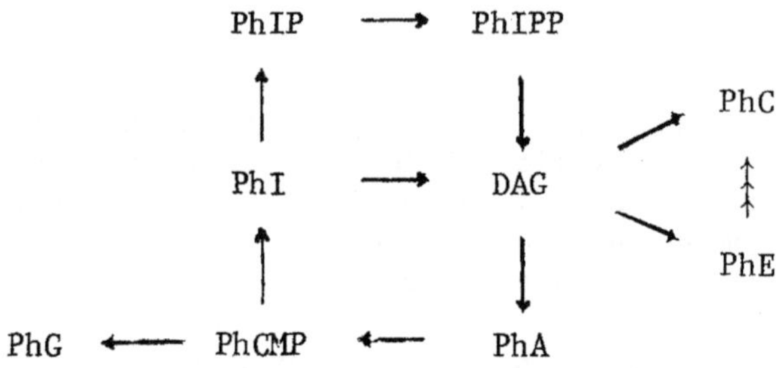

Fig. 3. Pathways of phospholipid metabolism. For the sake of clarity, only some of the enzymatic steps are shown. The conversion of PhA to PhI and PhG and that of DAG to neutral phospholipids occurs mainly in the endoplasmic reticulum. The other steps take place primarily in the plasma membrane.
PhCMP = phosphatidylCMP (CDP-diacylglycerol)
PhG = phosphatidylglycerol
PhIP = phosphatidylinositol-4-phosphate
PhIPP = phosphatidylinositol-4,5-bisphosphate

Dawson, 1980; Shukla, 1982) yielding DAG and inositol cyclic-1,2 phosphate (Dawson et al., 1971) and that the usual measurement of the events as elevated incorporation of ^{32}P into PhI and other acidic phospholipids reflects the secondary resynthesis of the starting material. Nonetheless this experimental approach is useful for analyzing many facets of the phenomenon, although whenever possible the more difficult determinations of PhI breakdown or disappearance are indicated. There are also suggestions that changes in the metabolism of the polyphosphoinositides, PhI-4-phosphate and PhI-4,5-bisphosphate, are intimately related to that of PhI following receptor activation, although the precise role of the more highly phosphorylated compounds has not yet been fully clarified (Abdel-Latif et al., 1977; Soukup et al., 1978; Soukup and Schanberg, 1982). The possibility has been raised that their degradation is the initial reaction and that PhI resynthesis follows its disappearance which is caused by phosphorylation for PhI-4-phosphate and PhI-4,5-bisphosphate replacement (Michell and Kirk, 1982). This alternative is also indicated in Figure 3.

Cerebral Cortex

Whichever of these pathways is eventually shown to be

initially stimulated, the receptors responsible have been identified in many tissues (see reviews cited above). We have used pharmacological agents as relatively specific antagonists to characterize the cholinergic and adrenergic receptor types in rat cerebral cortex in the course of studies on the interaction between agonist stimulation and K^+ depolarization (Table 2) (Majewska and Hauser, 1981). Stimulation of labeling of both PhI and PhA by NE was observed. The involvement of α_1-adrenergic receptors in the action of NE is demonstrated by the fact that neither the β-antagonist sotalol nor the α_2-antagonist yohimbine was

Table 2

Pharmacology of Stimulated Labeling of Phosphatidylinositol in Rat Brain Mince

Additions	Increase in Labeling	
	PhI	PhA
	percent	
None	0 ± 12.4	0 ± 11.5
NE	85.2 ± 10.1	168.7 ± 20.5
NE + prazosin	20.9 ± 11.8	17.4 ± 8.7
NE + WB4101	4.0 ± 17.4	4.0 ± 13.0
NE + yohimbine	111.3 ± 18.6	172.2 ± 11.7
NE + sotalol	93.0 ± 16.5	195.7 ± 21.2
CCh	74.1 ± 8.6	66.8 ± 3.9
CCh + atropine	2.2 ± 5.8	3.3 ± 8.6
CCh + d-tubocurarine	86.6 ± 11.2	72.4 ± 10.9

Mince was prepared from cerebral cortex slices of adult rats using a McIlwain tissue chopper. Incubations were for 1 h in 0.5 ml of Krebs-Ringer solution buffered to pH 7.4 with N-2-hydroxyethylpiperazine-N'-2-ethanesulfonic acid at 37° and contained, where indicated, NE, 0.1 mM; CCh, 1.0 mM; other drugs, 0.01 mM as well as 5 μCi of [^{32}P]orthophosphate. Reactions were stopped with chloroform-methanol, 2:1, and the extracted and washed lipids separated by thin-layer chromatography prior to scraping and counting as described previously (Hauser and Smith, 1981; Smith and Hauser, 1981).

able to affect the NE-induced labeling increases. On the other hand, both prazosin and WB4101, selective α_1-antagonists, prevented these effects. In the cholinergic system relatively high concentrations of added CCh were required to elicit an elevation of ^{32}P incorporation into PhI and PhA. This was precluded by the presence of atropine but not of d-tubocurarine, pointing to the sole participation of muscarinic rather than nicotinic receptors.

Anterior Pituitary Cells

We have also used the ability of agonists to change phospholipid metabolism as an indicator for the existence of the pertinent receptors. Thus, in order to test whether muscarinic cholinergic receptors are present in rat anterior pituitary gland, we examined the influence of cholinergic agonists on phospholipid labeling patterns in dissociated cells from this organ. Under the conditions used CCh increased the relative incorporation of ^{32}P into PhA, decreased that into PhI and left PhC labeling unchanged (Table 3). Similar results were obtained with pilocarpine and oxotremorine (Hauser and Parks, 1981). Atropine, scopolamine or quinuclidinyl benzilate but not d-tubocurarine abolished these effects. Although the action of these agonists did

Table 3

CCh-Induced Phospholipid Labeling Changes in Rat Anterior Pituitary Cells

Phospholipid	Incorporation of $^{32}P_i$ [1]	
	CCh	CCh + Atropine
PhA	1.60 ± 0.13	1.16 ± 0.16
PhI	0.77 ± 0.06	1.14 ± 0.09
PhC	1.13 ± 0.08	1.09 ± 0.07

Anterior pituitaries from adult rats were dissociated by digestion with collagenase and hyaluronidase (Hauser and Smith, 1981). Resuspended cells were incubated with or without 10^{-4} M CCh and 10^{-7} M atropine for 1 h (see legend to Table 2).

[1]Relative to appropriate control = 1.00.

not result in the classical elevation of PhI labeling, but rather in a shift from PhI to PhA, there is nevertheless a clear indication of a role for muscarinic cholinergic receptors. This information supplements results obtained by radioactive ligand binding (Mukherjee et al., 1980) and confirms the utility of the phospholipid effect in identifying receptor classes.

Pinealocytes

Similar findings in an adrenergic system were made with dissociated rat pineal cells, where in agreement with intact pineal glands (Smith et al., 1979) and other systems (Jones and Michell, 1978; Fain and García-Sáinz, 1980) α_1-receptors were shown to mediate the phospholipid effect (Table 4). The major stimulation in this preparation is in PhI and is inhibited by prazosin and WB4101 but not by sotalol or yohimbine.

Table 4

Phospholipid Effect in Rat Pinealocytes

Additions	Incorporation of $^{32}P_i$ PhC	Incorporation of $^{32}P_i$ PhI	PhI[1]/PhC
	cpm/h	cpm/h	
None	2495 ± 155	2320 ± 246	0.93
NE	2260 ± 340	6773 ± 963	3.00*
NE + prazosin	2687 ± 594	2762 ± 546	1.06
NE + WB4101	2509 ± 267	2785 ± 190	1.11
NE + sotalol	2810 ± 543	8197 ± 443	2.92*

Dissociated rat pinealocytes (Hauser and Smith, 1981) were incubated as described in the legend to Table 2 and contained, where indicated, NE 0.1 mM, prazosin 3.3 μM, WB4101 or sotalol 0.01 mM.

* = significant stimulation ($p < 0.05$).

[1]Incorporation ratio = radioactivity in PhI/radioactivity in PhC.

These cells and the phospholipid labeling approach have also been used to probe the nature of the receptors through the use of fluorinated NEs. These substances, prepared as ring-fluorinated analogues and made available through the cooperation of Dr. K. L. Kirk at NIAMDD-NIH, exert differential effects on α- and β-adrenergic systems of brain, as shown by stimulation of cyclic AMP formation and displacement of α- and β-specific ligands (Kirk et al., 1979; Daly et al., 1980). Using these two measurements the authors obtained data consistent with a selective agonist action at α- or β-adrenergic receptors when 6- or 2-fluoro-NE, respectively, was tested. In order to extend this type of study to the receptor-mediated phospholipid effect we compared the potency of NE with that of the two fluoro-derivatives. The effects on ^{32}P incorporation of NE and the 6-fluoro derivative were identical, whereas that of the 2-fluoro derivative was not only elicited solely at a higher concentration, but also gave a much lower plateau (Fig. 4). Very low concentrations of the α_1-adrenergic antagonist prazosin blocked these effects. Here, too, the stimulation by 2-fluoro-NE differed by being more susceptible to lower levels of the blocking agent (Table 5). It should be

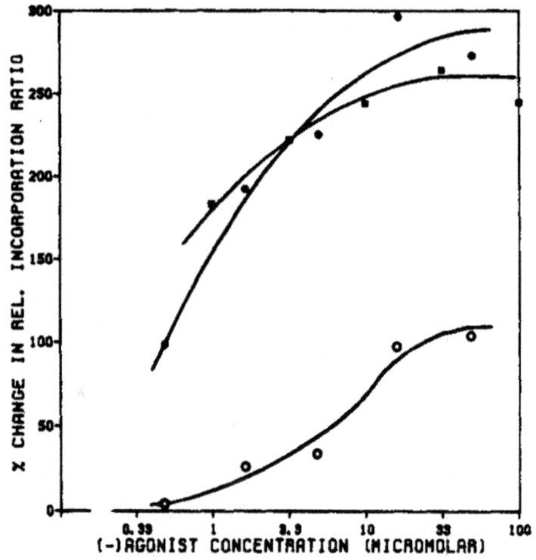

Fig. 4. Phospholipid effect elicited by fluoro-NEs in rat pinealocytes. Incubations of dissociated pinealocytes and determinations of incorporation of $^{32}P_i$ into phospholipid classes were done as indicated in the legend to Table 2. Incorporation ratio = radioactivity in PhI/radioactivity in PhC. ■ = NE; • = 6F-NE; o = 2F-NE.

Table 5

Inhibition by Prazosin of the Phospholipid Effect Elicited by Fluoro-NEs in Rat Pinealocytes

Concentration of prazosin	NE	2F-NE	6F-NE
μM		percent inhibition	
0.01	29.9	59.7	34.3
0.1	42.5	94.0	60.4
0.23	60.4	n.d.	74.6
1.0	84.3	n.d.	91.0
2.3	94.0	n.d.	94.8

Dissociated rat pinealocytes were incubated as described in the legend to Table 2. Agonists were 0.1 mM. n.d. = not determined.

illuminating to compare the agonist effects on the other, apparently closely related parameters, Ca^{2+} movements, cyclic GMP formation and arachidonate release. This should yield further information on their interrelationships.

The results with pineal cells support the order of potency of the fluorinated NEs at α_1-receptors as seen in cerebral cortex and other systems, although the specificity was not absolute. The same can be concluded for β-receptors where 6-fluoro-NE elicited cyclic AMP accumulation, albeit at concentrations two orders of magnitude greater than the unsubstituted or 5-substituted neurotransmitter. The 2-substituted compound had intermediate potency (Auerbach et al., 1981). Inasmuch as different responses are altered in a characteristic pattern depending only on the receptor type initially activated, it appears unlikely that the linkage to events on the opposite side of the membrane is affected by the molecular species of NE. Thus the coupling of the receptor-ligand complex to the cleavage of phosphoinositides does not seem to be responsible for the differential effects observed. It is more likely that receptor binding affinity is altered by the substituent, possibly owing to electronic factors. Whether the lipid environment plays a role in bringing about the very different relative effects at α- and β-receptor sites is not known, although fluorine substitution does in general increase the lipophilic character of the compounds.

EFFECTS OF PHOSPHOLIPASE TREATMENT

That the phospholipid composition of the cell surface may have a specific influence on the stimulation of phospholipid turnover by agonists is shown in our preliminary experiments on the modification of the membrane by enzyme treatment. When pinealocytes were incubated with phospholipases, the subsequent basal incorporation of ^{32}P into PhI remained unaltered (Table 6). Similarly the ability of NE to increase the labeling was not changed after phospholipase A_2 and only marginally reduced after PhI-specific phospholipase C treatment. In contrast, phospholipase D treatment, which results in PhA formation, markedly decreased the stimulation, so that the increment was less than half as great as in controls (Table 6). We plan to exploit this initial finding to test whether this and other enzymatic and chemical alterations of the plasma membrane affect ligand binding in either α_1-adrenergic or muscarinic cholinergic systems and have differential consequences on the ensuing membrane and intracellular events.

Table 6

Effect of Phospholipase Treatment on the Phosphatidylinositol Effect in Rat Pinealocytes

Treatment	Relative Incorporation into Phosphatidylinositol	
	Basal	NE-stimulated
None	1.00 ± 0.19	3.50 ± 0.39
Phospholipase A_2	0.90 ± 0.17	3.39 ± 0.28
Phospholipase C*	1.20 ± 0.30	2.72 ± 0.62
Phospholipase D	1.06 ± 0.27	2.07 ± 0.52†

Dissociated pinealocytes were incubated with or without enzyme for 10 min, recovered by centrifugation and reincubated with or without 30 μM NE for 60 min as described in the legend to Table 2. * from S. aureus (specific for PhI, courtesy of Dr. M. Low). † significantly reduced ($p < 0.05$).

CALCIUM REQUIREMENT AND MOBILIZATION

The most intriguing question that has plagued investigators in this field concerns the physiological significance of the receptor-mediated increase in phospholipid turnover. The most widely quoted heuristic suggestion is that this event controls Ca^{2+}-gates, thus permitting the entry of Ca^{2+} and therefore the elevation of cytosolic Ca^{2+} levels for further activation of intracellular events (Michell, 1975, 1979).

It is well established, as mentioned before, that the class of receptors being considered operates through Ca^{2+} as the output signal, but the relationship to the changes in phospholipid metabolism occurring in parallel is less clear. While the correlation exists in all systems so far investigated, there is controversy as to the sequence of events, namely whether hydrolysis of PhI or PhI-4,5-bisphosphate precedes Ca^{2+} gating or vice versa (Michell and Kirk, 1981, 1982).

The arguments revolve in part around the requirement for Ca^{2+} of the PhI-specific phospholipase C and have been advanced in detail by the protagonists in this polemic (Michell and Kirk, 1981, 1982; Cockcroft, 1981; Hawthorne, 1982; Michell, 1982). Even though Ca^{2+} is required for the cytoplasmic enzyme, there is uncertainty whether it responds to relatively small increases in intracellular Ca^{2+} and whether, therefore, the changes in cation concentration control phosphoinositide breakdown.

Another set of observations concerns the calcium sensitivity of the phospholipid effect in intact tissues or cells. It has generally been found that the increased turnover of PhI is independent of Ca^{2+} added to the incubation medium, but a number of exceptions have been reported (Michell and Kirk, 1981). Ca^{2+} omission and the addition of chelating agents reduced or abolished the response to adrenergic or muscarinic agonist stimulation in synaptosomes (Fisher and Agranoff, 1980) and cerebral cortex mince (Majewska and Hauser, unpublished observations) at least as measured by ^{32}P incorporation. These manipulations similarly modified the effect in rat pinealocytes (Fig. 5) although the addition of the ionophore A23187 did not restore or alter the response (Smith and Hauser, 1981). While these and other findings do not negate a role of PhI breakdown in

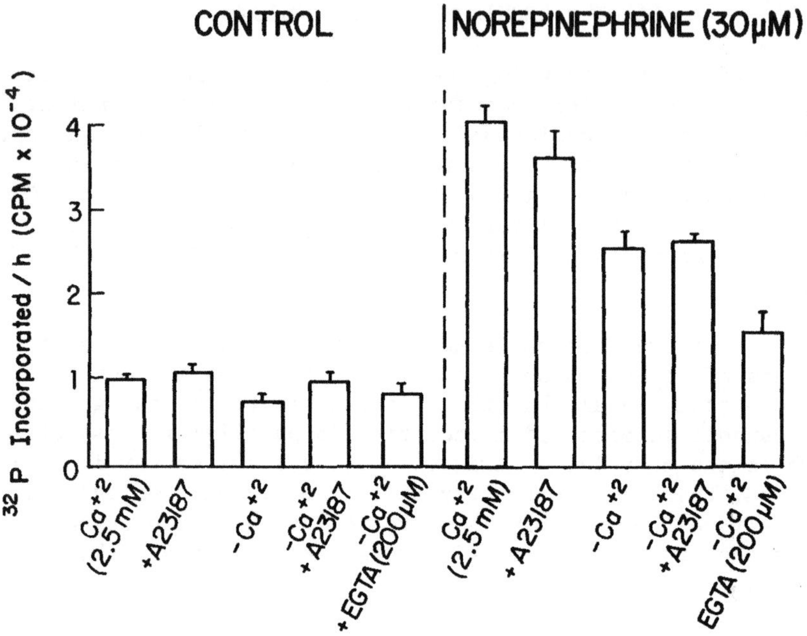

Fig. 5. Modification of basal and stimulated $^{32}P_i$ incorporation into PhI by changes in calcium availability in rat pinealocytes. Dissociated pinealocytes were incubated as described in the legend to Table 2. A23187 was 0.01 mM.

Ca^{2+} entry or mobilization, they do suggest caution in drawing universally applicable conclusions. It is conceivable that different mechanisms operate in nervous tissue than elsewhere although the postulate of a unified set of processes is certainly more attractive in visualizing the transfer of the receptor-mediated signal.

The manner in which the elevation of intracellular Ca^{2+} concentration could occur is through an increase in the calcium permeability of the membrane, possibly in conjunction with a release of Ca^{2+} from intracellular stores in the plasma membrane, mitochondria or endoplasmic reticulum. The participation of lipids in these putative events might involve several changes consequent to the cleavage of phosphoinositides. The most striking result of the removal of the phospholipid head group is the change in charge, and this might result in altered membrane fluidity. Specifically, if the highly anionic polyphosphoinositides are converted to DAG, bound Ca^{2+} would be liberated. Further, the removal of negative charges could affect the binding to specific proteins concerned with channel formation and similarly influence the mobility of proteins or complexes within the membrane, permitting the transient entry or release of cations.

One proposal which has received serious consideration deals with the role of PhA formed in the plasma membrane from the DAG liberated by the breakdown of PhI, especially through stimulated phospholipase C action. This compound has the ability to aid in the transfer of Ca^{2+} across a single bilayer and has been postulated to act in a similar manner in biological systems. Increased PhA formation as the result of receptor-mediated PhI degradation might provide a higher level of this putative Ca^{2+} ionophore and participate in the regulation of Ca^{2+} gating (reviewed in Putney, 1981).

CYCLIC GMP ELEVATION

One of the molecular events associated with receptor activation is increased cyclic GMP formation which has been shown in both the central and peripheral nervous systems (Goldberg and Haddox, 1977). This connection has been demonstrated for muscarinic receptors, most strikingly in mouse neuroblastoma cells, clone N1E-115 (Fig. 6) (Matsuzawa and Nirenberg, 1975), where stimulation of cyclic GMP production by CCh is prevented by atropine, but not by d-tubocurarine. This stimulation is parallel to the increase in PhI turnover in this cell line (Schmidt et al., 1980). A model for the interaction between the agonist, the muscarinic receptor and the effector has been proposed (Richelson and El-Fakahany, 1981), but does not address the role of lipids in complex formation or sensitivity changes. Analogous cyclic GMP increases also occur through adrenergic receptors, for example in rat pineal gland (O'Dea and Zatz, 1976), which is particularly relevant for our studies. Pharmacological characterization of the receptors involved has shown them to be of the α_1 type (Klein, D.C., personal communication) and therefore to be mediated by the same receptor class as the PhI response (Smith et al., 1979).

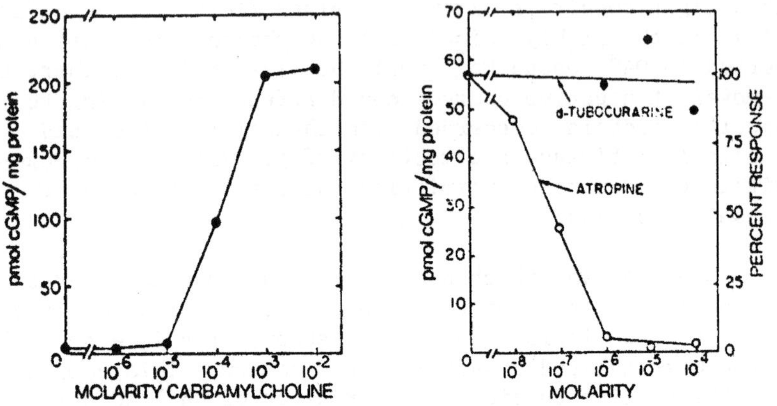

Fig. 6. Muscarinic elevation of cyclic GMP levels in neuroblastoma N1E-115 cells. Cells, grown for 9 days, were incubated for 30 sec. 0.5 mM 3-isobutyl-1 methyl-xanthine was present in the CCh dose-response experiments. CCh in the antagonist experiments was 1 mM. (From Matsuzawa and Nirenberg, 1975; Dr. M. W. Nirenberg kindly gave permission to reproduce these figures.)

Not only does there appear to be a relationship between cyclic GMP formation and phospholipid metabolism, but the guanylate cyclase itself may be regulated by lipids. This suggestion arises from experiments in vitro with neural tissues and cells where the activity of the enzyme is enhanced by lysolecithin and free, preferentially unsaturated fatty acids (Zwiller et al., 1976; Asakawa et al., 1978). These may arise as the result of phospholipase A_2 action (Sulakhe et al., 1976; Zwiller et al., 1982) and act in the form of the hydroperoxide produced by lipoxygenase (White et al., 1982). The guanylate cyclase appears also to be regulated by Ca^{2+}, and its level of activity may therefore depend on the increment in free Ca^{2+} in the cytosol brought about by receptor-mediated phospholipid breakdown.

The role of the increased cyclic GMP levels, as of intracellular cyclic GMP in general, is not well established. Most likely at least one of the functions is to regulate the activity of specific protein kinases which in turn phosphorylate specific proteins in brain as in other tissues (Glass and Krebs, 1980). However, very little detail on these enzymes or their substrates in the nervous system is known, let alone on the involvement of lipids in their modulation. In this context, a particular example of lipid participation in the events following activation of certain receptors comes from the studies of the group of Nishizuka (see Takai et al., 1982, for review). These investigators have discovered an enzyme designated protein kinase C which is activated by Ca^{2+} and dependent on the presence of phospholipids, particularly PhS. A most striking enhancement of enzyme activity is achieved by the addition of small amounts of DAG, but only if the lipid contains at least one unsaturated fatty acid group. Neither mono- nor triacylglycerol shows any capacity for activation, whereas DAG produces an increase in affinity of the enzyme for both phospholipid and Ca^{2+}.

It is intriguing to consider that the hydrolysis of PhI, increased following receptor activation, would provide additional amounts of DAG of the required fatty acid pattern. This could contribute to the phosphorylation of specific proteins and thus fulfill a second messenger function as well as serving as precursor for PhA formation, thus providing greater amounts of a Ca^{2+} ionophore as discussed earlier.

ARACHIDONIC ACID RELEASE

A further metabolic reaction which has recently gained prominent consideration is the liberation of the tetraenoic arachidonic acid from position 2 of the glycerol backbone of lipids. This reaction, catalyzed by a lipase and, like guanylate cyclase, requiring Ca^{2+}, is vital in providing free arachidonic acid as precursor for a number of physiologically important eicosanoids. Although little studied so far in the nervous system, activation of the appropriate receptors increases arachidonic acid release in parallel with the other phenomena mentioned so far (Berridge, 1981). While this close relationship has been established, the sequence of events is still uncertain, as is the exact source of the fatty acid.

PhC, PhE, PhI and PhS all contain arachidonic acid in different proportions and could thus serve as the metabolically important reservoir. Phospholipase A_2 is the enzyme responsible for the cleavage of the acyl group from these phospholipids and may therefore control the availability of the free fatty acid for further conversion to prostaglandins, prostacyclins, thromboxanes and leukotrienes (see Wolfe, 1982, for a review of these pathways). These molecules, which have a wide range of biological effects (Samuelsson et al., 1978), can indirectly be considered output signals and are, at the same time, lipid in nature.

Another candidate as the direct source of arachidonic acid is DAG rather than phospholipids with DAG lipase the responsible enzyme. This enzyme is also stimulated by Ca^{2+}. These alternatives are diagrammed in Figure 7. The pool of DAG involved in elevated arachidonic acid cleavage could at least in part arise from the agonist-stimulated PhI cleavage. It would need to fulfill this role in addition to its other postulated functions, namely as precursor for PhA and for neutral phospholipids and as cofactor for proteinase C, described elsewhere in this presentation.

It is clear that there is considerable uncertainty about the regulatory mechanisms operating to control the level of free arachidonic acid in the cell (Irvine, 1982). While it is known that the rate of its liberation is the limiting event in prostaglandin synthesis and that prosta-

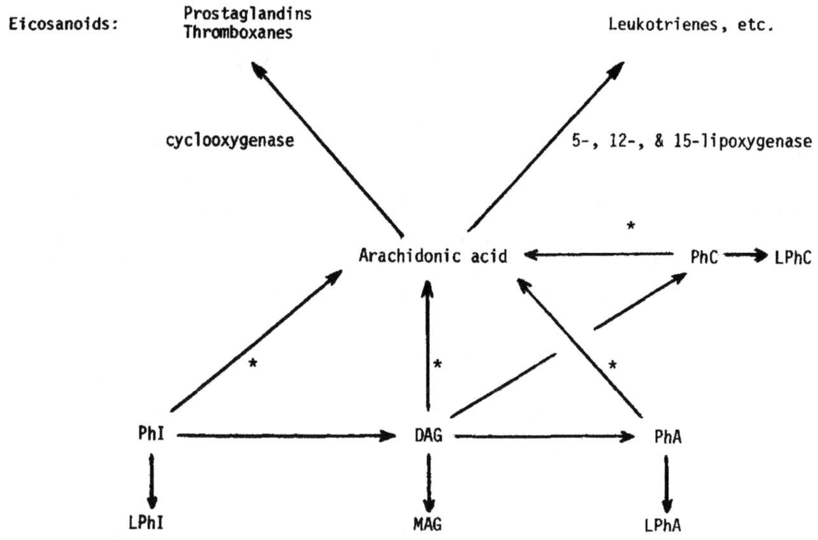

Fig. 7. Pathways of arachidonic acid metabolism. MAG = monoacylglycerol; LPhX = lysophosphatidylX. * = Ca^{2+}-activated lipases acting to liberate arachidonic acid (20:4) from C-2 of the glycerol moiety.

glandins are released in response to agonist action (e.g., in pineal gland [Ritta and Cardinali, 1982]), relatively little definitive information is available on the enzymes uniquely involved in this process, on their specificity for substrate pools, on the role of free intracellular Ca^{2+} or on the factors which affect the balance between acyl ester hydrolysis and reacylation catalyzed by acyltransferase(s). The connection between arachidonic acid turnover and Ca^{2+} entry has been more extensively studied in secretory cells (Rubin, 1982).

Information is sketchy, however, on the shift in the deacylation-reacylation cycle occurring upon activation of the receptors which also apparently initiate phosphoinositide breakdown via phospholipase C action in nervous tissue. Whether Ca^{2+} influx, membrane structure changes, enzyme activation or substrate availability is the critical factor must await further experimentation. This is made difficult inasmuch as the receptor-mediated phenomena occur only in intact cells, but not in broken cell preparations.

Our only experimental evidence that rapid degradation of several phospholipids occurs in brain upon additions of agonist in vitro is the disappearance of label from cerebral cortex mince prelabeled with ^{32}P (Fig. 8). In incubations with either NE or CCh, radioactivity disappeared from PhC, PhI and PhA during the first 20 sec after addition of the agonist, but increased again after the initial drop. We interpret the fact that all three of the phospholipids, virtually the only ones labeled, exhibited the same decrease, as being due to fatty acid loss rather than phospholipase C cleavage which should not have affected PhC. Clearly, further experiments with radioactive arachidonic acid will be necessary to confirm this hypothesis.

LIPID REQUIREMENTS FOR RECEPTOR ACTIVATION AND ENZYME ACTIVITIES

Several of the areas apart from metabolism changes, where lipids may be involved in the transfer of messages across cell membranes, have already been alluded to. They are in particular the binding of agonists to receptors, the coupling of the receptor to the effector and the generation of a messenger, in many cases through enzyme action (see Fig. 1). Our knowledge of the role of membrane lipids in

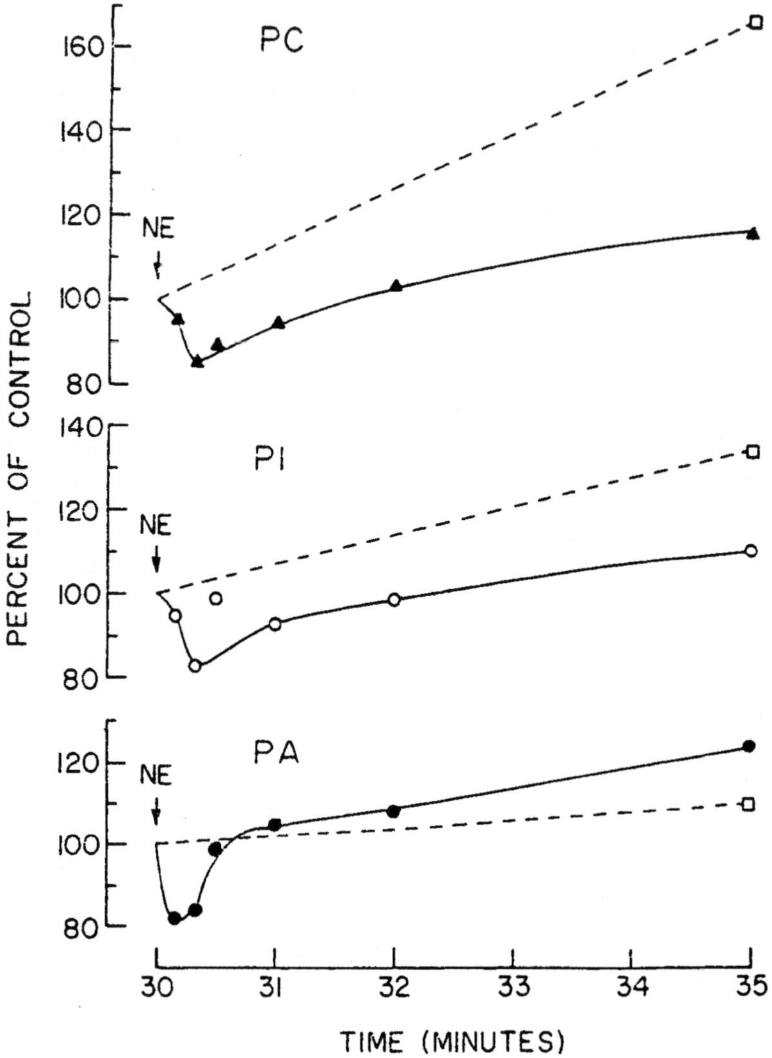

Fig. 8. Changes in ^{32}P content of prelabeled phospholipids in rat brain cortex mince following brief stimulation with NE. Aliquots of mince were incubated as described in the legend to Table 2 at 28°C. After 30 min NE was added (final concentration 0.1 mM) and the reaction was stopped at appropriate times with chloroform-methanol, 2:1. Lipids were separated and counted by standard techniques. Distribution of ^{32}P after prelabling for 30 min was (percent of total): PhA, 46; PhC, 15; PhI, 39. Dashed line, control; solid line, stimulated with NE.

receptor mechanisms has been reviewed by Loh and Law (1980). In addition to the changes in metabolism brought about by receptor activation, the binding of the ligand itself may depend on the presence of lipids, particularly acidic lipids. It is questionable whether the recognition sites of any receptor actually consist wholly or in part of lipid as some experiments _in vitro_ might suggest. However, the lipid matrix in which receptors are embedded can undoubtedly alter both the availability of binding sites and their affinity for agonists. Membrane fluidity and lipid-protein interactions are probably involved as regulators of the binding activity. Studies using phospholipase treatment prior to binding assays have shown, for example, reduction in quinuclidinyl benzilate binding to cholinergic receptors (Aronstam et al., 1977), supporting the assumption that the lipid environment affects the availability or affinity of receptor sites.

There is the further likelihood that the agonist-receptor complex must move within the membrane in order to couple to whatever effector may be pertinent. This can be adenylate cyclase where coupling occurs through the guanine nucleotide regulatory protein in the case of β-adrenergic receptors or possibly calcium channels in the case of receptors using Ca^{2+} influx as output signal. The information transfer could be controlled at this stage by the ability of the activated receptor to move laterally, as the result of conformation changes for which altered intermolecular forces affecting the structure and fluidity of the lipid bilayer would be some of the determining factors (see Boggs, 1980). In the context of the phospholipid effect the lateral movement of the agonist-receptor complex may favor the rearrangement of a particular PhI pool so that the molecules are caused to project from the cytoplasmic face of the membrane and thus become susceptible to hydrolysis by the soluble phospholipase C.

Finally, the enzymes involved in either stimulated phospholipid turnover or in messenger production may be modulated by lipids. Such influences can occur through specific binding of a lipid to a protein either through ionic or Van der Waals linkages so that the hydrophilic-hydrophobic relationships between relevant portions of the protein and the lipid assume an important role (Quinn, 1980).

The effect of lipids on guanylate cyclase has already

been mentioned. Similar considerations apply to adenylate cyclase (Ross and Gilman, 1980). In a mitochondrial fraction from rat cerebral cortex, for example, added PhS or PhI materially enhanced both basal and hormone-stimulated enzyme activity, perhaps because the functional lipid environment of the enzyme was disrupted during isolation of the preparation (Von Hungen and Roberts, 1973). As in other cells (e.g., Sinensky et al., 1979), cholesterol content and the ordering of lipid acyl chains could play important roles in adenylate cyclase activity in the nervous system.

The enzymes responsible for initiating the turnover of components of the PhI molecule, phospholipases A_2 and C, as well as those concerned with its resynthesis are all presumably subject to similar effects, and even cytoplasmic forms may be influenced by lipid cofactors. Acidic phospholipids and free unsaturated long chain fatty acids appear the most effective modulators in most of the systems examined, but space does not permit a detailed description of these studies in this discussion.

The most carefully studied enzyme of relevance in the context of signal transmission is the protein kinase C (Katoh and Kuo, 1982; Takai et al., 1982). Its dependence on PhS and Ca^{2+} and activation by DAG have already been mentioned and constitute a clear example of the importance of lipids in the function of enzyme proteins.

SUMMARY

The foregoing considerations make it evident that the transmission of information from the extracellular to the intracellular environment can involve lipids in several ways, some of which are better understood than others. The activation of receptors by agonists, the coupling of the receptor-ligand complex to effectors and the function of the effector in generating a messenger within the cell are three stages which can be profoundly influenced directly or indirectly by lipids within the membrane bilayer or generated in the course of metabolic conversions. Local or generalized changes in viscosity and hence protein mobility may be involved in regulating signal transmission (Harris and Hitzemann, 1981).

The two receptor classes considered in this presentation appear intimately connected to specific and characteristic metabolic changes, PhE methylation and phosphoinositide turnover, which play as yet imperfectly defined roles in the cascade of events initiated by ligand binding. Some of the lipid molecules produced may themselves act as messengers and facilitate subsequent events.

In the case of α_1-adrenergic and certain muscarinic cholinergic receptors, these events include, in addition to phospholipid metabolism changes, Ca^{2+} influx and mobilization, cyclic GMP formation and cleavage of lipids to liberate arachidonic acid. The complex relationships among them will require considerable effort using different approaches for complete elucidation. Two lines of experimentation are the modification of the cell surface through enzyme treatment and the specific inhibition of one of the events (Fig. 9) which can profoundly influence the balance in the agonist-initiated sequence. They should aid in clarifying the mechanism of signal transmission where the essential participation of lipids is only beginning to be fully appreciated.

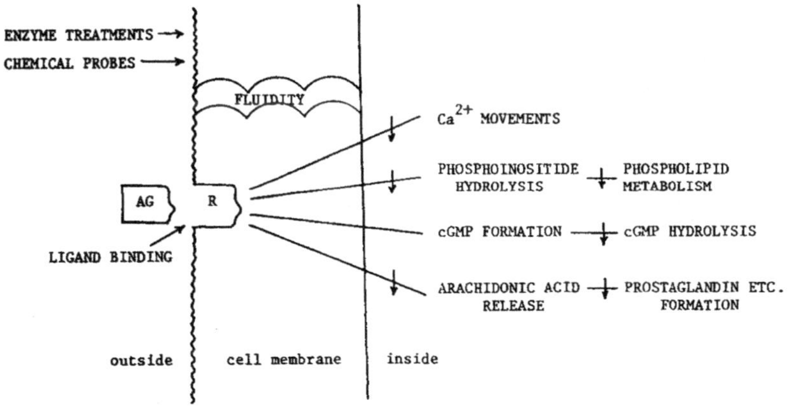

Fig. 9. Approaches to the study of signal transmission across cell membranes. Tissue or cell preparations may be treated with enzymes or chemical reagents before assay of the effect of agonists on the indicated parameters, or the incubation itself may be carried out in the presence of a specific inhibitor before the addition of an agonist to test its effects. AG = agonist; R = receptor; ↓ = site of specific inhibitor action.

ACKNOWLEDGMENTS

The experimental work discussed in this presentation was done in collaboration with Drs. M. Dorota Majewska, Joe Parks and Thomas L. Smith. Ms. Johanne H. Khan provided excellent secretarial services. Financial support came from research grants NS06399 (National Institutes of Health) and PCM 7824387 (National Science Foundation) and Biomedical Research Support Grant RR05484 awarded by the Division of Research Resources, NIH, to McLean Hospital.

REFERENCES

Abdel-Latif A.A., Akhtar R.A., and Hawthorne, J.N. (1977) Acetylcholine increases the breakdown of triphosphoinositide of rabbit iris muscle prelabeled with [^{32}P] phosphate. Biochem. J. 162, 61-73.

Aronstam R.S., Abood L.G., and Baumgold J. (1977) Role of phospholipids in muscarinic binding by neural membranes. Biochem. Pharmacol. 26, 1689-1695.

Asakawa T., Takenoshita M., Uchida S., and Tanaka S. (1978) Activation of guanylate cyclase in synaptic plasma membranes of cerebral cortex by free fatty acids. J. Neurochem. 30, 161-166.

Auerbach D.A., Klein D.C., Kirk K.L., Cantacuzene D., and Creveling C.R. (1981) Effects of fluorine analogs of norepinephrine on stimulation of cyclic adenosine 3',5'-monophosphate and binding to β-adrenergic receptors in intact pinealocytes. Biochem. Pharmacol. 30, 1085-1089.

Berridge M.J. (1981) Phosphatidylinositol hydrolysis: a multifunctional transducing mechanism. Mol. Cell. Endocrinol. 24, 115-140.

Blusztajn J.K., Zeisel S.H., and Wurtman R.J. (1979) Synthesis of lecithin (phosphatidylcholine) from phosphatidylethanolamine in bovine brain. Brain Res. 179, 319-327.

Boggs J.M. (1980) Intermolecular hydrogen bonding between lipids: influence on organization and function of lipids in membranes. Can. J. Biochem. 58, 755-770.

Cockcroft S. (1981) Does phosphatidylinositol breakdown control the Ca^{2+}-gating mechanism? Trends Pharmacol. Sci. 2, 340-342.

Crews F.T., Hirata F., and Axelrod J. (1980) Identification and properties of methyltransferases that synthesize phosphatidylcholine in rat brain synaptosmes. J. Neurochem. 34, 1491-1498.

Daly J.W., Padgett W., Nimitkitpaisan Y., Creveling C.R., Cantacuzene D., and Kirk K.L. (1980) Fluoronorepinephrines: specific agonists for the activation of alpha and beta adrenergic-sensitive cyclic AMP-generating systems in brain slices. J. Pharmacol. Exp. Ther. 212, 382-389.

Dawson R.M.C., Freinkel N., Jungalwala F.B., and Clarke N. (1971) The enzymic formation of myo-inositol 1:2-cyclic phosphate from phosphatidylinositol. Biochem. J. 122, 605-607.

Fain J.N. and García-Sáinz J.A. (1980) Role of phosphatidylinositol turnover in $alpha_1$ and of adenylate cyclase inhibition in $alpha_2$ effects of catecholamines. Life Sci. 26, 1183-1194.

Fisher S.K. and Agranoff B.W. (1980) Calcium and the muscarinic synaptosomal phospholipid labeling effect. J. Neurochem. 34, 1231-1240.

Friedel R.O. and Schanberg S.M. (1972) Effects of carbamylcholine and atropine on incorporation in vivo of intracisternally injected $^{33}P_i$ into phospholipids of rat brain. J. Pharmacol. Exp. Ther. 183, 326-332.

Glass D.B. and Krebs E.G. (1980) Protein phosphorylation catalyzed by cyclic AMP-dependent and cyclic GMP-dependent protein kinases. Annu. Rev. Pharmacol. Toxicol. 20, 363-388.

Goldberg N.D. and Haddox M.K. (1977) Cyclic GMP metabolism and involvement in biological regulation. Annu. Rev. Biochem. 46, 823-896.

Harris R.A. and Hitzemann R.J. (1981) Membrane fluidity and alcohol actions, in Currents in Alcoholism, Vol. VIII (Galanter M., ed.), pp. 379-404. Grune & Stratton, New York.

Hauser G. and Parks J.M. (1981) Muscarinic receptors on pituitary cells: phospholipid metabolism changes. Transact. Am. Soc. Neurochem. 12, 194.

Hauser G. and Smith T.L. (1981) Characteristics of the norepinephrine-stimulated phosphatidylinositol turnover in rat pineal cell dispersions. Neurochem. Res. 6, 1067-1079.

Hawthorne J.N. (1982) Is phosphatidylinositol now out of the calcium gate? Nature 295, 281-282.

Hawthorne J.N. and Pickard M.R. (1979) Phospholipids in synaptic function. J. Neurochem. 32, 5-14.

Hirata F. and Axelrod J. (1980) Phospholipid methylation and biological signal transmission. Science 209, 1082-1090.

Hokin L.E. and Hokin M.R. (1955a) Effects of acetylcholine on phosphate turnover in phospholipids of brain cortex in vitro. Biochim. Biophys. Acta 16, 229-237.
Hokin L.E. and Hokin M.R. (1955b) Effects of acetylcholine on the turnover of phosphoryl units in individual phospholipids of pancreas slices and brain cortex slices. Biochim. Biophys. Acta 18, 102-110.
Hokin L.E. and Hokin M.R. (1958) Acetylcholine and the exchange of inositol and phosphate in brain phosphoinositide. J. Biol. Chem. 233, 818-822.
Irvine R.F. (1982) How is the level of free arachidonic acid controlled in mammalian cells? Biochem. J. 204, 3-16.
Irvine R.F. and Dawson R.M.C. (1980) The control of phosphatidylinositol turnover in cell membranes. Biochem. Soc. Transact. 8, 27-30.
Irvine R.F., Dawson R.M.C., and Freinkel N. (1982) Stimulated phosphatidylinositol turnover--a brief appraisal, in Contemporary Metabolism, Vol. 2 (Freinkel N., ed.), pp. 301-342. Plenum Medical Book Co., New York.
Johnson G.L., Kaslow H.R., Farfel C.Z., and Bourne H.R. (1980) Genetic analysis of hormone sensitive adenylate cyclase. Adv. Cyclic Nucleotide Res. 13, 2-38.
Jones L.M. and Michell R.H. (1978) Stimulus-response coupling at α-adrenergic receptors. Biochem. Soc. Transact. 6, 673-688.
Karnovsky M.J., Kleinfeld A.M., Hoover R.L., and Klausner R.D. (1982) The concept of lipid domains in membranes. J. Cell Biol. 94, 1-6.
Katoh N. and Kuo J.F. (1982) Subcellular distribution of phospholipid-sensitive calcium-dependent protein kinase in guinea pig heart, spleen and cerebral cortex, and inhibition of the enzyme by Triton X-100. Biochem. Biophys. Res. Commun. 106, 590-595.
Kirk K.L., Cantacuzene D., Nimitkitpaisan Y., McCullough D., Padgett W.L., Daly J.W., and Creveling C.R. (1979) Synthesis and biological properties of 2-,5-, and 6-fluoro-norepinephrines. J. Med. Chem. 22, 1493-1497.
Lefkowitz R.J., Caron M.G., Michel, T., and Stadel J.M. (1982) Mechanism of hormone receptor-effector coupling: the β-adrenergic receptor and adenylate cyclase. Fed. Proc. 41, 2664-2670.
Loh H.H. and Law P.Y. (1980). The role of membrane lipids in receptor mechanisms. Annu. Rev. Pharmacol. Toxicol. 20, 201-234.
Majewska M.D. and Hauser G. (1981) Effects of neurotransmitters and depolarizing agents on phospholipid metabolism

in rat cerebral cortex mince. Abstr. 8th Internat. Meeting, Internat. Soc. Neurochem., p. 331.

Margolis R.U. and Heller A. (1966) The effect of cholinergic and other pharmacologic agents on brain monophosphoinositide turnover in vivo. J. Pharmacol. Exp. Ther. 151, 307-312.

Matsuzawa H. and Nirenberg M. (1975) Receptor-mediated shifts in cGMP and cAMP levels in neuroblastoma cells. Proc. Natl. Acad. Sci. USA 72, 3472-3476.

Michell R.H. (1975) Inositol phospholipids and cell surface receptor function. Biochim. Biophys. Acta 415, 81-147.

Michell R.H. (1979) Inositol phospholipids in membrane function. Trends Biochem. Sci. 4, 128-131.

Michell R.H. (1982) Is phosphatidylinositol really out of the calcium gate? Nature 296, 492-493.

Michell R.H. and Kirk C.J. (1981) Why is phosphatidylinositol degraded in response to stimulation of certain receptors? Trends Pharmacol. Sci. 2, 86-89.

Michell R.H. and Kirk C.J. (1982) The unknown meaning of receptor-stimulated inositol lipid metabolism. Trends Pharmacol. Sci. 3, 140-141.

Mukherjee A., Snyder G., and McCann S.M. (1980) Characterization of muscarinic cholinergic receptors on intact rat anterior pituitary cells. Life Sci. 27, 475-482.

O'Dea R.F. and Zatz M. (1976) Catecholamine-stimulated cyclic GMP accumulation in the rat pineal: apparent presynaptic site of action. Proc. Natl. Acad. Sci. USA 73, 3398-3402.

Putney Jr. J.W. (1981) Recent hypotheses regarding the phosphatidylinositol effect. Life Sci. 29, 1183-1194.

Quinn P.J., ed. (1980) Colloquium on Membrane Enzymes in their Lipid Environment. Biochem. Soc. Transact. 8, 25-48.

Richelson E. and El-Fakahany E. (1981) The molecular basis of neurotransmission at the muscarinic receptor. Biochem. Pharmacol. 30, 2887-2891.

Ritta M.N. and Cardinali D.P. (1982) Involvement of α-adrenoceptors in norepinephrine-induced prostaglandin E_2 release by rat pineal gland in vitro. Neurosci. Lett. 31, 307-311.

Ross E.M. and Gilman A.G. (1980) Biochemical properties of hormone sensitive adenylate cyclase. Annu. Rev. Biochem. 49, 533-564.

Rubin R.P. (1982) Calcium-phospholipid interactions in secretory cells: a new perspective on stimulus-secretion coupling. Fed. Proc. 41, 2181-2187.

Samuelsson B., Goldyne M., Granström E., Hamberg M., Hammarström S., and Malmsten C. (1978) Prostaglandins and thromboxanes. Annu. Rev. Biochem. 47, 997-1029.

Schacht J. and Agranoff B.W. (1972) Effects of acetylcholine on labeling of phosphatidate and phosphoinositides by [^{32}P]orthophosphate in nerve ending fractions of guinea-pig cortex. J. Biol. Chem. 247, 771-777.

Schmidt D.M., McGlennen R.C., and Klein W.L. (1980) Receptor-mediated increase in phosphatidylinositol turnover in cloned cell lines. Soc. Neurosci. Abstr. 6, 254.

Shukla S.D. (1982) Phosphatidylinositol specific phospholipases C. Life Sci. 30, 1323-1335.

Sinensky M., Minneman K.P., and Molinoff P.B. (1979) Increased membrane acyl chain ordering activates adenylate cyclase. J. Biol. Chem. 254, 9135-9141.

Singer S.J. and Nicolson G.L. (1972) The fluid mosaic model of the structure of cell membranes. Science 175, 720-731.

Smith, T.L., Eichberg J., and Hauser G. (1979) Postsynaptic localization of the alpha receptor-mediated stimulation of phosphatidylinositol turnover in pineal gland. Life Sci. 24, 2179-2184.

Smith T.L. and Hauser G. (1981) Effects of changes in calcium concentration on basal and stimulated ^{32}P incorporation into phospholipids in rat pineal cells. J. Neurochem. 37, 427-435.

Soukup J. and Schanberg S. (1982) Involvement of alpha noradrenergic receptors in mediation of brain polyphosphoinositide metabolism in vivo. J. Pharmacol. Exp. Ther. 222, 209-214.

Soukup J.F., Friedel R.O., and Schanberg S.M. (1978) Cholinergic stimulation of polyphosphoinositide metabolism in brain in vivo. Biochem. Pharmacol. 27, 1239-1243.

Stadel J.M., DeLean A., and Lefkowitz R.J. (1982) Molecular mechanisms of coupling in hormone receptor-adenylate cyclase systems. Adv. Enzymol. 53, 1-43.

Sulakhe S.J., Leung N.L., and Sulakhe P.V. (1976) Properties of particulate, membrane-associated and soluble guanylate cyclase from cardiac muscle, skeletal muscle, cerebral cortex and liver. Biochem. J. 157, 713-719.

Takai Y., Kishimoto A., and Nishizuka Y. (1982) Calcium and phospholipid turnover as transmembrane signaling for protein phosphorylation, in Calcium and Cell Function, Vol. II, (Cheung W.Y., ed.), pp. 385-412. Academic Press, New York.

Von Hungen K. and Roberts S. (1973) Adenylate cyclase receptors for adrenergic neurotransmitters in rat cerebral cortex. Eur. J. Biochem. 36, 391-401.

White A.A., Karr D.B., and Patt C.S. (1982) Role of lipoxygenase in the O_2-dependent activation of soluble guanylate cyclase from rat lung. Biochem. J. 204, 383-392.

Wolfe L.S. (1982) Eicosanoids: prostaglandins, thromboxanes, leukotrienes, and other derivatives of carbon-20 unsaturated fatty acids. J. Neurochem. 38, 1-14.

Yagihara Y. and Hawthorne J.N. (1972) Effects of acetylcholine on the incorporation of $^{32}P_i$ *in vitro* into the phospholipids of nerve-ending particles from guinea-pig brain. J. Neurochem. 19, 355-367.

Zwiller J., Ciesíelskí-Treska J., and Mandel P. (1976) Effect of lysolecithin on guanylate and adenylate cyclase activities in neuroblastoma cells in culture. FEBS Lett. 69, 286-290.

Zwiller J., Ciesíelskí-Treska J., Ulrich G., Revel M.O., and Mandel P. (1982) Activation of brain guanylate cyclase by phospholipase A_2. J. Neurochem. 38, 856-858.

INVOLVEMENT OF PHOSPHOLIPID METABOLITES IN NEURONAL MEMBRANE FUNCTIONS

Grace Y. Sun, Wilson Tang, M. Dorota Majewska, David W. Hallett, Laurie Foudin and Steven Huang[1]

Sinclair Comparative Medicine Research Farm and Biochemistry Department, University of Missouri, Columbia, MO 65201 USA, and [1]Neuroscience Institute, National Yang Ming Medical College, Taipei, Taiwan, R.O.C.

INTRODUCTION

Since the proposal of the fluid mosaic model of membrane structure by Singer and Nicolson (1972), new research interests were initiated towards understanding the structure and functions of biological membrane. Modern biochemical and biophysical techniques have been developed for probing the membrane properties, and through these studies, unique features of membrane topography have been revealed. Indeed, we are only beginning to realize the complexity of the structural organization of biomembranes and the intricate metabolic relationships among the membrane components. Consequently, the modern concept of biological membrane has evolved from what used to be a simple lipid bilayer structure to include more detailed considerations such as presence of pools, hydrophobicity, charges, channels, pores, asymmetry and microenvironment.

Phospholipids not only form an important part of the structural component of biological membrane, they also provide the membranes with a suitable environment, fluidity, and divalent cation binding sites for expression of their functional activities (Sun and Sun, 1976). Phospholipid molecules are largely synthesized *de novo* by enzymes in the microsomal fraction and are subsequently transported to other membranes througn a carrier mechanism. Within the membrane, enzymes are present for metabolic turnover of the phospholipid molecules (Sun, 1982; and Porcellati, this volume). There is increasing evidence that metabolism of membrane phospholipids is a highly regulated process and that the lipid metabolites probably play an important role in maintaining the functional integrity of the membrane. In the central nervous system, unique functions such as neurotransmitter release, ion transport and receptor binding activities are attributed to the neuronal membranes. The studies described in this chapter are attempts to reveal the role of phospholipids and their metabolites in neuronal membrane functions. Attention is focused mainly on the metabolic relationship among FFA[1], acyl-CoA, lysophospholipids and neuronal membrane phospholipids as depicted in Fig 1.

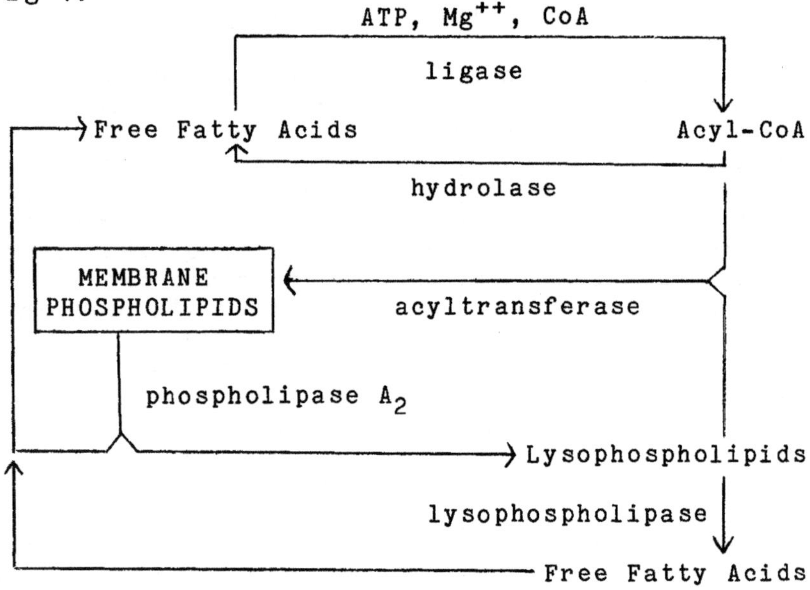

Fig. 1 Scheme depicting the metabolic relationship of phospholipid metabolites.

Phospholipids and their metabolites in neuronal membranes

It is generally known that mature neurons in brain have little ability to replicate themselves. Therefore, intricate biochemical mechanisms probably exist in order to maintain the structural integrity and functional activities of the neuronal membrane. One of the mechanisms for maintaining the cellular integrity is to prevent unnecessary breakdown of membrane components and to re-utilize existing molecules as much as possible. An example of this is the mechanisms available for recycling of the acyl groups of membrane phospholipids. Experiments with labeled fatty acids which were injected intracerebrally into mouse brain indicated that phospholipids in brain subcellular membranes are undergoing metabolic turnover with slow and fast turnover pools (Sun and Horrocks, 1973; Sun, 1973; Sun and Su, 1979). Furthermore, results of these studies also show that phospholipid turnover is a complex mechanism mediated by multi-enzymic pathways. Many of the phospholipid metabolizing enzymes are localized inside the membrane matrix, and they require a lipid environment for proper functioning. Due to their topographic arrangement, many of these enzymes are not readily accessible to exogenous lipid substrates, which poses some difficulties with regard to elucidation of their mode of action.

The metabolism of membrane phospholipids as mediated by enzymes depicted in Fig 1 is known to give rise to lipid intermediates such as lysophospholipids, acyl-CoA and FFA (Sun et al., 1979b). These phospholipid intermediates are normally found in trace amounts in tissues and body organs, but may accumulate under pathological conditions. Most of these compounds are

---

[1]Abbreviations used: FFA, free fatty acids; LPC, lysophosphatidylcholine; LPE, lysophosphatidylethanolamine; 1-acyl-GPC, 1-acyl-glycerophosphocholine; BSA, bovine serum albumin.

amphipathic; consequently, their presence in excess amounts may cause physical and metabolic disturbances to the membrane. Interaction of amphipathic phospholipid intermediates with the hydrophobic region of the membrane lipid bilayer can cause a disordering effect which is correlated to an increase in membrane fluidity. Amphiphiles which interact with the phospholipid polar head groups may cause a displacement of the divalent cations and thus alter the ion transport mechanism and receptor binding activity. On the other hand, presence of a regulated amount of these lipid metabolites within the membrane matrix is probably important in mediating the overall membrane functions, because these compounds can act as naturally occurring detergents for facilitating the metabolic activity. Due to the unique physical properties of these lipid metabolites, their involvement in neuronal membrane functions should not be underestimated.

Stimulated phospholipid deacylation activity during neuronal membrane-stimulation

One research interest in our laboratory is to demonstrate the involvement of phospholipids and their metabolites in neuronal membrane functions. In one series of experiments, phospholipids in rat synaptosomes were labeled with ^{14}C-arachidonate (Majewska et al., 1981). The prelabeled synaptosomes were suspended in Krebs-Ringer-bicarbonate buffer supplemented with glucose and Ca^{++} (2.5 mM), and were subsequently depolarized by raising the K^+ concentration in the incubation medium from 5 to 55 mM. This procedure is known to elicit a calcium-dependent neurotransmitter release from the synaptosomes (Blaustein et al., 1972). The results from this study showed that K^+-induced depolarization of arachidonoyl-labeled synaptosomes was positively correlated to an increase in deacylation of the phosphatidylcholine and phosphatidylinositol (Majewska and Sun, 1982). This calcium-mediated deacylation activity was observed in the synaptic plasma membranes, as well as synaptic vesicles, but not in the intra-synaptic mitochondrial fraction.

However, using this procedure, we were not able to completely recover the labeled arachidonate released from the phosphoglycerides. This may be due to the fact that FFA tended to form a complex with calcium, and the complex was recovered in the interphase of the extraction medium instead of in the organic phase.

The release of labeled arachidonate from pre-labeled synaptosomes was further investigated by the superfusion procedure in which synaptosomes were superfused with Krebs-Ringer-bicarbonate buffer containing BSA which was used to bind the FFA (Lazarewicz et al., 1983). This type of superfusion technique has been successfully used by other investigators for eliciting the release of neurotransmitters from synaptosomes (Mulder et al., 1975). K^+-depolarization of arachidonoyl-labeled synaptosomes in the presence of calcium elicited an increase in arachidonate release, albeit the amount released was very small. It is concluded from the results of these experiments that a small pool of the membrane phospholipids is involved in the neuronal stimulation associated with neurotransmitter release.

The role of calcium in membrane phospholipid metabolism

A number of the biochemical and physiological events associated with the neuronal membrane functions are modulated by calcium and calmodulin (DeLorenzo, 1982). These processes include depolarization of synaptosomal membranes leading to subsequent neurotransmitter release as well as binding of receptors to the neuronal membrane. In turn, the binding of Ca^{++} to synaptic plasma membranes and subsequent flux through a channel-mediated mechanism may require the cooperation of specific membrane phospholipids or metabolites. For example, Serhan et al. (1982) showed that leukotriene B_2 and phosphatidic acids can act as calcium ionophores. Calcium is needed for membrane or vesicle fusion activity, and in the presence of calcium binding protein, only micromolar amount of Ca^{++} is needed to stimulate the fusion of lipid vesicles (Zimmerberg et al., 1980). There is increasing evidence that the presence of a small amount of acidic phospholipids in the phospholipid bilayer is important in mediating the calcium-complex formation as well as the membrane fusion activity (Sundler et al., 1981).

Calcium also forms complexes with FFA and acyl-CoA, and it inhibits the brain acyl-CoA hydrolase activity (Strosznajder et al., 1981). The inhibition of acyl-CoA hydrolase by calcium is probably due to formation of a calcium complex between acyl-CoA and membrane phospholipids, and the complex can immobilize the acyl-CoA for the hydrolase action. Since acyl-CoA is an important lipid intermediate for phospholipid biosynthesis and for mediating the incorporation of FFA into membrane phospholipids (Fig 1), its interaction with calcium is likely to affect the overall membrane phospholipid metabolism. Not only is acyl-CoA a potent inhibitor of the mitochondrial adenosine transport system (Vignais et al., 1976), it is also shown to inhibit many membrane-bound enzymes and ion transport mechanisms (Asimakis and Sordahl, 1977; Wood et al., 1977). Obviously, this type of lipid intermediate can impose a dramatic effect on neuronal membrane functions.

Almost all the phospholipase A_2 present in plasma membranes require calcium for activity (Van den Bosch, 1980). The physiological implications of the enzyme and its metabolic products in neuronal membrane functions are described in the next section.

Involvement of phospholipase A_2 in deacylation of membrane phospholipids

There is little doubt that phospholipase A_2 is present in brain (Webster and Cooper, 1968; Cooper and Webster, 1970; 1972), as well as in brain synaptosomal membranes (Woelk et al.,

1974b). At least two types of phospholipase A_2 are present in brain, one exhibiting a pH optimum at 4-5 and the other at 8-9. However, only the pH 8-9 enzyme shows a Ca^{++}-dependency. Woelk et al. (1981) showed that the phospholipase A_2 in neuronal cells preferred hydrolyzing the arachidonoyl groups to the linoleoyl groups. The specificity of this enzyme towards different types of substrates has not been fully revealed. In fact, it is not surprising to find that a number of phospholipase A_2 are present in the membrane, each acting on a different type of phospholipid substrate. A phospholipase A_2 with high specificity towards phosphatidylinositol has been isolated from brain microsomes (Gray and Strickland, 1982). The acyl group specificity suggests that this enzyme may be involved in remodeling the acyl groups of phosphatidylinositol.

Studies to elucidate the properties of phospholipase A_2 in neuronal membranes are frequently confronted with problems regarding substrate selection and its accessibility to exogenous lipid substrates. One method to circumvent these limitations is to directly measure the FFA released from the membrane. We found that synaptosomes which were first "pre-washed" with BSA to remove endogenous FFA elicited a time-dependent release of FFA upon incubation in an isotonic solution at 37°C (Fig 2). However, exogenously added calcium did not elicit and increase in the FFA release, although addition of EDTA or EGTA resulted in partial inhibition of the FFA release, especially the unsaturated fatty acids (Fig 3). Results of the study indicated that (1) multiple enzymic mechanisms are involved in the FFA release process, (2) at least a portion of the unsaturated fatty acids are released from synaptosomes through a calcium-dependent process, and (3) there is apparently enough endogenous calcium in the native synaptosomes (isolated in the absence of calcium chelating agents) for mediating the calcium-dependent FFA release.

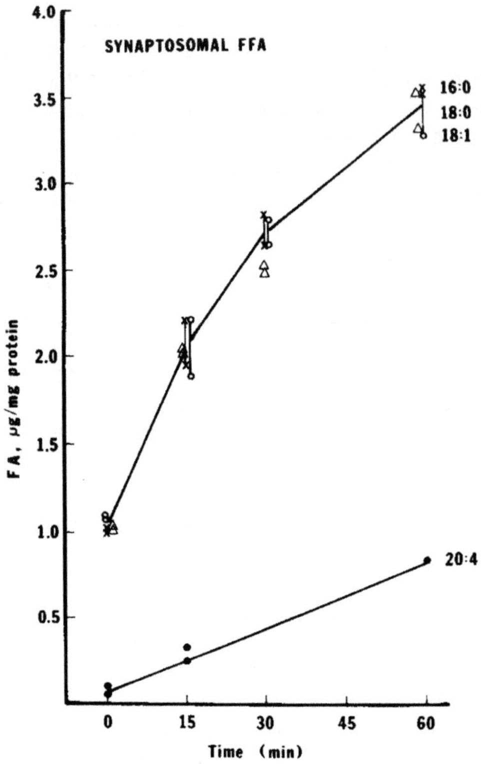

Fig 2. FFA release due to incubation of BSA-washed synaptosomes at 37°C. After incubation, lipids were extracted with chloroform-methanol (2:1, v/v). The FFA were separated by TLC and quantitatively assayed by GLC.

When synaptosomal FFA release was examined with respect to pH, an increase in the release of arachidonate was shown at pH 5 and pH 8-9, whereas the release of saturated fatty acids was more active at the acidic pH range (Fig 4).

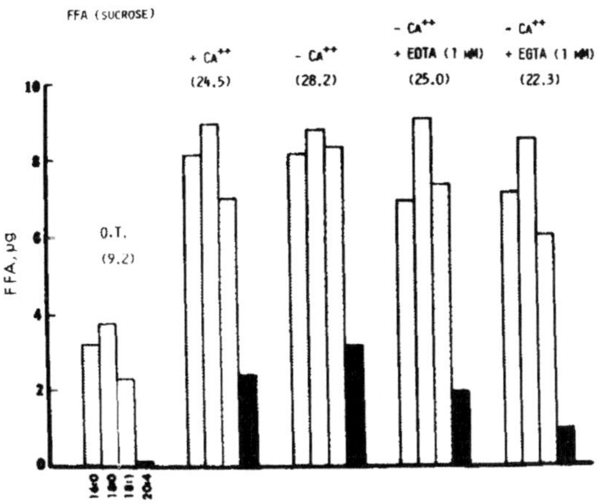

Fig 3. Effect of Ca^{++} and EGTA on synaptosomal FFA release. In this preparation, rat brain synaptosomes were isolated in 0.32 M sucrose-50 mM Tris-HCl (pH 7.4) in the absence of calcium chelating agents($-Ca^{++}$). Synaptosomes were incubated at 37°C for 1 hr in the presence of (1) 2.5 mM Ca^{++}, (2) without exogenous Ca^{++}, (3) Ca^{++} (2.5 mM) and EDTA (1 mM), or (4) Ca^{++} (2.5 mM) and EGTA (1 mM).

Elevated phospholipase A_2 activity is implicated in pathogenesis of degenerative diseases, especially those involving demyelination (Webster, 1973; Woelk et al., 1974a). Unlike other types of phospholipases, phospholipase A_2 is known to exert a potent inhibitory effect on synaptosomal (Na^+,K^+)-ATPase activity (Sun and Sun, 1976), and on the norepinephrine and serotonin uptake processes (Rotman, 1977). Recently, Moskowitz et al. (1982) gave evidence that the calcium-dependent phospholipase A_2 in synaptic vesicles is responsive to stimulation by a number of modulators including prostaglandin E_2, ATP and adenosine. In their study, stimulation of the endogenous phospholipase A_2 was also correlated to vesicle aggregation activity.

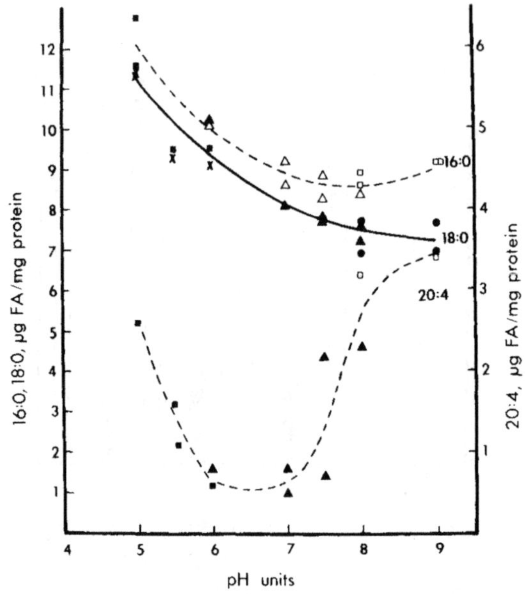

Fig 4. Release of arachidonic acid (20:4) and saturated fatty acids (16:0 and 18:0) due to incubation of rat brain synaptosomes as a function of pH. Three types of buffers were used: sodium acetate, pH 5-6; sodium phosphate, pH 6-8; and Tris-HCl, pH 8-9.

Release of FFA from membrane phospholipids

The action of phospholipase A_2 on membrane phospholipids is known to give rise to FFA and lysophospholipids as products. In brain, the FFA are normally present in a greater quantity than the lysophospholipids. Therefore, other mechanisms must be present to contribute to this non-stoichiometric release of FFA. It is realized that the FFA level in brain is increased under almost all stimulated conditions examined, including ischemia (Bazan, 1970; 1976; De Medio et al., 1980; Shiu et al., 1981; Tang and Sun, 1982), electroconvulsive shock (Bazan and Rakowski, 1970), drug-induced stimulation (Seisjo et al., 1982) and edema (Chan and Fishman, 1983).

On the other hand, FFA level in brain is depressed after administration of depressant drugs including acute ethanol administration (Sun et al., 1982). Brain ischemia induced by decapitation is correlated to a biphasic mode of increase in FFA. During the early time period, arachidonate and stearate were preferentially released (Tang and Sun, 1982). The brain FFA release due to post-decapitation (2 min treatment) was not obvious in immature brain from birth up to 2 weeks of age (Fig 5). After 2 weeks of age, there was a 2-3 fold increase due to the ischemic treatment. This developmental profile suggests that enzyme(s) involved in the release mechanism is not present in the newborn brain but becomes obvious at the time of synaptogenesis.

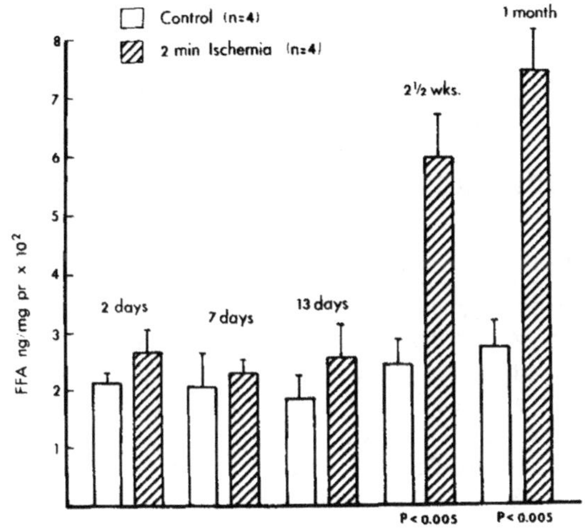

Fig 5. FFA release from rat brain due to a 2 min post-decapitative ischemic treatment. Comparison with age during brain development. No significant difference in FFA release was observed in 2, 7 and 13 day-old animals.

The physiological significance of FFA on neuronal membrane functions has not been fully assessed. Since these molecules are largely generated in the hydrophobic region of the lipid bilayer, their presence in increasing amounts is likely to interfere with the packing of the phospholipid acyl chains and to cause a disordering effect. FFA are potent uncouplers of mitochondrial oxidative-phosphorylation (Wojtezak, 1976). In the sarcoplasmic reticulum vesicles, low concentration of FFA can inhibit the calcium efflux activity (Katz et al., 1981). FFA was also shown to inhibit the Na^{+}-dependent amino acid uptake system (Rhoads et al., 1982). When polyunsaturated fatty acids were added to cortical slices, a disturbance of the ion transport activity and subsequently tissue edema was observed (Chan and Fishman, 1978). The polyunsaturated fatty acids are excellent substrates for peroxidation reactions. The superoxide radicals formed would, in turn, lead to tissue and cellular damage (Chan and Fishman, 1980; 1982). From this and other studies, FFA is shown to play an important role in regulating neuronal membrane functions, although the exact mode of their interaction with specific systems has not yet been elucidated.

Metabolism of lysophospholipids in neural membranes

Lysophospholipids, the other metabolic product derived from phospholipase A action, are also ubiquitously found in body organs and tissues including the brain (Gallai-Hatchard et al., 1962; Webster and Alpern, 1964; Hallett, 1981). Analysis of the acyl groups of LPC in brain indicated a profile comprised mainly of 16:0, 18:0, and 18:1 (Hallett, 1981). This profile reflects that the major source of the LPC is probably derived from the action of phospholipase A_2. This type of acyl-metabolite can exert a potent cytolytic effect on tissue and membranes due to their detergent properties (Weltzien, 1979). Its action in mediating membrane fusion activity is well known (Howell and Lucy, 1969). A small amount of the LPC delivered into the brain white matter is known to induce focal demyelination (Foster et al., 1980). The lysophospholipids may play an important role in neuronal membrane receptor functions through modulation of the cyclic nucleotide metabolism. For example, LPC is shown to activate the guanylate cyclase but inhibit the adenylate cyclase activity (Shier et al., 1976; Zwiller et al., 1976; Houslay and Palmer, 1979). It also exerts a stimulatory effect on the 3'5'-nucleotide phosphodiesterase in brain cytoplasm (Pichard and Cheung, 1977). Besides the above mentioned enzymic systems, an increasing number of membrane bound enzymes are also affected by the detergent properties of the lysophospholipids. Consequently, the physiological importance of the lysophospholipids in modulating neurotransmitter release, receptor response, as well as other neuronal membrane transport systems, cannot be denied (Baker et al., 1975).

In the past, technical difficulties existed in the isolation of lysophospholipids from tissues and membranes. Bjerve et al. (1974) showed that some of the lysophosphatidates were poorly extracted by conventional organic solvent system. Extraction of tissue lipids by acidified organic solvent system was shown to cause a degradation of the plasmalogens, and thus may result in an erroneously high estimation of the lyso-compounds. Recently, a sensitive enzymic procedure was developed in our laboratory for assaying the 1-acyl-GPC in tissues, body fluids and subcellular membrane fractions (Hallett, 1981). With this procedure, only small amounts of tissues or body fluids are needed for the assay. The level of LPC in various brain regions, except the brain stem, is around 170 nmole/gm weight of tissue. The brain stem has a higher level of 280 nmole/gm wt. Unlike the FFA, which increased dramatically during ischemia and stimulation, post-decapitative ischemia elicited only a small increase in the LPC level (Fig 6). The lack of an obvious accumulation of lysophospholipids in the tissue, even during the ischemic treatment, strongly suggests that these compounds are metabolically active and that their metabolic activities are probably highly regulated.

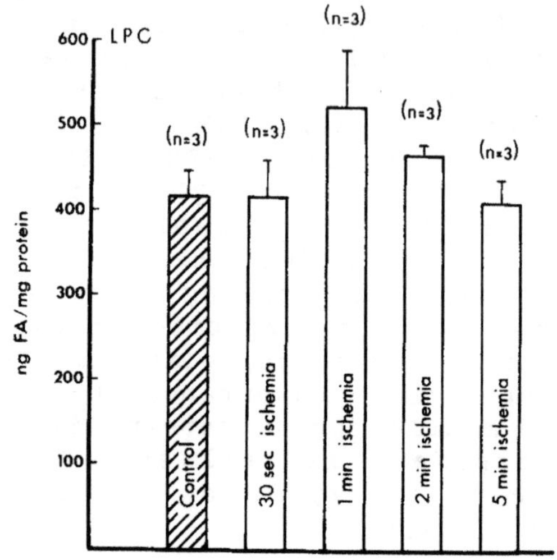

Fig 6. LPC level in rat brain due to post-decapitative ischemia. LPC was extracted from brain with chloroform-methanol (2:1, v/v) together with internal standard. LPC content was determined by a combined TLC-GLC procedure.

As depicted in the scheme in Fig 7, the lysophospholipids are maintained in a dynamic equilibrium with membrane phospholipids by several enzymic pathways, which are important in the intracellular formation as well as removal of the lysophospholipids (Van den Bosch and van den Besselaar, 1978).

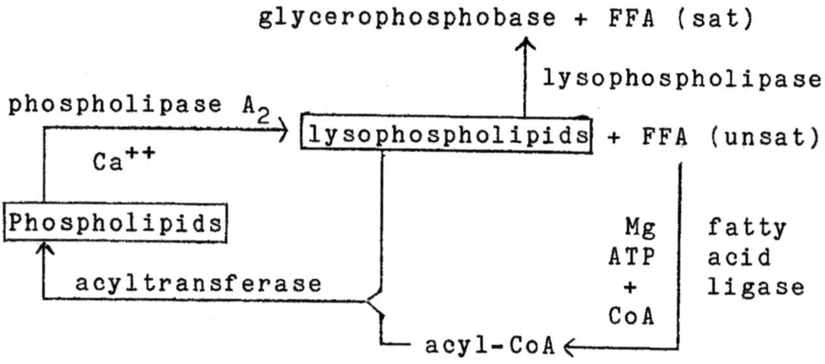

Fig. 7 Metabolism of lysophospholipids.

Although lysophospholipids in brain are formed as a result of phospholipase A_1 or A_2 action, their acyl group profiles suggest that contribution by the phospholipase A_2 is probably much greater than that by the phospholipase A_1 action (Hallett, 1981). It is realized that lysophospholipids which are formed by phospholipase A_2 action can be catabolized by two routes: (1) hydrolysis by lysophospholipase to form glycerophospho-base and FFA, or(2) they may be rapidly reacylated and returned to the phospholipid pool via the lysophospholipid: acyl-CoA acyltransferase. Recently, experiments were carried out in our laboratory to investigate the metabolic significance of lysophospholipids in brain subcellullar membranes and the contribution by these two pathways for catabolizing the compounds. When the subcellular membranes were incubated in the absence of added cofactors (ATP, Mg and CoA), it is surprising to find that the labeled lysophospholipids (^{32}P-LPC) were largely degraded by the lysophospholipase, with practically no reacylation activity (Fig 8). This result also shows that subcellular membranes isolated from brain homogenates are active in lysophospholipid degradation, but lack the ability to reacylate the FFA released. However, when Mg^{++}-ATP and CoA were added to the incubation system, a large portion of the LPC was reacylated to form phosphatidylcholine, whereas similar amounts were degraded by the lysophospholipase (Fig 8). In this regard, addition of exogenous fatty acids did not further enhance the reacylation activity, suggesting that enough endogenous FFA are probably present in the membrane for the reacylation process. Since the entire reacylation process, i.e. starting from the CoA and ATP-dependent activation of FFA to acyl-CoA to the transfer of acyl-CoA to phospholipids, is probably occurring within the membrane matrix, it is not surprising that exogenous lipid substrates may have difficulty in equilibrating with the endogenous pool.

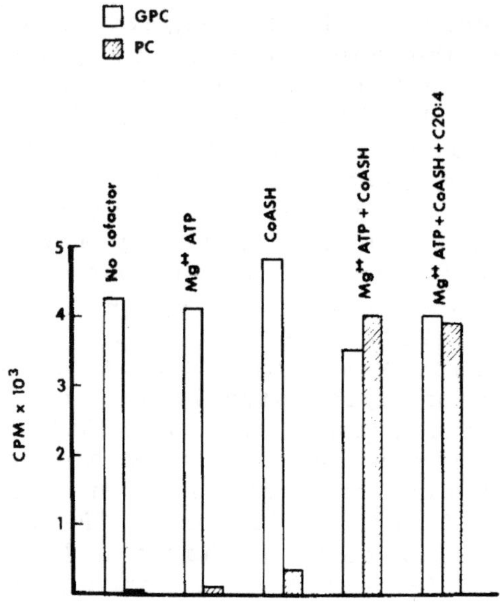

Fig 8. Incubation of [^{32}P]-LPC with brain microsomes. Assay of the lysophospholipase activity (GPC) and cofactor requirements for the lysophospholipid acyltransferase activity (PC).

Inability of the exogenous FFA pool to exchange with the membrane endogenous pool is a known phenomenon with a number of enzymic processes involving fatty acid transformation (Gan-Elepano et al., 1981; Aeberhard et al., 1981). Our results here have provided additional evidence to the presence of discrete lipid pools for metabolism of the FFA. Apparently, the lysophospholipase is only loosely bound to the membrane, whereas the fatty acid ligase and the lysophospholipid acyltransferase are well situated within the membrane lipid bilayer. The fatty acid ligase thus plays an important role in linking the deacylation event, in which FFA are released from the phospholipids, with the reacylation event, in which the FFA are returned to the membrane. Fatty acid ligase activity is found in most membranes exhibiting the acyltransferase activity and evidence was obtained from our laboratory that it may even be coupled to the acyltransferase. In the ischemic condition, an obvious decrease in the reacylation activity was observed, indicating that the enzymic process for recyling the FFA is curtailed. Under this condition, the lysophospholipids would be mainly hydrolyzed by the lysophospholipase and, in turn, more FFA are released (probably the saturated type). The observation of a nearly parallel increase in stearic and arachidonic acids due to post-decapitative ischemia may the result of such a biochemical disturbance.

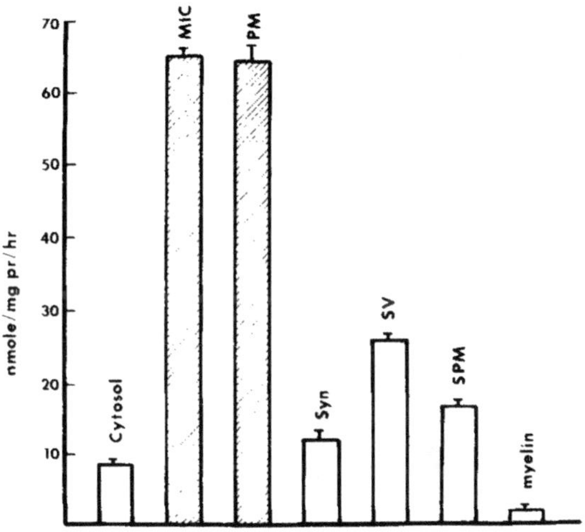

Fig 9. Distribution of lysophospholipase activity in brain subcellular fractions. Mic, microsomes; PM, plasma membranes; SV, synaptic vesicles; SPM, synaptic plasma membranes.

With the exception of myelin, most subcellular membranes exhibit various levels of the lysophospholipase activity. As shown in Fig 9, specific activity of this enzyme is highest in the microsomal and plasma membrane fractions.

Although the synaptosomes and synaptic membranes have a lower level of the enzymic activity, significant amount of the hydrolysis is found due to the large amount of membrane attributed to these fractions. The cytosolic fraction shows only a low level of the enzyme activity. The absence of an active catabolic pathway for the lysophospholipids in the myelin may be an important consideration related to the pathogenisis of diseases involving demyelination. This may also explain why the myelin phospholipids are more susceptible to degradation by invasive cells from non-neural origin. Since most of the brain subcellular membranes contain both lysophospholipase and acyltransferase activities, a comparison of the two enzymic processes can be made using labeled lysophospholipids as substrate. As shown in Fig 10, brain microsomes contain lysophospholipase and acyltransferase in approximately equal proportions. In the plasma membrane fraction,

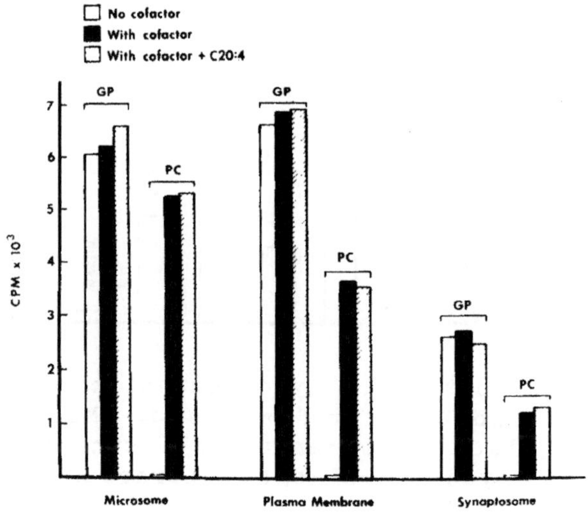

Fig 10. A comparison of the lysophospholipase (GPC formed) and acyltransferase (PC formed) activities in different subcellular membrane fractions from rat brain.

however, activity of the lysophospholipase is higher than the acyltransferase. In the synaptosomes, activity of both enzymes are low, but the lysophospholipase activity is two-fold higher than the acyltransferase.

The lysophospholipid acyltransferase in brain prefers transferring arachidonate to lysophosphatidylinositol and lysophosphatidylcholine (Corbin and Sun, 1978; Baker and Thompson, 1973). This enzyme is tightly bound to the membrane and its activity is shown to be affected by most detergents and agents which may perturb the membrane structure (Sun et al., 1979a; Shier, 1977).

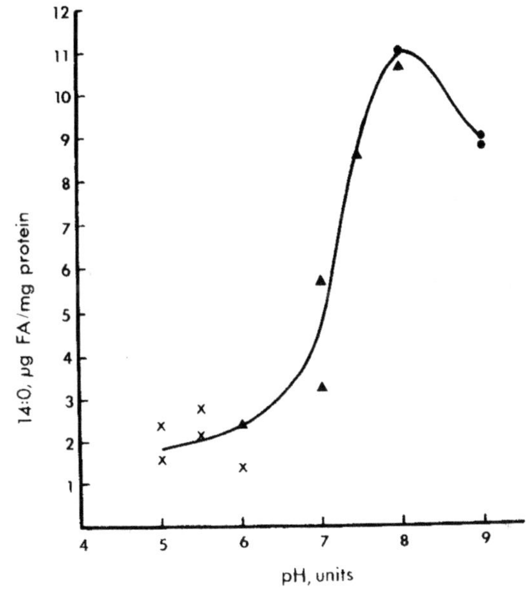

Fig 11. Lysophospholipase activity in rat brain synaptosome as a function of pH. Assays were based on the release of 14:0 from 14:0-LPC after incubation with synaptosomes in 0.32M sucrose with three different types of buffers. pH 5-6, acetate; pH 6-8, phosphate; and pH 8-9, Tris-HCl.

Previous studies of the lysophospholipase in brain also indicated that the enzyme is largely present in the particulate fraction but is apparently only loosely bound to the membrane (Leibovitz-BenGersohn, 1972). From our determinations, the pH optima for the synaptosomes is around 8 (Fig 11). This enzyme does not require calcium for activity, but is inhibited by other divalent cations such as Zn^{++}, Cu^{++} and Hg^{++} (Fig 12). The lysophospholipase in brain synaptosomes seems to show little selectivity towards lysophospholipids with different base groups, but lysophospholipid substrates with an alkenylether bond are apparently not hydrolyzed by this enzyme (Fig 13).

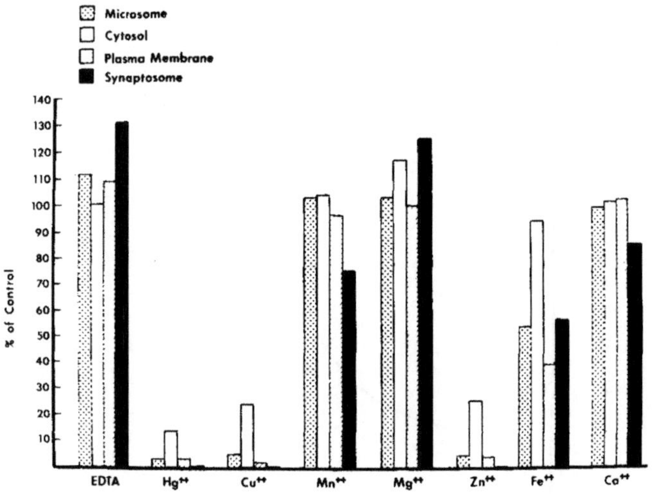

Fig 12. Effect of Ca^{++} and other divalent cations on brain subcellular lysophospholipase activity.

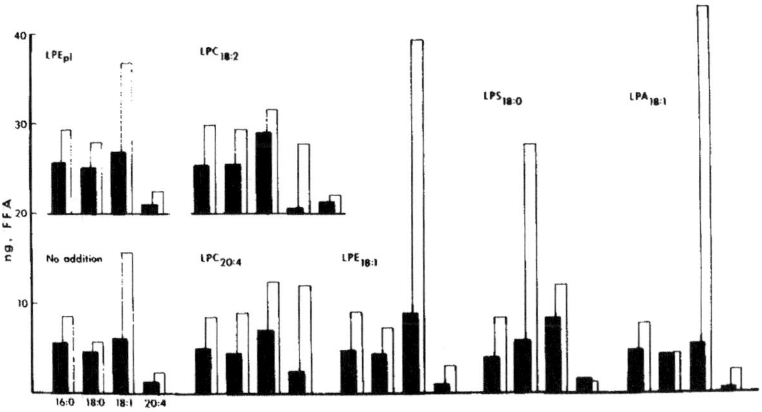

Fig 13. Substrate specificity of lysophospholipase in brain synaptosomes. Synaptosomes were incubated with different lysophospholipids. Open bars represent fatty acids released due to incubation, solid bars represent the fatty acids present in the fraction before incubation.

SUMMARY

There is increasing evidence indicating that phospholipid "pools" in neuronal membranes are actively engaged in metabolic turnover which is mediated by a number of phospholipid-metabolizing enzymes. These enzymic reactions yield metabolites such as free fatty acids (FFA), acyl-CoA, and lysophospholipids, which are amphiphiles and exhibit potent detergent properties. In brain, phospholipid metabolites may play an important role in regulating neuronal membrane functions such as neurotransmitter release, ion transport, calcium homeostasis, receptor binding and membrane fusion activity. Since these compounds are normally maintained in a dynamic equilibrium with the membrane phospholipids, a disturbance of the equilibrium may result in a change in the metabolite level. An increase in FFA in brain is shown in ischemia, edema, and many forms of stimulation. Studies *in vitro* with arachidonoyl-labeled synaptosomes indicated that synaptosomal stimulation is associated with an increase in phospholipid deacylation activity and arachidonate release. The transient release of arachidonate and lysophospholipids may, in turn, alter neuronal membrane functions.

Among the phospholipid metabolites, the lysophospholipids may play a most relevant role in membrane functions on account of their potent detergent properties. Although only present in trace amounts, these compounds are ubiquitous in biological membranes. Since many membrane-bound enzymes are modulated by lysophospholipids, their role as naturally occurring detergents has been well accepted. On the other hand, excessive accumulation of lysophospholipids in membrane would produce undesirable effects leading to abnormal membrane functions, demyelination and tissue pathogenesis. The acyl group profile of lysophospholipids in brain revealed that they are generated mainly from phospholipase A_2 action. The lysophospholipids which are generated in membranes are catabolized either by reacylation back to the membrane phospholipids or further degraded by lysophospholipase to yield FFA and a water soluble product. In the presence of ATP, Mg^{++} and CoA, synaptosomal membranes indicate a higher lysophospholipase than acyl-transferase activity. The lysophospholipase in neural membranes does not require calcium for activity, but is inhibited by divalent metal ions such as Zn^{++}, Hg^{++} and Cu^{++} (1 mM). Results of these and other studies have provided evidence that phospholipids and their metabolites are involved in neuronal membrane functions, although the underlying mechanisms for their involvement may be complex and may depend on many factors regulating the enzymes involved in metabolism of individual membrane phospholipid "pools."

Acknowledgement: Special thanks are due to Ms. D. Torres for her help in preparation of the manuscript. The research work is supported in part by research grants NS 16715 from Department of Human Health and Service and BNS 76-24338 from National Science Foundation.

REFERENCES

Aeberhard, E.E., Gan-Elepano, M. and Mead, J.F. (1981) Metabolism of fatty acids in rat brain membranes. Lipids 16, 705-713.

Asimakis, G.K. and Sordahl, L.A. (1977) Effects of atractyloside and palmitoyl coenzyme A on calcium transport in cardiac mitochondria. Arch. Biochem. Biophys. 179, 200-210.

Baker, R.R. and Thompson, W. (1973) Selective acylation of 1-acylglycerophosphorylinositol by rat brain microsomes. Comparison with 1-acylglycerophosphorylcholine. J. Biol. Chem. 248, 7060-7065.

Baker, R.R., Dowdall, M.J. and Whittaker, V.P. (1975) The involvement of lysophosphoglycerides in neurotransmitter release: composition and turnover of phospholipids of synaptic vesicles of guinea pig cerebral cortex and *Torpedo* electric organ and the effect of stimulation. Brain Res. 100, 629-644.

Bazan, N.G. (1970) Effects of ischemia and electroconvulsive shock on free fatty acid pool in the brain. Biochim. Biophys. Acta 218, 1-10.

Bazan, N.G. (1976) Free arachidonic acid and other lipids in the nervous system during early ischemia and after electroshock. In: *Function and Metabolism of Phospholipids in the Central and Peripheral Nervous Systems* (Porcellati, G., Amaducci, L and Galli, C., eds.) pp. 317-335, Plenum Press, New York.

Bazan, N.G. and Rakowski, H. (1970) Increased levels of brain free fatty acids after electroconvulsive shock. Life Sci. 9, 501-507.

Bjerve, K.S., Daae, L.N.W. and Bremer, J. (1974) The selective loss of lysophospholipids in some commonly used lipid-extraction procedures. Anal. Biochem. 58, 238-245.

Blaustein, M.P., Johnson, E.M. and Needleman, P. (1972) Calcium dependent norepinephrine release from presynaptic nerve endings *in vitro*. Proc. Natl. Acad. Sci. 69, 2237-2240.

Chan, P.H. and Fishman, R.A. (1978) Brain edema: induction in cortical slices by polyunsaturated fatty acids. Science 201, 358-360.
Chan, P.H. and Fishman, R.A. (1980) Transient formation of superoxide radicals in polyunsaturated fatty acids-induced brain swelling. J. Neurochem. 35, 1004-1007.
Chan, P.H. and Fishman, R.A. (1982) Alterations of membrane integrity and cellular constituents by arachidonic acid in neuroblastoma and glioma cells. Brain Res. 248, 151-157.
Chan, P.H. and Fishman, R.A. (1983) Phospholipid degradation and the early release of polyunsaturated fatty acids in the evolution of brain edema. 5th Int. Symp. on Brain Edema. (in press)
Cooper, M.F. and Webster, G.R. (1970) The differentiation of phospholipase A_1 and A_2 in rat and human nervous tissues. J. Neurochem. 17, 1543-1554.
Cooper, M.F. and Webster, G.R. (1972) On the phospholipase A_2 activity of human cerebral cortex. J. Neurochem. 19, 333-340.
Corbin, D.R. and Sun, G.Y. (1978) Characterization of the enzymic transfer of arachidonoyl groups to 1-acyl-phosphoglycerides in mouse synaptosome fraction. J. Neurochem. 30, 77-82.
DeLorenzo, R.J. (1982) Calmodulin in neurotransmitter release and synaptic function. Fed. Proc. 41, 2265-2272.
De Medio, G.E., Goracci, G., Horrocks, L.A., Lazarewicz, J.W., Mazzari, S., Porcellati, G., Strosznajder, J. and Trovarelli, G. (1980) The effect of transient ischemia on fatty acid and lipid metabolism in the gerbil brain. Ital. J. Biochem. 29, 412-432.
Foster, R.E., Kocsis, J.D., Malenka, R.C. and Waxman, S.G. (1980) Lysophosphatidylcholine-induced focal demyelination in the rabbit corpus callosum. J. Neurol. Sci. 48, 221-231.
Gallai-Hatchard, J., Magee, W.L., Thompson, R.H.S. and Webster, G.R. (1962) The formation of lysophosphatides from diacyl phosphatides

by brain preparations. J. Neurochem. 9, 545-554.

Gan-Elepano, M., Aeberhard, E. and Mead, J.F. (1981) On the mechanisms of fatty acid transformations in membranes. Lipids 16, 790-795.

Gray, N.C.c. and Strickland, K.P. (1982) On the specificity of a phospholipase A_2 purified from the 106,000 x g pellet of bovine brain. Lipids 17, 91-96.

Hallett, D.W. (1981) A sensitive enzymic assay for lysolecithin. Thesis, University of Missouri-Columbia.

Houslay, M.D. and Palmer, R.W. (1979) Lysophosphatidylcholines can modulate the activity of the glucagon-stimulated adenylated cyclase from rat liver plasma membranes. Biochem. J. 178, 217-221.

Howell, J.I. and Lucy, J.A. (1969) Cell fusion induced by lysolecithin. FEBS Lett. 4, 147-150.

Katz, A.M., Messineo, F., Miceli, J. and Nash-Adler, P.A. (1981) Low concentrations of fatty acids can inhibit calcium efflux from sarcoplasmic reticulum vesicles. Life Sci. 28, 1103-1107.

Lazarewicz, J., Leu, V., Sun, G.Y. and Sun, A.Y. (1983) Arachidonic acid release from K^+-evoked depolarization of brain synaptosomes. Neurochem. Int. (in press)

Leibovitz-BenGershon, Z., Kobiler, H. and Gatt, S. (1972) Lysophospholipases of rat brain. J. Biol. Chem. 247, 6840-6817.

Majewska, M.D., Manning, R. and Sun, G.Y. (1981) Effects of postdecapitative ischemia on arachidonate release from brain synaptosomes. Neurochem. Res. 6, 567-576.

Majewska, M.D. and Sun, G.Y. (1982) Activation of arachidonoyl-phosphatidylinositol and phosphatidylcholine turnover by K^+-evoked stimulation of brain synaptosomes. Neurochem. Int. 4, 427-433.

Moskowitz, N., Schook, W. and Puszkin, S. (1982) Interaction of brain synaptic vesicles induced by endogenous Ca^{2+}-dependent phospholipase A_2. Science 216, 305-307.

Mulder, A.H., van den Berg, W.B. and Stoot, J.G. (1975) Calcium-dependent release of radio-labeled catecholamines and serotonin from rat brain synaptosomes in a superfusion system. Brain Res. 99, 419-429.

Pichard, A.L. and Cheung, W.Y. (1977) Cyclic 3':5'-nucleotide phosphodiesterase. Stimulation of bovine brain cytoplasmic enzyme by lysophosphatidylcholine. J. Biol. Chem. 252, 4872-4875.

Rhoads, D.E., Kaplan, M.A., Peterson, N.A. and Raghupathy, E. (1982) Effects of free fatty acids on synaptosomal amino acid uptake systems. J. Neurochem. 38, 1255-1260.

Rotman, A. (1977) The effect of phospholipase C, phospholipase A_2 and neuraminidase on the uptake of [^{3}H]norepinephrine and [^{3}H]-serotonin by rat brain synaptosomes. J. Neurochem. 28, 1369-1372.

Seisjo, B.K., Martin Ingvar, and Westerberg, E. (1982) The influence of bicuculline-induced seizures on free fatty acid concentrations in cerebral cortex, hippocampus, and cerebellum. J. Neurochem. 39, 796-802.

Serhan, C.N., Fridovich, J., Goetzl, E.J., Dunham, P.B. and Weissmann, G. (1982) Leukotriene B_2 and phosphatidic acid are calcium ionophores. Studies employing arsenazo III in liposomes. J. Biol. Chem. 257, 4746-4752.

Shier, W.T. (1977) Inhibition of acyl-CoA: lysolecithin acyltransferases by local anesthetics, detergents and inhibitors of cyclic nucleotide phosphodiesterases. Biochem. Biophys. Res. Commun. 75, 186-193.

Shier, W.T., Baldwin, J.H., Nilsen-Hamilton, M., Hamilton, R.T. and Thanassi, N.M. (1976) Regulation of guanylate and adenylate cyclase activities by lysolecithin. Proc. Natl. Acad. Sci. USA 73, 1586-1590.

Shiu, G.K., Nemoto, E.M. and Alexander, H.L. (1981) Brain free fatty acid changes during global ischemia with barbiturate anesthesia and hypothermia. Brit. J. Anaesthesia 53, 304.

Singer, S.J. and Nicolson, G.L.(1972) The fluid mosaic model of the structure of cell membranes. Science 175, 720-724.

Strosznajder, J., Tang, W., Manning, R., Lin, A.Y.-T., MacQuarrie, R. and Sun, G.Y. (1981) Metabolism of oleoyl-CoA in rat brain synaptosomes: Effects of calcium and post-decapitative ischemia. Neurochem. Res. 6, 1231-1240.

Sun, A.Y. and Sun, G.Y. (1976) Functional roles of phospholipids of synaptosomal membrane. In: Function and Metabolism of Phospholipids in the Central and Peripheral Nervous Systems (Porcellati, G., Amaducci, L. and Galli, C., eds.) pp. 169-197, Plenum Press, New York.

Sun, G.Y. (1973) The turnover of phosphoglycerides in the subcellular fractions of mouse brain. A study using [^{14}C]-oleic acid as precursor. J. Neurochem. 21, 1083-1092.

Sun, G.Y. (1982) Metabolic turnover of arachidonoyl groups in brain membrane phosphoglycerides. In: Phospholipid Metabolism in the Nervous System, Metabolism (Horrocks, L.A., Ansell, G.B., and Porcellati, G., eds.) Vol. 1, pp. 75-89, Raven Press, New York.

Sun, G.Y. and Horrocks, L.A. (1973) The metabolism of palmitic acid in the subcellular fractions of mouse brain. J. Lipid Res. 14, 206-214.

Sun, G.Y. and Su, K.L. (1979) Metabolism of arachidonoyl phosphoglycerides in mouse brain subcellular fractions. J. Neurochem. 32, 1053-1059.

Sun, G.Y., Corbin, D.R., Wise, R.W. and MacQuarrie, R. (1979a) Effects of lipid intermediates, lyso-glycerophospholipids and detergents on arachidonate transfer to 1-acyl-glycerophospholipids by brain synaptosomes. Int. J. Biochem. 10, 557-563.

Sun, G.Y., Su, K.L., Der, O.M. and Tang, W. (1979b) Enzymic regulation of arachidonate metabolism in brain membrane phosphoglycerides. Lipids 14, 229-235.

Sun, G.Y., Tang, W. and Sun, A.Y. (1982) Changes in free fatty acids and diacylglycerols in rat brain due to acute ethanol administration. Fed. Proc. 41, 4270.

Sundler, R., Duzgunes, N. and Papahadjopoulos, D. (1981) Control of membrane fusion by phospholipid head groups. II. The role of phosphatidylethanolamine in mixtures with phosphatidate and phosphatidylinositol. Biochim. Biophys. Acta 649, 751-758.

Tang, W. and Sun, G.Y. (1982) Factors affecting the free fatty acids in rat brain cortex. Neurochem. Int. 4, 269-273.

Van den Bosch, H. (1980) Intracellular phospholipase A. Biochim. Biophys. Acta 604, 191-246.

Van den Bosch, H. and van den Besselaar, A.M.H.P. (1978) Intracellular formation and removal of lysophospholipids. In: Advances in Prostaglandin and Thromboxane Research (Galli, C. et al., eds.) Vol. 3, pp. 69-75, Raven Press, New York.

Vignais, P.V., Lauquin, G.J.M. and Vignais, P.M. (1976) In: Mitochondria (Packer, L. and Gomez-Puyon, A., eds.) pp. 109-125, Academic Press, New York.

Webster, G.R. (1973) Phospholipase A activites in normal and sectioned rat sciatic nerve. J. Neurochem. 21, 873-876.

Webster, G.R. and Alpern, R.J. (1964) Studies on the acylation of lysolecithin by rat brain. Biochem. J. 90, 35-41.

Webster, G.R. and Cooper, M. (1968) On the site of action of phosphatide acylhydrolase activity of rat brain homogenates on lecithin. J. Neurochem. 15, 795-802.

Weltzien, H.U. (1979) Cytolytic and membrane-perturbing properties of lysophosphatidylcholine. Biochim. Biophys. Acta 559, 259-287.

Woelk, H., Kanig, K. and Peiler-Ichikawa, K. (1974a) Phospholipid metabolism in experimental allergic encephalomyelitis: activity of mitochondrial phospholipase A_2 of rat brain towards specifically labelled 1,2-

diacyl-1-alk-1'-enyl-2-acyl- and 1-alkyl-2-acyl-_sn_-glycero-3-phosphorylcholine. J. Neurochem. 23, 745-750.
Woelk, H., Peiler-Ichikawa, K., Binaglia, L., Goracci, G. and Porcellati, G. (1974b) Distribution and properties of phospholipases A_1 and A_2 in synaptosomes and subsynaptosomal fractions of rat brain. Hoppe Seyler's Z. Physiol. Chem. 355, 1535-1542.
Woelk, H., Ariente, G., Gaiti, A., Kanig, K. and Porcellati, G. (1981) Action of phospholipase A_2 of rabbit neuronal and glial cells on 1,2-diacyl-, 2-acyl-1-alk-1'-enyl-, and 2-acyl-1-alkyl-glycerophosphatides. Neurochem. Res. 6, 23-32.
Wojtezak, L. (1976) Effect of long-chain fatty acid and acyl-CoA on mitochondrial permeability, transport, and energy coupling process. J. Bioenerg. Biomemb. 8, 293-311.
Wood, J.M., Bush, B., Pitts, B.J.R. and Schwartz, A. (1977) Inhibition of bovine heart Na^+,K^+-ATPase by palmitylcarnitine and palmityl-CoA. Biochem. Biophys. Res. Commun. 74, 677-684.
Zimmerberg, J., Cohen, F.S. and Finkelstein, A. (1980) Micromolar Ca^{++} stimulates fusion of lipid vesicles with planar bilayers containing a calcium-binding protein. Science 210, 906-908.
Zwiller, J., Cresielski-Treska, J. and Mandel, P. (1976) Effect of lysolecithin on guanylate and adenylate cyclase activities in neuroblastoma cells in culture. FEBS Lett. 69, 286-290.

EFFECTS OF CATECHOLAMINES AND THEIR METABOLITES AND OTHER PHARMACOLOGICAL AGENTS ON ARACHIDONIC ACID METABOLISM AND PROSTAGLANDIN RELEASE BY RABBIT IRIS MUSCLE AND IRIS MICROSOMES

Ata A. Abdel-Latif, Jack P. Smith,
Sardar Y.K. Yousufzai and Richard K. Dover

Department of Cell and Molecular Biology

Medical College of Georgia

Augusta, GA 30912, U.S.A.

SUMMARY

Comparative studies were conducted on the effects of catecholamines and their metabolites and other pharmacological agents on arachidonic acid (AA) metabolism and prostaglandin (PG) release by rabbit iris smooth muscle and iris microsomes. Norepinephrine (Nor) and related compounds were found to stimulate significantly the conversion of 1-^{14}C-AA into $PGF_{2\alpha}$ and PGE_2. Nor also stimulated the basal release of PGE_2 from the iris in a dose-dependent manner. The Nor-stimulation of PG synthesis was blocked by indomethacin (1.5 μM). The studies on the mechanism(s) underlying the action of catecholamines on PG synthesis revealed the following: (1) The structural requirement for maximal catecholamine stimulation of PG synthesis by the iris and iris microsomes is a catechol nucleus and ethylamine polar side-chain. Thus the deaminated metabolites of Nor had little effect on PG synthesis by the iris and iris microsomes; in contrast normetanephrine stimulated significantly PG synthesis. Furthermore, catechol inhibited PG synthesis by the iris in a dose-dependent manner. (2) In order for the amine to activate PG synthesis by the iris, the amine should be permeable to the cell membrane. Permeability studies revealed that ^{14}C-Nor is taken up rapidly by the iris. (3) α-Adrenergic agonists stimulated PG synthesis and release by the iris, and this was blocked by phentolamine.

It is concluded that in the iris, catecholamines stimulate PG synthesis in two ways: (a) They are taken up by the tissue where they act as cofactors for the cyclo-oxygenation of AA. (b) They stimulate PG synthesis through adrenoreceptor-mediated mechanisms. Participation of an adrenoreceptor-linked phospholipase A_2 in the expression of the catecholamine stimulation of PG synthesis remains to be defined.

INTRODUCTION

There is a relationship between prostaglandins (PGs) of the E series and the sympathetic nerve-effector transmission. Stimulation of the sympathetic nerve leads to an increased formation and release of $PGF_{2\alpha}$ and PGE_2 in a variety of tissues (Ferreira and Vane, 1967; Gilmore et al, 1968; Davis et al, 1968; Hedqvist and Euler, 1972; Ferreira et al, 1973; Pipili and Poyser, 1981; Petkov and Radomirov, 1980). Further, PGs of the E series have an inhibitory effect on norepinephrine (Nor) release in response to sympathetic nerve stimulation (for review see Hedqvist, 1977). On the basis of these observations, Hedqvist (1977) and others (Westfall, 1980; Starke, 1977; Kahn and Malik, 1982) proposed that PGs exert a negative feedback control of sympathetic neuroeffector transmission.

Catecholamines have been reported to stimulate PG biosynthesis in a variety of tissues including spleen (Ferreira and Vane, 1967; Gilmore et al, 1968; Davis et al, 1968), heart (Wennmalm and Brundin, 1978), phrenic diaphragm (Ramwell et al, 1965) kidney (Needleman et al, 1974), brain (Seregi et al., 1982), rabbit mesenteric blood vessels (Simmit and Hertting, 1980), rabbit iris smooth muscle (Bhattacherjee et al., 1979; Abdel-Latif et al., 1981; Abdel-Latif and Smith, 1982) and cultured cells (Levine and Moskowitz, 1979; Bruckner-Schmidt et al., 1981). The mechanism underlying the stimulatory effect of these amines

---

Abbreviations used: AA, arachidonic acid; PG, prostaglandin; Nor, norepinephrine; RIA, radioimmunoassay.

on PG synthesis is poorly understood. It has been suggested that: (a) They act as cofactors for the cyclooxygenation of arachidonic acid (AA), as shown by their stimulatory effects on PG synthetase activity (Takaguchi et al, 1971; Baumann et al, 1979). (b) They stimulate PG synthesis through receptor-mediated mechanisms; thus treatment with the α-adrenergic receptor blocking agent phenoxybenzamine inhibits the appearance of PGs from dog spleen (Davis et al, 1968) and rabbit kidney (Needleman et al, 1974). Furthermore, PG synthesis in cell cultures from dog kidney cells (Levine and Moskowitz, 1979) and from rabbit splenic pulpa (Bruckner-Schmidt et al, 1981) is mediated through α-adrenoreceptors. (c) Finally, PG release could occur during muscle contraction following receptor activation by the catecholamines.

In a previous communication from this laboratory (Abdel-Latif et al, 1981), we reported that in the rabbit iris smooth muscle Nor stimulates PG synthesis in a dose-dependent manner; it was also suggested that following activation of α-adrenergic receptors by Nor, there could be an increase in AA release, mediated via Ca^{2+}, and consequently an increase in PG biosynthesis. The iris is innervated by adrenergic and cholinergic nerve fibers. It is extremely rich in Nor, which can be demonstrated histochemically; stimulation of the cervical sympathetic depletes the iris of Nor.

In this tissue $PGF_{2\alpha}$ and PGE_2 are the major PGs formed from AA (Abdel-Latif et al, 1981). To throw more light on the molecular mechanism underlying Nor-stimulation we have compared the effects of various catecholamines and their metabolites and several chemically related compounds on conversion of AA into PGs by rabbit iris muscle and iris microsomes, and on the basal release of PGE_2 (measured by RIA) from the iris. Our aims from this study are: (a) to define the essential molecular structures necessary for the action of catecholamines as cofactors in the cyclooxygenase pathway in the rabbit iris and (b) to answer the question whether adrenoreceptors are involved in the Nor-stimulation of AA metabolism and PG release in this tissue.

EFFECTS OF NOR ON PG RELEASE

Effect of Nor Concentration on Conversion of 1-^{14}C-AA into PGs by Iris and Iris Microsomes

Comparative studies on the effects of different concentrations of Nor on PG synthesis from 1-^{14}C-AA in rabbit iris muscle and iris microsomes showed that the catecholamine-induced release of $PGF_{2\alpha}$ by microsomes is considerably higher than that of the whole iris at all concentrations studied (Fig. 1). Thus at 0.5 mM Nor, a concentration at which the catecholamine-induced release of PGs levelled off, $PGF_{2\alpha}$ synthesis in muscle increased by 80% and that of microsomes increased by 300%. Similarly at lower concentrations of Nor (0.1 mM), $PGF_{2\alpha}$ synthesis by muscle and microsomes increased by 40% and 120%, respectively. It can be concluded that in the rabbit iris Nor-stimulation of PG synthesis by microsomes is three times as high as that of the muscle.

Effect of Nor Concentration on Basal Release of PGE_2

The results on the effect of Nor concentration on the basal formation from endogenous AA pools of PGE_2 are given in Fig. 2. PGE_2 was quantified using the RIA method of Jaffe et al (1973). PGE_2 ^{3}H-RIA kit was obtained from Seragen Inc., 54 Clayton Street, Boston, MA. The mean levels of PGE_2 in media from 12 experiments incubated without Nor was 828 ng/g wet irides. Incubation with various concentrations of Nor produced during 20 min of incubation a steady increase in PGE_2 concentration from 54% increase at 0.05 mM to 230% at 0.75 mM of the neurotransmitter. In mammalian tissues AA, the precursor of PGs, is found mainly esterified to phospholipids and triacylglycerols; there is very little free AA. It is esterified almost exclusively at the second position of glycerol of phospholipids, from which it could be released by the action of phospholipase A_2 and other lipases. It is now generally agreed that PG formation must be preceded by a lipolytic process to release free AA from the tissue glycerolipids. The data presented could suggest that the increment in PG synthesis associated with Nor treatment was probably caused by release of AA from a glycerolipid pool after stimulation of the adrenoreceptors.

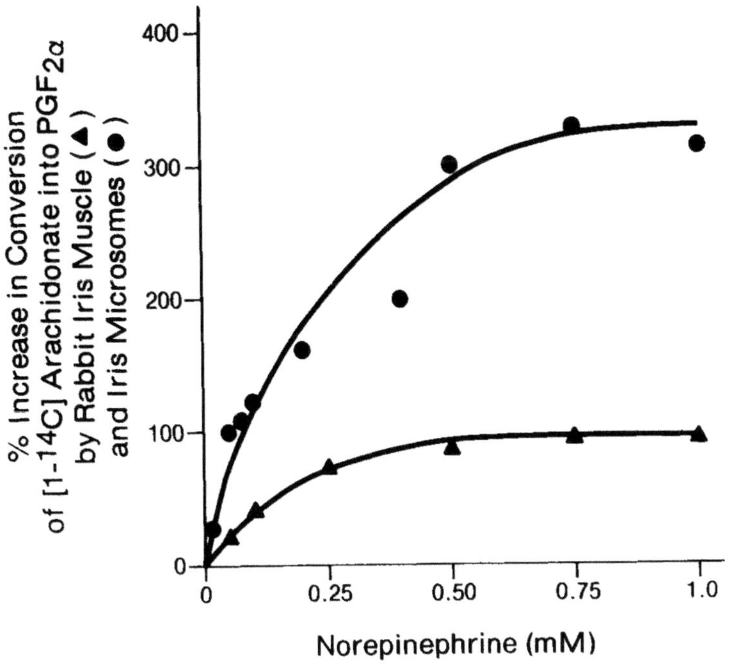

Fig. 1. Effect of Nor concentration on conversion of 1-^{14}C-AA into $PGF_{2\alpha}$ by rabbit iris muscle (▲) and iris microsomes (●). In these experiments 2 irides from pairs were incubated in 1 ml of an isoosmotic medium that contained 0.25 μCi of 1-^{14}C-AA (specific activity 56.9 mCi/mmole) bound to 0.1 mg albumin in absence and presence of various concentrations of Nor for 1 hr at 37°C. At the end of incubation the medium was analyzed for PGs as previously described (Abdel-Latif and Smith, 1982). The data presented are means of two separate experiments and each experiment was carried out in duplicate, with variability less than 7%. Assay conditions for conversion of 1-^{14}C-AA into PGs by iris microsomes were as previously described (Abdel-Latif and Smith, 1982). Each point represents means of 3 experiments. S.E.M. were less than 5% (Taken from Abdel-Latif et al, 1982, with permission).

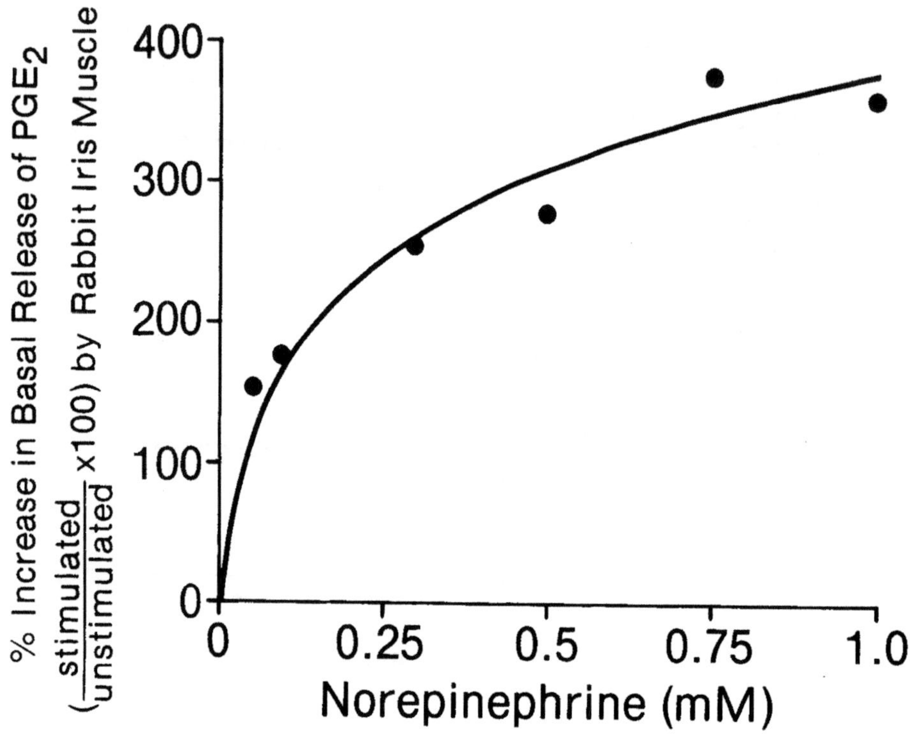

Fig. 2. Effect of Nor concentration on basal release of PGE_2. Conditions of incubation were the same as described in Fig. 1, except that no ^{14}C-AA was added, and time of incubation was reduced to 20 min. PGE_2 in the medium was quantified by RIA. Mean levels of PGE_2 in media from 12 experiments incubated without Nor was 828 ng/g wet irides. Each point represents means of 3 experiments.

Effects of Nor, Glutathione and Indomethacin on PG Synthesis by Rabbit Iris and Iris Microsomes

In contrast to Nor which stimulated PG synthesis by both muscle and microsomes, glutathione, which does not penetrate the plasma membrane of the cell, stimulated only the microsomal conversion of 1-^{14}C-AA into PGs (Table I). The stimulatory effects of both Nor and glutathione were abolished by 1.5 μM indomethacin. These data could suggest either that Nor acts on the surface of the plasma membrane to activate the PG synthetase or the catecholamine is taken up by the cell where it may stimulate PG

TABLE I. Comparative Studies on the Effect of Nor, Glutathione and Indomethacin on Conversion of 1-^{14}C-AA into PGs by Rabbit Iris and Iris Microsomes*

Additions	Radioactivity released into medium			
	Whole iris (cpm/2 irides)		Iris microsomes (cpm/mg protein)	
	PGE_2	% of control	PGE_2	% of control
None	23971	100	12295	100
1 mM-Nor	38765	162	44741	364
1 mM-Glutathione	24208	100	34522	281
Nor + glutathione	37900	158	93421	760
Nor + glutathione + 1.5 μM-indomethacin	5993	25	7825	64

*Conditions of incubation were the same as described under Fig. 1. Data given are averages of three separate experiments (Taken from Abdel-Latif and Smith, 1982, with permission).

synthetase of the cellular organelles. The first possibility is supported by the fact that PG biosynthesis has recently been demonstrated in plasma membranes and mitochondrial and microsomal fractions from rabbit kidney medulla (Erman and Raz, 1981; Erman et al, 1982). Furthermore, Marion et al (1979), working with rat cerebral cortex slices, suggested that Nor could stimulate $PGF_{2\alpha}$ formation at the surface of the slice by stimulating directly the PG synthetase. The latter possibility is supported by the finding that in the iris ^{14}C-Nor was taken up by the tissue (Fig. 3). Thus after 1 hr of incubation,the ^{14}C-Nor found in the tissue was about 20% of that recovered in the medium.

EFFECTS OF CATECHOLAMINES AND THEIR METABOLITES AND OTHER AMINES ON PG RELEASE

Effects of Catecholamines and their Precursors

Results of studies on the effects of catecholamines and their precursors on PG synthesis are given in Table II. Tyrosine had little effect on PG synthesis by both muscle and microsomes. DOPA, with two hydroxyl groups substituted in the 3 and 4 positions of the benzene ring, increased PGE_2 synthesis in iris and iris microsomes by 63% and 36%, respectively. In contrast dopamine had little effect on PGE_2 synthesis by the iris, but increased that of microsomes by 215%. This could suggest that dopamine but not DOPA has an appreciable stimulatory effect on PG synthesis. The effect of DOPA observed in the iris is probably due to the dopamine formed from DOPA via decarboxylation. The finding that dopamine had little effect on PG synthesis in the iris but exerted an appreciable effect on that of microsomes is probably due to the fact that this catecholamine is impermeable to the plasma membranes of the smooth muscle. Nor and epinephrine increased PGE_2 synthesis in the iris and iris microsomes by about 60% and 140%, respectively. These results show that the minimum structural requirement for an activation is a catechol nucleus (a benzene ring with two adjacent hydroxyl substituents) containing an aliphatic polar sidechain. Hydroxylation of the ethylamine side chain, e.g. in Nor and epinephrine, appears either to increase the permeability of the catecholamine or it facilitates its binding to the adrenoreceptor. Lack of alcoholic hydroxyl

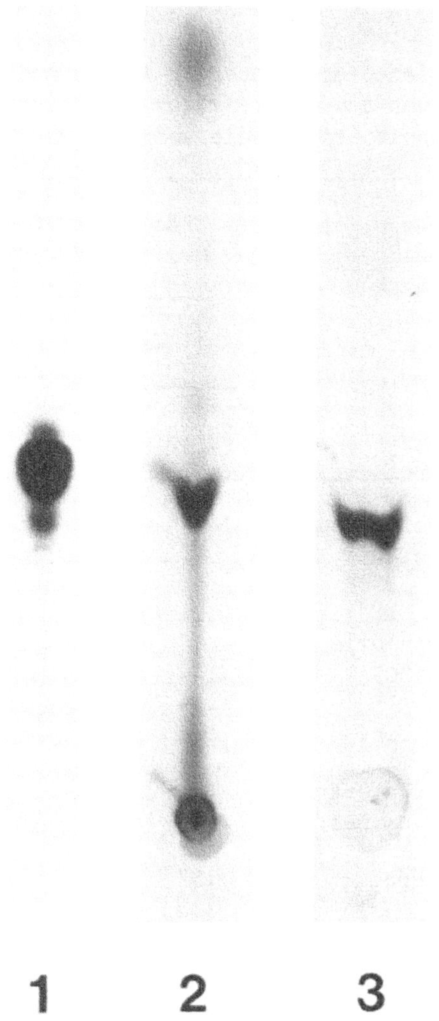

Fig. 3. An autoradiograph showing the metabolism of DL-[7-^{14}C]-Nor by rabbit iris. Two irides were incubated in 1 ml Krebs-Ringer medium that contained 2 μCi of ^{14}C-Nor (Specific radioactivity 54.8 mCi/mmol, obtained from New England Nuclear, Boston, Mass.). Nor metabolites were extracted from the tissue with 0.05 N HCl. The extract was lyophilized, then separated by means of one-dimensional thin-layer chromatography with cellulose plates in a solvent system of n-butanol-acetic acid-water (25:4:10 by vol.). 1, standard [7-^{14}C]-Nor; 2, ^{14}C-labeled metabolites extracted from the tissue; 3, ^{14}C-labeled metabolites obtained from the medium.

TABLE II. Effects of Adrenergic Agents and their Metabolites and Other Amines on PG Synthesis by Rabbit Iris Muscle and Iris Microsomes.

Additions (0.1 mM)	Effects on PG synthesis (% of control)			
	Iris		Microsomes	
	$PGF_{2\alpha}$	PGE_2	$PGF_{2\alpha}$	PGE_2
A. Catecholamines and Precursors:				
Tyrosine	121	119	103	114
Dihydroxyphenyl-alanine (DOPA)	146	163	125	136
Dopamine	104	90	236	315
Norepinephrine	141	157	197	271
B. Adrenergic agents and other amines:				
Isoproterenol	157	163	285	193
Epinine	133	174	184	208
Phenylephrine	129	131	157	137
Serotonin	120	102	211	326
C. Amine metabolites:				
Normetanephrine	141	148	153	198
3-Methoxy, 4-hydroxy mandelic acid	125	127	110	138
Adrenochrome	77	92	143	197

*Conditions of incubation were the same as described under Fig. 1.

group on the side chain is known to reduce receptor affinity and so to reduce activity.

Effect of Various Adrenergic Agents and other Amines

Other amines such as isoproterenol, a β-adrenergic agonist, and epinine, both of which possess a catechol nucleus and an ethylamine side-chain with an alkyl substituent on the terminal amino group, exerted a stimulatory effect on PGE_2 synthesis in both iris and iris microsomes, comparable to that of Nor and epinephrine (Table II). In contrast phenylephrine, an α-adrenergic agonist lacking the catechol ring hydroxyl in the 4 position, had a weak effect on PGE_2 synthesis by iris and iris microsomes. Unexpectedly, serotonin (5-hydroxytryptamine), which is impermeable to the plasma membrane of the muscle had no effect on PGE_2 synthesis by the iris, however it increased its microsomal synthesis by 226% (Table II). It has been suggested that the catecholamines and serotonin stimulate the cyclooxygenase directly by supplying reducing equivalents (Takaguchi et al, 1971).

Effects of Catecholamine Metabolites

Normetanephrine, an O-methylated metabolite of Nor, stimulated significantly PG synthesis by both iris and iris microsomes (Table II). In contrast, the deaminated metabolites of Nor, products of monoamine oxidase action, such as 3-methoxy-4-hydroxy-mandelic acid and 3-methoxy-4-hydroxyphenylacetic acid had little effect on PG synthesis by the iris and iris microsomes. This indicates that the aliphatic polar side-chain ethylamine in catecholamines is essential for the catecholamine-stimulated PG biosynthesis by the iris. Seregi et al (1982), working with rat brain homogenates, suggested that catecholamines stimulate $PGF_{2\alpha}$ formation in brain in two ways: (a) Monoamine oxidase deaminates catecholamines and produces H_2O_2; the latter then stimulates endoperoxide synthesis. Hydroperoxides, including H_2O_2, have been reported to activate PG synthesis (Smith and Lands, 1972; Panganamala et al, 1974; Polgar and Taylor, 1980). (b) Catecholamines, as hydrogen donors, promote the non-enzymatic transformation of cyclic endoperoxides into $PGF_{2\alpha}$.

The important role played by the ethylamine side-chain of catecholamines in PG synthesis is further supported by the finding that adrenochrome, an orthoquinone product of Nor oxidation, stimulated PGE_2 synthesis by iris microsomes. However, in the iris this product of Nor oxidation inhibited formation of $PGF_{2\alpha}$ and PGE_2 by 23% and 8%, respectively (Table II).

Effects of Catechol

Since catechol lacks the ethylamine polar side-chain, it was of interest to investigate its effect on AA metabolism by iris and iris microsomes. We found that in muscle, catechol inhibited PG formation in a dose-dependent manner (Fig. 4). Thus dose-response curves for inhibition of PGE_2 and $PGF_{2\alpha}$ biosynthesis by catechol gave IC_{50} (inhibitory $concentration_{50}$) values of 0.03 mM and 0.05 mM, respectively. In contrast, in microsomes (Fig. 5) there was up to 50% stimulation in PGE_2 formation at lower concentrations of catechol (0.05 mM), and up to 67% inhibition at higher concentrations (0.1 mM). These data indicate that the ethylamine polar side-chain is essential to catecholamine-activation of the cyclooxygenase pathway in the muscle. The inhibitory effect observed with catechol on AA metabolism in muscle could be due to oxidation of catechol into o-benzoquinone or other metabolites. To test this possibility we prepared o-benzoquinone from catechol according to the method of Erickson and Dechary (1952). Results of comparative studies on the effects of catechol and quinones, at two concentrations, on PGE_2 synthesis by iris muscle and iris microsomes are given in Table III. Catechol and o-benzoquinone inhibited PGE_2 synthesis by muscle at both concentrations, in contrast, at 0.05 mM they stimulated its biosynthesis by microsomes. Hydroquinone and p-quinone had some inhibitory effects on PGE_2 synthesis by muscle but stimulated significantly its biosynthesis by microsomes. Hydroquinone and p-quinone have been reported to stimulate PG synthesis by microsomes from bovine seminal vesicles (Takaguchi et al, 1971; Miyamoto et al, 1976; Egan et al, 1978). These data suggest that other metabolites of catechol, in addition to o-benzoquinone, could be involved in the pronounced inhibitory effect of this compound on PG synthesis by the iris.

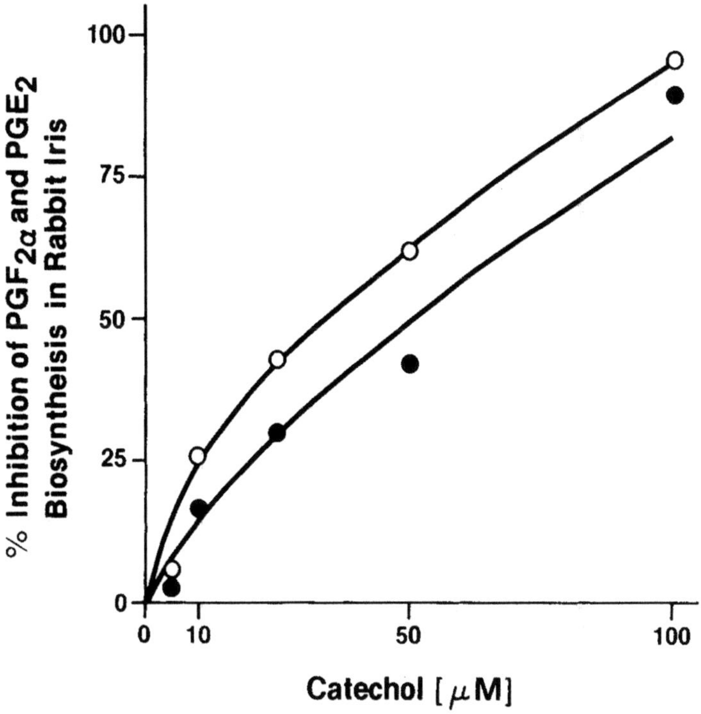

Fig. 4. Effects of different concentrations of catechol on formation of $PGF_{2\alpha}$ (●) and PGE_2 (○) from AA by iris muscle. Conditions of incubation were as described under Fig. 1.

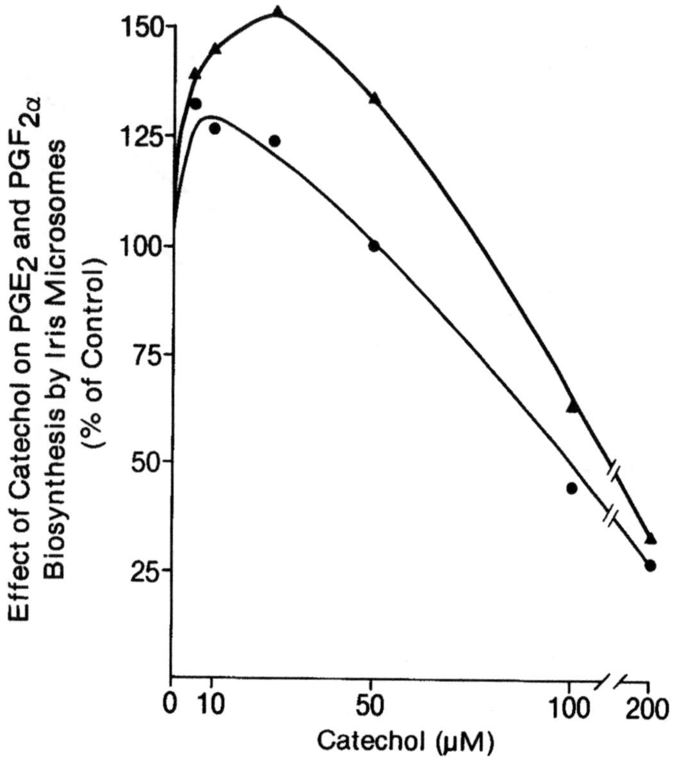

Fig. 5. Effects of different concentrations of catechol on formation of $PGF_{2\alpha}$ (●) and PGE_2 (▲) from AA by iris microsomes. Conditions of incubation were as described under Fig. 1.

TABLE III. Effects of Catechol and Quinones on PGE_2 Synthesis by Iris Muscle and Iris Microsomes*

Additions	PGE_2 (percent of control) at catechol or quinone concentration (mM)			
	Muscle		Microsomes	
	0.05	0.1	0.05	0.1
Catechol	38	5	134	63
o-Benzoquinone	81	41	231	218
Hydroquinone	92	69	287	282
p-Quinone	98	95	194	117

*Conditions of incubation were the same as described under Fig. 1.

ARE α-ADRENORECEPTORS INVOLVED IN PG BIOSYNTHESIS BY THE IRIS?

To answer this question we investigated the effect of phentolamine, an α-adrenergic antagonist, on Nor-stimulated AA metabolism by rabbit iris and iris microsomes. Phentolamine reduced the Nor-stimulated $PGF_{2\alpha}$ and PGE_2 synthesis in the iris by 52% and 68%, respectively, and in iris microsomes by 22% and 19%, respectively (Table IV). Propranolol, a β-adrenergic antagonist, had no effect (data not shown). These data could suggest some possible involvement of α-adrenoreceptors in PG formation and release. This conclusion is supported by the finding that 5–20 μM phentolamine inhibited the Nor-stimulated basal release of PGE_2 in the iris muscle (Table V). Nor had little effect on the release of ^{14}C-AA from iris muscle prelabeled with the isotope (Table VI). It is possible that endogenous AA and ^{14}C-AA have different pools.

DISCUSSION AND CONCLUSIONS

In summary, Nor and other related compounds were found to stimulate significantly the conversion of 1-^{14}C-AA into $PGF_{2\alpha}$ and PGE_2 by both rabbit iris and iris microsomes. Nor also stimulated the basal release of PGE_2 by the iris. The Nor-stimulation of PG synthesis was completely blocked by low concentrations of indomethacin (1.5 μM). The data presented show that: (1) The structural requirement for maximal catecholamine stimulation of PG synthesis by the iris and iris microsomes is a catechol nucleus and ethylamine polar side-chain, e.g. Nor and epinephrine. (2) The amine should be taken up by the muscle in order for activation to occur. Thus DOPA, but not dopamine, stimulated PG biosynthesis by the iris muscle, In contrast dopamine, but not DOPA, stimulated PG biosynthesis by microsomes. Also serotonin and glutathione stimulated PG biosynthesis by microsomes, but not by the muscle. We found that ^{14}C-Nor is taken up rapidly by the iris smooth muscle. This indicates that catecholamines can act both at the plasma membrane and at intracellular sites. (3) In the rabbit iris catecholamines stimulate PG synthesis in two ways (Fig. 6): (a) They act as cofactors for the cyclooxygenation of AA (Fig. 6-2). This is supported by the comparative studies on their activation of PG synthesis by iris muscle and iris microsomes. (b) They

TABLE IV. Effect of Phentolamine on Nor-Stimulated Formation of PGs by Rabbit Iris and Iris Microsomes*

Additions	Radioactivity in PGs (% of control)			
	Whole iris		Iris microsomes	
	$PGF_{2\alpha}$	PGE_2	$PGF_{2\alpha}$	PGE_2
0.1 mM-Nor	160 ± 17	166 ± 18	210 ± 12	197 ± 7
0.05 mM-phentolamine	101 ± 8	103 ± 8	103 ± 3	115 ±11
0.1 mM-Nor + 0.05 mM-phentolamine	129 ± 9	121 ± 7	186 ± 11	179 ± 2

*Conditions of incubation were the same as described under Fig. 1. Results are means ± SEM of three experiments and each experiment was run in triplicate (Taken from Abdel-Latif et al, 1982, with permission).

TABLE V. Effect of Phentolamine on Nor-stimulated Basal Release of PGE_2 by Rabbit Iris*

Additions	PGE_2 (ng/g tissue)	Effect of drug (% of control)
None	720	100
0.1 mM-Nor	1400	194
10 μM phentolamine	712	99
0.1 mM-Nor + 5 μM phentolamine	874	121
0.1 mM-Nor + 10 μM phentolamine	845	119
0.1 mM-Nor + 20 μM phentolamine	658	91

*Conditions of incubation were as described under Fig. 2. PGE_2 was quantified by RIA.

TABLE VI. Effect of Nor on Release of 1-^{14}C-AA from Rabbit Iris Prelabeled with the Isotope In Vitro*

Lipids	Radioactivity released (cpm x 10^{-2})		
	Control	0.1 mM Nor	% of control
Tissue:			
Phosphatidic acid	18	22	122
Phosphatidyl-inositol	60	63	105
Phosphatidyl-choline	263	255	97
Phosphatidyl-ethanolamine	56	53	95
Diacylglycerol	155	141	91
Triacylglycerol	362	320	88
Arachidonic acid	196	182	93
Medium:			
Arachidonic acid	145	157	108

*Conditions of incubation were the same as described under Fig. 1 with the following modifications: Irides were incubated with 1-^{14}C-AA for 1 hr, washed twice with non-radioactive medium that contained 1 mg/ml bovine serum albumin, then transferred into tubes that contained iso-osmotic medium without 1-^{14}C-AA. Incubations were then carried out in the absence and presence of 0.1 mM Nor for 1 hr. Phospholipids and neutral lipids were analyzed as described previously (Abdel-Latif and Smith, 1982). All values are means of two separate experiments and each experiment was run in duplicate.

act through α-adrenoreceptor-mediated mechanisms (Fig. 6-1); thus treatment with the α-adrenergic blocking agent phentolamine inhibited both the Nor-stimulated conversion of 1-^{14}C-AA into PGs and the Nor-stimulated basal release of PGE_2 by the iris muscle.

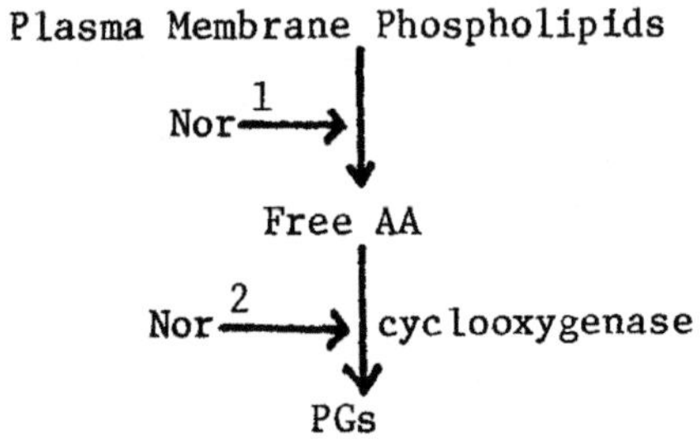

Fig. 6. Possible sites of Nor effects on AA release and metabolism by iris muscle.

The molecular mechanism underlying the action of catecholamines as cofactors in the cyclooxygenase pathway is still unclear at the present time. Thus it has been suggested that catecholamines and hydroquinones stimulate cyclooxygenase directly be supplying reducing equivalents (Miyamoto et al, 1976). Peterson et al (1980) concluded that epinephrine and other activators of PG endoperoxide synthetase act by reducing Fe^{3+} heme to Fe^{2+}-heme of the cyclooxygenase enzyme. Seregi et al (1982), working with rat brain homogenates, proposed that monoamine oxidase deaminates catecholamines and produces H_2O_2, the latter then stimulates endoperoxide synthesis. In the present work the deaminated metabolites of Nor, products of monoamine oxidase action, had little effect on PG synthesis by the iris and iris microsomes. In contrast, normetanephrine, an o-methylated metabolite of Nor, stimulated significantly PG synthesis by both muscle and microsomes. These findings emphasize the important role played by the ethylamine polar side-chain of catecholamines in stimulation of PG synthesis. Since catecholamine-activation of PG synthesis by microsomes and by purified PG synthetase has been reported in bovine seminal vesicles (Takaguchi et al, 1971; Baumann et al, 1979; and Miyamoto et al, 1976) and in iris microsomes (the present study), where there is no monoamine oxidase activity, one may conclude that both the catechol nucleus and ethylamine side chain are required for activation. This conclusion is further supported by the finding in the present work that catechol, which lacks the ethylamine side chain, inhibited significantly PG synthesis by both muscle and microsomes. This inhibition does not appear to be entirely due to oxidation of catechol into o-benzoquinone. Other metabolites of catechol are probably involved in this inhibition. This conclusion is supported by the finding that adrenochrome, an oxidation product of Nor, and o-benzoquinone exerted an inhibitory effect on PG synthesis by the muscle, but they stimulated appreciably PG synthesis by microsomes. Since the hydrogens are lacking in these compounds, one may conclude that activation of PG synthesis by catecholamines requires not only the reducing equivalents but also the catechol nucleus.

It can be concluded from the above studies that both the catechol nucleus and the ethylamine side chain are required for maximal activation of PG synthesis by catecholamines.

Both α-adrenergic and β-adrenergic agonists were found to stimulate PG synthesis by the rabbit iris. The finding that phentolamine does block Nor-induced PG release could suggest the involvement of α-adrenoreceptors. Nor has been reported to increase the release of Ca^{2+} from intracellular sites in the iris (Akhtar and Abdel-Latif, 1979). Thus activation of adrenoreceptors by catecholamines could bring about a release of AA from membrane phospholipids (Fig. 6-1), mediated via Ca^{2+}-activation of phospholipase A_2, and consequently an increase in PG biosynthesis. There is no conclusive experimental evidence that shows which phospholipid is the source for AA in PG synthesis. Furthermore, the precise lipase(s) activated upon hormonal stimulation of the cell remains to be established.

In conclusion there is compelling experimental evidence which indicates that adrenoreceptors are probably involved in AA release and PG biosynthesis; however, the molecular mechanism for participation of an adrenoreceptor-linked phospholipase A_2 in the expression of the catecholamine stimulation of PG biosynthesis remains to be defined.

Stimulation of PG synthesis by Nor may be of importance in normal or pathological situations. In the eye the increased release of catecholamines from neurons may serve as a stimulus for increased AA release and a subsequent PG synthesis. Furthermore catecholamines and adrenergic drugs are routinely employed therapeutically to lower intraocular pressure in the eye. A catecholamine-induced increase in PG synthesis may play a role in mediating the effects of these therapeutic agents. In brain, during carotid ligation, there is a large increase in brain catecholamine release (Akiguchi et al, 1980) and free AA levels (Bazan, 1970), and PG synthesis is stimulated following re-perfusion of the ischemic tissue (Gaudet et al, 1980).

ACKNOWLEDGEMENTS

This work was supported by NIH grant EY-04387 from the National Eye Institute. This is contribution No. 0720 from the Department of Cell and Molecular Biology, Medical College of Georgia.

REFERENCES

Abdel-Latif, A.A. and Smith, J.P. (1982) Studies on the incorporation of [1-^{14}C]arachidonic acid into glycerolipids and its conversion into prostaglandins by rabbit iris. Effects of anti-inflammatory drugs and phospholipase A_2 inhibitors. Biochim. Biophys. Acta 711, 478-489.

Abdel-Latif, A.A., Smith, J.P. and Dover, R.K. (1982) Norepinephrine and prostaglandin biosynthesis by iris smooth muscle and iris microsomes. Biochem. Pharmacol. Accepted for publication.

Abdel-Latif, AA., Smith, J.P. and Mitra, R. (1981) Glycerolipids and prostaglandin biosynthesis in the rabbit iris. Prog. in Lipid Res. 20, 183-188.

Akhtar, R.A. and Abdel-Latif, A.A. (1979) Effects of acetylcholine and Nor on ^{45}Ca uptake and efflux in rabbit iris smooth muscle. Gen. Pharmac. 10, 445-450.

Akiguchi, I., Horie, R. and Yamori, Y. (1980) Role of central aminergic fibers in experimental cerebral ischemia in stroke-prone SHR. Stroke 11, 383-389.

Baumann, J., Bruchhausen, F.V. and Wurm, G. (1979) A structure-activity study on the influence of phenolic compounds and bioflavonoids on rat renal prostaglandin synthetase. Naunyn-Schmiedeberg's Arch. Pharmacol. 307, 73-78.

Bazan, N.G. (1970) Effects of ischemia and electroconvulsive shock on free fatty acid pool in the brain. Biochim. Biophys. Acta 118, 1-10.

Bhattacherjee, P., Kulkarni, P.S. and Eakin, K.E. (1979) Metabolism of arachidonic acid in rabbit ocular tissues. Invest. Ophthalmol. and Visual Sci. 18, 172-178.

Bruckner-Schmidt, R., Jackisch, R. and Hertting, G. (1981) Stimulation of prostaglandin E_2-synthesis by noradrenaline in primary cell cultures from rabbit pulpa is mediated by atypical α-adrenoreceptors. Naunyn-Schmiedeberg's Arch. Pharmacol. 316, 1-7.

Davis, B.N., Horton, E.W. and Withrington, P.G. (1968) The occurrence of prostaglandin E_2 in splenic venous blood of the dog following splenic nerve stimulation. Br. J. Pharmac. Chemother. 32, 127-135.

Egan, R.W., Humes, J.L. and Kuehl, F.A. (1978) Differential effects of prostaglandin synthetase stimulators on inhibition of cyclooxygenase. Biochem. 17, 2230-2234.

Erickson, J.L.E. and Dechary, J.M. (1952) The addition of diphenylketone to o-benzoquinone. J. Am. Chem. Soc. 74, 2644-2646.

Erman, A., Azuri, R. and Raz, A. (1982) Enzymic coupling of acylhydrolase and prostaglandin synthase activities in subcellular fractions from rabbit renal medulla. Biochem. J. 201, 635-640.

Erman, A. and Raz, A. (1981) Prostaglandin biosynthesis and lipolysis in subcellular fractions from rabbit kidney medulla. Biochem. J. 194, 957-961.

Ferreira, S.H., Moncada, S. and Vane, J.R. (1973) Some effects of inhibiting endogenous prostaglandin formation on the responses of the cat spleen. Br. J. Pharmac. 47, 48-58.

Ferreira, S.H. and Vane, J.R. (1967) Prostaglandins: their disappearance from release into the circulation. Nature (Lond.) 216, 868-873.

Gaudet, R.J., Alan, I. and Levine, L. (1980) Accumulation of cyclooxygenase products of arachidonic acid metabolism in gerbil brain during reperfusion after bilateral common carotid artery occlusion. J. Neurochem. 35, 653-658.

Gilmore, N., Vane, J.R. and Wyllie, J.H. (1968) Prostaglandins released by the spleen. Nature (Lond.) 218, 1135-1140.

Hedqvist, P. (1977) Basic mechanisms of prostaglandin action on autonomic neurotransmission. Annu. Rev. Pharmacol. Toxicol. 17, 259-279.

Hedqvist, P. and Euler, U.S. (1972) Prostaglandins control neuromuscular transmission in guinea-pig vas deferens. Nature New Biol. 236, 113-115.

Jaffe, B.M. Behrman, H.A. and Parker, C.W. (1973) Radioimmunoassay measurement of prostaglandins E, A, and F in human plasma. J. Clin. Investig. 52, 398-405.

Khan, M.T. and Malik, K.U. (1982) Modulation by prostaglandins of the release of [^{3}H] noradrenaline evoked by potassium and nerve stimulation in the isolated rat heart. European J. Pharmacol. 78, 213-218.

Levine, L. and Moskowitz, M.A. (1979) α- and β-adrenergic stimulation of arachidonic acid metabolism in cells in culture. Proc. Natl. Acad. Sci. USA 76, 6632-6636.

Marion, J., Pappiusm H.M. and Wolfe, L.S. (1979) Evidence for the use of a pool of the free arachidonic acid in rat cerebral cortex tissue for prostaglandin $F_{2\alpha}$ synthesis _in vitro_. Biochim. Biophys. Acta 573, 229-237.

Miyamoto, T., Ogino, N., Yamamoto, S. and Hayaishi, O. (1976) Purification of prostaglandin endoperoxide synthetase from bovine vesicular gland microsomes. J. Biol. Chem. 251, 2629-2636.

Needleman, P., Douglas, J.R., Jakschik, B., Stoecklein, P.B. and Johnson, E.M. (1974) Release of renal prostaglandin by catecholamines: relationship to renal endocrine function. J. Pharmacol. Exp. Ther. 188, 453-460.

Panganamala, R.V., Sharma, H.M., Sprecher, H., Geer, J.C. and Cornwell, D.G. (1974) A suggested role for hydrogen peroxide in biosynthesis of prostaglandins. Prostaglandins 10, 3-11.

Peterson, D.A., Gerard, J.M., Rao, G.H.R. and White, J.G. (1980) Epinephrine and other activators of prostaglandin endoperoxide synthetase can reduce Fe^{3+}-heme to Fe^{2+}-heme. Prostag. and Med. 5, 357-364.

Petkov, V. and Radomirov, R. (1980) On the origin of prostaglandin and its role in the sympathetic nerve transmission in vas deferens. Gen. Pharmac. 11, 275-282.

Pipili, E. and Poyser, N.L. (1981) Effects of nerve stimulation and of administration of noradrenaline or potassium chloride upon the release of prostaglandins I_2, E_2 and $F_{2\alpha}$ from the perfused mesenteric arterial bed of the rabbit. Br. J. Pharmac. 72, 89-93.

Polgar, P. and Taylor, L. (1980) Stimulation of prostaglandin synthesis by ascorbic acid via hydrogen peroxide formation. Prostaglandins 19, 693-700.

Ramwell, P.W., Shaw, J.E. and Kucharski, J. (1965) Prostaglandin release from the rat phrenic nerve-diaphragm preparation. Science 149, 1390-1391.

Seregi, A., Serfozo, P., Mergl, Z. and Schaefer, A. (1982) On the mechanism of the involvement of monoamine oxidase in catecholamine-stimulated prostaglandin biosynthesis in particulate fraction of rat brain homogenates: role of hydrogen peroxide. J. Neurochem. 38, 20-27.

Simmit, T. and Hertting, G. (1980) On the relation between contraction and prostaglandin release in rabbit mesenteric blood vessels. Eur. J. Pharmacol. 65, 325-331.

Smith, W.L. and Lands, W.E.M. (1972) Stimulation and blockade of prostaglandin biosynthesis. J. Biol. Chem. 246, 6700-6704.

Starke, K. (1977) Regulation of noradrenaline release by presynaptic receptor systems. Rev. Physiol. Biochem. Pharmacol. 77, 1-124.

Takaguchi, C., Kohno, E. and Sih, C.J. (1971) Mechanism of prostaglandin biosynthesis. I. Characterization and assay of bovine prostaglandin synthetase. Biochem. 10, 2372-2376.

Wennmalm, A. and Brundin, T. (1978) Prostaglandin-mediate inhibition of noradrenaline release: Prostagladin synthesis is stimulated by myocardial adrenoreceptors differing from the α- and β-type. Acta Physiol. Scand. 102, 374-381.

Westfall, T.C. (1980) Local regulation of adrenergic neurotransmission. Annu. Rev. Physiol. 42, 383-397.

FREE ARACHIDONIC AND DOCOSAHEXAENOIC ACID ACCUMULATION IN THE CENTRAL NERVOUS SYSTEM DURING STIMULATION

Nicolas G. Bazan[1], Susana G. Morelli de Liberti[2], Elena B. Rodriguez de Turco[2], and Maria F. Pediconi[2]

[1]LSU Eye Center, Louisiana State University Medical Center, School of Medicine, 136 South Roman Street, New Orleans, LA 70112 and [2]Instituto de Investigaciones Bioquimicas, Universidad Nacional del Sur-Consejo Nacional de Investigaciones Cientificas y Tecnicas, Bahia Blanca, Argentina.

INTRODUCTION

A single seizure promotes the accumulation of free fatty acids (FFA), especially arachidonic, stearic, docosahexaenoic, palmitic, and oleic acids, in the central nervous system (Bazan, 1970; Bazan and Rakowski, 1970). The stimulation of deacylation pathways of excitable membrane phospholipids and the decreased ability to activate-reacylate the released acyl groups leads to a transient accumulation of FFA (Bazan, 1970; Bazan and Rakowski, 1970; Rodriguez de Turco et al., 1977). Diglycerides (DG) also are accumulated and display a fatty acid composition similar to the FFA (Aveldano and Bazan, 1979). Electroconvulsive shock (Bazan, 1970; Bazan and Rakowski, 1970), pentylenetetrazol (Cenedella et al., 1975) and bicuculline (Seisjo et al., 1982; Rodriguez de Turco et al., 1983; Bazan et al., 1982b) are some of the convulsive-producing treatments that have been found to increase brain FFA. It seems that although hypoxia (Porcellati et al., 1978), ischemia (Bazan, 1970, 1971a, 1976; Galli and Spagnuolo, 1976; Bazan et al., 1971, 1982a; Marion and Wolfe, 1979; Rehncrona et al., 1982; Kuwashima et al., 1978; Majewska et al., 1981; Aveldano and Bazan, 1975) and carotid artery ligation (Porcellati et al., 1978) yield lipid effects that are similar to those seen with induced seizure activity, hypoxia is not involved during seizures inasmuch as the same effect was seen when experiments were performed on well-oxygenated and paralyzed animals (Seisjo et al., 1982). It has been postulated that seizures overstimulate a set of reactions including a regulatory phospholipase A_2 that operate under basal conditions for the maintenance of permeability properties in excitable membranes (Bazan and Rakowski, 1970; Bazan, 1971b). Moreover, a phospholipase A_2 has been implicated in neurotransmission (Bazan and Rakowski, 1970; Bazan et al., 1971; Bazan, 1971b). Thus, the mechanism involved in the accumulation of these brain lipids has been explored using drug treatments that modify the brain content of neurotransmitters prior to the induction of seizures. Pretreatment with α-methyl-p-tyrosine (α-MPT) has been shown to inhibit the release of FFA and DG induced by electroconvulsive shock (Aveldano and Bazan, 1979). Pretreatment with α-MPT also affects the lipid effect in bicuculline-induced seizures (Rodriguez de Turco et al., 1983) and p-chlorophenylalanine (PCPA) by itself stimulates FFA release (Rodriguez de Turco et al., 1983). Alpha-MPT depletes catecholamines by acting as a competitive inhibitor of tyrosine hydroxylase, the key enzyme in the pathway leading to the formation of catecholamines. PCPA inhibits tryptophane hydroxylase the key enzyme in the synthesis of 5-hydroxy-tryptamine.

This paper presents data on 1) the accumulated quantities of FFA and DG during bicuculline-induced seizures and 2) the effect of pretreatment with α-MPT and PCPA. It shows for the first time the nature of PCPA pretreatment on cerebral and cerebellar DG. Also, the effect of electroconvulsive shock and ischemia on the removal of intracerebrally injected radiolabeled arachidonic acid is shown.

MATERIALS AND METHODS

Electroshock and Ischemia Experiments

Male and female mice weighing approximately 20 g were used. Animals were anesthetized lightly 30 to 60 min prior to injection and an incision was made in the skin exposing the left parietal bone. Unanesthetized mice were injected intracerebrally in the area of the right frontal horn of the lateral ventricle. Each animal received 6 μL [1-^{14}C]arachidonic acid (specific activity, 58 mCi/mmol) complexed with lipid-free bovine serum albumin (labeled fatty acid : protein, 4:1). EEG platinum needle electrodes were placed under the skin on the temporal bones of the head and unidirectional rectangular pulses (0.5 msec at 45 V) were applied at 7 msec intervals over a one sec period.

Bicuculline-induced Seizure Experiments

Female Wistar rats weighing 240-280 g were injected intraperitoneally with 10 mg/Kg bicuculline (Sigma Chemical Co., Saint Louis, MO). Animals were decapitated and heads were placed in liquid nitrogen and swirled vigorously. Lipids were extracted from a frozen powder of cerebrum or cerebellum with chloroform-methanol, 2:1 by volume (Folch et al., 1957). FFA and DG were separated by gradient-thickness thin-layer chromatography and quantified after methanolysis with 14% BF_3 in methanol by gas-liquid chromatography (Aveldano and Bazan, 1975; Bazan and Bazan, 1975).

RESULTS

Effect of Pretreatment with Catecholamine- and Serotonin-Depleting Drugs on the Bicuculline-Induced Release of Free Fatty Acids and Diglycerides

Fig. 1 shows the difference between controls and the three groups of experimental animals. Note that the

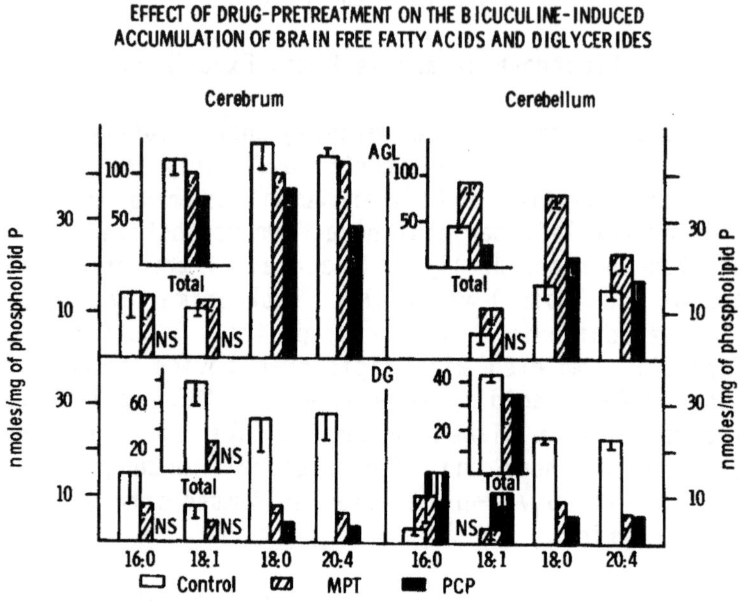

Figure 1. Effect of pretreatment with α-methyl-p-tyrosine or p-chlorophenylalanine on the accumulation of brain free fatty acids and diglycerides by bicuculline. Values are nmoles of fatty acid methyl esters per mg of lipid P, calculated by subtracting basal concentrations from experimental concentrations. Values are averages ± S.D. of 4-6 rats in each case. NS = no significant difference between basal and experimental values.

nmoles of individual fatty acids in FFA and DG are presented as accumulated quantities, and changes in docosahexaenoic acid are given (Table 1). The bicuculline stimulated accumulation of FFA and DG is shown in the cerebrum and cerebellum. The fatty acids that showed the greatest accumulation in both cerebellar and cerebral FFA and DG were arachidonic and stearic acids. Also, the cerebrum exhibited a greater overall lipid effect than the cerebellum when data were expressed in terms of total phospholipid content. Alpha-MPT pretreatment did not modify

Table 1

Accumulation of Free Docosahexaenoic Acid and Docosahexaenoyl-Diglycerides in the Brain of Animals Treated with Bicuculline and With or Without Pretreatment with α-Methyl-p-tyrosine and P-chlorophenylalanine

	Free 22:6		Docosahexaenoyl-DG	
	Cerebrum	Cerebellum	Cerebrum	Cerebellum
Bicuculline	1.07 ± 0.37	1.35 ± 0.7	2.22 ± 0.64	2.06 ± 0.19
α-MPT	2.05 ± 0.97	2.52 ± 1.7	0.90 ± 0.26	2.10 ± 0.70
PCPA	1.10 ± 0.6	NS	NS	1.64 ± 0.70

NS, no significant difference between basal and experimental values; α-MPT, α-methyl-p-tyrosine; PCPA, p-chlorophenylalanine. Values are nmoles of fatty acid methyl esters per mg of lipid P and are the averages ± S.D. of 4-6 rats in each case.

the changes induced by bicuculline in the cerebral FFA. However, it greatly decreased the accumulated DG in both the cerebrum and cerebellum. FFA accumulation in the cerebellum was potentiated by α-MPT pretreatment. We have shown that PCPA by itself increases brain FFA (Rodriguez de Turco et al., 1983) and DG content (present work). Hence, all the nonsignificant differences presented in Fig. 1 are due to the PCPA-stimulated accumulation of FFA and DG in controls, except oleate and palmitate in cerebellar DG.

Electroconvulsive Shock and Postdecapitation Ischemia Decreases the Removal of Intracerebrally Injected [1-^{14}C]Arachidonic Acid from the Free Fatty Acid Pool

When radiolabeled arachidonic acid is injected intracerebrally, a rapid decrease of radioactivity occurs in the FFA pool (Yau and Sun, 1974). This is caused by the activation-acylation of the fatty acid, mainly in phospholipids, and to the loss of labeled fatty acid from the brain, possibly towards the blood stream. Ischemia markedly inhibits the removal of radiolabeled arachidonic acid from the FFA pool and a single electroconvulsive shock greatly reduces this process (Fig. 2).

DISCUSSION

We have shown that α-MPT potentiates the bicuculline-stimulated release of FFA in the cerebellum and that PCPA increases basal levels of FFA. Thus, bicuculline-injected animals that were pretreated with PCPA failed to show the expected bicuculline-induced lipid changes. Cerebellar and cerebral DG were decreased by α-MPT pretreatment. Although we cannot rule out the existence of extraneous secondary drug effects, we suggest that brain catecholamines are required for the control of cellular free arachidonic acid and other FFA in the central nervous system and that the stimulation-induced depletion of them results in the loss of a vital component in this system.

This agrees well with the suggestion that phospholipases A_1 and A_2, primarily the latter, are regulatory enzymes involved in maintenance of the properties of excitable neuronal membranes playing a central role in neurotransmission (Bazan and Rakowski, 1970; Bazan, 1971b; Bazan et al., 1971).

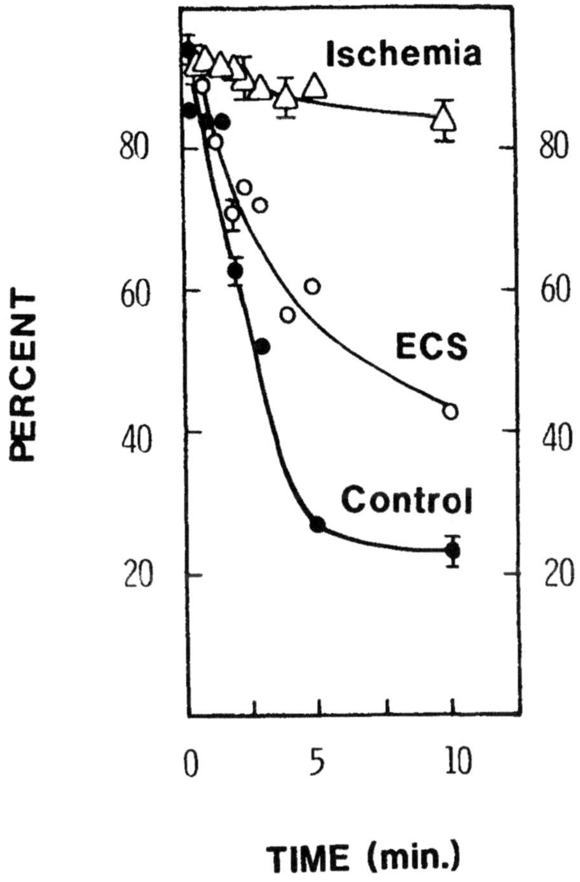

Figure 2. Effect of a single electroconvulsive shock and postdecapitation ischemia on the labeling of the brain free fatty acid pool by intracerebrally injected [1-^{14}C]-arachidonic acid. Values are averages of at least 5 mice when ± S.D. is given and of 2 mice when S.D. is not given.

This potentiation of the bicuculline effect was seen in the cerebellum. Perhaps the noradrenergic innervation of the cerebellum provides an inhibitory factor in the deacylation of membrane phospholipids. Alpha-MPT treatment of animals lacking catecholamines resulted in an overstimulatory effect. Previous studies with electroshock and α-MPT-pretreated animals showed an inhibition of the electroshock-induced release of FFA when measurements were taken of the entire brain (Aveldano and Bazan, 1979). We interpreted those results to mean that catecholamines were necessary for the electroshock effect to be produced in FFA, perhaps through the activation of a phospholipase A_2. It is likely that in various cell types different mechanisms may be operating and catecholamines may act as neuromodulators for lipid breakdown in some of these membranes. Inasmuch as serotonin has been implicated in seizure activity (Munkenbeck and Schwark, 1982), the experiments with PCPA were aimed at determining the relationship between the lipid changes and the neurotransmitter content. However, an unexpected effect of PCPA (by itself and not additive to bicuculline) was observed. At this stage it is difficult to conclude with any certainty about the mechanism of action of PCPA, because nonspecific effects cannot be ruled out. This observation is of interest because, although PCPA does not produce seizures, it does increase FFA. Further studies of these differences may help to elucidate the detailed biochemical pathways of neuronal lipid metabolism that have been modified by convulsions and/or drugs.

The decreased removal of radiolabeled arachidonic acid after its intracerebral injection may be due to an electroshock- or ischemia-induced inhibition of the diffusion of the fatty acid though the ventricular system (Pediconi et al., 1982). In addition, a decreased activation-reacylation (reactions I and II) may be the result of either an altered enzymatic activity or a dilution of the exogenous fatty acid with endogenously accumulated arachidonic acid, thus producing a competitive inhibition of the enzyme.

Free docosahexaenoic acid also is accumulated during seizures induced by electroshock (Bazan, 1970) or bicuculline (Bazan et al., 1982b; Rodriguez de Turco et al., 1983; Siesjo et al., 1982). Docosahexaenoate is enriched in phospholipids of synaptic plasma membranes (Sun and Sun, 1972; Cotman et al., 1969; Breckenridge et al., 1971), and may be involved in excitable membrane functions through the influence of its six double bonds on membrane fluidity properties. However, there is little available information about the metabolism of this fatty acid.

It is likely that the activation-acylation pathway of arachidonic acid is altered after electroconvulsive shock (Bazan and Rakowski, 1970; Rodriguez de Turco et al., 1977) by affecting the following reactions:

Free arachidonic acid + CoA ---> arachidonoyl-CoA (I)

Arachidonoyl-CoA + lysophospholipid --->
arachidonoyl phospholipid (II)

The importance of these reactions for the maintenance of the brain FFA pool was discussed recently (Sun, 1982). Labeling experiments support the hypothesis that the arachidonate metabolism of brain synaptosomes is altered during post-decapitation ischemia (Aveldano and Bazan, 1979; Siesjo et al., 1982; Majewska et al., 1981; Banschbach and Geison, 1974; Der and Sun, 1981).

A similar set of reactions may activate-reacylate docosahexaenoate:

Free docosahexaenoic acid + CoA --->
docosahexaenoyl - CoA (III)

Docosahexaenoyl - CoA + lysophospholipid --->
docosahexaenoyl - phospholipid (IV)

Diglycerides hydrolyzed by a diglyceride lipase still remain as an additional source that may be involved in releasing FFA in brain ischemia (Aveldano and Bazan,

1975b) and during convulsions (Aveldano and Bazan, 1979; Rodriguez de Turco et al., 1983). During ischemia (Bazan, 1970; Aveldano and Bazan, 1975b; Banschbach and Geison, 1974) and convulsions (Aveldano and Bazan, 1979; Rodriguez de Turco et al., 1983), diglycerides accumulate in brain, and a decrease in the stearoyl-arachidonoyl groups occurs during repeated convulsions (Rodriguez de Turco et al., 1983; Bazan et al., 1982b).

Reactions V and VI may be involved in the release of polyenoic fatty acids in brain. The following formulae represent reactions where such enzymes are involved in the release of arachidonic and docosahexaenoic acids:

Arachidonoyl-phospholipid ---> free arachidonic acid + lysophospholipid (V)

Docosahexaenoyl-phospholipid ---> free docosahexaenoic acid + lysophospholipid (VI)

Because the polyenoic acyl groups are esterified to position 2 of the glycerol backbone, at least two different phospholipid classes are degraded by convulsions, each containing one of these fatty acids. Arachidonate in the brain is enriched in phosphatidylinositol and phosphatidylethanolamine; these phospholipids, in addition to phosphatidylcholine, have been implicated as sources of the brain FFA pool (Bazan, 1976; Bazan et al., 1982a, 1982b; Marion and Wolfe, 1979; Porcellati et al., 1978; Rodriguez de Turco et al., 1977, 1983; Aveldano and Bazan, 1975a, 1979; Sun et al., 1980; Der and Sun, 1981). The activation of an ethanolamine phospholipase A_2 has been reported with brain ischemia (Edgar et al., 1980). Moreover, the activation of a plasmalogenase also has been proposed, because plasmalogens in the central nervous system contain high proportions of arachidonate (Horrocks et al., 1978).

Various enzymatic systems and their substrates have been considered as sources of the released FFA (Bazan et al., 1980). More recently, it has been suggested that the reversal of the CDP-choline and -ethanolamine phosphotransferases are an additional pathway leading to the release of FFA during brain ischemia through the splitting of the accumulated diglycerides by a diglyceride lipase (Goracci et al., 1981; Horrocks et al., 1981). When the electroconvulsive shock-induced release of brain FFA was first observed, it was suggested that phospholipases A_1 and A_2 (predominantly the latter) were activated and that excitable membrane phospholipids were deacylated (Bazan, 1970,1971b; Bazan and Rakowski, 1970). This hypothesis was based on the predominance of free unsaturated fatty acids (Bazan, 1970), the simultaneous increase of diglycerides (Bazan, 1970), and the relatively insignificant lipolytic decrease of triglycerides (Bazan, 1970). The inhibition of the release of brain FFA produced by head-focused microwave irradiation supports the theory that highly active phospholipases are involved in the production of brain FFA during ischemia (Cenedella et al., 1975).

At least five different routes may be involved in the further metabolism of FFA. First, biologically active molecules may be converted through lipoxygenase and cyclooxygenase catalyzed reactions. This avenue was explored (Marion and Wolfe, 1978; Spagnuolo et al., 1978; Steinhauer et al., 1979; Wolfe, 1979; Galli et al., 1980), and it was concluded that the release of prostaglandins is due to seizure activity, rather than to hypoxia (Steinhauer et al., 1979). In ischemia, as during arterial occlusion, the lack of oxygen may inhibit the formation of prostaglandins from accumulated free arachidonic acid, and this may increase during recirculation (Gaudet and Levine, 1979). A similar increase in eicosanoid formation may be responsible for the damage that occurs in humans during reperfusion after stroke. It is not known how extensively the cerebral microvasculature is affected by the formation of eicosanoids, especially during seizure activity.

The second possible route for FFA is that they are reacylated into phospholipids. This is dependent on the enzymes that catalyze steps I,II,III and IV for arachidonic and docosahexaenoic acids having all the required cofactors, such as ATP, and that no other modulatory changes occur.

The third possible pathway is that the FFA may be bound to cellular proteins. A number of proteins with high affinity binding sites for fatty acids are present in different cells. However, there is no information available about the participation of such binding activity during the stimulation-induced release of brain FFA. The accumulation of these FFA may be harmful to the nerve cells and emergency mechanisms may be activated to decrease, bind or transform the accumulated FFA.

Another possible pathway for the removal of FFA is the activation of enzymes that methylate these lipids. Several years ago it was found that this does occur in the brain and retina during anoxia (Bazan, 1976). A confirmation and extension of this work (Giusto and Bazan, 1983) suggests that the following reactions may take place in the central nervous system:

Free arachidonic acid + s-adenosyl methionine ---> arachidonoyl-methyl ester + s-adenosyl homocysteine (VII)

Free docosahexaenoic acid + s-adenosyl methionine ---> docosahexaenoyl-methyl ester + s-adenosyl homocysteine
Fatty acid methyl ester ---> free fatty acid (VIII)

A fifth alternative pathway is the peroxidation of the accumulated free polyenoic fatty acids, leading to possible brain damage (Cenedella et al., 1975; Bazan, 1976; Aveldano and Bazan, 1975a; Chan and Fishman, 1980; Seisjo, 1981).

The chemical pathogenesis of brain damage that leads to irreversible loss of function may involve the same changes in the FFA pool that have been seen during ischemia and convulsions. Newborn mammalians (Bazan, 1976; Bazan et al., 1971), and adult poikilotherms (Aveldano and Bazan, 1975a), both known to survive prolonged periods of oxygen deprivation, release brain FFA at a much slower rate than the brain of mature homeothermic animals. Further studies into the lipid changes that occur in membranes during convulsions will increase our understanding of the dynamics of excitable membrane components, as well as the chemical pathogenesis of the lesions that complicate epilepsy.

REFERENCES

Aveldano M. I. and Bazan N. G. (1975a) Differential lipid deacylation during brain ischemia in a homeotherm and a poikilotherm. Content and composition of free fatty acids and triacylglycerols. Brain Res. 100, 99-110.

Aveldano M. I. and Bazan N. G. (1975b) Rapid production of diacylglycerols enriched in arachidonate and stearate during early brain ischemia. J. Neurochem. 25, 919-920.

Aveldano de Caldironi M. I. and Bazan N. G. (1979) Alpha--methyl-p-tyrosine inhibits the production of free arachidonic acid and diacylglycerols in brain after a single electroconvulsive shock. Neurochem. Res. 4, 213-221.

Banschbach M. W. and Geison R. L. (1974) Post-mortem increase in rat cerebral hemisphere diglyceride pool size. J. Neurochem. 23, 875-877.

Bazan N. G. (1970) Effects of ischemia and electroconvulsive shock on free fatty acid pool in the brain. Biochim. Biophys. Acta 218, 1-10.

Bazan N. G. (1971a) Free fatty acid production in cerebral white and grey matter of the squirrel monkey. Lipids 6, 211-212.

Bazan N. G. (1971b) Changes in free fatty acids of the brain by drug-induced convulsions, electroshock and anesthesia. J. Neurochem. 18, 1379-1385.

Bazan N. G. (1971c) Modifications in the free fatty acids of developing rat brain. Acta Physiol. Latinoamer. 21, 101-123.

Bazan N. G. (1976) Free arachidonic acid and other lipids in the nervous system during early ischemia and after electroshock. Adv. Exp. Med. Biol. 72, 317-335.

Bazan N. G. and Bazan H. E. P. (1975) Analysis of free and esterified fatty acids in neural tissues using gradient-thickness thin-layer chromatography. In: Research Methods in Neurochemistry. N. Marks and R. Rodnight, eds., vol. 3, Plenum Press, New York, pp. 309-324.

Bazan N. G. and Rakowski H. (1970) Increased levels of brain free fatty acids after electroconvulsive shock. Life Sci. 9, 501-507.

Bazan N. G. and Rodriguez de Turco E. B. (1980) Membrane lipids in the pathogenesis of brain edema: Phospholipids and arachidonic acid, the earliest membrane components changed at the onset of ischemia. Adv. Neurol. 28, 197-205.

Bazan N. G., Aveldano de Caldironi M. I., Cascone de Suarez G. D. and Rodriguez de Turco E. B. (1980) Transient modifications in brain free arachidonic acid in experimental animals during convulsions. In: Neurochemical and Clinical Neurology. L. Batistin, G. Hashim, and A. Lajtha, eds., Alan R. Liss, New York, pp. 167-169.

Bazan N. G., Aveldano de Caldironi M. I. and Rodriguez de Turco E. B. (1982a) Rapid release of free arachidonic acid in the central nervous system due to stimulation. Prog. Lipid Res. 20, 523-529.

Bazan N. G., Bazan H. E. P., Kennedy W. G. and Joel C. D. (1971) Regional distribution and rate of production of free fatty acids in rat brain. J. Neurochem. 18, 1387-1393.

Bazan N. G., Rodriguez de Turco E. B. and Morelli di Liberti S. M. (1982b) Arachidonic acid and arachidonoyl-diglycerols increase in rat cerebrum in bicuculline-induced status epilepticus. Neurochem. Res. 7, 123-128.

Breckenridge W. C., Gombos G. and Morgan I. G. (1971) The docosahexaenoic acid of the phospholipids of synaptic membranes, vesicles and mitochondria. Brain Res. 33, 581-583.

Cenedella R. J., Galli C. and Paoletti R. (1975) Brain free fatty acid levels in rats sacrificed by decapitation versus focused microwave irradiation. Lipids 10, 290-293.

Chan P. H. and Fishman R. A. (1980) Transient formation of superoxide radicals in polyunsaturated fatty acid-induced brain swelling. J. Neurochem. 35, 1004-1007.

Cotman C., Blank M. L., Moehl A. and Snyder F. (1969) Lipid composition of synaptic plasma membranes isolated from rat brain by zonal centrifugation. Biochemistry 8, 4606-4610.

Der O. M. and Sun G. Y. (1981) Degradation of arachidonoyl-labeled phosphatidylinositols by brain synaptosomes. J. Neurochem. 36, 355-362.

Edgar A. D., Strasznajder J. and Horrocks L. A. (1982) Activation of ethanolamine phospholipase A_2 in brain during ischemia. J. Neurochem. 39, 1111-1116.

Folch J., Lees M. and Sloane-Stanley G. H. (1957) A simple method for the isolation and purification of total lipids from animal tissue. J. Biol. Chem. 226, 497-509.

Galli C. and Spagnuolo C. (1976) The release of brain free fatty acids during ischemia in essential fatty acid-deficient rats. J. Neurochem. 26, 401-404.

Galli C., Spagnuolo C. and Petroni A. (1980) Factors affecting brain prostaglandin formation. In: Advances in Prostaglandin and Thromboxane Research. B. Samuelsson, P. W. Ramwell, and R. Paoletti, eds., vol. 8, Raven Press, New York, pp. 1235-1239.

Gardiner M., Nilsson B., Rehncrona S. and Siesjo B. K. (1981) Free fatty acids in the rat brain in moderate and severe hypoxia. J. Neurochem. 36, 1500-1505.

Gaudet R. J. and Levine L. (1979) Transient cerebral ischemia and brain prostaglandins. Biochem. Biophys. Res. Comm. 86, 893-901.

Giusto N. M. and Bazan N. G. (1983) Anoxia-induced production of methylated and free fatty acids in retina, cerebral cortex and white matter in vitro; comparison with triacylglycerol and with other tissues. Neurochem. Pathol. 1, (in press).

Goracci G., Francescangeli E., Horrocks L. A. and Porcellati G. (1981) The reverse reaction of cholinephosphotransferase in rat brain microsomes. A new pathway for degradation of phosphatidylcholine. Biochim. Biophys. Acta 664, 373-379.

Horrocks L. A., Spanner S., Mozzi R., Chun Fu S., D'Amato R. A. and Krakowka S. (1978) Plasmalogenase is elevated in early demyelinating lesions. In: Myelination and demyelination, J. Palo, ed., Plenum Publishing Corp., New York, pp. 423-438.

Kuwashima J., Nakamura K., Fujitani B., Kadokawa T., Yoshida K. and Shimizu M. (1978) Relationship between cerebral energy failure and free fatty acid accumulation following prolonged brain ischemia. Jpn. J. Pharmacol. 28, 277-287.

Majewska M. D., Manning R. and Sun G. Y. (1981) Effects of postdecapitative ischemia on arachidonate release from brain synaptosomes. Neurochem. Res. 6, 567-576.

Marion J. and Wolfe L. S. (1978) Increase in vivo of unesterified fatty acids, prostaglandin $F_{2\alpha}$ but not thromboxane B_2 in rat brain during drug induced convulsions. 37. Prostaglandins 16, 99-110.

Marion J. and Wolfe L. S. (1979) Origin of the arachidonic acid released post-mortem in rat forebrain. Biochim. Biophys. Acta 574, 25-32.

Munkenbeck K. E. and Schwark W. S. (1982) Serotonergic mechanisms in amygdaloid-kindled seizures in the rat. Exper. Neurol. 76, 246-253.

Pediconi M. F., Rodriguez de Turco E. B. and Bazan N. G. (1982) Diffusion of intracerebrally injected [1-^{14}C]arachidonic acid and [2-^{3}H]glycerol in the mouse brain. Effects of ischemia and electroconvulsive shock. Neurochem. Res. 7, 1453-1456.

Porcellati G., DeMedio G. E., Fini C., Floridi A., Gorracci G., Horrocks L. A., Lazarewicz J. W., Palmerini C. A., Strosznajder J. and Trovarelli G. (1978) Phospholipid and its metabolism in ischemia. In: Proceedings of the European Society for Neurochemistry, vol. 1, V. Neuhoff, ed., Verlag Chemie, New York, pp. 285-302.

Rehncrona S., Westerberg E., Akesson B. and Siesjo B. K. (1982) Brain cortical fatty acids and phospholipids during and following complete and severe incomplete ischemia. J. Neurochem. 38, 84-93.

Rodriguez de Turco E. B., Cascone G. D., Pediconi M. F. and Bazan N. G. (1977) Phosphatidate, phosphatidylinositol, diacylglycerols, and free fatty acids in the brain following ECS, anoxia or ischemia. Adv. Exp. Med. Biol. 83, 389-396.

Rodriguez de Turco E. B., Morelli di Liberti S. and Bazan N. G. (1983) Stimulation of free fatty acids and diacylglycerol accumulation in cerebrum and cerebellum during bicuculline-induced status epilepticus. Effect of pretreatment with alpha-methyl-p-tyrosine and p-chlorophenylalanine. J. Neurochem. 40, 252-259.

Siesjo B. K. (1981) Cell damage in the brain: A speculative synthesis. J. Cereb. Blood Flow Metab. 1, 155-185.

Siesjo B. K., Ingvar M., and Westerberg E. (1982) The influence of bicuculline-induced seizures on free fatty acid concentrations in cerebral cortex, hippocampus, and cerebellum. J. Neurochem. 39, 796-802.

Spagnuolo C., Terzi C. and Galli C. (1978) Differential response of brain $PGF_{2\alpha}$ synthesis to methionine sulfoximine in respect of other convulsant drugs. Pharmacol. Res. Commun. 10, 541-544.

Steinhauer H. B., Anhut H. and Hertting G. (1979) The synthesis of prostaglandins and thromboxane in the mouse brain in vivo. Influence of drug induced convulsions, hypoxia and the anticonvulsants trimethadione and diazepam, Naynyn-Schmeidberg's Arch. Pharmacol. 310, 53-58.

Sun G. Y. (1982) Metabolic turnover of arachidonoyl groups in brain membrane phosphoglycerides. In: Phospholipids in the Nervous System, vol. 1: Metabolism, L. A. Horrocks, G. B. Ansell, and G. Porcellati, eds., Raven Press, New York, pp. 75-89.

Sun G. Y. and Sun A. Y. (1972) Phospholipids and acyl groups of synaptosomal and myelin membranes isolated from the cerebral cortex of squirrel monkey (Saimiri sciureus). Biochim. Biophys. Acta 280, 306-315.

Sun G. Y., Manning R., and Stroaznajder J. (1980) Effects of postdecapitative ischemia and hypoxia on the phosphoglyceride acyl groups of rat brain membranes. Neurochem. Res. 5, 1211-1219.

Wolfe, L. S. (1979) The role of prostaglandins in the central nervous system. Ann. Rev. Physiol. 41, 669-684.

Yau T. M. and Sun G. Y. 1974. The metabolism of [1-^{14}C]arachidonic acid in the neutral glycerides and phosphoglycerides of mouse brain. J. Neurochem. 23, 99-104.

THE ROLE OF ARACHIDONIC ACID IN CELLULAR BRAIN EDEMA

Pak Hoo Chan, Robert A. Fishman,
Susan Longar, Sylvia Chen and Susan Chew

Brain Edema Clinical Research Center
Department of Neurology
University of California
School of Medicine
San Francisco, CA 94143 USA

Cerebral edema is defined as an increase in brain volume due to an increase in its water and sodium content. Based on neuropathological and experimental observations, brain edema has been divided into three major categories (Klatzo, 1967; Fishman, 1975, 1980, 1981; Fishman and Chan, 1980): 1) Vasogenic edema indicates increased permeability in capillary endothelial cells associated with an expanded extracellular volume. 2) Cellular (cytotoxic) edema indicates the increase in intracellular volume of brain cells associated with a decreased extracellular volume. 3) Interstitial or hydrocephalic edema indicates the increase in brain water that characterizes obstructive hydrocephalus.

Cerebral edema accompanies a wide variety of pathological processes in the brain and it contributes to the morbidity and mortality of many neurological diseases. It plays a major role in head injury, stroke, and brain tumor, as well as in cerebral infections, including brain abscess, encephalitis and meningitis, lead encephalopathy, hypoxia, hypo-osmolality, the disequilibrium syndromes associated with dialysis and diabetic keto-acidosis. Extensive clinical investigations and laboratory studies of brain edema using neurochemical, physiological, and ultrastructural techniques have been able to clarify many pathological and clinical uncertainties. However, the fundamental mechanisms of brain swelling in various pathological conditions are poorly understood.

Cellular edema induced by polyunsaturated fatty acids

Using single cortical slices of the rat brain as a bioassay system *in vitro*, our earlier work has demonstrated that metabolism was altered by incubation of the slices with a crude preparation of membranes from granulocytic leukocytes obtained from glycogen-induced rat peritoneal exudates. Such leukocytic preparations induced brain edema characterized by increased water content, cellular swelling (decreased inulin space), increased intracellular sodium, and decreased intracellular potassium. Glucose oxidation and lactate production were also stimulated, and the energy charge was reduced in the edematous brain tissue (Fishman et al., 1977).

We have further demonstrated that arachidonic acid and other polyunsaturated fatty acids which are intrinsic constituents of the phospholipids of cellular membranes are potent inducers of cellular (cytotoxic) edema (Chan and Fishman, 1978; Chan et al., 1980). In these experiments, the edematous brain slices were characterized by increased sodium and decreased potassium contents. Cellular metabolism was affected; lactic acid levels were increased and high energy nucleotides were reduced. Both saturated fatty acids including nonanoic acid (9:0), lauric acid (12:0) and palmitic acid (16:0) and monounsaturated oleic acid (18:1) were not effective (Table I). The swelling of cortical slices induced by 0.5 mM arachidonic acid (20:4) reached 12.6% at 15 minutes and continued to increase during the 90 minutes of incubation to reach a maximum of 35.2%. From the studies of various

Table I. Effects of Fatty Acid on Cellular Edema of Cortical Slices

Fatty Acid	Swelling (%)	Inulin Space (%)	Lactate (mmol/kg dry wt.)	Na+ (meq/kg dry wt.)	K+ (meq/kg dry wt.)
Control	11.3	46.11	236	599	393
Nonanoic acid (9:0)	12.6	49.42	269*	519	312
Lauric acid (12:0)	11.6	49.22	320*	651	315
Palmitic acid (16:0)	9.4	46.12	258	838*	387
Oleic acid (18:1)	11.0	41.07*	278**	694**	419
Linoleic acid (18:2)	32.5*	33.47*	519*	1340*	92*
Linolenic acid (18:3)	26.0*	42.88	391*	949*	188*
Arachidonic acid (20:4)	35.2*	32.6*	527*	1239*	140*
Docosahexaenoic acid (22:6)	33.5*	41.47*	543*	1215*	77*

Data obtained from Chan and Fishman (1978).
*$p<0.001$, **$p<0.02$, Student's t-test.

molar ratios between 20:4 and bovine serum albumin (BSA), we have concluded that free or unbound arachidonate activates the induction of brain swelling. The swelling processes induced by arachidonic acid in this system were irreversible; i.e., they were not modulated by BSA, although the latter could extract 46% of the transported arachidonic acid from brain slices (Chan et al., 1980). Arachidonic acid also caused a significant increase in intracellular sodium levels concomitant with a decrease in intracellular potassium in both C-6 glioma cells and N18TG2 neuroblastoma cells. Both $(Na^{+}+K^{+})$-ATPase and p-nitrophenylphosphatase of glioma cells were inhibited by arachidonic acid, whereas only the p-nitrophenylphosphatase of neuroblastoma cells was inactivated. Low concentrations of arachidonic acid stimulated lactic acid release, whereas high concentrations had an opposite effect (Chan and Fishman, 1982). These data suggest that arachidonic acid *per se* or its metabolites served to induce cellular swelling.

The role of arachidonic acid and free radicals in cellular edema

The basic mechanisms underlying PUFA-induced cellular edema were investigated further. PUFA readily undergo auto-oxidation to form lipid peroxides chemically (Mead, 1976, Demopoulos et al., 1979; Fridovich and Porter, 1981) or readily convert to radical intermediates by cyclooxygenase and lipoxygenase pathways (Samuelsson et al., 1979; Wolfe, 1982). Furthermore, the metabolism of arachidonic acid by cytochrome P-450 could lead to the formation of superoxide anions and other arachidonic acid hydroperoxides (Capdevila et al., 1981). These processes suggest that free radicals are likely candidates capable of mediating the cellular swelling. When rat brain cortical slices were incubated with PUFA, transient formation of superoxide radicals and lipid peroxide was observed (Chan and Fishman, 1980). Furthermore, free radicals generated with the xanthine oxidase-Fe^{3+} system induced cellular swelling and increased lactic acid levels in rat brain slices (Chan et al., 1982). In addition, the lipid peroxide content of glioma cells was increased abruptly by 50 uM arachidonic acid, whereas only a slight increase of malondialdehyde was observed in neuroblastoma cells. Exogenous arachidonic acid free radicals caused a significant release of endogenous labeled arachidonic acid from cellular membrane phospholipids. The role of arachidonic acid and free radicals in cellular edema is summarized in Table II.

Apart from these metabolic disturbances, free radicals also induced structural disturbances of membrane phospholipids. When purified synaptosomal plasma membranes were incubated with the xanthine oxidase-Fe^{3+} system, phosphatidylethanolamine and phosphatidylcholine were decreased concomitant with increases of lysophosphatidylethanolamine and lysophosphatidylcholine (Chan and Fishman, 1982). Furthermore, free PUFA, especially arachidonic acid and docosahexaenoic acid, were released from synaptic plasma membranes (Chan et al., 1982). These data indicate that free radicals stimulate the membrane bound enzyme phospholipase A_2. Other phospholipases, like phospholipase C, may also be stimulated by free radicals. Bazan has proposed that hydrolysis of phospholipids by phospholipases A_2 and C and the subsequent release of arachidonic acid are the first changes occurring in membrane at the onset of ischemia (Bazan and Turco, 1980). Our cold-injury models in rat have confirmed these possibilities (Fishman and Chan, 1981; Chan and Fishman, 1982).

Arachidonic acid, once released from the cellular membrane, rapidly intercalates into membrane domain enriched in polyunsaturated acyl moieties and produces significant change in the packing of the lipid molecules (Klausner et al., 1980). The intercalation of arachidonic acid into membranes also lowers the transition temperature and distorts the crystal lattice of the

Table II. Effects of arachidonic acid and free radicals on cellular edema

Factors	Tissue or cells	Na+	K+	Lactate	Lipid Peroxidation	Arachidonic Acid Release	(Na++K+)-ATPase	p-nitrophenylphosphatase
Arachidonic Acid (50 uM)	brain slices	+	-	+	+	+	-	-
	neuroblastoma	+	-	+	+	ND	NS	-
	glioma	+	-	+	+	+	-	-
Free Radicals (xanthine oxidase -Fe3+)	brain slices	+	-	+	+	+	-	-
	neuroblastoma	+	-	+	+	ND	ND	ND
	glioma	+	-	+	+	+	ND	ND

+, increased significantly; -, decreased significantly; NS, no significant change; ND, not determined.

gel states of phospholipids (Usher et al., 1978). Furthermore, increased synaptosomal membrane $(Na^{+}+K^{+})$-ATPase activity could be induced by the deficiency of arachidonic acid in the membranes (Sun and Sun, 1974). It has also been demonstrated by Ahmed and Thomas (1971) that unsaturated long chain fatty acids strongly inhibited rat brain $(Na^{+}+K^{+})$-ATPase. These studies prompted our investigation of the direct effect of arachidonic acid on $(Na^{+}+K^{+})$-ATPase activity and neurotransmitter uptake of both cortical slices and synaptosomes.

Recently, we have demonstrated that the initial rate of uptake of the neurotransmitter amino acids, glutamic acid and GABA, was reduced by arachidonic acid in both cortical slices and synaptosomes (Figs. 1, 2). This reduction in neurotransmitter uptake was concomitant with the inactivation of $(Na^{+}+K^{+})$-ATPase activity. The uptake of amino-isobutyric acid (AIBA), an inert non-metabolized amino acid without neurotransmitter properties was not affected (Chan et al., 1982). The reduction of synaptosomal $(Na^{+}+K^{+})$-ATPase activity was followed by an early elevation of intrasynaptosomal sodium induced by arachidonic acid. These data further suggest that membrane perturbations induced by arachidonic acid may cause unfavorable functional consequences. These processes may lead to the increased movement of water molecules into the cells. Other metabolic and functional disturbances like increased glycolysis (Chan and Fishman, 1978; Chan et al., 1980; Chan and Fishman, 1982) and uncoupled mitochondrial phosphorylation (Wojtezak, 1976) induced by arachidonic acid also may be associated with cellular edema. Further elucidation of the biochemical mechanisms of arachidonic acid and free radical-induced cellular edema may allow the development of more rational therapeutic approaches toward brain edema in patients (Fishman, 1982).

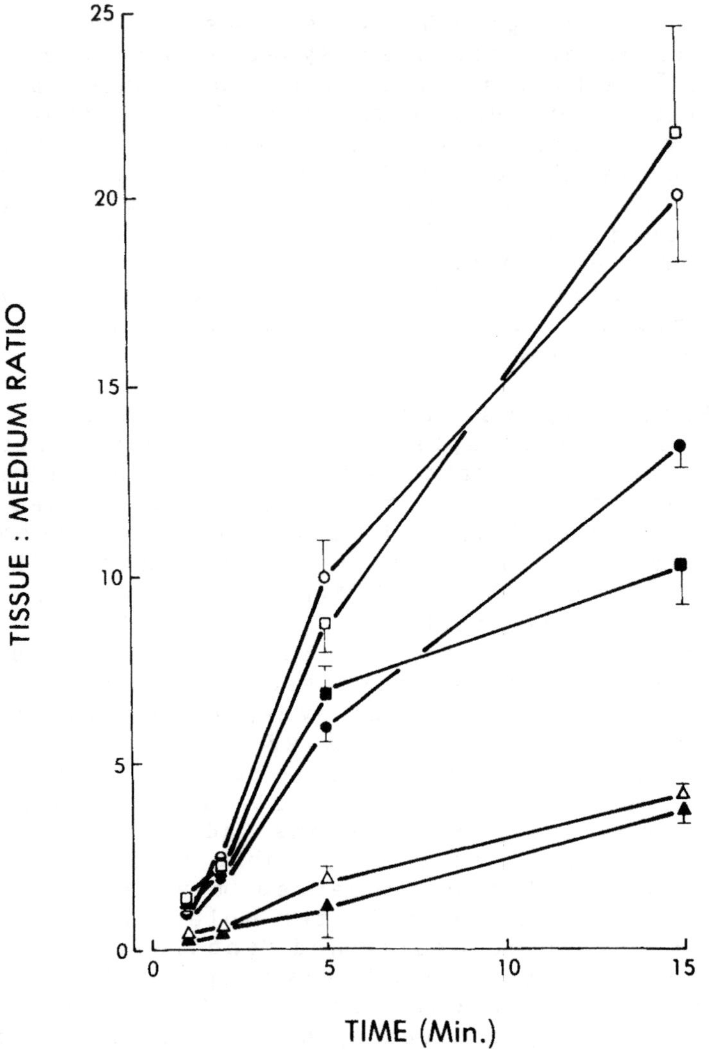

Fig 1. Time-course of the effect of arachidonic acid on the uptake of GABA, glutamate and aminoisobutyric acid (AIBA) in rat brain cortical slices. Cortical slices were reconstituted in Krebs-Ringer buffer (pH 7.4) for 20 min at 37°C and transferred to a fresh medium containing 11.2 uM [^{14}C] GABA or 22 uM [^{14}C] L-glutamic acid, or 55.5 uM [^{14}C] AIBA for various times of incubation. Square, glutamic acid; circle, GABA; triangle, AIBA. Open symbols, control brain slices. Values are the mean ± SEM of four cortical slices.

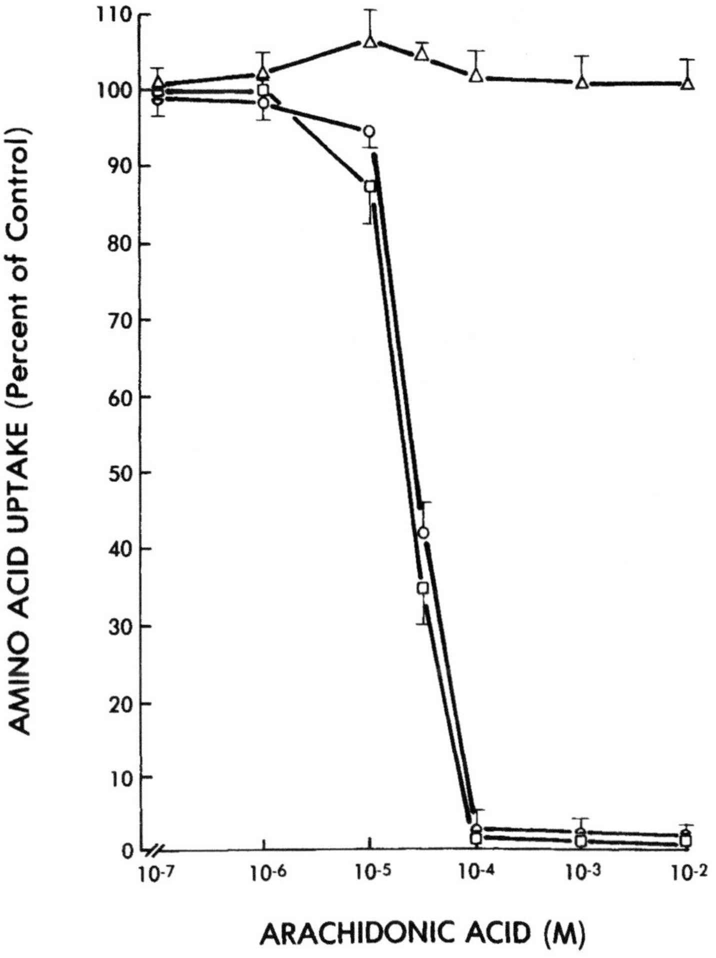

Fig 2. Dose-response studies of synaptosomal uptake of GABA, glutamate and AIBA. Synaptosomes (6 mg protein per ml) were pretreated with arachidonic acid for 10 min followed by incubation with various radiolabelled amino acids for another 5 min. Circle, GABA; square, glutamic acid; triangle, AIBA. Values are the mean ± S.E.M. of four different experiments.

ACKNOWLEDGEMENTS

This work was supported by grant NS-14543 from the National Institutes of Health, Public Health Services, Department of Health and Human Services, U.S.A.

REFERENCES

Ahmed, K. and Thomas, B.S. (1971) The effects of long chain fatty acids on sodium plus potassium ion-stimulated adenosine triphosphatase of rat brain. J. Biol. Chem. 246, 103-109.

Bazan, N.G. and Turco, E.B.R. (1980) Membrane lipids in the pathogenesis of brain edema - phospholipids and arachidonic acid, the earliest membrane components changed at the onset of ischemia. Adv. Neurol. 28, 197-205.

Capdevila, J., Chacoa, N., Werringloer, J., Prough, R.A. and Estabrook, R.W. (1981) Liver microsomal cytochrome P-450 and the oxidative metabolism of arachidonic acid.

Chan, P.H. and Fishman, R.A. (1978) Brain edema: induction in cortical slices by polyunsaturated fatty acids. Science 201, 358-360.

Chan, P.H. and Fishman, R.A. (1980) Transient formation superoxide radicals in polyunsaturated fatty acids-induced brain swelling. J. Neurochem. 35, 1004-1007.

Chan, P.H. and Fishman, R.A. (1982) Alterations of membrane integrity and cellular constituents by arachidonic acid in neuroblastoma and glioma cells. Brain Res. 248, 151-157.

Chan, P.H. and Fishman, R.A. (1982) Phospholipid degradation and the early release of polyunsaturated fatty acids in the evolution of brain edema. 5th International Symposium on Brain Edema (in press).

Chan, P.H., Fishman, R.A., Lee, J.L. and Quan, S.C. (1980) Arachidonic acid-induced swelling in incubated rat brain cortical slices: Effect of bovine serum albumin. Neurochem. Res. 5, 629-640.

Chan, P.H., Kerlan, R. and Fishman, R.A. (1982) Reductions of GABA and glutamate uptake and

$(Na^{+}+K^{+})$-ATPase activity in brain slices and synaptosomes by arachidonic acid. J. Neurochem. (in press)

Chan, P.H., Yurko, M. and Fishman, R.A. (1982) Phospholipid degradation and cellular edema induced by free radicals in brain cortical slices. J. Neurochem. 38, 525-531.

Demopoulos, H.B., Flamm, E.S., Seligman, M.L., Mitamauir, J.A. and Ransohoff, J. (1979) Membrane perturbations in central nervous system injury: Theoretical basis for free radical damage and a review of the experimental data. In: Neural Trauma (Popp, J., Bourke, R.S., Nelson, L.R. and Kimelberg, H.K., eds.) pp. 63-78, Raven Press, New York.

Fishman, R.A. (1975) Brain edema. N. Eng. J. Med. 293, 706-711.

Fishman, R.A. (1980) Brain edema. In: Cerebrospinal Fluid in Diseases of the Nervous System. pp. 107-128, Saunders, Philadelphia.

Fishman, R.A. (1981) Brain edema. In: Basic Neurochemistry pp. 681-691, Little, Brown and Co., Boston.

Fishman, R.A. (1982) Steroids in the treatment of brain edema (editorial). N. Eng. J. Med. 306, 359-360.

Fishman, R.A. and Chan, P.H. (1980) Metabolic basis of brain edema. In: Advances in Neurology (Cervos-Navarro, J. and Ferszt, R., eds.) Vol. 28, pp. 207-214, Raven Press, New York.

Fishman, R.A. and Chan, P.H. (1981) Hypothesis: Membrane phospholipid degradation and polyunsaturated fatty acids play a key role in the pathogenesis of brain edema. Ann. Neurol. 10, 75.

Fishman, R.A., Sligar, K. and Hake, R.B. (1977) Effects of leukocytes on brain metabolism in granulocytic brain edema. Ann. Neurol. 2, 89-94.

Fridovich, S.E. and Porter, N.A. (1981) Oxidation of arachidonic acid in micelles by superoxide and hydrogen peroxide. J. Biol. Chem. 256, 260-265.

Klatzo, I. (1967) Neuropathological aspects of brain edema. J. Neuropathol. Exp. Neurol. 26, 1-14.

Klausner, R.D., Kleinfeld, A.M., Hoover, Karnovsky (1980) Lipid domains in membrane. Evidence derived from structural perturbations induced by free fatty acids and lifetime heterogeneity analysis. J. Biol. Chem. 255, 1286-1295.

Mead, J.F. (1976) Free radicals mechanisms of lipid damage and consequences for cellular membranes. In: Free Radicals in Biology (Pryor, W.A., ed.) Vol. 1, pp. 51-68, Academic Press, New York.

Samuelsson, B., Hammarstrom, S. and Borgeat, P. (1979) Pathway of arachidonic acid metabolism. Adv. Inflam. Res. 1, 405-411.

Sun, G.Y. and Sun, A.Y. (1974) Synaptosomal plasma membranes - acyl group composition of phosphoglycerides and $(Na^{+}+K^{+})$-ATPase activity during fatty acid deficiency. J. Neurochem. 22, 15-18.

Usher, J.R., Epand, R.M. and Papahadjopoulos, D. (1978) The effect of free fatty acids on the thermotropic phase transition of dimyristoyl glycrophosphocholine. Chem. Phys. Lipids 22, 245-253.

Wojtezak, L. (1976) Effect of long-chain fatty acids and acyl-CoA on mitrochondrial permeability, transport, and energy-coupling processes. J. Bioenerg. Biomemb. 8, 293-311.

Wolfe, L.S. (1982) Eicosanoids: prostaglandins, thromboxanes, leukotrienes, and other derivatives of carbon-20 unsaturated fatty acids. J. Neurochem. 38, 1-14.

METABOLISM OF DOCOSAHEXAENOIC-ACID-CONTAINING SPECIES OF ETHER PHOSPHOLIPID IN RAT BRAIN

Yasuo Masuzawa, Yoshiaki Onuma, Keizo Waku and [a]Yoshio Ishima

Faculty of Pharmaceutical Sciences, Teikyo University, Sagamiko, Tsukuigun, Kanagawa 199-01, Japan; [a]Tomobe Hospital Medical Centre Tomobe, Nishiibarakigun, Ibaraki, 654, Japan

ABSTRACT

The metabolism of hexaene species of ethanolamine plasmalogen (1-alk-1'-enyl-2-acylglycero-3-phosphoethanolamine, alkenylacyl-GPE) and 1-alkyl-2-acylglycero-3-phosphoethanolamine (alkylacyl-GPE) was studied by estimating the incorporation rates of two kinds of the precursor into brain phospholipids *in vivo*. In the first experiment, 1-^{3}H-glycerol was intracerebrally injected into brain of 18-day-old rats. After selected periods, the lipids were extracted and 1-acyl-, 1-alkyl- and 1-alkenyl-2-acyl-3-acetylglycerols were prepared from the ethanolamine phosphoglycerides (EPG). Six molecular species were fractionated by $AgNO_3$-impregnated TLC. Since it has been established that ethanolamine plasmalogen is biosynthesized from alkylacyl-GPE, the turnover time, synthetic rates and synthetic rate constant for each of the species of ethanolamine plasmalogen were calculated from the kinetic data of specific radioactivities for both ether-linked EPG. The synthetic rate constant of hexaene species was found to be approximately two times that of the other species. The high content of docosahexaenoic-acid-containing species in plasmalogen may be partially due to the rapid synthesis of this species from alkylacyl-GPE. In the second experiment, the incorporation rate of ^{14}C-docosahexaenoic acid (22:6) was compared with that of ^{3}H-arachidonic acid (20:4). Labelled 22:6 was distributed in the EPG fraction in much higher amount than ^{3}H-20:4. Among the three types of EPG, the distribution of ^{14}C-22:6 in both ether-linked EPG was much higher than in 1-^{3}H-glycerol. The metabolic pathways which possibly contributed to the special retention of 22:6 are discussed.

INTRODUCTION

Brain contains ethanolamine plasmalogen at much higher levels than most other healthy tissues (Wykle, 1977). It has been established that the ethanolamine plasmalogen is synthesized from alkylacyl-GPE by a desaturase in mammalian tissues (Horrocks and Radominska-Pyrek, 1972; Wykle et al., 1972; Paltauf and Holasek, 1973; Wykle, 1977). Although there is evidence that the fatty acid composition of ether-linked phospholipids can be regulated distinctively from diacylglycerophospholipids (Lands and Crawford, 1976; Holub and Kuksis, 1978), little information has been obtained about the selective synthesis of molecular species of ether-linked EPG in nervous tissue.

One of the most effective tools to know the specificity of individual reactions in vivo is to estimate the synthetic rate constant of individual species using the equation according to Zilversmit et al. (1942). Although we (Waku and Nakazawa, 1978; 1979) have estimated the incorporation rate of 1-^{3}H-glycerol into the molecular species of ether-linked ethanolamine phospholipids in Ehrlich ascites tumor cells, the turnover time or synthetic rate constant for the molecular species of ether-linked lipid has never been calculated by the equation of Zilversmit's.

In this paper, we estimated the incorporation rate of 1-^{3}H-glycerol into each molecular species in developing brain to discover how the molecular species of ether-linked lipids turns over in developing brain, and how the composition of molecular species is governed under physiological conditions. Also, the turnover time and turnover rate constant for each of the species of ethanolamine plasmalogen were calculated. By this analysis, we found the extremely rapid turnover of plasmalogen, compared to that previously reported (Horrocks et al., 1975; Miller et al., 1977), and the preferential synthesis of hexaene species in the formation of ethanolamine plasmalogen from alkylacyl-GPE. In addition, the incorporation rate of

^{14}C-22:6 into the developing rat brain was compared with that of ^{3}H-20:4 to examine the selectivity of the synthetic enzymes for polyunsaturated fatty acids.

MATERIALS AND METHODS

All chemicals were of reagent grade and solvents were distilled before use. 1-^{3}H-glycerol (2.5 Ci/mmole), 1-^{14}C-glycerol (51.3 mCi/mmole) and 5,6,8,9,11,12,14,15-^{3}H-arachidonic acid (120 Ci/mmole) were purchased from Amersham International (Amersham, England). The uniformly labelled ^{14}C-22:6 (60.0 mCi/mmole) was purchased from New England Nuclear (Boston, MA). The ^{3}H-arachidonic acid was purified by thin-layer chromatography (TLC) before use.

Male Wistar rats were obtained from Sankyo Lab. Service, Inc. (Tokyo, Japan). Eighteen-day-old rats were given intracerebral injections of 70 ul of Ringer's balanced salt solution (buffered to pH 7.4 with 50 mM Tris-HCl) containing 200 uCi/ml of ^{3}H-glycerol. To estimate the incorporation rate of ^{14}C-22:6 and ^{3}H-20:4, each of the fatty acids was dissolved in Ringer's balanced salt solution containing bovine serum albumin (1 mg/ml), essentially fatty acid free, Sigma Chemical Co., St. Louis) at the concentration of 86.7 uCi/ml and 28.9 uCi/ml, respectively, and was injected in the same manner. After selected periods, groups of 4 rats were decapitated, and the brains were quickly removed. The brains were immediately homogenized in ice cold saline, and the lipids were extracted as described by Bligh and Dyer (1959). The phosphoglycerides were fractionated by DEAE cellulose and silicic acid column chromatography as described previously (Waku et al., 1974). Throughout the fractionation procedure, a small amount of butyrated hydroxytoluene was added to the solvents.

The ethanolamine phosphoglyceride fraction was hydrolyzed by phospholipase C (B. cereus) and acetylated as described previously (Waku et al., 1974). 1-radyl-2-acyl-3-acetylglycerols prepared by this treatment were separated into 1-alkenyl-2-acyl-3-acetylglycerols, 1-alkyl-2-acyl-3-acetylglycerols and 1,2-diacyl-3-acetylglycerols by TLC, according to Renkonen and Varo (1976). The individual spots were visualized under ultraviolet light by spraying with primuline, a non-destructive fluorescent reagent. Three types of diradylacetylglycerol were eluted from silica gel with chloroform:methanol:water (1:2:0.8) and extracted as described by Bligh and Dyer (1959). The purity of each of the fractions was confirmed by TLC. For further analysis of molecular species, each of the diradylacetylglycerols was fractionated by 20% $AgNO_3$-impregnated TLC developed with 0.8% methanol in chloroform. Individual spots were visualized by 2',7'-dichlorofluorescein under ultraviolet. Each of the molecular species was eluted from $AgNO_3$-containing silica gel with 2 ml of 1% NaCl in 90% methanol and extracted with hexane. A part of the lipid fraction was used for the measurement of the radioactivity, and the remaining sample was used for the analysis of fatty acyl chains. The recovery of lipids and radioactivity from the plate was over 90%.

Fatty acyl moieties of the separated molecular species were transmethylated with sodium methoxide, and the quantities were estimated by GLC using methyl heptadecanoate as an internal standard. A 2 m column containing 15% diethyleneglycolsuccinate on chromosorb W was run at 195 C with nitrogen as a carrier gas in a Hitachi 163 gas-liquid chromatograph.

The radioactivity was measured in a Packard 3320 scintillation counter, using the scintillation fluid described previously (Waku and Nakazawa, 1978).

To determine the pool size of EPG, the total lipid extracted from rat brain was subjected to two-dimensional TLC for separation of phospholipid classes, using the solvent system described previously (Masuzawa et al., 1973). The phosphorus of each fraction was determined as described by Rouser et al. (1966). The pool size of each molecular species of EPG was determined from the phosphorus content of EPG and the quantity of the fatty acyl moiety of each species.

From the time dependent change of specific radioactivity of each molecular species of alkylacyl-GPE and alkenylacyl-GPE, the turnover time, synthetic rate and synthetic rate constant of brain alkenylacyl-GPE was calculated by the following equation, according to Zilversmit et al. (1942):

$$T_{alkenyl} = \frac{\int_{t_1}^{t_2} S_{alkyl}\,dt - \int_{t_1}^{t_2} S_{alkenyl}\,dt}{\int_{t_1}^{t_2} dS_{alkenyl}}$$

$$V_{alkenyl} = \frac{(ALKENYL)}{T_{alkenyl}}$$

$$V_{alkyl \rightarrow alkenyl} = V_{alkenyl}$$

$$k_{alkyl \rightarrow alkenyl} = \frac{V_{alkyl \rightarrow alkenyl}}{(ALKENYL)}$$

S_{alkyl}, $S_{alkenyl}$: specific activity of molecular species of alkylacyl-GPE or alkenylacyl-GPE (cpm/nmole)

$T_{alkenyl}$: turnover time of molecular species of alkenylacyl-GPE (min)

$V_{alkyl \rightarrow alkenyl}$: synthetic rate of molecular species of alkenylacyl-GPE from alkylacyl-GPE (μmole/min/g wet weight)

$V_{alkenyl}$: turnover rate (degradation rate) of molecular species of alkenylacyl-GPE (μmole/min/g wet weight of brain)

(ALKENYL), (ALKYL) : pool size of molecular species of alkenylacyl-GPE or alkylacyl-GPE (μmole/g wet weight)

$k_{alkyl \rightarrow alkenyl}$: synthetic rate constant from alkylacyl-GPE to alkenylacyl-GPE (min^{-1})

t_1, t_2 : time after injection of radioactive precursor (min)

RESULTS

Fig 1-a. shows the time-dependent change of specific radioactivities of molecular species of rat brain diacyl-GPE at 18 days of age. The curves for hexaene and monoene were similar, and the specific activities were much higher than for diene and tetraene species. On the other hand, the specific activity of tetraene species was the lowest, although it gradually increased. This phenomenon indicated that a synthetic route other than de novo must contribute to the formation of this molecular species. A reacylation mechanism of 1-acyl-lysophospholipids (Lands and Hart, 1965) could be involved in the formation of the tetraene species. Deacylation-reacylation, or other

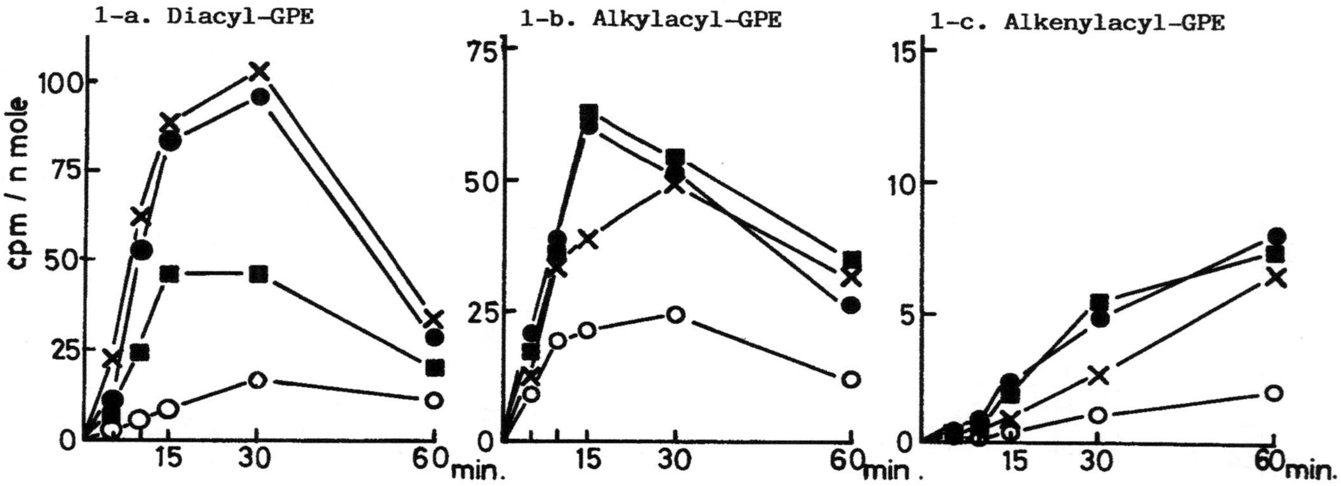

Fig. 1. The specific activities of molecular species of the three types of 1-radyl-2-acyl-3-acetylglycerol derived from ethanolamine phospholipids of 18 day old rat brains as a function of time. 1-a, diacyl compound; 1-b, alkyl-acyl compound; 1-c, alkenylacyl compound. 1-^{3}H-glycerol (14 uCi) was injected intracerebrally into the brain of 18 day old rats. After selected periods, the brains were removed and the lipids fraction extracted. 1-radyl-2-acyl-3--acetylglycerol was prepared from ethanolamine phospholipd fraction. Three types of diradylacetylglycerols were separated by TLC. Each of the molecular species was separated by $AgNO_3$-TLC, and the specific radioactivity estimated, as described in Materials and Methods. (●-●)Monoene; (■-■)diene; (O-O)tetraene; (X-X)hexaene.

routes differing from the de novo pathway, may be partly involved in the formation of diene species since the time-dependent curve for this species was located between those of the other species.

The time-dependent change of specific radioactivity of each molecular species of both types of ether-linked EPG was also estimated, and the results are shown in Fig 1-b. and Fig 1-c. The order of specific radioactivity at 5 min after injection was: diene > monoene > hexaene > pentaene > tetraene (the data for pentaene species is not shown). This order of specific radioactivity was the same for both alkylacyl-GPE and alkenylacyl-GPE during the first 30 min after injection. In alkylacyl-GPE, the curves for monoene and diene species rose rapidly and reached a maximum at 15 min after injection. On the other hand, the tetraene species showed behavior similar to the corresponding species of diacyl compounds; the specific activity was low and the time-dependent increase was slow. The time-dependent change of each molecular species of alkenylacyl-GPE appeared to reflect that of the corresponding species of alkylacyl-GPE. This result indicates that each molecular species of alkenylacyl-GPE is synthesized from the corresponding molecular species of alkylacyl-GPE.

From the above data, it is possible to estimate the turnover time of each molecular species of alkenylacyl-GPE according to Zilversmit's equation (Zilversmit et al., 1942). Furthermore, synthetic rate and synthetic rate constant can be calculated from the observed turnover time and tissue content of alkenylacyl-GPE and alkylacyl-GPE. The details of calculation have been described in the section of Materials and Methods. The results, using the brain of 18-day-old rats are shown in Table 1. The changes of specific activities at earlier times (5, 10 min) were used for the calculation to exclude the effect of re-utilization of radioactive precursors and of influx of radioactivity from pathways other than de novo as much as possible. The pool size of each molecular species of both alkenylacyl- and alkylacyl-GPE was determined from the phosphorus content of the EPG fraction, and the quantity of the fatty acyl moiety of each molecular species was determined as described in Materials and Methods. The calculated turnover time of total ethanolamine plasmalogen was found to be approximately 5 hr and much shorter than the previous data for the half-life as determined from graphs of specific activity (logarithmic scale) against times (half-life: 4-20 days in microsomes, 11-34 days in myelin (Miller et al., 1977). The calculated turnover time of molecular species of alkenylacyl-GPE were in the order: tetraene ≥ hexaene > pentaene ≥ monoene ≥ diene. The differences of synthetic rate among molecular species reflected the difference of the pool size of particular species. The synthetic rate constant of each species showed the following order: hexaene > pentaene > tetraene > monoene > diene. The value for the hexaene species was 3.8-fold higher than that of the diene species. This result suggests that the hexaene species is preferentially synthesized from alkylacyl-GPE among all species. The preferential synthesis of hexaene species of alkenylacyl-GPE in the desaturation of alkylacyl-GPE may explain the difference of the composition of this species, between alkylacyl-GPE and alkenylacyl-GPE.

Table 1. The turnover time, synthetic rate and synthetic rate constant of each of the molecular species of ethanolamine plasmalogen synthesized *de novo*. The pool size and specific radioactivity of 1-^{3}H-glycerol of each of the molecular species were determined as described in Materials and Methods. Turnover rates, synthetic rates and synthetic rate constants were calculated according to Zilversmit et al. (7). The time dependent changes of specific activities from 5 min to 10 min were used for the calculation.

molecular species	turnover time (min)	synthetic rate (μmole/g/min)	synthetic rate constant (min^{-1})
monoene	250	4.0	0.021
diene	210	1.7	0.017
tetraene	360	9.5	0.037
pentaene	270	3.0	0.038
hexaene	310	14	0.064
total	320	32	0.037

Since rat brain contained a large amount of 22:6-containing species (data not shown), and it has been shown that this molecular species is preferentially synthesized from alkylacyl-GPE in developing rat brain, we studied the incorporation rate of radioactive 22:6 or 20:4 into the phospholipids in the same tissues in vivo. Table 2 shows the incorporation of 20:4 and 22:6 into various types of lipids in rat brain. As shown in the column of the mole percent distribution, 97% of fatty acids are esterified in the form of phospholipids. Most of the radioactive fatty acids were found to be esterified and only 1-2% of each of the radioactivities were distributed in the free fatty acid (FFA) fraction at 1 hr after injection.

Table 3 shows the distribution of both of the radioactive fatty acids among the various types of phospholipids. The distribution of radioactivity at 60 min after injection was almost similar to that at 15 min, apart from slight changes found in EPG, inositolphosphoglycerides (IPG) and serinephosphoglycerides (SPG) fractions. Arachidonate was incorporated at the highest amount in cholinephosphoglycerides (CPG) followed by IPG. In comparison with the mole percent, radioactive 20:4 was found to be most rapidly incorporated into IPG. On the other hand, this polyunsaturated fatty acid was less esterified in EPG and SPG, compared with the mass distribution. Radioactive 22:6 was distributed at the highest amount in CPG followed by EPG. Only in EPG fraction, the percent distribution of labelled 22:6 was significantly ($p<0.001$) higher than that of labelled 20:4. The ratio of these two values (distribution of radioactive 22:6/distribution of radioactive 20:4) was 3.2 in the EPG fraction. This result may be explained by the difference of the metabolic pathway for each of the phospholipids and/or the selectivity of the synthetic enzymes for 20:4 and 22:6. The lower proportion of radioactive 20:4 and 22:6 in the SPG fraction, in comparison with the mass distribution, can be explained by the synthesis of SPG through the base exchange pathway. The rapid incorporation of 20:4 into CPG and IPG may be due to the distinctive selectivity of acyltransferases, such as the acyl-CoA:1-acyl-GPC acyltransferase (Lands and Hart, 1965; Okuyama et al., 1975) or the acyl-CoA:1-acyl-glycerophosphoinositol (1-acyl-GPI) acyltransferase (Baker and Thompson, 1973). Labelled 22:6 may be a poorer substrate than 20:4 for the 1-acyl-GPI

Table 2. The distribution of radioactivities among lipid classes. The designated labeled fatty acids were intracerebrally injected into 18-day –old rat brain. After 60 min, brains were removed and the lipid extract was separated into cholesterol ester (ChE), triacylglycerols (TG), free fatty acids (FFA), diacylglycerol (DG) and phospholipids (PL) by TLC using the solvent system, petroleum ether: diethylether: acetic acid , 80:20:1. Percentage of fatty ester of each of the fractions was determined by estimating the quantity of the total fatty esters as described in Materials and Methods. The radioactivities of the individual fractions were determined in a liquid scintillation counter. Each value is the mean ± S.D. of six separate experiments.

	ChE	TG	FFA	DG	PL
% of fatty ester	0.3	1.0	0.9	0.8	97.0
			% ± S.D.		
20:4	0.3 ±0.3	0.8 ±0.7	1.9 ±1.4	1.9 ±0.9	95.1 ±2.3
22:6	0.3 ±0.2	5.8 ±2.0	1.0 ±1.0	1.2 ±0.6	91.7 ±2.5

Table 3. The distribution of esterified radioactivities among various phospholipids. Phospholipids were fractionated by two-dimensional TLC. Percentage of each of the phospholipid classes was determined by estimating the phosphorus content. Each value is the avarage of three separate experiments. The % distribution of radioactivities is the mean ± S.D. of three (15 min) or six (60 min) separate experiments. CPG, choline-phosphoglycerides; EPG, ethanolaminephosphoglycerides; SPG, serinephosphoglycerides; IPG, inositolphosphoglycerides; Sph, sphingomyelin; CL, cardiolipin; PA, phosphatidic acid.

		CPG	EPG	SPG	IPG	Sph	CL	PA
mole %		42.4	34.3	12.0	3.9	3.6	2.6	1.2
		% ± S.D.						
20:4	15min	57.7±1.9	15.7±0.7	3.3±0.4	20.4±0.8	0.3 ±0.1	0.3 ±0.3	2.3±0.2
	60min	60.2±2.6	13.3±2.0	3.8±0.5	19.5±2.8	0.5 ±0.4	0.4 ±0.4	2.3±0.8
22:6	15min	52.3±1.7	36.0±1.1	2.9±0.3	5.5±0.4	0.1 ±0.1	0.3 ±0.3	3.0±0.3
	60min	50.2±3.7	41.2±3.0	4.6±0.9	2.6±0.3	0.2 ±0.1	0.1 ±0.1	1.1±0.2

acyltransferase because only 2.6% of the radioactivity was incorporated into the IPG fraction. The high distribution of radioactive 22:6 in the EPG fraction indicates that a specific reaction of the ethanolamine phosphotransferase for the molecular species containing 22:6 (Kanoh and Ohno, 1975; Holub, 1978) may also occur in brain.

Table 4 shows the distribution of radioactivities among diacyl, alkyl and alkenyl compounds of EPG. In addition to the incorporation of radioactive polyunsaturated fatty acids, the percent distribution of 1-^{3}H-glycerol among three types of EPG were re-calculated from the data shown in Fig 1 to evaluate the net synthesis of each of the compounds by _de novo_ pathway. Both of the radioactive fatty acids were found more in diacyl and alkylacyl compounds and less in alkenylacyl compounds in comparison with the mass distribution. This observation revealed that the high distribution of radioactive 22:6 in the EPG fraction, as shown in Table 3, was mainly due to the rapid incorporation of 22:6 into diacyl and alkylacyl compounds. The distribution of radioactive 20:4 or 22:6 differed from that of radioactive glycerol. In diacyl compounds, the percent distribution of both radioactive polyunsaturated fatty acids was less than that of radioactive glycerol. However, radioactive 20:4 and 22:6 were much more abundant in both of the ether-linked EPG fractions, compared to radioactive glycerol. The difference in incorporation of the three radioactive precursors in the three species of EPG was clearer when the ratios of relative distributions of radioactive 20:4 or 22:6 to radioactive glycerol (distribution of radioactive 20:4 or 22:6/distribution of radioactive glycerol) were compared. The ratios for both ether-linked EPG were higher than for diacyl-GPE, with respect to 20:4 and 22:6. For example, the values for alkenylacyl-GPE at 15 min after injection were 10-fold higher than that for diacyl-GPE (20:4 - 12 vs 0.8; 22:6 - 9.3 vs 0.8). This result indicates that the selective synthesis of molecular species containing 20:4 or 22:6 effectively functions in the formation of ether-linked EPG rather than in diacyl-GPE. Furthermore, the relative distribution of 20:4 and 22:6 differed between alkylacyl-GPE and alkenylacyl-GPE; the value for radioactive 22:6 was higher than for radioactive 20:4 in alkylacyl-GPE, but the reverse was true for alkenylacyl-GPE. The distributions of both radioactive 20:4 and 22:6 relative to

Table 4. The distribution of radioactivities in diacyl-, alkylacyl- and alkenylacylglycerophosphoethanolamine. 1-radyl-2-acyl-3-acetylglycerols were prepared from the ethanolamine glycerophospholipid fraction. The diacyl, alkylacyl and alkenylacyl acetates were separated by TLC. Each value is the mean ± S.D. of three (15 min) or six (60 min) separate experiments. The mole % of each fraction was determined by estimating the quantity of the total fatty esters.

		Diacyl -GPE			Alkylacyl -GPE			Alkenylacyl -GPE		
	mole %	48.5			4.1			47.4		
					% ± S.D.					
20:4	15 min	72.9	±2.1	(0.8[a])	7.9	±0.8	(1.4)	19.3	±2.0	(12.0)
	60 min	62.7	±4.1	(0.8)	4.8	±1.0	(1.4)	32.4	±3.8	(1.8)
22:6	15 min	72.2	±0.9	(0.8)	12.9	±0.5	(2.2)	14.9	±0.4	(9.3)
	60 min	64.3	±3.1	(0.8)	9.1	±1.1	(2.6)	26.6	±3.6	(1.5)
glycerol[b]	15 min	92.7	±3.0		5.7	±1.8		1.6	±0.5	
	60 min	78.3			3.5			18.3		

a. Figures in parentheses show relative distribution of radioactive fatty acids to that of radioactive glycerol (% distribution of ^{3}H-20:4 or ^{14}C-22:6 / % distribution of ^{3}H-glycerol).

b. The distribution of radioactive glycerol was re-calculated from the data shown in Fig 1.

radioactive glycerol were much higher in alkenylacyl-GPE than in alkylacyl-GPE, a precursor of alkenylacyl-GPE, at 15 min after injection.

There are four enzymic activities possibly contributing to the different distribution of three radioactive precursors in the three EPG species: 1) 1-radyl-GP acyltransferase(s). Okuyama and Lands (1972) have revealed that 1-acyl-glycerophosphate (1-acyl-GP) is esterified with high selectivity for oleate or linoleate at low 1-acyl-GP concentration. On the other hand, Fleming and Hajra (1977) demonstrated that the 1-alkyl-GP acyltransferase selected arachidonyl- and docosahexaenoyl-CoA over more saturated acyl-CoAs. Higher distributions of radioactive polyunsaturated fatty acids relative to radioactive glycerol in ether-linked lipids than in diacyl-GPE may be partially due to the different selectivity between these two enzymic activities. 2) Ethanolamine phosphotransferase. Kanoh and Ohno (1975) and Holub (1978) have provided evidence that the CDP-ethanolamine:diacylglycerol ethanolamine phosphotransferase has a specificity for the molecular species containing 22:6. Our data, shown in Table 3, indicate that the same selective action of this enzyme also occurs in brain. Masuzawa et al. (1982) and Nakagawa and Waku (1982) have suggested that in Ehrlich ascites tumor cells, ethanolamine phosphotransferase may have a strong specificity for molecular species containing 22-carbon acids of 1-alkyl-2-acylglycerol. In the present data, the distribution of radioactive 22:6 relative to radioactive glycerol in alkylacyl-GPE was higher than both of the values for radioactive 20:4 in the same compounds, and for the labeled 22:6 in diacyl-GPE. This result may be due to a unique selectivity of ethanolamine phosphotransferase in the synthesis of alkylacyl-GPE as described above. 3) 1-radyl-GPE acyltransferase(s). Lands and Hart (1965) indicated that 1-acyl-GPE acyltransferase has a selectivity for arachidonyl-CoA. A number of isotopic studies supported this original recognition (Lands and Crawford, 1976; Holub and Kuksis, 1978). In addition, Wykle et al. (1973) have shown that 20:4 is incorporated into the 2-position of 1-alkenyl-GPE by a microsomal acyltransferase of rat testis. The higher ratio of ^{3}H-20:4 to ^{3}H-glycerol in alkenylacyl-GPE compared to alkylacyl-GPE, the precursor of alkenylacyl-GPE (Horrocks and Radominska-Pyrek, 1972; Wykle et al., 1972; Paltauf and Holasek,

1973; Wykle, 1977), indicated that a large amount of tetraene species of ethanolamine plasmalogen is synthesized by a deacylation-reacylation mechanism in rat brain. Also, the same mechanism may contribute to the retention of 22:6 in alkenylacyl-GPE. However, previously the selective action for docosahexaenoyl-CoA by the 1-radyl phospholipids acyltransferase(s) has never been described. 4) Desaturation enzyme for the synthesis of ethanolamine plasmalogen. It has been established that alkenylacyl-GPE is synthesized by the desaturation of alkylacyl-GPE (Horrocks and Radominska-Pyrek, 1972; Wykle et al., 1972; Paltauf and Holasek, 1973; Wykle, 1977). We have calculated the synthetic rate constant for each of the molecular species of alkenylacyl-GPE in rat brain. The synthetic rate constant of hexaene species was highest among the five species (Table 1). This result indicates that the preferential synthesis of hexaene species occurs in the formation of alkenylacyl-GPE from alkylacyl-GPE. The specificity of this enzyme may contribute to the higher relative distribution of ^{14}C-22:6 to ^{3}H-glycerol in alkenylacyl-GPE compared to alkylacyl-GPE.

Although further investigation is necessary to reveal which enzyme would most effectively function for the retention of 22:6 in ether-linked lipids, it seems that each of the steps for the synthesis of ether-linked phospholipids tends to form hexaene species preferentially. The special retention of 22:6 may be related to the development of central nervous system and myelination (Sun et al., 1975; Tinoco et al., 1979; Mekkena and Campagnini, 1979). Furthermore, the high level of 22:6 may influence either the release of prostaglandin precursors or the formation of prostaglandins (Lands et al., 1973). In the next experiment, we hope to ascertain whether polyunsaturated fatty acids are selectively incorporated into alkylacyl-GPE _in vivo_, and whether the 1-alkenyl-GPE acyltransferase has a specificity for the docosahexaenoyl-CoA _in vitro_.

REFERENCES

Baker, R.R. and Thompson, W. (1973) Selective acylation of 1-acylglycerophosphorylinositol by rat brain microsomes. Composition with 1-acylglycerophosphorylcholine. J. Biol. Chem. 248, 7060-7065.

Bligh, E.G. and Dyer, W.J. (1959) A rapid method of total lipid extraction and purification. Can. J. Biochem. Physiol. 37, 911-917.

Fleming, P.J. and Hajra, A.K. (1977) 1-alkyl-sn-glycero-3-phosphate: acyl-CoA acyltransferase in rat brain microsomes. J. Biol. Chem. 252, 1663-1672.

Holub, B.J. (1978) Differential utilization of 1-palmitoyl and 1-stearoyl homologues of various unsaturated 1,2-diacyl-sn-glycerols for phosphatidylcholine and phosphatidylethanolamine synthesis in rat liver microsomes. J. Biol. Chem. 253, 691-696.

Holub, B.J. and Kuksis, A. (1978) Metabolism of molecular species of diacylglycerophospholipids. Adv. in Lipid Res. 16, 1-125.

Horrocks, L.A. and Radominska-Pyrek, A. (1972) Enzymic synthesis of ethanolamine plasmalogens from 1-alkyl-2-acyl-sn-glycero-3-(^{32}P)-phosphorylethanolamines by microsomes from rat brain. FEBS Lett. 22, 190-192.

Horrocks, L.A., Toews, A.D., Tompson, D.K. and Chin, J.Y. (1975) Synthesis and turnover of brain phosphoglycerides. Results, methods of calculation and interpretation. In: Function and Metabolism of Phospholipids in the Central and Peripheral Nervous System (G. Porcellati, L. Amaducci and C. Galli, eds.) pp. 37-54, Plenum Press, New York.

Kanoh, H. and Ohno, K. (1975) Substrate-selectivity of rat liver microsomal 1,2-diacylglycerol: CDP choline (ethanolamine) choline (ethanolamine) phosphotransferase in utilizing endogenous substrates. Biochim. Biophys. Acta 380, 199-207.

Lands, W.E.M. and Crawford, C.G. (1976) Enzymes of membrane phospholipid metabolism in animals. In: The Enzymes of Biological Membranes (A. Martonosi, ed.) Vol. 2, pp. 3-85, Plenum Press, New York.

Lands, W.E.M. and Hart, P. (1965) Metabolism of glycerolipids. VI. Specificities of acyl coenzyme A: phospholipid acyltransferase. J. Biol. Chem. 240, 1905-1911.

Lands, W.E.M., LeTellier, P.R., Rome, L.H. and Vanderhoek, J.Y. (1973) In: Advances in the Biosciences (S. Bergstrom, ed.) Vol. 9, pp. 15-28.

Masuzawa, Y., Osawa, T., Inoue, K. and Nojima, S. (1973) Effects of various mitogens on the phospholipid metabolism of human peripheral lymphocytes. Biochim. Biophys. Acta 326, 339-344.

Masuzawa, Y., Nakagawa, Y., Waku, K. and Lands, W.E.M. (1982) Biochim. Biophys. Acta (in press)

Mekkena, M.C. and Campagnoni, A.T. (1979) Effect of pre- and postnatal essential fatty acid deficiency on brain development and myelination. J. Nutr. 109, 1195-1204.

Miller, S.L., Benjamines, J.A. and Morell, P. (1977) Metabolism of glycerophospholipids of myelin and microsomes in rat brain. J. Biol. Chem. 252, 4025-4037.

Nakagawa, Y. and Waku, K. (1982) Abstract for 12th International Congress of Biochemistry, p. 353, Perth, Australia.

Okuyama, H. and Lands, W.E.M. (1972) Variable selectivities of acyl coenzyme A: Monoacylglycerophosphate acyltransferase in rat liver. J. Biol. Chem. 247, 1414-1423.

Okuyama, H., Yamada, K. and Ikezawa, H. (1975) Acceptor concentration effect in the selectivity of acyl coenzyme A: 1-acylglycerylphosphorylcholine acyltransferase system in rat liver. J. Biol. Chem. 250, 1710-1713.

Paltauf, F. and Holasek, A. (1973) Enzymatic synthesis of plasmalogens. Characterization of the 1-O-alkyl-2-acylglycero-3-phosphorylethanolamine desaturase from mucosa of hamster small intestine. J. Biol. Chem. 248, 1609-1615.

Renkonen, O. and Varo, P. (1976) Thin-layer chromatography of phosphatides and glycolipids. In: Lipid chromatographic Analysis (G.v. Marinetti, ed.) Vol. 1, pp. 41-98, Marcel Dekker, New York.

Rouser, G., Siakotos, A.N. and Fleischer, S. (1966) Quantitative analysis of phospholipids by thin-layer chromatography and phosphorus analysis of spots. Lipids 1, 85-87.

Sun, G.Y., Winniczek, H., Go, J. and Sheng, S.L. (1975) Essential fatty acid deficiency: metabolism of 20:3(n-9) and 22:3(n-9) of major phosphoglycerides in subcellular fractions of developing and mature mouse brain. Lipids 10, 365-373.

Tinoco, J., Babcock, R., Hincenbergs, J., Medwadowski, B., Miljanich, P. and Williams, M.A. (1979) Linolenic acid deficiency. Lipids 14, 166-173.

Waku, K. and Nakazawa, Y. (1978) Incorporation rates of [1-^{14}C] glycerol into the molecular species of alkyl ether phospholipids of Ehrlich ascites tumor cells *in vivo*. Eur. J. Biochem. 88, 489-494.

Waku, K. and Nakazawa, Y. (1979) Turnover rates of the molecular species of alkenyl ether phospholipids of Ehrlich ascites tumor cells. Eur. J. Biochem. 100, 317-320.

Waku, K., Ito, H., Bito, T. and Nakazawa, Y. (1974) Fatty chains of acyl, alkenyl, and alkyl phosphoglycerides of rabbit sarcoplasmic reticulum. The metabolic relationship considered on the basis of structural analysis. J. Biochem. (Tokyo) 75, 1307-1312.

Wykle, R.L. (1977) Brain. In: Lipid Metabolism in Mammals (F. Snyder, ed.) Vol. 1, pp. 317-367, Plenum Press, New York.

Wykle, R.L., Blank, M.L., Malone, B. and Snyder, F. (1972) Evidence for a mixed function oxidase in the biosynthesis of ethanolamine plasmalogens from 1-alkyl-2-acyl-sn-glycero-3-phosphorylethanolamine. J. Biol. Chem. 247, 5442-5447.

Wykle, R.L., Blank, M.L. and Snyder, F. (1973) The enzymic incorporation of arachidonic acid into ether-containing choline and ethanolamine phosphoglycerides by deacylation-reacylation reactions. Biochim. Biophys. Acta 326, 26-33.

Zilversmit, D.B., Entenman, C. and Fischler, M.C. (1942) On the calculation of "Turnover time" and "Turnover rate" from experiments involving the use of labeling agents. J. Gen. Physiol. 26, 325-331.

THE EFFECT OF DIETARY CHOLINE ON BRAIN PHOSPHOLIPID CONTENT

Charles E. Pendley II, Lloyd A. Horrocks and Ronald F. Mervis
Departments of Physiological Chemistry and Pathology
The Ohio State University
Columbus, OH 43210, U.S.A.

INTRODUCTION

Considerable interest in dietary choline was generated following the demonstration that it could increase brain acetylcholine (Cohen and Wurtman, 1975). However, the variability of this effect and its small magnitude have led to questions about the mechanism of choline action. In addition to its effects upon acetylcholine synthesis, choline is a phospholipid precursor, and may also act as a weak cholinergic agonist (Krnjevic and Reinhardt, 1979). Despite the controversy regarding the effect of choline on acetylcholine synthesis, choline may be clinically beneficial in treating those conditions which are hypocholinergic.

Senescence in the normal aging rodent may be due to cholinergic dysfunction (Sherman *et al*. 1981). Bartus *et al*. (1980) showed that 4 months of dietary choline treatment increased the performance of senescent animals on single trial passive avoidance tests. The length of choline treatment required for an effect and the minimal changes effected in acetylcholine levels suggest to us that phospholipids may be important in this model system. We therefore chose to study the effect of chronic dietary choline on brain phospholipid. We have utilized both morphological and biochemical methods to more completely characterize our model system.

Aging in Mouse Brain Membranes

Membrane function is largely dependent upon the maintenance of membrane fluidity (Hegner, 1980). Membrane fluidity varies with many factors, some of which include the content of cholesterol, fatty acid (length and unsaturation), and phospholipid polar head groups. Changes in the composition of membrane lipid components should ultimately change membrane fluidity, and eventually membrane function.

Age-related changes in membrane composition can be induced by changes in the rate of synthesis of different compounds or may be caused by lipid degradation. Brunetti et al. (1979) have studied the effect of aging on the microsomal synthesis of brain phospholipids starting with CDPcholine and CDPethanolamine precursors. The rates of synthesis of choline and ethanolamine plasmalogens, as well as the synthesis of diacylglycerophosphocholine and diacylglycerophosphoethanolamine, were significantly reduced in 18-month rats relative to 56-day old controls. The differences in activity were probably a result of decreased availability of diacylglycerol in the aged microsomes. The diacylglycerols from aged microsomes contained proportionally more mono and di-unsaturated fatty acids compared to the microsomes from young rats. Similar experiments performed on homogenates of glial and neuronal enriched fractions yielded similar results (Gaiti et al., 1981). Aged neuronal enriched homogenates had a decreased rate of synthesis, whereas the rate of synthesis in glial fractions was unchanged with age. Collectively, these results suggest that a major effect of aging is the decrease in the neuronal choline and ethanolamine phosphotransferase reactions.

The phospholipid composition of whole brain membranes does not reflect the enzyme differences noted above. Gaiti et al. (1979; 1981) found no difference in the concentration of cellular lipid phosphorus in the above studies. No changes in total phospholipid are seen in brains of young versus aged mice if the data are expressed as a concentration (Horrocks et al., 1981). However, if the data are expressed per brain, a 10% decrease in total phospholipid is seen.

Part of the discrepancy between phospholipid content and enzyme rates may be a result of mass action. The studies of Gaiti et al. (1979) suggest that the decreased rate of lipid synthesis occurs only in neurons, and not in the myelin derived from oligodendroglia. Thus, the very small changes in total phospholipid may be difficult to detect. The content of myelin in the mouse brain increases during aging whereas that of the human brain decreases from the age of 30-40 years onward.

The activities of cholinergic enzymes in the brain are also functionally decreased with age. The synthesis of acetylcholine decreases with age (Gibson et al., 1981). This decreased acetylcholine synthesis is a result of several factors such as the decreased sodium-dependent high-affinity choline uptake (Sherman et al., 1981), a generalized decrease in oxidative metabolism (Gibson and Peterson, 1981, Leong et al., 1981), and decreased choline acetyltransferase (Lippa et al., 1980). Aging also results in a loss of muscarinic cholinergic receptors (Freund, 1980). Thus, a loss of cholinergic tone may be manifested in a variety of ways.

Metabolism of Choline

The pathways of choline metabolism in brain are, for the most part, well known (Fig. 1). However the relative flux through each of these pathways remains undetermined. The bulk of free brain choline is utilized for phospholipid synthesis. The phospholipid-bound choline pool is 270 times greater than the acetylcholine-bound pool (Ansell and Spanner, 1982), and thus would be expected to account for a much greater utilization of choline.

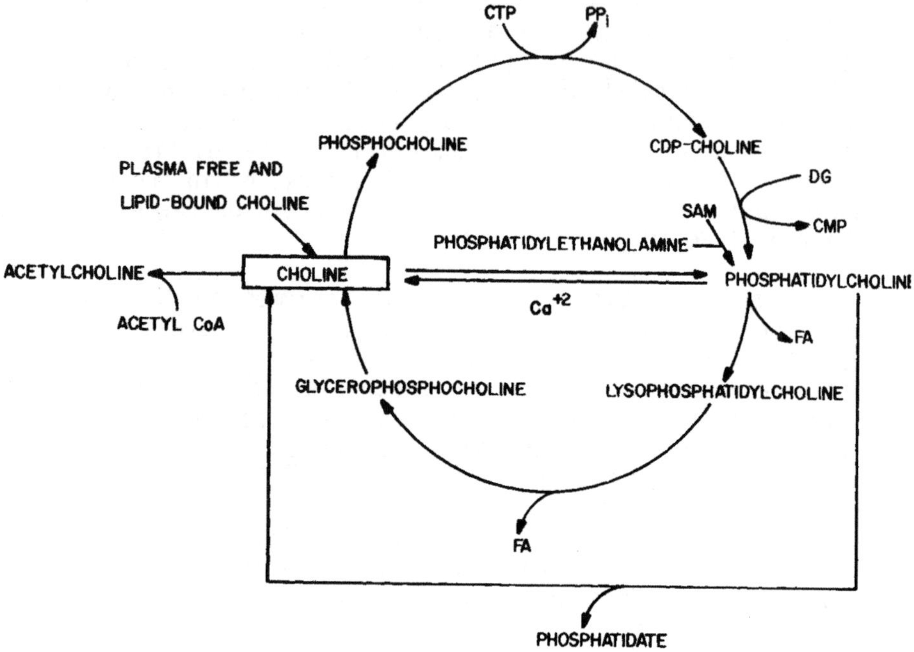

Figure 1. Metabolic pathways for the synthesis of acetylcholine and phosphatidylcholine from choline. Diacylglycerol, fatty acid, and S-adenosyl methionine are abbreviated as DG, FA, and SAM, respectively.

The origin of free choline in the brain is rather complicated. A number of different sources contribute to the free choline pool. A major source of choline is the bloodstream. Choline may be transported across the blood brain barrier in a free (Cornford et al., 1978) or bound form (Ansell and Spanner, 1979). Other contributions to the free choline pool are made by the breakdown of acetylcholine and choline-containing lipids. A small amount of bound choline is synthesized by phospholipid methylation in the brain (Mozzi and Porcellati, 1979).

The free choline pool serves as the source of precursor for the synthesis of neurotransmitter and choline-containing lipids. Thus an equilibrium exists between the free and bound choline compounds. This equilibrium can be affected by increasing the level of choline in the diet.

Increased dietary choline leads to increased brain choline, and finally greater bound choline. This hypothesis is based upon data which show that the transport of choline into the brain is not saturated, even at high plasma concentrations of choline (Freeman et al., 1975).

The slow rate of choline uptake is most likely a consequence of the concentration gradient for choline that exists between brain and blood. In fact, a significant amount of free choline is generated by the brain. Comparisons of the arterial input to the brain and the venous output from the brain have shown a small net efflux of free choline from the brain (Choi et al., 1975; Dross and Kewitz, 1972). Thus choline uptake into the brain is probably a combination of a variety of factors which include the blood and brain concentrations of choline, the form of the transported choline, and the metabolic equilibrium between the choline-requiring and choline-releasing reactions.

Phospholipid Metabolism

The major route for choline glycerophospholipid biosynthesis in brain was first described by Kennedy and Weiss (1956). Choline phosphorylated by choline kinase reacts with CTP to form CDPcholine as an intermediate. CDPcholine then reacts with diacylglycerol to form diacylglycerophosphocholine.

The synthesis of choline appears to be regulated at the CTP:phosphocholine cytidylyltransferase step (Porcellati and Arienti, 1970). The active liver enzyme is located in the soluble cell fraction and requires lysophosphatidylethanolamine for maximum activity (Choy et al., 1977). Diacylglycerol causes enzyme aggregation. Lysophosphatidylcholine and, to a lesser extent, phosphatidylcholine are enzyme inhibitors. CDPcholine: 1, 2-diacylglycerol phosphocholinetransferase is another potential regulatory site. Phosphocholinetransferase is a freely reversible enzyme depending on the relative concentrations of products and reactants (Goracci et al., 1981). Therefore, addition of CDPcholine to the reaction should stimulate the formation of phosphatidylcholine.

An alternative synthetic pathway has recently been seen in brain (Blusztajn et al., 1979, Crews et al., 1980, Mozzi and Porcellati, 1979). Phosphatidylethanolamine is methylated to phosphatidylcholine by the action of two N-methyltransferases utilizing S-adenosyl methionine. In this process, the phosphatidylcholine "flip-flops" to the outer side of the membrane. The activity of these enzymes, however, is far lower than those of the Kennedy pathway. The reaction does not contribute substantially to net phosphatidylcholine synthesis, but may account for the small net efflux of choline from the brain.

The breakdown of phosphatidylcholine to release choline can be accomplished in a number of ways. Free choline may be released in a single step by phospholipase D hydrolysis (Saito and Kanfer, 1973) or by base exchange between phosphatidylcholine and phosphatidylethanolamine (Porcellati et al., 1971). This pathway is not thought to be a major route for choline release (Sundler and Akesson, 1975). Like the methylation reactions above, base exchange activity leads to differences in phospholipid composition. The major route of choline release from phosphatidylcholine is via enzymic hydrolysis of glycerophosphocholine (Jope and Jenden, 1979). Glycerophosphocholine is derived from phosphatidylcholine by deacylation through the sequential action of a phospholipase A_1 or A_2, followed by a lysophospholipase.

Effect of Choline Supplementation

Morphology

Structure and function in the nervous system are intimately related. With changes in neuronal connectivity, changes in information processing, and ultimately function, may be inferred.

The Golgi stained cortical pyramidal cell has been extensively utilized to study morphological changes representative of the effects of aging. In the aging pyramidal cell, structural changes have been reported in cell number, cell body, dendrites, and dendritic spines (Peters and Vaughan, 1981).

We have investigated the effect of chronic (4 months) choline treatment on pyramidal cell dendritic spines. Brains from choline-enriched, choline-deficient, and control mice (Bartus et al., 1980) were perfused through the left ventricle with a solution of 2% glutaraldehyde and 2% paraformaldehyde. Small coronally cut brain slices were Golgi stained in a Golgi-Kopsch modification of the method of Riley (Cupp and Uemura, 1980). Sections were cut 120, 150 and 180 µm thick and counter stained with cresyl violet. Spine counts were made on randomly selected layer V pyramidal neurons.

Data for the terminal tip spine content for the basilar and apical dendritic trees are shown in Table 1. The content of apical terminal tip spines from 19-month control animals is significantly reduced relative to 8-month old control animals. Choline deficient 19-month old mice had slightly fewer spines then age-matched controls. Choline supplemented 19-month old mice, however, had either no loss or an increase in spines relative to 8-month old controls.

Table 1. Effect of Chronic Choline Treatment on Spine Density in Aging C57BL/6J Mice

	Apical	Basilar
Control, 8 month	941	748
Control, 19 month	781	732
Choline deficient, 19 month	760	718
Choline enriched, 19 month	925	838

Values are spines per mm. Mice were treated as described by Bartus et al. (1980). Dendritic spines were counted on layer V pyramidal neurons in the cortex.

Effect of Choline on Phospholipid Content

The results of choline treatment on neuronal spine content suggests that choline prevents the age-related loss of dendritic spines. Such an effect should result in differences in total phospholipid content.

Total brain phospholipids were extracted with hexane and isopropyl alcohol (Hara and Radin, 1978). Extracted lipids were separated by two-dimensional thin layer chromatography (Horrocks and Sun, 1972). Phosphorus analyses utilized the method of Rouser et al. (1970), and cholesterol was assayed by the method of Bowman and Wolf (1962).

The effect of choline on the phospholipid content of the mouse striatum and brainstem is shown in Table 2. Choline deficient animals contained somewhat less total phospholipid, and substantially less phosphatidylcholine. Aged controls and enriched brains did not differ in content of either total phospholipid or phosphatidylcholine. The content of phospholipid and cholesterol of forebrain and cerebellum was not affected by dietary choline.

Table 2. Total Phospholipid and Phosphatidylcholine Contents of Striatum and Brainstem From Mice Maintained on Choline Enriched, Deficient, and Control Diets for Four Months.

Diet (n)	Lipid P	Phosphatidylcholine
Control (3)	104.2	26.7
Enriched (3)	105.1	25.2
Deficient (4)	95.5	16.9

The data are expressed as median values in units of μmol/g fresh tissue. Diets and drinking water of the mice were as described by Bartus et al. (1980).

Proposed Mechanism of Choline Action

The results of our morphological studies suggest that the primary effect of chronic dietary choline is maintenance of synaptic structure. The maintenance of dendritic spines could be a function of increased membrane synthesis under continuing conditions of spine loss or might simply be the prevention of spine loss. In any event, more phospholipid-containing dendritic spines are present in the choline enriched animals.

Choline is not degraded in brain, and the only other pathway for choline utilization, other than acetylcholine synthesis, is choline glycerophospholipid synthesis (Francescangeli et al., 1977). Elevated choline levels may stimulate the synthesis of a normal membrane containing a variety of lipid and phospholipid components. Possmayer (1974; Possmayer and Mudd, 1971) showed that CDPcholine stimulates the synthesis of phosphatidate by stimulating the fatty acid acylation of glycerol-3-phosphate. Stimulation of diglyceride levels by increasing the synthesis of phosphatidate could then account for the synthesis of a variety of phospholipids and neutral lipids by increasing the supply of precursor. In support of the work of Possmayer we have previously shown that CDPcholine exerts profound effects on brain lipid metabolism in vivo (Dabrowiecki et al., 1980). Intracerebral injection of CDPcholine and CDPethanolamine together (Dorman et al., 1981) caused more than a 4-fold increase in diglyceride labeling, and a 40-50% increase in the labeling of choline and ethanolamine glycerophospholipids.

No definitive evidence exists regarding the effect of chronic choline administration on levels of CDPcholine. Millington and Wurtman (1982) have shown that a single oral dose of choline chloride (20 mmol/kg) significantly elevated phosphocholine from 2 to more than 34 hours, whereas the acetylcholine increase was significant at only one time point. Phosphocholine represents the first committed step in phosphatidylcholine synthesis, although enzymes do exist for the dephosphorylation. In the presence of an increasing concentration of phosphocholine substrate, the rate limiting cytidylyltransferase reaction should result in elevated CDPcholine, and thus greater phosphatidylcholine. The activity of cytidylyltransferase is modified by a variety of effectors (lysophosphatidyl-choline, lysophosphatidylethanolamine, CTP, diacylglycerol) and thus the conversion of phosphocholine to CDPcholine may not be strictly dependent upon substrate concentrations. Ultimately the effectiveness of choline is in direct proportion to its conversion to CDPcholine.

The efficacy of choline in reversing the loss of dendritic spines in the aging animal is probably related to the effect of aging on brain membranes. The flux of metabolites through the oxidative and lipogenic pathways is slowed. A catabolic rate which is relatively greater than the synthetic rate results in dissolution of the synaptic membrane. The increased brain choline leads to the formation of increased phosphocholine and CDPcholine. The relatively greater CDPcholine in enriched animals stimulates the flow of diacylglycerol through its synthetic pathway by reacting with substrate diacylglycerol to form phosphatidylcholine, and by allosterically stimulating the *de novo* synthesis of diacylglycerol. The branch point position of diacylglycerol and phosphatidate makes these compounds precursors of many other lipids.

The flow of metabolites through a pathway at a normal rate is predicated on the functional integrity of enzymes. Clearly, the synthetic machinery for phospholipid synthesis is not intact in the aging animal. The *in vitro* rate of choline and ethanolamine glycerophospholipid synthesis is decreased 38 and 41 percent, respectively, for 18 month versus 8 week-old rat brains (Brunetti *et al*., 1979).

An effective strategy for the treatment of aging animals would be the stimulation of the metabolic flux through a given pathway by stimulating the synthesis of precursor. Piracetam (2-oxo-pyrolidine-1-acetamide) has been used in geriatric conditions to stimulate metabolism, possibly by facilitating the synthesis of ATP (Nickolson and Wolthuis, 1976).

Piracetam in combination with choline has been shown to synergistically improve the retention of a passive avoidance task in the rat (Bartus *et al*., 1981). Choline in single or week-long dosages had no effect on behavior, while the same dosages of piracetam alone increased retention non-significantly.

Piracetam may also affect the synthesis of lipids directly. The incorporation of labeled phosphate into phosphatidylcholine and phosphatidylinositol in isolated neurons and glial cells was stimulated by piracetam (Woelk, 1979). Microsomal ethanolamine plasmalogen synthesis is also stimulated by piracetam (Woelk and

Peiler-Ichikawa, 1978). These results suggest that the efficacy of piracetam treatment may be mediated by changes in the rate of phospholipid synthesis.

In several studies the effect of choline on brain choline and acetylcholine did not correlate with the positive effects of choline. Bartus et al. (1981) reported mimimal increases in brain acetylcholine following choline administration, while the acetylcholine content of the choline plus piracetam animals did not differ from saline controls. Schmidt and Wecker (1981), in agreement with the above results, have shown a difference in the temporal dependence for acetylcholine synthesis and maximum brain choline levels. It is also interesting to note that the effect of choline on memory is observed only after chronic choline administration. If the modulation of phospholipid biosynthesis is responsible for the behavior effects of choline, then chronic administration would probably be the rule since membranes are composed of a number of components with widely different turnover rates.

Effect of Choline on Membrane Protein Function

If elevated dietary choline results in increases in memory and phosphatidylcholine, then it is important to examine the relationship between phosphatidylcholine and memory. Since phosphatidylcholine is a major membrane component the answer is simplified by examining the relationship between membrane changes and neural function. The information processing capability of the brain is a plasma membrane phenomenon. Thus an examination of the effect of choline on membrane protein-lipid interaction should be valuable.

Membrane function is critically dependent upon membrane fluidity (Hegner, 1980). Hirata, Axelrod and coworkers have shown that the methylation of phosphatidylethanolamine to phosphatidylcholine causes a marked increase in membrane fluidity (Hirata and Axelrod, 1978) and receptor function (Strittmatter et al., 1979; Hirata and Axelrod, 1980).

Despite their low activity the methyltransferases may also substantially contribute to regional lipid synthesis (Pfenninger and Johnson, 1981). Dorsal root ganglia cultures challenged with nerve growth factor showed a

4-fold stimulation in phospholipid methylation in the nerve neurites but not in perikarya. The results of the present investigation are similar in that choline caused the greatest inhibition of spine loss at the terminal tips of the apical and basilar branches.

A number of other membrane bound enzymes are affected by their lipid environment. Lipid depletion and fluidity changes have profound effects on the activity of Na^+, K^+-ATPase (Wheeler and Walker, 1975), adenylate cyclase (Axelrod et al., 1976), and acetylcholine receptor (Schiebler and Hucho, 1977; Briley and Changeux, 1978). The activity of liver mitochondrial β-hydroxybutyrate dehydrogenase was decreased 50% in rats fed a choline deficient diet for just 3 days (Leelavathi et al., 1974). The phospholipid composition and the recovery of activity following preincubation of mitochondria with phosphatidylcholine or increased dietary choline suggests that changes in enzyme activity were a result of changes in choline as a phospholipid precursor.

The effect of choline-induced membrane alterations on enzyme activity and brain function can be mimicked by the administration of CDPcholine. Qualitatively CDPcholine is more indicative of the effect of choline on phospholipid synthesis since this rapidly metabolized compound represents a firm commitment to lipid synthesis. Hayaishi et al. (1961) first noticed that the primary site of CDPcholine action was upon membranes. CDPcholine was particularly effective in reversing the neural cell membrane damage caused by cerebral edema. Using various pathological model systems, CDPcholine has beneficial effects upon the mitochondrial ATPase (Alberghina et al., 1981), Na^+, K^+-ATPase (Rigoulet et al., 1979), Ca^{2+}, Mg^{2+}-ATPase (Clendenon et al., 1981), free fatty acid levels (Trovarelli et al., 1981), protein synthesis (Serra et al., 1981), and Ca^{2+} permeability in myocytes (Dorman et al., unpublished observations). The clinical antiParkinson effectiveness of CDPcholine is a function of its stimulation of tyrosine hydroxylase (Martinet et al., 1981) and dopamine uptake (Martinet et al., 1978, 1979). The water solubility of CDPcholine and its effects upon lipid-requiring intrinsic membrane proteins would preclude a direct action. Instead, CDPcholine action is probably mediated by the synthesis of phosphatidylcholine and its association with membranes.

Summary

We have studied the effect of chronic dietary choline on brain phospholipids as a potential explanation for the increased memory of choline-treated aging mice. They showed a complete retention of cortical dendritic spines as contrasted with aging controls. Untreated senescent control mice had significantly fewer terminal dendritic spines. The total phospholipid phosphorus content and phosphatidylcholine content of choline supplemented mice was significantly greater than that of deficient mice, but did not differ from control mice. The lack of complete correspondence between the morphologic and chemical changes is not surprising because only the terminal tips of the apical and basilar dendrite branches were significantly affected by choline treatment.

A possible mechanism of choline action is proposed. Choline is effective in the aging model because of its ability to stimulate phospholipid synthesis (membrane formation). Aging animals have a decreased phospholipid synthetic capability relative to young animals. The increased phospholipid synthesis is due to increased precursor choline and to the stimulatory effects of CDPcholine on the synthesis of phosphatidic acid and other phospholipids. The enhanced memory is then a result of the lipid effects upon membrane-bound protein function.

REFERENCES

Alberghina, M., Viola, M., Serra, I., Mistretta, A. and Guiffrida, A.M. 1981. Effect of CDPcholine on the biosynthesis of phospholipids in brain regions during hypoxic treatment. J. Neurosci. Res. 6: 421-433.

Ansell, G.B. and Spanner, S. 1979. Sources of choline for acetylcholine synthesis in the brain. In: Choline and Lecithin in Brain Disorders. (A. Barbeau, J.H. Growdon, R.J. Wurtman, eds.), Plenum Press, New York, pp. 35-46.

Ansell, G.B. and Spanner, S. 1982. Choline transport and metabolism in the brain. In: Phospholipids in the Nervous System. Vol. I Metabolism. (L.A. Horrocks, G.B. Ansell, and G. Porcellati, eds.), Raven Press, New York, pp. 137-144.

Axelrod, D., Koppel, D.E., Schlessinger, J., Elson, E. and Webb, W.W. 1976. Mobility measurement by analysis of fluorescence photobleaching recovery kinetics. Biophys. J. 16: 1055-1069.

Bartus, R.T., Dean, R.L., Goas, J.A. and Lippa, A.S. 1980. Age-related changes in passive avoidance retention: modulation with dietary choline. Science 209: 301-303.

Bartus, R.T., Dean, R.L., Sherman, K.A., Friedman, E. and Beer, B. 1981. Profound effects of combining choline and piracetam on memory enhancement and cholinergic function in aged rats. Neurobiol. Aging 2: 105-111.

Blusztajn, J.K., Zeisel, S.H. and Wurtman, R.J. 1979. Synthesis of lecithin (phosphatidylcholine) from phosphatidylethanolamine in bovine brain. Brain Res. 179: 319-327.

Bowman, R.E. and Wolf, R.C. 1962. A rapid and specific ultramicro method for total serum cholesterol. Clin. Chem. 8: 302-309.

Briley, M.S. and Changeux, J. 1978. Recovery of some functional properties of the detergent extracted cholinergic receptor protein from Torpedo marmorata after reintegration into a membrane environment. Eur. J. Biochem. 84: 429-439.

Brunetti, M., Gaiti, A. and Porcellati, G. 1979. Synthesis of phosphatidylcholine and phosphatidylethanolamine at different ages in the rat brain in vitro. Lipids 14: 925-931.

Choi, R.L., Freeman, J.J. and Jenden, D.J. 1975. Kinetics of plasma choline in relation to turnover of brain choline and formation of acetylcholine. J. Neurochem. 24: 735-741.

Choy, P.C., Lim, P.M. and Vance, D.E. 1977. Purification and characterization of CTP: cholinephosphate cytidyltransferase from rat liver cytosol. J. Biol. Chem. 252: 7673-7677.

Clendenon, N.R., Palayoor, S.T. and Gordon, W.A. 1981. Effect of CDPcholine treatment on ATPase activity post-spinal cord trauma. Trans. Am. Soc. Neurochem. 12: 93.

Cohen, E.L. and Wurtman, R.J. 1975. Brain acetylcholine: increase after systemic choline administration. Life Sci. 16: 1095.

Cornford, E.M., Braun, L.D. and Oldendorf, W.H. 1978. Carrier mediated blood-brain barrier transport of choline and certain choline analogs. J. Neurochem. 30: 299-308.

Crews, F.T., Hirata, F. and Axelrod, J. 1980. Identification and properties of methyltransferases that synthesize phosphatidylcholine in rat brain synaptosomes. J. Neurochem. 34: 1491-1498.

Cupp, C.J. and Uemura, E. 1980. Age-related changes in prefrontal cortex of _Macaca mulatta_: synaptic density. Exp. Neurol. 69: 143-163.

Dabrowiecki, Z., Dorman, R.V. and Horrocks, L.A. 1980. Alteration of rat brain lipid metabolism _in vivo_. Trans. Am. Soc. Neurochem. 11: 136.

Dorman, R.V., Dabrowiecki, Z., DeMedio, G.E., Porcellati, G. and Horrocks, L.A. 1982. Effects of cytidine nucleotides on CNS membranes during ischemia. In: _Neural Trauma_ (R.G. Grossman and P. L. Gildenberg, eds.), Raven Press, New York, pp. 123-135.

Dross, K. and Kewitz, H. 1972. Concentration and origin of choline in the rat brain. Naunyn-Schmiedebergs Arch. Pharmakol. 274: 91-106.

Francescangeli, E., Goracci, G., Piccinin, G.L., Mozzi, R., Woelk, H. and Porcellati, G. 1977. The metabolism of labelled choline in neuronal and glial cells of the rabbit _in vivo_. J. Neurochem. 28: 171-176.

Freeman, J.J., Choi, R.L. and Jenden, D.J. 1975. Plasma choline: Its turnover and exchange with brain choline. J. Neurochem. 24: 729-734.

Freund, G. 1980. Cholinergic receptor loss in brains of aging mice. Life Sci. 26: 371-375.

Gaiti, A., Brunetti, M. and Porcellati, G. 1979. Base exchange reaction and lipid composition of liver and brain membranes during aging. Bull. Mol. Biol. Med. 4: 24-36.

Gaiti, A., Sitkievicz, D., Brunetti, J. and Porcellati, G. 1981. Phospholipid metabolism in neuronal and glial cells during aging. Neurochem. Res. 6: 13-22.

Gibson, G.E. and Peterson, C. 1981. Aging decreases oxidative metabolism and the release and synthesis of acetylcholine. J. Neurochem. 37: 978-984.

Gibson, G.E., Peterson, C. and Jenden, D.J. 1981. Whole brain acetylcholine synthesis declines with senescence. Science 213: 674-676.

Goracci, G., Francescangeli, E., Horrocks, L.A. and Porcellati, G. 1981. The reverse reaction of cholinephosphotransferase in rat brain microsomes. A new pathway for degradation of phosphatidylcholine. Biochim. Biophys. Acta 664: 373-379.

Hara, A. and Radin, N.S. 1978. Lipid extraction of tissues with a low-toxicity solvent. Anal. Biochem. 90: 420-426.

Hayaishi, O., Ozawa, A., Araki, S., Ishii, S. and Kondo, H. 1961. Biochemistry of brain injury and cerebral edema. Jpn. J. Med. Prog. 48: 519-27.

Hegner, D. 1980. Age dependence of molecular and functional changes in biological membrane properties. Mech. Aging Develop. 14: 101-118.

Hirata, F. and Axelrod, J. 1978. Enzymatic methylation of phosphatidylethanolamine increases erythrocyte membrane fluidity. Nature 275: 219-220.

Hirata, F. and Axelrod, J. 1980. Phospholipid methylation and biological signal transmission. Science 209: 1082-1090.

Horrocks, L.A. and Sun, G.Y. 1972. Ethanolamine plasmalogens. In: Research Methods of Neurochemistry. (R. Rodnight and N. Marks, eds.), Plenum Press, New York, Vol. I, 223-231.

Horrocks, L.A., Van Rollins, M. and Yates, A.J. 1981. Lipid changes in the aging brain. In: The Molecular Basis of Neuropathology. (A.N. Davison, and R.H.S. Thompson, eds.), Edward Arnold Publishers, London, pp. 601-630.

Jope, R.S. and Jenden, D.J. 1979. Choline and phospholipid metabolism and the synthesis of acetylcholine in rat brain. J. Neurosci. Res. 4: 69-82.

Kennedy, E.P. and Weiss, S.B. 1956. The function of cytidine coenzymes in the biosynthesis of phospholipids. J. Biol. Chem. 222: 193-214.

Krnjevic, K. and Reinhardt, W. 1979. Choline excites cortical neurons. Science 206: 1321-1323.

Leelavathi, D.E., Katyal, S.L. and Lombardi, B. 1974. Lecithin depletion in lower mitochondria of rats fed a choline-deficient diet. Effect on β-hydroxybutyrate dehydrogenase. Life Sci. 14: 1203-1210.

Leong, S.F., Lai, J.C.K., Lim, L. and Clark, J.B. 1981. Energy-metabolizing enzymes in brain regions of adult and aging rats. J. Neurochem. 37: 1548-1556.

Lippa, A.S., Pelham, R.W., Beer, B., Critchett, D.J., Dean, R.L. and Bartus, R.T. 1980. Brain cholinergic dysfunction and memory in aged rats. Neurobiol. Aging 1: 13-14.

Martinet, M., Fonlupt, P. and Pacheco, H. 1978. Interaction of CDPcholine with synaptosomal transport of biogenic amines and their precursors in vitro and in vivo in the rat corpus striatum. Experientia 34: 1197-1199.

Martinet, M., Fonlupt, P. and Pacheco, H. 1979. Effects of cytidine-5'-diphosphocholine on norepinephrine, dopamine and serotonin synthesis in various regions of the rat brain. Arch. Int. Pharmacodyn. Ther. 239: 52-61.

Martinet, M., Fonlupt, P. and Pacheco, H. 1981. Activation of soluble tyrosine hydroxylase in the rat brain after CDPcholine administration. Biochem. Pharmacol. 30: 539-41.

Millington, W.R. and Wurtman, R.J. 1982. Choline administration elevates brain phosphorylcholine concentrations. J. Neurochem. 38: 1748-1752.

Mozzi, R. and Porcellati, G. 1979. Conversion of phosphatidylethanolamine to phosphatidylcholine in rat brain by the methylation pathway. FEBS Lett. 100: 363-366.

Nickolson, V.J. and Wolthuis, O.L. 1976. Effect of the acquisition - enhancing drug piracetam on rat cerebral energy metabolism: Comparison with naftidrofuryl and methamphetamine. Biochem. Pharmacol. 25: 2241-2244.

Peters, A. and Vaughan, D.W. 1981. Central nervous system. In: Aging and Cell Structure. (J. E. Johnson, ed.), Plenum Press, New York, pp. 1-34.

Pfenninger, K.H. and Johnson, M.P. 1981. Nerve growth factor stimulates phospholipid methylation in growing neurites. Proc. Nat. Acad. Sci. 78: 7797-7800.

Porcellati, G. and Arienti, G. 1970. The cytidine triphosphate: choline phosphate cytidylyltransferase of normal and degenerating peripheral nerve: A study of its activation by phospholipids. Brain Res. 19: 451-464.

Porcellati, G., Arienti, G., Pirotta, M. and Giorgini, D. 1971. Base-exchange reactions for the synthesis of phospholipids in nervous tissue: the incorporation of serine and ethanolamine into the phospholipids of isolated brain microsomes. J. Neurochem. 18: 1395-1417.

Possmayer, F. 1974. CDPcholine reversal of the CMP and CTP inhibtion of phosphatidic acid synthesis by rat brain preparations. Biochem. Biophys. Res. Comm. 61: 1415-26.

Possmayer, F. and Mudd, J.B. 1971. The regulation of sn-glycerol-3-phosphate acylation by cytidine nucleotides in rat brain cerebral hemispheres. Biochim. Biophys. Acta. 239: 217-33.

Rigoulet, M., Guerin, K.B., Cohadon, F. and Vandendreissche, M. 1979. Unilateral brain injury in the rabbit: reversible and irreversible damage of the membrane ATPases. J. Neurochem. 32: 535-541.

Rouser, G., Fleischer, S. and Yamamoto, A. 1970. Two-dimensional thin layer chromatographic separation of polar lipids and determination of phopholipids by phosphorus analysis of spots. Lipids 5: 613-620.

Saito, M. and Kanfer, J. 1973. Solubilization and properties of a membrane-bound enzyme from rat brain catalyzing a base-exchange reaction. Biochem. Biophys. Res. Comm. 53: 391-398.

Schiebler, N. and Hucho, F. 1978. Membranes rich in acetylcholine receptor: characterization and reconstitution to excitable membranes from exogenous lipids. Eur. J. Biochem. 85: 55-63.

Schmidt, D.E. and Wecker, L. 1981. CNS effects of choline administration: Evidence for temporal dependence. Neuropharmacology 20: 535-539.

Serra, I., Alberghina, M., Viola, M., Mistretta, A. and Giuffrida, A.M. 1981. Effect of CDPcholine on the biosynthesis of nucleic acids and proteins in brain regions during hypoxia. Neurochem. Res. 6: 607-18.

Sherman, K.A., Kuster, J.E., Dean, R.L., Bartus, R.T. and Friedman, E. 1981. Presynaptic cholinergic mechanisms in brain of aged rats with memory impairments. Neurobiol. Aging 2: 99-104.

Strittmatter, W.J., Hirata, F. and Axelrod, J. 1979. Phospholipid methylation unmasks cryptic beta-adrenergic receptors in rat reticulocytes. Science 204: 1205-1207.

Sundler, R. and Akesson, B. 1975. Regulation of phospholipid biosynthesis in isolated rat hepatocytes. Effect of different substrates. J. Biol. Chem. 250: 3359-67.

Trovarelli, G., DeMedio, G.E., Dorman, R.V., Piccinin, G. L., Horrocks, L.A. and Porcellati, G. 1981. Effect of cytidine diphosphate choline, CDPcholine, on ischemia induced alterations of brain lipid in the gerbil. Neurochem. Res. 6: 821-833.

Wheeler, K.P. and Walker, J.A. 1975. Differential effects of lipid depletion on membrane sodium-plus-potassium ion-dependent adenosine triphosphatase and potassium ion dependent phosphatase. Biochem. J. 146: 723-727.

Woelk, H. 1979. Effects of piracetam on the incorporation of ^{32}P into the phospholipids of neurons and glial cells isolated from rabbit cerebral cortex. Pharmakopsychiat. 12: 251-256.

Woelk, H. and Peiler-Ichikawa, K. 1978. Action of piracetam on formation of ethanolamine plasmalogen by neuronal microsomes of developing rat brain. Arzneimittel-Forschung 28: 1752-1756.

SOLUBILIZATION, PURIFICATION AND PROPERTIES OF MEMBRANE-BOUND BRAIN ENZYMES WHICH BIOSYNTHESIZE PHOSPHOINOSITIDES

Joseph Eichberg, J. Robert Bostwick and Abdallah Ghalayini

Department of Biochemical and Biophysical Sciences, University of Houston,
Houston, TX 77004

INTRODUCTION

For some years, evidence has steadily accumulated that cellular inositol phospholipid metabolism is stimulated in response to hormones or neurotransmitters in a variety of tissues. These are now generally considered to be receptor-mediated, cyclic AMP-independent phenomena which involve an initial accelerated breakdown of one or more of the phosphoinositides followed by their resynthesis. The nature of the molecular linkage between receptor activation, stimulated phosphoinositide degradation and subsequent cellular events, such as mobilization of cytosolic Ca^{2+} or liberation of arachidonic acid as precursor to prostaglandin formation, has become a significant problem in membrane biochemistry and is discussed in another chapter in this volume (Hauser 1983).

In nervous tissue, changes in the metabolism of phosphatidylinositol and the polyphosphoinositides, phosphatidylinositol-4-phosphate and phosphatidylinositol-4,5-bisphosphate, as well as phosphatidic acid, occur when synaptosomes are exposed to acetylcholine (Yagihara and Hawthorne 1972; Schacht and Agranoff 1972; Fisher and Agranoff 1981). Current evidence indicates that these effects are mediated by muscarinic cholinergic receptors at a postsynaptic location (Fisher _et al_ 1981). There is also circumstantial evidence that the turnover of polyphosphoinositides is specifically affected during propagation of electrical signals along the axon. (White _et al_

1974; Killian and Schacht 1981). One approach to gaining further understanding of the factors which regulate the metabolic alterations, as a step toward appreciating their functional significance, is to isolate and characterize the enzymes which biosynthesize and degrade phosphoinositides. To this end, we initiated studies to solubilize, purify and study the properties of selected membrane-bound enzymes of phosphoinositide metabolism.

Previous efforts to purify enzymes which metabolize phosphoinositides have been most successful in the case of soluble enzymes. Thus cytosolic Ca^{2+} -dependent phosphatidylinositol-specific phospholipase C has been purified over 4000 fold to homogeneity from rat liver and sheep seminal vesicles (Takenawa and Nagai 1981; Hoffman and Majerus 1982), and appreciably purified from bacterial sources as well (Low and Finean 1977, Sundler et al 1978). Another degradative enzyme, phosphatidylinositol-4,5-bisphosphate phosphatase was purified 430 fold from rat brain and was also found to be active against phosphatidylinositol-4-phosphate (Nijjar and Hawthorne 1977). For the most part, the purification of these enzymes was achieved by means of conventional multistep column fractionation techniques.

In contrast, efforts to purify membrane-associated enzymes by similar methodology have been much less successful. A procedure to enrich CDP-diacylglycerol: inositol phosphatidyl transferase (phosphatidylinositol synthetase) 60 fold from rat brain was reported, but little supporting data was given (Rao and Strickland 1974). The same enzyme was stated to be purified to homogeneity from rat liver microsomes in 3% yield but with a specific activity increase of only 28-fold (Takenawa and Egami 1977). In the case of phosphatidylinositol kinase, only sketchy information was provided by Harwood and Hawthorne (1969) concerning its partial purification from rat liver, and Buckley (1977) obtained negligible enrichment of the enzyme from human erythrocyte membrane.

A general characteristic of these molecules, in common with many other membrane-bound enzymes, is that they require the presence of detergent to become and remain solubilized. However detergents, especially nonionic compounds, often interfere with the usual procedures for enzyme purification which rely on ionic or other hydrophilic interactions. Therefore we have attempted the purification of detergent solubilized phosphatidylinositol kinase and phosphatidylinositol synthetase by application of methodology which promises to take advantage of the particular properties of these molecules, namely their hydrophobic nature and their substrate specificity. Thus we have examined the usefulness of hydrophobic interaction chromatography and affinity chromatography for purification of these enzymes. In addition, we have examined some properties not heretofore investigated of these detergent solubilized preparations.

DETERGENT ACTIVATION AND SOLUBILIZATION OF PHOSPHATIDYLINOSITOL KINASE

Previous investigators have shown that phosphatidylinositol kinase activity is markedly stimulated by detergents and has previously been assayed in the presence of sodium deoxycholate, Triton X-100, and Cutscum (Kai *et al* 1966; Harwood and Hawthorne 1969; Shaikh and Palmer 1977; Buckley 1977). In our hands, increasing levels of sodium deoxycholate or octylglucoside produced sharp peaks of enzyme activation at detergent concentrations of 3 mM and 30 mM, respectively (Bostwick and Eichberg 1981). The non-ionic detergent Triton X-100 produced activation over a broader range of concentration but with a maximum at 2% (w/v).

The ability of those detergents which most effectively activated the enzyme to solubilize it was investigated (Figure 1). Up to 80% of total protein, but no more than 45% of phosphatidylinositol kinase, was present in the 105,000 g supernatant obtained by centrifugation after treatment of the microsomal pellet with varying concentrations of sodium deoxycholate (Figure 1A). The recovery of enzyme activity in pellet plus supernatant never exceeded 54%, indicating that loss of activity occurred during extraction.

In the presence of 1% Triton X-100, 105% of microsomal enzyme activity and about 50% of microsomal protein were solubilized with a 2-fold enhancement in specific activity. At higher Triton levels, the appearance of enzyme activity in the supernatant declined sharply (Figure 1B). Octylglucoside also proved effective in solubilizing the enzyme (Figure 1C). Protein extraction increased linearly over a detergent concentration range of 10 to 40 mM and reached a maximum of 55%. Total phosphatidylinositol

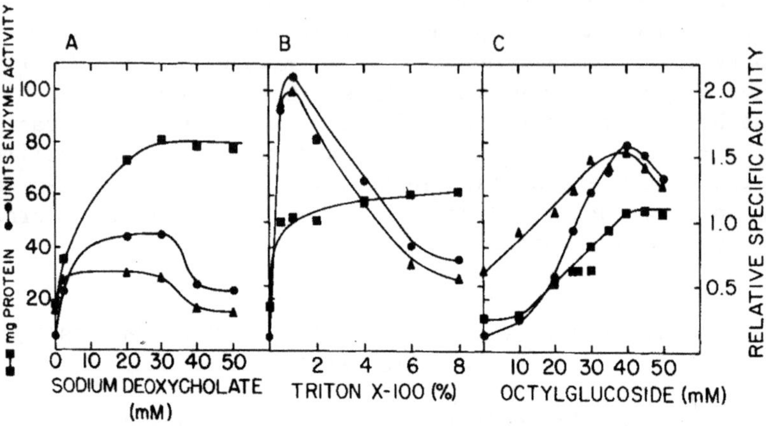

Figure 1. Effect of detergent concentration on the solubilization of microsomal phosphatidylinositol kinase. The enzyme was solubilized from microsomes and assayed as described in the text. (-■-) % total protein. (-●-) % total enzyme activity, (-▲-) relative specific activity (microsomes = 1.0). Each point is the average of triplicate assays. A. sodium deoxycholate; B. Triton X-100; C. octylglucoside (From Bostwick and Eichberg 1981).

kinase activity in the supernatant reached a maximum of 80% at 40 mM octylglucoside. The extraction of enzyme as a function of octylglucoside concentration followed a sigmoidal curve with an inflection point in the region of the critical micelle concentration of the detergent.

SOME PROPERTIES OF PHOSPHATIDYLINOSITOL KINASE

Nature of phosphatidylinositol and kinetic constants

The effect of exogenous phosphatidylinositol on microsomal phosphatidylinositol kinase activity in the absence and presence of 3 mM sodium deoxycholate is shown in Table 1. The enzyme failed to utilize added substrate unless detergent was present, in which case the activity rate was increased 7-fold as compared to the rate when only endogenous substrate was available.

TABLE 1
EFFECT OF EXOGENOUS PHOSPHATIDYLINOSITOL ON MICROSOMAL PHOSPHATIDYLINOSITOL KINASE IN THE ABSENCE AND PRESENCE OF SODIUM DEOXYCHOLATE

PhI Added	Sodium Deoxycholate	
	None	3.0 mM
	nmoles PhIP formed/mg protein/minute	
None	1.4	0.7
1.0 mM	1.3	4.8

The enzyme assay mixture contained 50 mM Tris buffer, pH 8.3, 30 mM $MgCl_2$ 2 mM EDTA, 0.5 mM dithiothreitol, 5 mM soybean phosphatidylinositol, 5 mM γ-[^{32}P]-ATP and 100 μg microsomal protein in a final volume of 0.2 ml. Incubations were performed for 5 minutes at 37°. Preparation of substrates and quantification of product were carried out as described elsewhere (Bostwick and Eichberg 1981). PhI: phosphatidylinositol; PhIP: phosphatidylinositol-4-phosphate.

The acyl group composition of exogenous phosphatidylinositol may play an important role in determining the rate of phosphorylation. In these studies, we utilized both brain phosphatidylinositol, which contains mostly arachidonic acid at the 2-position of the glycerol backbone, and soybean phosphatidylinositol which has mainly linoleic acid at this position and hence is substantially more saturated. Prior to detergent solubilization of microsomal phosphatidylinositol kinase, the apparent K_m for both brain and soybean phosphatidylinositol was 0.4 mM and the apparent V_{max} values were 6.1 and 6.8 nmoles min^{-1} mg $protein^{-1}$ respectively. Following solubilization in 1% Triton X-100, the enzyme exhibited the same affinity for brain phosphatidylinositol, but the V_{max} was lowered to 2.6 nmoles min^{-1} mg $protein^{-1}$. For soybean phosphatidylinositol, the apparent K_m increased to 1.9 mM and the V_{max} fell to 2.9 nmoles min^{-1} mg $protein^{-1}$. The solubilized enzyme displayed an apparent K_m for ATP of 0.7 mM and a V_{max} of 5.8 nmoles min^{-1} mg $protein^{-1}$.

TABLE 2
EFFECT OF CYCLITOLS ON PHOSPHATIDYLINOSITOL KINASE ACTIVITY

Compound	Concentration(mM)			
	0.5	2.0	2.0	4.0
	% of activity with no additions			
myo-inositol	100	111	-	93
D-*myo*-inositol-1-phosphate	100	86+	86	82*
myo-inositol-2-phosphate	96	82+	71*	64*
myo-inositol hexaphosphate	-	82	86	96
1,2-isopropylidene *myo*-inositol	-	125*	121*	129*
2-*myo*-inosose-2-oxime	-	121*	129*	125*

Assay conditions were identical to those described in Table 1 except that bovine brain phosphatidylinositol (1.0 mM) was used. +$p < 0.05$; *$p < 0.01$ different from activity in absence of added cyclitol.

Inhibition of the reaction.

Although phosphatidylinositol kinase was active in the absence of sulfhydryl protecting reagents, a 30% increase in activity was observed when dithiothreitol was included in all buffers used. In agreement with the findings of Benjamins and Agranoff (1969), p-chloromecuribenzoate and N-ethymaleimide (1.0 mM) abolished enzyme activity. Dithiothreitol did not prevent this inhibition.

Cyclitol derivatives at concentrations ranging from 0.5 to 4.0 times that of phosphatidylinositol were at most weak inhibitors (Table 2). Only inositol-2-phosphate (4.0 mM) produced an appreciable decrease in activity. In contrast, 1,2-isopropylidene-*myo*-inositol and 2-*myo*-inosose appeared to activate phosphatidylinositol kinase slightly.

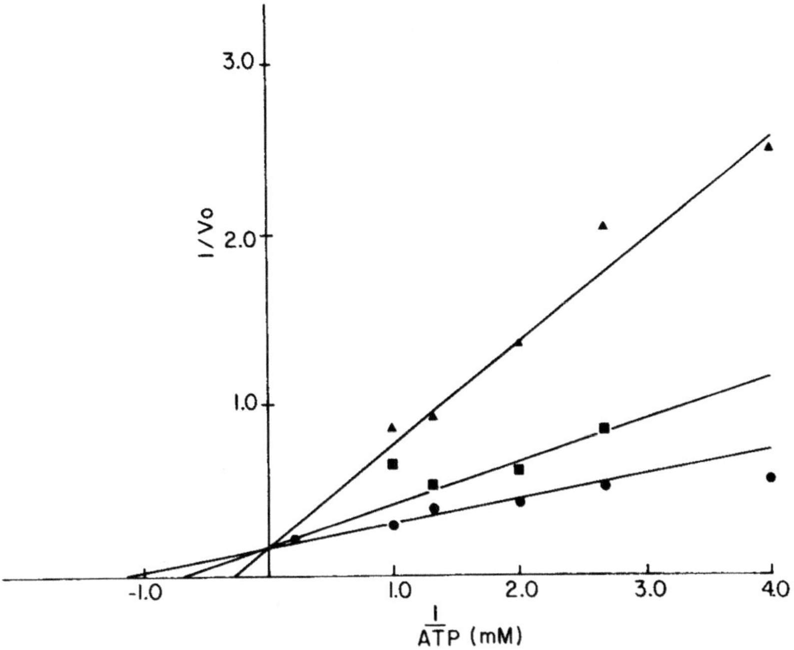

Figure 2: Effect of adenosine on the ATP dependence of phosphatidylinositol kinase

Solubilized brain microsomal enzyme was assayed under standard conditions except that the concentration of ATP was varied (-●-) in the absence of adenosine, (-■-) in the presence of 0.25 mM adenosine, (-▲-) in the presence of 0.50 mM adenosine. Each point represents the average of triplicate values.

Compounds which contain adenosine were moderately effective inhibitors of phosphatidylinositol kinase (Table 3). At concentrations twice that of ATP, ADP and 2', 3'-cyclic AMP inhibited enzyme activity up to 65% whereas 3',5'-cyclic AMP did not inhibit activity nearly as well and 5'-AMP was even less effective. Adenosine itself reduced enzyme activity to 64% of control at a concentration of 2.5 mM in the presence of 5 mM ATP. The insolubility of adenosine prevented the use of higher concentrations. However, when the ATP concentration was varied in the presence of 0.25 mM or 0.5 mM adenosine, double reciprocal plots of velocity versus ATP concentration showed competitive inhibition by adenosine with

TABLE 3
EFFECT OF ADENINE CONTAINING COMPOUNDS ON PHOSPHATIDYLINOSITOL KINASE ACTIVITY

Compound	Concentration(mM)		
	2.5	4.0	10.0
	% of activity with no additions		
Adenosine	64*	-	-
5'-AMP	87	74+	73+
2',3'-cyclic AMP	94	57*	37*
3',5'-AMP	69*	66*	66*
ADP	71*	53*	35*

Assay conditions were the same as those described in Table 1.
*$p < 0.001$; $^+p < 0.001$.

an apparent K_i of approximately 0.2 mM (Figure 2).

In view of the inhibition by adenosine, the effects of an ATP analog, 5'-p-fluorosulfonylbenzoyladenosine (FSBA), were investigated. This compound has been used as an irreversible affinity label at the adenine nucleotide binding sites of a number of enzymes where the sulfonyl fluoride moiety is readily able to form covalent bonds with amino acids such as lysine and tyrosine (Colman et al 1977). Since FSBA is not soluble in water, it was necessary to use dimethylformamide to dissolve the compound at concentrations high enough to have an inhibitory effect. Table 4 shows the results obtained after preincubating solubilized enzyme with and without added FSBA.

After a 15 minute preincubation, dimethylformamide alone at concentrations of 7% and 14% reduced enzyme activity by 29% and 46% respectively. However, inclusion of FSBA at concentrations of 10 and 20 mM in the preincubation mixture resulted in additional losses of activity to 26% and 3% of control, respectively. A longer preincubation did not increase the inhibition. The action of FSBA was not affected by the presence of Mg^{2+} during the preincubation. Competition experiments using ATP in an

TABLE 4
INHIBITION OF PHOSPHATIDYLINOSITOL KINASE ACTIVITY BY 5'-p-FLUOROSULFONYLBENZOYLADENOSINE

Preincubation Condition	15 minutes preincubation		30 minutes preincubation	
	^{32}P incor-poration into PhIP	Enzyme activity	^{32}P incor-poration into PhIP	Enzyme activit
	cpm	%	cpm	%
No additions	4675	100	3705	100
7% DMF	3651	71	2289	62
10mM FSBA in 7% DMF	1204	26	562	15
14% DMF	2506	54	1272	34
20mM FSBA in 14% DMF	133	3	123	3

Triton X-100 solubilized enzyme was preincubated at 0° in the presence of the appropriate amounts of DMF or FSBA in DMF for the times indicated. An aliquot corresponding to 100 µg protein was then withdrawn for enzyme assay under conditions described in Table 1. DMF: dimethylformamide

effort to block the inhibition by FSBA were tried, but these were inconclusive because ATP itself inhibited activity at the necessarily high concentrations of nucleotide used. Hence, although covalent binding of FSBA at the active site of phosphatidylinositol kinase seems likely, as occurs with other nucleotide-requiring enzymes, conclusive proof that the inhibitor reacted specifically in this manner must await further experimentation.

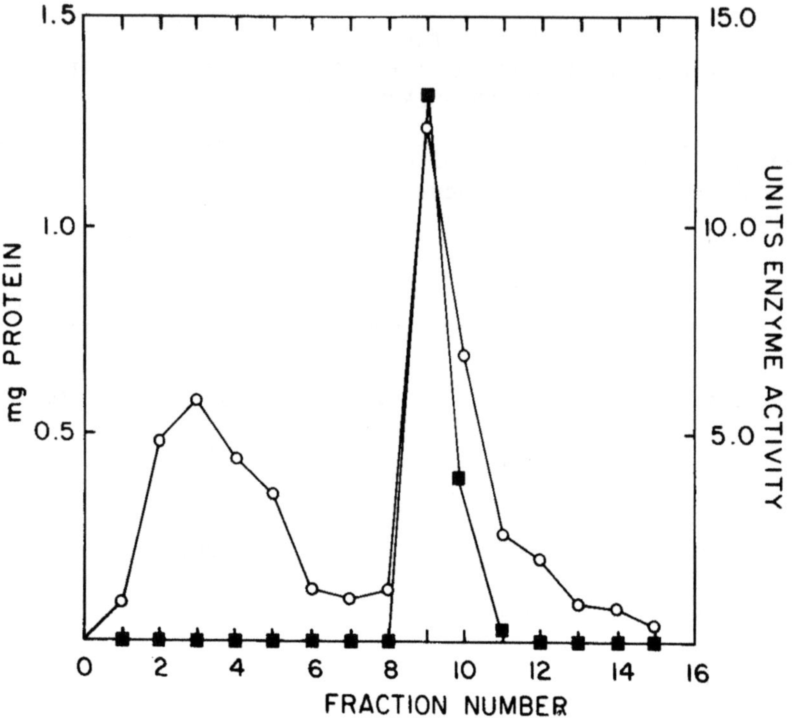

Figure 3. Chromatography of Triton X-100 solubilized phosphatidylinositol kinase on uncharged dodecyl-Sepharose 4B. Solubilized enzyme (1.5 ml containing 6.3 mg protein and 22.1 units enzyme activity) was loaded onto a 7.0 ml resin bed volume. The first 5 fractions (3.0 ml each) were eluted with buffer. Elution with 10 mM Tris buffer containing 8 mM sodium deoxycholate and 0.5 mM dithiothreitol began with fraction 6 and fraction volumes were reduced to 1.5 ml. The unit of enzyme activity protein eluted is expressed as amount per fraction (-0-) protein; (-■-) units enzyme activity one unit is defined as 1 nmole product formed per minute (From Bostwick and Eichberg 1981).

PURIFICATION OF PHOSPHATIDYLINOSITOL KINASE

Hydrophobic chromatography

In initial experiments, solubilized phosphatidylinositol kinase was found not to be adsorbed on ATP-Sepharose affinity columns and could not be purified on hydrophobic resins synthesized using cyanogen bromide and containing charged sites. However solubilized enzyme could be adsorbed on an uncharged hydrophobic resin in which dodecyl chains were attached to Sepharose 4B by ether bonds and eluted in a characteristic and reproducible pattern (Figure 3; Bostwick and Eichberg 1981). When solubilized preparations in 1% Triton X-100 were applied to a lightly substituted resin (16 nmole alkyl chains/mole galactose), from 30-45% of protein and at most a small amount of enzyme activity passed through the column. Following a wash to remove Triton X-100, elution in the presence of 8 mM sodium deoxycholate yielded a sharp peak which contained 35-40% of protein and from 44-85% of enzyme activity loaded on the column. A 2- to 3-fold increase in specific activity was obtained in the most active column fraction. Further elution of the column with up to 100 mM sodium deoxycholate failed to elute additional protein or enzyme.

Inositol affinity chromatography

Inositol-Sepharose 4B containing 3.5 mM cyclitol in ester linkage per ml of packed resin was prepared (Bostwick and Eichberg, unpublished experiments). In preliminary experiments, phosphatidylinositol kinase, extracted from microsomes with 40 mM octylglucoside, was applied to an inositol column in the presence of ATP and Mg^{2+}. A previous column run had established that little protein or enzyme activity was recovered if detergent was absent or in low concentration in the eluting Tris buffer, probably because of strong hydrophobic interactions between protein and the hydrocarbon spacer arms linking the inositol to the column matrix. When the column was washed with Tris buffer containing 8 mM deoxycholate, 1 mM ATP and 1 mM Mg^{2+}, approximately 60% of the enzyme and 95% of the protein was recovered. Continued elution with buffer which contained detergent, but no ATP or Mg^{2+}, failed to yield more enzyme. When the column was then eluted with the deoxycholate containing buffer in which 1.8 mM inositol was present, another 37% of the phosphatidylinositol kinase activity was obtained purified at least 25-fold.

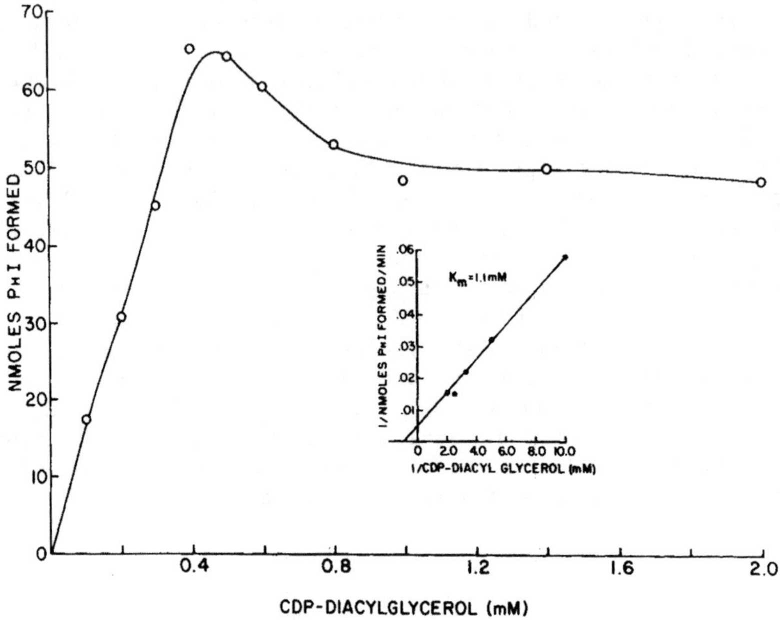

Figure 4. Effect of CDP-diacylglycerol concentration on the activity of detergent-solubilized phosphatidylinositol synthetase. The enzyme assay mixture contained 1.0 mM CDP-diacylglycerol, 50 mM glycylglycine buffer, pH 8.6, 48 mM $MgCl_2$, 0.6% Triton-X-100, 6.0 mM [^3H] myo-inositol (ca. 50,000 cpm) and 0.3 mg microsmal protein in a final volume of 0.2 ml. Incubations were performed for 8 minutes at 37°. Lipids were extracted and chloroform-soluble radioactivity was measured.

SOLUBILIZATION AND PROPERTIES OF PHOSPHATIDYLINOSITOL SYNTHETASE

Phosphatidylinositol synthetase from several mammalian tissues, as well as from a number of other eukaryotic organisms, has been well described (Paulus and Kennedy 1960; Benjamins and Agranoff 1969; Takenawa and Egami 1977; Sexton and Moore 1978; Carman and Feder 1980; Carman and Matas 1981). Earlier efforts at solubilizing the enzyme have shown the most effective detergents are

TABLE 5
EFFECT OF INOSITOL DERIVATIVES ON PHOSPHATIDYLINOSITOL SYNTHETASE ACTIVITY

Derivative	Concentration	Activity
	mM	%
None	-	100
epi-inosose	1.0	89
myo-inosose	1.0	89
scyllo-inositol	1.0	101
myo-inositol-2-oxime	1.0	98
1,2-O,C-methylene *myo*-inositol	1.0	80
	2.0	78
1,2-isopropylidene *myo*-inositol	1.0	85
	2.0	79
	4.0	60

Incubations were performed as described in the Legend to Figure 4 using solubilized enzyme and in the presence of 2.0 mM *myo*-inositol.

Renex 690, Cutscum and Triton-X-100 (Bishop and Strickland 1970; Rao and Strickland 1974; Takenawa *et al* 1977; Carman and Matas 1981). In our hands, treatment of rat brain microsomes with either 0.3% or 0.6% Triton-X-100 solubilized practically all phosphatidylinositol synthetase activity together with about 70% of microsomal protein. Since the activation of the enzyme was near maximal at 0.6% Triton-X-100, this concentration was adopted for the solubilization procedure.

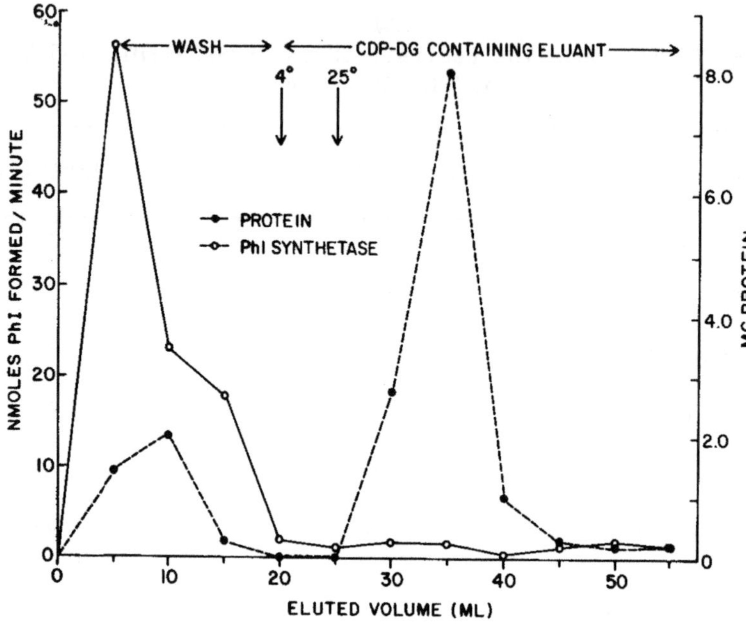

Figure 5 Purification of phosphatidylinositol synthetase by CDP-diacylglycerol affinity chromatography. Solubilized brain microsomal protein was applied to a column of CDP-diacylglycerol-Sepharose 4B. The column was washed and phosphatidylinositol synthetase was eluted as described in the text. Enzyme activity was assayed as described in the Legend to Figure 4.

The solubilized enzyme displayed optimal activity with a CDP-diacylglycerol concentration of 0.5 mM, whereas higher concentrations were somewhat inhibitory (Figure 4). The apparent K_m for the liponucleotide was 1.1 mM. The optimal myo-inositol concentration was 4 to 6 mM and the enzyme displayed an apparent K_m of 1.8 mM for this substrate. These kinetic constants are in generally good agreement with the results of other studies on mammalian phosphatidylinositol synthetase, and indicate a lower affinity for both substrates than is exhibited by the enzyme from Saccharomyces cerevisiae (Carman and Matas 1981).

Other workers have reported the inability of many inositol and cyclitol derivatives to affect the activity of phosphatidylinositol synthetase (Benjamins and Agranoff 1969; Takenawa et al 1977). We have tested additional compounds and found that both 1, 2-O,C-methylene-myo-inositol and 1,2 isopropylidene myo-inositol partially inhibited the reaction (Table 5). The slight effect observed with inosose derivatives is consistent with previous suggestions that these compounds can serve as poor substrates in the reaction (Paulus and Kennedy 1960; Benjamins and Agranoff 1969).

PURIFICATION OF PHOSPHATIDYLINOSITOL SYNTHETASE BY CDP-DIACYLGLYCEROL AFFINITY COLUMN CHROMATOGRAPHY

CDP-diacylglycerol was chemically synthesized according to published procedures (Agranoff and Suomi 1963; Dowhan and Larson 1981) starting with egg phosphatidylcholine-derived phosphatidic acid. The product was covalently linked to Sepharose 4B through the ribose moiety as described by Larson et al (1976). Solubilized phosphatidylinositol synthetase was adsorbed onto CDP-diacylglycerol-Sepharose 4B and the column washed at 4° with 50 mM glycylglycine buffer, pH 8.6, containing 0.3% Triton-X-100. The enzyme was then eluted at 25° using the same buffer containing in addition 1 mM CDP-diacylglycerol and 10 μg P/ml Asolectin (Figure 5).

This procedure yielded an enzyme preparation purified nearly 100-fold in an overall yield of 3-4% from brain homogenate (Table 6). However, based on the amount of enzyme activity solubilized from microsomes, 34% of the enzyme applied to the column was recovered. More extensive work has shown that from 50 to 70% of solubilized enzyme activity can be recovered by modification of the elution procedure. We have also obtained enzyme with substantially higher specific activity, probably due at least in part to an improved means of protein determination which caused measurements of the extent of enzyme purification in earlier experiments most likely to be underestimated.

TABLE 6

PURIFICATION OF RAT BRAIN PHOSPHATIDYLINOSITOL SYNTHETASE

Fraction	Total Enzyme Units	Total Protein	Specific Activity	Fold Purification	% Yield
	nmole PhI formed/min	mg	nmole/min/ mg protein		
Rat brain homogenate	1546	537	2.9	1.0	100
Microsomal Triton X-100 extract	236	21.5	11.0	3.7	5
CDP-diacylglycerol affinity column:					
Void volume + wash	25	15.1	1.6	–	2
CDP-diacylglycerol elution at 25°	54	0.2	270.0	93	4

*Another less highly purified fraction was eluted preceding the major peak and accounted for 10% of activity applied to the column.

Purified phosphatidylinositol synthetase revealed three Coomassie Blue positive bands when examined by polyacrylamide gradient slab gel electrophoresis. The preparation had an apparent K_m for myo-inositol of 4.6 mM. The presence of CDP-diacylglycerol in the eluting buffer made it difficult to obtain accurate kinetic measurements for this substrate, but the reaction rate increased up to 0.4 mM CDP-diacylglycerol and decreased somewhat at higher concentrations. The rate of phosphatidylinositol formation was linear for 10 minutes and the optimal pH for the reaction was 8.5-9.0, although appreciable activity was evident at physiological pH. When the quantity of $[^3H]$-inositol converted to lipid product was compared to the amount of $[^3H]$-CDP-diacylglycerol transformed into water soluble product, a ratio of 1.00 to 1.13 was obtained, which was close to the expected stoichiometry in the reaction. The synthesis of phosphatidylinositol was absolutely dependent on the presence of Mg^{2+} and optimal activity was achieved with 10 mM cation. Substitution of Mn^{2+} for Mg^{2+} reduced the reaction rate by 75%. The inclusion of EGTA in the assay approximately doubled the reaction rate.

The purified phosphatidylinositol synthetase preparation was also examined for the presence of contaminating enzyme which utilize CDP-diacylglycerol. Using the assay conditions of Possmayer et al (1973), no CDP-diacylglycerol: sn-3-glycerophosphate phosphatidyl transferase activity was detected, and at most a low activity of CDP-diacylglycerol hydrolase was manifested when evidence for this enzyme was sought according to the conditions of Rittenhouse et al (1981).

DISCUSSION

The results described in this paper support findings of previous investigators concerning the solubilization of phosphatidylinositol kinase and phosphatidylinositol synthetase. Our observations also illustrate the usefulness of hydrophobic interaction chromatography and affinity chromatography for the purification of these phosphoinositide metabolizing enzymes.

Complete solubilization of phosphatidylinositol kinase activity from microsomes occurred in the presence of 1% Triton-X-100, whereas 80% solubilization of enzyme activity was achieved using 40 mM octylglucoside. In previous investigations, Cutscum, another polyoxyethylene detergent, was reported to solubilize the enzyme partially (Harwood and Hawthorne 1969; Tou et al 1969). Maximal solubilization by by Triton-X-100 and octylglucoside took place at similar detergent; protein ratios on a weight: weight basis (approximately 20:1), but at quite dissimilar molar ratios. Thus the mass of detergent rather than the number of detergent molecules is important in assuring a sufficiently hydrophobic environment to bring about the extraction of phosphatidylinositol kinase. It seems likely that the solubilized enzyme exists in micellar form with Triton-X-100 and possibly other membrane components.

The solubilization curve for octylglucoside was sigmoidal with an inflection point at the critical micelle concentration for this detergent. Thus it may be postulated that there are two population of enzyme molecules which may differ in the strength of their hydrophobic interactions with other membrane components. The change in K_M and V_{max} for phosphatidylinositol following solubilization could result from an alteration in enzyme conformation as a result of association with detergent.

In confirmation of earlier studies, phosphatidylinositol kinase was shown to be inhibited by sulfhydryl reagents. Several cyclitol derivatives were not effective inhibitors of the enzyme, suggesting that the inositol moiety of phosphatidylinositol is not primarily involved in binding of substrate to the enzyme. The reduction of enzyme activity by adenine-containing compounds and especially the competitive inhibition by adenosine may indicate that this portion of the ATP molecule binds to the kinase. Buckley (1977) observed similar inhibitory effects of adenine-containing substances on detergent solubilized phosphatidylinositol kinase from human erythrocytes. Moreover, the blockade of activity by ADP and adenosine is compatible with the concept of Talwalker and Lester (1974) that polyphosphoinositide metabolism may be regulated by the adenylate energy charge within the cell. The striking inhibition by FSBA points to the potential usefulness of this reagent as an affinity label for the purified enzyme and appears to implicate the involvement of sites other than sulfhydryl groups in the catalytic mechanism.

Since extracted microsomal membrane constituents are capable of strong mutual hydrophobic interactions, hydrophobic chromatography was tried as a step to dissociate phosphatidylinositol kinase from these macromolecular complexes. Homcy et al (1977, 1978), using uncharged alkyl-Sepharose resins, earlier demonstrated the value of this chromatographic technique for the purification of adenylate cyclase from more hydrophobic proteins. In our hands, the bulk of phosphatidylinositol kinase was bound to uncharged lightly substituted dodecyl Sepharose, retained during removal of excess Triton-X-100 and largely eluted with sodium deoxycholate. Thus the adsorbent behaved as a solid phase detergent and its properties enabled us to separate phosphatidylinositol kinase from proteins of differing hydrophobicity. While use of the column yielded only a 2-3 fold purification of the enzyme, we were able to change the nature of the bulk detergent associated with the enzyme, and this procedure may find more general application in devising optimal conditions for purification of other membrane-bound enzymes following solubilization. Phosphatidylinositol kinase eluted from the hydrophobic column still failed to bind to an ATP affinity resin, but the partial purification obtained using an inositol affinity column suggests that this method may be combined with hydrophobic chromatography to effect appreciable purification of the enzyme.

In contrast to our experience with phosphatidylinositol kinase, solubilized phosphatidylinositol synthetase was substantially purified by a simple and rapid procedure employing CDP-diacylglycerol affinity chromatography. The recovery of active enzyme required the presence of phospholipid throughout the procedure, presumably to maintain the molecule in a relatively native hydrophobic environment. In general, the ion requirements and pH optimum for purified phosphatidylinositol synthetase resembled those of the crude microsomal enzyme. However, the purified preparation displayed a higher K_m for inositol than did solubilized enzyme, such that a modest drop in the level of myo-inositol generally prevailing in tissues could result in a significant decrease in phosphatidylinositol biosynthesis. Indeed a 25% reduction in peripheral nerve myo-inositol concentration in diabetic neuropathy has been reported to be associated with diminished phosphatidylinositol synthetase activity (Greene et al 1975; Whiting et al 1979) and this may be partially responsible for depressed phosphatidylinositol levels in the disorder (Palmano et al 1977; Natarajan et al 1981).

The availability of purified phosphatidylinositol synthetase should considerably facilitate biochemical and cytochemical studies of this enzyme. In particular, the identification of an inhibitor for the enzyme should prove useful in studies of the mechanism which underlies stimulated phosphoinositide metabolism.

Acknowledgements: This work was supported by NIH grant NSI2493 and by grant E-675 from the Robert A. Welch Foundation.

REFERENCES

Agranoff, B. W. and Suomi, W. D. (1963) Cytidine diphosphate-DL-dipalmitin. Biochem. Prep. 10, 47-51.

Benjamins, J. A. and Agranoff, B. W. (1969) Distribution and properties of CDP-diglyceride: inositol transferease from brain. J. Neurochem. 16, 513-517.

Bishop, H. H. and Strickland, K. P. (1970) On the specificity of cytidine diphosphate diglycerides in monophosphoinositide biosynthesis by rat brain preparations. Canad. J. Biochem. 48, 269-277.

Bostwick, J. R. and Eichberg, J. (1981) Detergent solubilization and hydrophobic chromatography of rat brain phosphatidylinositol kinase. Neurochem. Res. 6, 1053-1065.

Buckley, J. T. (1977) Properties of human erythrocyte phosphatidylinositol kinase and inhibition by adenosine, ADP and related compounds. Biochim. Biophys. Acta. 498, 1-9.

Carman, G. M. and Matas, J. (1981) Solubilization of microsomal associated phosphatidylserine synthetase and phosphatidylinositol synthase from Saccharomyces cerevisiae. Canad. J. Microbiol. 27, 1140-1149.

Colman, R. F., Pal, P. K. and Wyatt, J. (1977) Adenosine derivatives for dehydrogenases and kinases. Methods Enzymol. 46, 240-249.

Dowhan, W. and Larson, T. (1981) Phosphatidylserine synthase from Escherchia coli. Methods in Enzymol. 71, 561-571.

Fisher, S. K. and Agranoff, B. W. (1981) Enhancement of the muscarinic synaptosomal phospholipid labeling effect by the ionophor A23187. J. Neurochem. 37, 968-977.

Fisher, S. K., Frey, K. A. and Agranoff, B. W. (1981) Loss of muscarinic receptors and of stimulated phospholipid labeling in ibotenate-treated hippocampus. J. Neurosci 1, 1407-1413.

Greene, D. A., DeJesus, P. V. and Winegrad, A. I. (1975) Effects of insulin and dietary myo-inositol on impaired peripheral nerve motor nerve conduction velocities in acute streptozotocin diabetes. J. Clin, Invest. 55, 1326-1336.

Harwood, J. L. and Hawthorne, J. N. (1969) The properties and subcellular distribution of phosphatidylinositol kinase in mammalian tissues. Biochim. Biophys. Acta 171, 75-88.

Hauser, G. (1983) The involvement of lipids in signal transmission across cell membranes of the nervous system, this volume.

Hoffman, S. L. and Majerus, P. W. (1982) Identification and properties of two distinct phosphatidylinositol specific phospholipase C enzymes from sheep seminal vesicular glands. J. Biol Chem. 257, 6461-6469.

Homcy, C. J., Wrenn, S. M. and Haber, E. (1977) Demonstration of the hydrophobic character of adenylate cyclase following hydrophobic resolution on immobilized alkyl residues. J. Biol. Chem. 252, 8957-8964.

Homcy, C. J., Wrenn, S. and Haber, E. (1978) Affinity purification of cardiac adenylate cyclase. Dependence on prior hydrophobic resolution. Proc. Nat. Acad. Sci. USA 75, 59-63.

Kai, M., White, G. L. and Hawthorne, J. N. (1966) The phosphatidylinositol kinase of rat brain. Biochem. J. 101, 328-377.

Killian P. and Schacht, J. (1980) Sound stimulates labeling of polyphosphoinositides in the auditory organ of the noctuid moth. J. Neurochem. 34, 709-712.

Larson, T. J., Hirabayashi, T. and Dowhan, W. (1976) Phosphatidylglycerol biosynthesis in Bacillus licheniformis. Resolution of membrane-bound enzymes by affinity chromatography on cytidine diphospho-sn-1,2-diacyglycerol Sepharose. Biochemistry 15, 974-980.

Low, M. B. and Finean, J. B. (1976) The action of phosphatidylinositol-specific phospholipase C on membranes. Biochem. J. 154, 203-208.

Natarajan, V., Dyck, P. J. and Schmid, H. H. O. (1981) Alterations of inositol lipid metabolism of rat sciatic nerve in streptozotocin-induced diabetes. J. Neurochem. 36, 413-419.

Nijjar, M. S. and Hawthorne, J. N. (1977) Purification and properties of polyphosphoinositide phosphomonoesterase. Biochim. Biophys. Acta. 480, 390-402.

Palmano, K. P., Whiting, P. H. and Hawthorne, J. N. (1977) Free and lipid myo-inositol in tissues from rats with acute and less severe streptozotocin-induced diabetes. Biochem. J. 167, 229-235.

Paulus, H. and Kennedy, E. P. (1960) The enzymatic synthesis of inositol monophosphatide. J. Biol. Chem. 235, 1303-1311.

Possmayer, F., Meiners, B. and Mudd, J. B. (1973) Regulation by cytidine nucleotides of acylation of sn-C^{14}-glycerol-3-phosphate: regional and subcellular distribution of enzymes responsible for phosphatidic acid synthesis de novo in the central nervous system of the rat. Biochem. J. 132, 381-388.

Rao, R. H. and Strickland, K. P. (1974) On the solubility, stability and partial purification of CDP diacyl-sn-glycerol: inositol transferase from rat brain. Biochim. Biophys. Acta 348, 306-314.

Rittenhouse, H., Seguin, E. G., Fisher, S. K. and Agranoff, B. W. (1981) Properties of CDP-diglyceride hydrolase from guinea pig brain. J. Neurochem. 36, 991-999.

Robinson, M. L. and Carman, G. M. (1982) Solubilization of microsomal-associated phosphatidylinositol synthase from germinating soybeans. Plant Physiol. 69, 146-149.

Schacht, J. and Agranoff, B. W. (1972) Effect of acetylcholine on labeling of phosphatidate and phosphoinositides by ^{32}P orthophosphate in nerve ending fractions of guinea pig cortex. J. Biol. Chem. 247, 771-777.

Sexton, J. C. and Moore, J. S. (1978) Phosphatidylinositol synthesis in castor bean endosperm. Plant Physiol. 62, 978-980.

Shaikh, N. A. and Palmer, F. B. St. C. (1977) Phosphoinositide kinases in chick brain and sciatic nerve: a developmental study. J. Neurochem. 28, 395-402.

Sundler, R., Alberts, A. W. and Vagelos, P. R. (1978) Enzymatic properties of phosphatidylinositol inositolphosphohydrolase from *Bacillus cereus*. J. Biol Chem. 253, 4175-4179.

Takenawa, T. and Egami, K. (1977) CDP-diglyceride: inositol transferase from rat liver: purification and properties. J. Biol Chem. 252, 5419-5413.

Takenawa, T. and Nagai, Y. (1981) Purification of phosphatidylinositol specific phospholipase-C from rat liver. J. Biol. Chem. 256, 6769-6775.

Talwalker, R. T. and Lester, R. L. (1973) The response of diphosphoinositide and triphosphoinositide to perturbations of the adenylate energy charge in cells of *Saccharomyces cerevisiae*. Biochim. Biophys. Acta. 306, 412-421.

Tou, J-S., Hurst, M. W. and Huggins, C. B. (1969) Phosphatidylinositol kinase in rat kidney cortex: subcellular distribution and kinetic properties. Arch. Biochem. Biophys. 131, 596-602.

White, G. L. Schellhase, U. and Hawthorne, J. N. (1974) Phosphoinositide metabolism in rat superior cervical ganglion, vagus and phrenic nerve: effects of electrical stimulation and various blocking agents J. Neurochem. 22, 149-158.

Whiting, P. H., Palmano, K. P. and Hawthorne, J. N. (1979) Enzymes of *myo*-inositol and inositol lipid metabolism in rats with streptozotocin-induced diabetes. Biochem. J. 179, 549-553.

Yagihara, Y. and Hawthorne, J. N. (1972) Effects of acetylcholine on the incorporation of $^{32}P_i$ *in vitro* into phospholipids of nerve ending particles from guinea pig brain. J. Neurochem. 19, 355-367.

Part II: Receptors

DEMONSTRATION OF GABA-SYNTHESIZING ENZYME, GAD, BENZODIAZEPINE RECEPTOR AND COATED VESICLES IN THE VERTEBRATE NERVOUS SYSTEM

Jang-Yen Wu[1,3], Ching-Tarng Lin[1], and Y.Y.T Su[2]

Dept. of Cell Biology[1] Cullen Eye Institute[2] &
Program in Neuroscience[3]

Baylor College of Medicine, Texas Medical Center
Houston, TX 77030

INTRODUCTION

In vertebrate central nervous system (CNS), the communication between neurons is conducted virtually via chemical transmission. At least three components, namely transmitter enzymes for transmitter synthesis and degradation, receptors for binding to transmitters and modulators and coated vesicles for the recycling of synaptic vesicle membrane and transport of various substances are involved in chemical synapses. This review discusses mainly the progress to date dealing with γ-aminobutyric acid (GABA) including its synthetic enzyme, L-glutamate decarboxylase (GAD), and receptor for benzodiazepine (BDZP), a group of substances that facilitate the action of GABA. In addition, the site of synthesis and distribution of coated vesicles protein, clathrin, in mammalian brain is also discussed.

IDENTIFICATION OF GABAERGIC NEURONS AND PATHWAYS

Evidence from physiological studies has established that GABA is an inhibitory neurotransmitter at the crustacean neuromuscular junction and that it probably also is the major inhibitory neurotransmitter in the vertebrate CNS (Florey, 1957; Kravitz, 1967; Krnjevic; 1974). One of the important criteria of a neurotransmitter is the presence of the substance and its synthesizing enzyme in the nerve terminal of the suspected neuron. The approach that we have been using in the identification of GABA-ergic neurons and pathways in the vertebrate CNS is to purify the GABA-synthesizing enzyme, GAD, to homogeneity, to prepare monospecific antibody against GAD and to visualize its cellular and subcellular locations by indirect immunocytochemical techniques. The rationale for our approach is that there is a good correlation between GABA levels and GAD activity in the nervous tissues. (Kravitz and Potter, 1965; Kuriyama et al., 1966; Baxter, 1970). Hence GAD is a better marker for GABA-ergic neurons than GABA per se, which may redistribute or be metabolized during the preparation of the tissue (Baxter, 1970; Fonnum, 1975).

In the last few years, we have purified GAD to homogeneity from mouse brain (Wu et al., 1973; Wu, 1976; Wu et al., 1981b), catfish brain (Su et al., 1979; Wu et al., 1981) and bovine brain (Wu, 1982a) The purity of GAD preparations was established based on the following criteria: (a) 25-59 μg of the purified enzyme migrated as a single protein band with the location of the enzyme activity corresponding to the location of the protein band on polyacrylamide gel electrophoresis (Wu, 1976); (b) the purified mouse brain GAD preparation appeared to be homogeneous in size as judged from the linear plot of the logarithm of concentrations against the squares of the distances from the center of rotation to points of interest in high-speed sedimentation equilibrium runs in H_2O and D_2O solutions and in dissociating conditions, e.g., guanidine HCl and 2-mercapto-ethanol (Wu, 1976); (c) on 3.7 to 15% gradient polyacrylamide gel electrophoresis, the purified catfish enzyme preparation also migrated as a sharp, single protein band which contained all the enzyme activity (Su et al., 1979; Wu et al., 1981) and (d) antibodies against the purified GAD preparations appear to be specific only to GAD, as demonstrated in immunodiffusion and immunoelectrophoresis experiments in which a sharp, single precipitin band was obtained with GAD antisera and crude enzyme preparations (Matsuda et al., 1973; Saito et al., 1974; Wong et al., 1975; Wu, 1976). When antibodies against partially purified GAD were used, a sharp, single band was obtained with the purified GAD preparations, whereas the crude preparations showed multiple bands (Matsuda et al., 1973). Furthermore, the precipitin band also contained GAD activities, while comparable piece of gel did not show any enzyme activity, suggesting that the precipitin band was indeed GAD-anti-GAD complex. Once the purity of GAD preparations had been established, rabbits were immunized with various amounts of purified GAD for the production of polyclonal antibodies. The specificity of polyclonal GAD antibodies was established from extensive immunochemical tests, e.g. immunodiffusion, immunoelectrophoresis, enzyme inhibition test, microcomplement fixation test and enzyme immunoassay (Saito et al., 1974; Wong et al., 1975; Wu 1982b; Wu et al., 1982a). Recently, we have employed the hybridoma technique originally developed by Köhler and Milstein (1975, 1976) for the production of monoclonal antibody against GAD (Wu et al., 1982a,b). The well-characterized and specific GAD antibodies have been used extensively in our own laboratoy as well as others for the identification of GABA-ergic neurons and their pathways. GAD has been shown immunocytochemically to be present in the suspected GABA-ergic neurons and their terminals such as Purkinje cells and basket cells in cerebellum (Saito et al., 1974; McLaughlin et al., 1974, 1975b; Chan-Palay et al., 1979, 1981) amacrine cells in retina (Brandon et al., 1979, 1980; Wu et al., 1981a, 1982a), horizontal cells in goldfish retina (Lam et al., 1979; Wu et al., 1981a) neurons projected from stria medullaris to lateral nuclei of habenula (Gottesfeld et al., 1980, 1981; Wu et al., 1982a) and GABA-ergic neurons in spinal cord (McLaughlin et al., 1975; Hunt et al., 1981), striate cortex (Hendrickson et al., 1981), hypothalamus and the pituitary gland (Vincent et al., 1982a), dentate gyrus (Goldowitz et al., 1982) and pancreas (Vincent et al., 1982b).

CHARACTERIZATION OF BENZODIAZEPINE SYSTEM

Benzodiazepines (BDZP) are the most widely prescribed psychoactive drugs in current therapeutic use. A variety of neurotransmitter systems (Guidotti, 1978) including acetylcholine,lcatecholamines (Fuxe et al., 1975), serotonin (Cook and Sepinwall, 1975; Stein et al., 1975), glycine (Snyder and Enna, 1975) and GABA (Costa and Guidotti, 1979; Guidotti, et al., 1979; Olsen, 1981), have been suggested in the anxiolytic, muscle relaxation and anticonvulsant action of BDZP. Recently, a possible interaction with the GABA system has received most attention. The following observations support the concept that most of the neuropharmacological and neurochemical actions of the BDZP may be mediated via a primary action on GABA transmission. First of all, BDZP potentiate the actions of both synaptically released and exogenously administered GABA on mammalian neuronal preparations (Choi, et al., 1977; MacDonald and Barker, 1978); secondly, binding studies suggest that BDZP combine with highly specific sites in the neuronal membrane (Squires and Breastrup, 1977; Möhler and Okada, 1977; Braestrup and Squires, 1978) and these BDZP receptor bindings are selectively enhanced by GABA and GABA-like drugs (Tallman et al., 1978; Wastek et al., 1978; Karobath and Sperk, 1979; Supavilai and Karobath, 1980); thirdly, BDZP stimulate GABA binding by removal of an endogenous inhibitor (Guidotti et al., 1978); fourthly, intravenous administration of BDZP potentiate hippocampal recurrent inhibition (Tsuchiya and Fukushina, 1978) which is believed to be mediated by GABA (McNamara et al., 1980); fifthly, the agonistic action of BDZP on postsynaptic GABAergic mechanism has been demonstrated in a number of brain regions or in cultured neuronal systems (Dray and Straughn, 1976; Kozhechkin and Ostovskaya, 1977; Gallager, 1978) and sixthly, the BDZP receptor has been shown to be in the vicinity of the GABAergic synapse by a combination of autoradiographic localization of BDZP receptor and immunocytochemical localization of GAD in a GABAergic terminal (Möhler et al., 1981)

In this communication, we would like to focus on the isolation and characterization of BDZP receptor and the search for an endogeneous substrate(s) for BDZP receptor.

Isolation and Characterization of Benzodiazepine Receptor

1. Subunit Structure of BDZP Receptor. We took the advantage that BDZP receptor became covalently linked to its ligand, ^{3}H-flunitrazepam (FNZP)) upon UV irradiation and hence the subunit structure of BDZP receptor could be analyzed by SDS-polyacrylamide gel electrophoresis (SDS-PAGE) following the distribution of ^{3}H-FNZP. The photoaffinity labeling of BDZP receptor was carried out according to the procedure of Möhler et al. (1981) with some modifications (Wu et al., 1982b).

The photoaffinity labeled BDZP receptors, [^{3}H] FNZP-BDZP receptor, were treated with 1% SDS at 100°C for 3 min

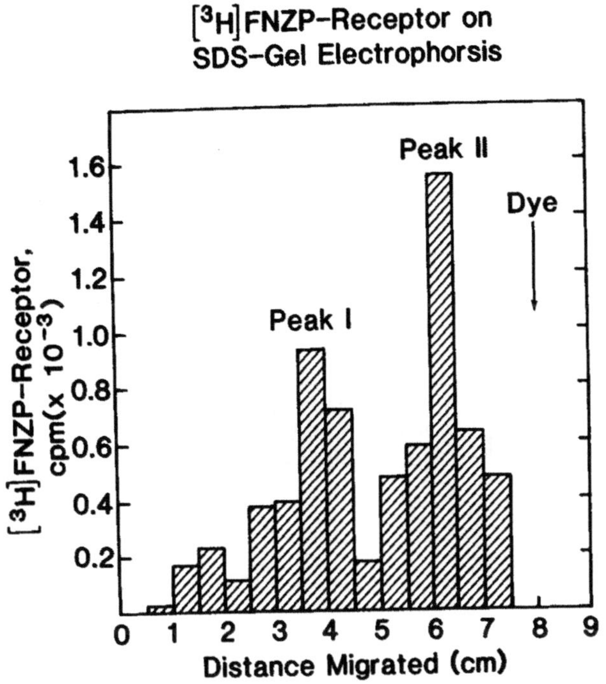

Fig. 1. SDS gel electrophoresis of [^{3}H]FNZP-labeled benzodiazepine receptor. The specific binding expressed as counts per min (cpm) was plotted against the distance of migration on SDS gel in cm.

and then applied to 15% SDS gel. The SDS-PAGE was carried out at 15 mA for 1 hr and 20 mA for an additional 2 hr. Gel was sliced at 0.5 cm intervals and counted for radioactivity. A parallel gel was stained with Coomassie Blue for protein pattern. When the specific binding was plotted against the distance of migration of SDS-PAGE, two peaks, peak I and II, corresponding to a molecular weight of 50,000 ± 5,000 and 12,000 ± 2,000, respectively, were obtained (Fig. 1). Similar results were obtained with SDS gel filtration column chromatography suggesting that BDZP

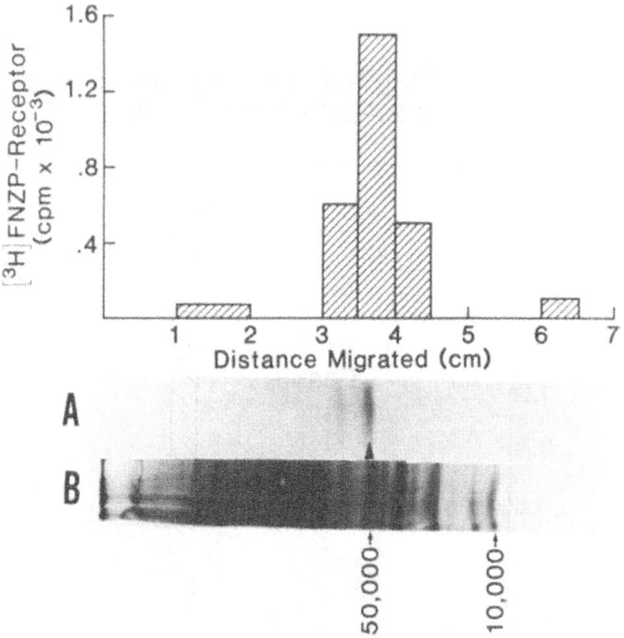

Fig. 2. A: SDS-gel electrophoresis of benzodiazepine receptor subunit. Benzodiazepine receptor subunit, 50,000-dalton, was first purified by preparative SDS-gel electrophoresis as described in text. The arrow head indicates the position of 50,000-dalton subunit.

B: SDS-gel electrophoresis of [^{3}H]-FNZP labeled benzodiazepine receptor. The conditions were the same as those described in Fig. 1. The first arrow indicates the position of 50,000-dalton subunit and the second arrow indicates the position of 12,000-dalton subunit. The distribution of [^{3}H]-FNZP labeled benzodiazepine receptor subunit, 50,000-dalton, expressed as cpm is shown on the top figure.

receptor may consist of two different types of subunit, 50,000 and 12,000 daltons. Peak II, the 12,000-dalton subunit, is probably not the degradation product of peak I, because the native BDZP receptor obtained by solubilization with Triton X-100 eluted as a macromolecule with a molecular weight greater than 200,000 (Fig. 3).

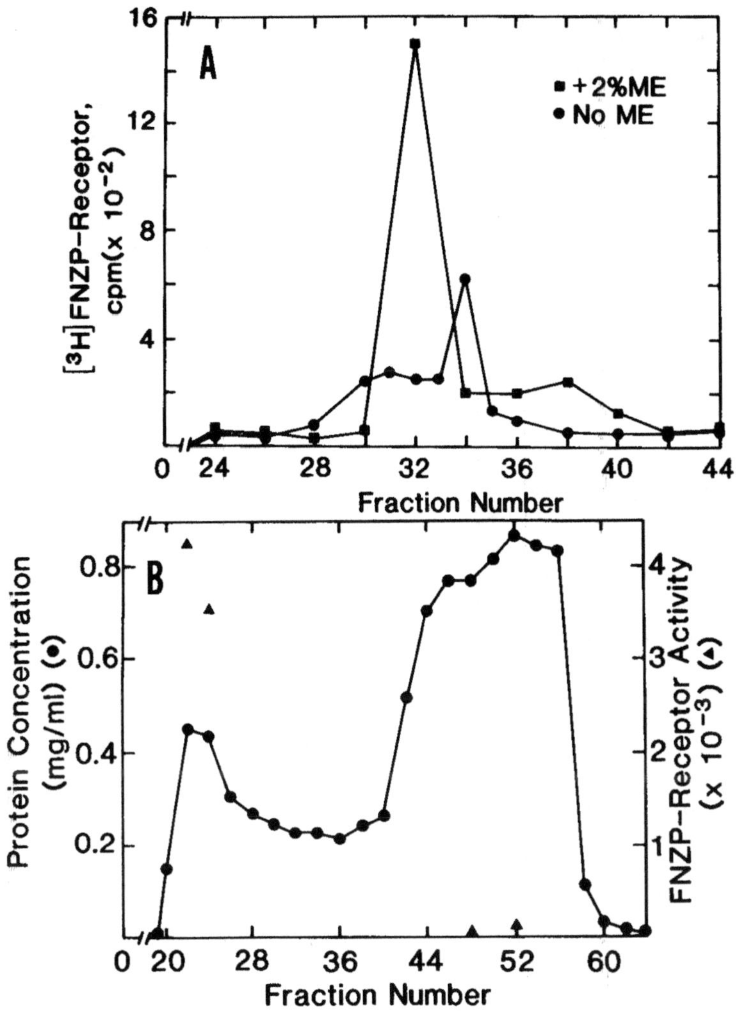

Fig. 3. A: Gel filtration of Triton X-100 solubilized benzodiazepine receptor. 1.2 ml per fraction was collected. 100 μl aliquot was counted.

B: Gel filtration of Triton X-100 solubilized benzodiazepine receptor. [^{3}H]-FNZP was used as ligand in receptor (▲). 1.8 ml per fraction was collected. Protein concentration was expressed as mg/ml (●).

2. Purification of BDZP Receptor Subunit. The 50,000-dalton subunit was purified by preparative SDS-PAGE. The

condition was the same as above except that the gel thickness was increased from 1 mm to about 2.7 mm and the current was increased to 30 mA. The gel strip which contained the 50,000-dalton subunit was frozen overnight at -20°C and extracted by homogenizing the gel in SDS buffer (400 µl/cm^2 gel). Gel was removed by brief centrifugation and the supernatant thus obtained was concentrated by vacuum centrifugation and analyzed on analytical SDS-PAGE as described above. The protein pattern of Coomassie Blue stained gel of the purified 50,000-dalton subunit is shown in Fig. 2A. The protein pattern of the starting material is included for comparison (Fig. 2B). Although it is difficult to estimate the purity of the 50,000-dalton subunit, it is reasonable to conclude that substantial purification has been achieved by SDS-PAGE.

3. Solubilization of BDZP Receptor. The photoaffinity-labeled P_2 fraction was incubated with 10 volumes of 0.2% Triton X-100 in 50 mM Tris-citrate buffer, pH 7.1 at 4°C for 30 min. The supernatant thus obtained was concentrated and applied to a Sephadex G200 column (2 x 30 cm) which had been equilibrated with the same Triton X-100 solution. The column was eluted with the same Triscitrate buffer containing 0.2% Triton X-100 at a flow rate of 12 ml/hr. 1.2 ml per fraction was collected and 100 µl aliquot was counted. The elution profile is shown in Figure 3A. We have also fractionated the unlabeled Triton X-100 solubilized BDZP receptor on the same column. In this case, the position of BDZP receptor was determined by assaying each fraction for the receptor binding activity using [^{3}H]-FNZP as ligand (Fig. 3B). Similar results were obtained with an elution volume of about 38 ml (in Fig. 3B, 1.8 ml/fraction was collected while in Fig. 3A the fraction size was 1.2 ml), which was in the void volume, suggesting that the native BDZP receptor was probably larger than 200,000 daltons. The Triton X-100 solubilized receptor preparation had the same properties as the membrane-bound receptor preparations as shown in the Scatchard plot in which a K_d, 1.8 nM and Bmax, 440 fmoles/mg protein and K_d, 2.0 nM and Bmax, 255 fmoles/mg protein were obtained with membrane-bound and solubilized receptor preparations, respectively (Fig. 4A,B).

4. Endogenous Benzodiazepine-Like Substance. The strategy that we have used in the search for endogenous BDZP-like substances is to test the effect of the substance of interest on the binding of FNZP to its receptor. In a typical

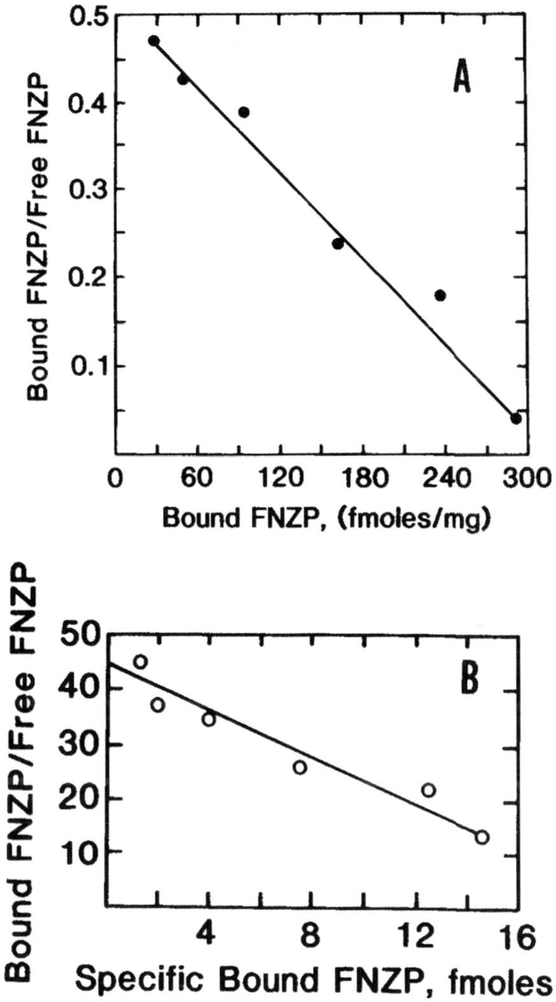

Fig. 4. A: Scatchard analysis of membrane-bound benzodiazepine receptor. K_d, 1.8 nM and B_{max}, 440 fmoles/mg protein were obtained.

B: Scatchard analysis of Triton X-100 solubilized benzodiazepine receptor. K_d, 2.0 and Bmax, 255 fmoles/mg protein were obtained.

experiment, 8.5 grams of rat brain were homogenized in 90 ml of water containing 1 mM AET, 0.2 mM pyridoxal phosphate, 1 mM EDTA, 1 mM benzamidine - HCl, 0.1 mM phenylmethane sulfonyl fluoride, 10 μg/ml of leupeptin, 10 μg/ml of pepstatin A, and 20 μg/ml of soybean trypsin inhibitor, pH7.2. The homogenate was centrifuged at 29,000 rpm for 1 hour. The supernatant was filtered through Diaflo Ultrafiltration membrane PM10 which has a cut-off limit of a molecular weight, 10,000. The filtrate was further concentrated either by vacuum centrifugation or lyophilization and was referred as F-I. The pellet from the first centrifugation was re-extracted with the same volume of the homogenizing buffer. The supernatant from the re-extraction was filtered, concentrated as before and the concentrated solution was referred as F-II. F-I was found to be a potent inhibitor for BDZP binding, inhibiting 63% of FNZP binding to BDZP receptor with 5 μl of F-I (Fig. 5), while no inhibition was observed with F-II at comparable concentration. The inhibition of FNZP binding by F-I appeared to be competitive since the extent of inhibition decreased with the increase of FNZP concentration. For instance, the inhibition by F-I decreased from 63 to 38% when the concentrations of FNZP increased from 1.5 to 24 nM. Currently we are in the process of purifying F-I to homogeneity, so that the identity of F-I or the endogenous BDZP-like substances can be unequivocally established.

CHARACTERIZATION of COATED VESICLE

Clathrin, a single polypeptide chain with a molecular weight of about 175,000 has been found as a major protein component of the lattice-like coated vesicles (Pearse, 1975, 1976; Garbern and Wu, 1981). This protein is believed to be capable of assembling to form coated vesicles and coated pits. Coated pits and coated vesicles function in a variety of processes such as receptor-mediated protein uptake (Anderson, Goldstein and Brown, 1976; Anderson, Brown and Goldstein, 1977), protein uptake in the proximal tubule (Miller, 1960; Miller and Palade, 1964; Ericsson, 1965; Maunsbach, 1966), synaptic vesicle membrane recycling (Heuser and Reese, 1973, 1977) and membrane transport processes (Friend and Farquhar, 1967; Holtzmann, Novikoff and Villaverde, 1967). Recently, we have purified the clathrin subunit by SDS-PAGE to homogeneity and specific antiserum to clathrin has also been obtained (Garbern and

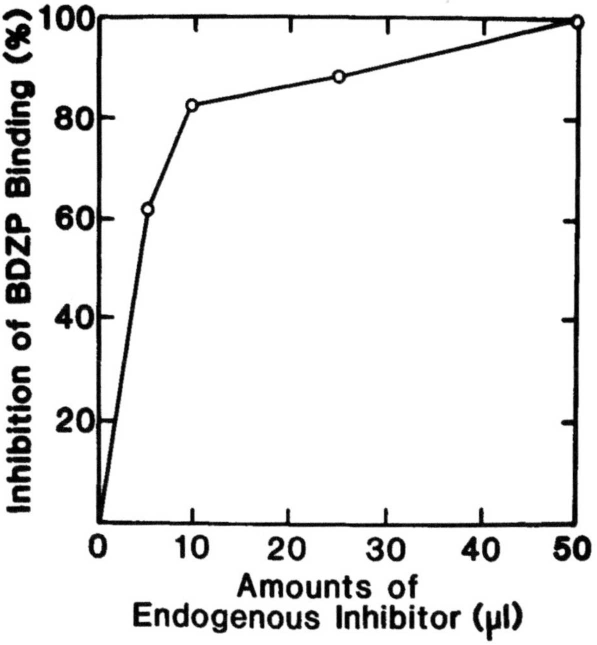

Fig. 5. Inhibition of benzodiazepine receptor binding by endogenous brain substances. The binding assay consisted of 10 μl crude mitochondrial pellet membrane, P_2 membrane (15.67 mg/ml), 15 μl 0.5 M Tris-citrate buffer, pH 7.0, 10 μl of 118 nM [^{3}H]-FNZP and various amounts of endogenous brain substances. The final volume was 200 μl. The extent of inhibition of benzodiazepine receptor binding, in %, is plotted against various amounts of concentrated endogenous inhibitor, in μl.

Wu, 1981). Furthermore, the site of synthesis and the distribution of clathrin in brain and kidney have also been visualized by immunocytochemical probes (Lin *et al.*, 1982).

Characterization of Clathrin Antiserum

The specificity of clathrin antibodies have been established from extensive immunochemical tests, e.g., double

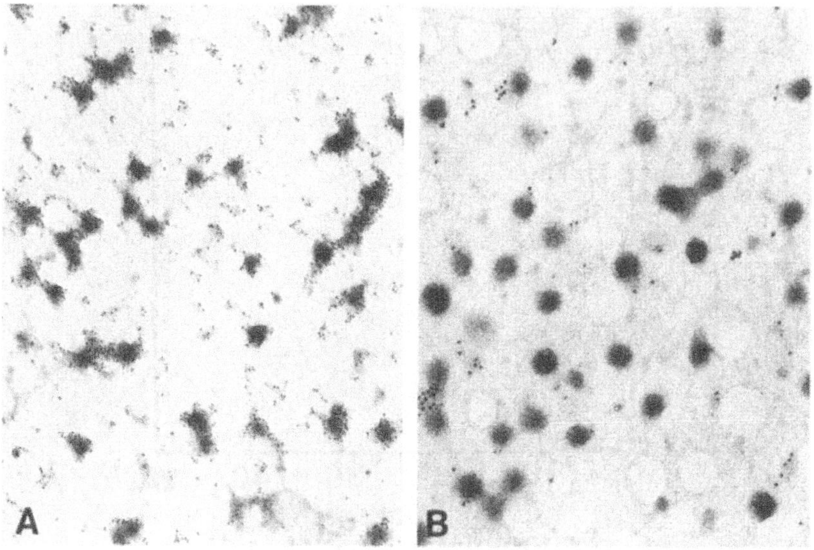

Fig. 6. Immunocytochemistry of reassociated coats. Electron micrographs of coats treated with (A) anticlathrin IgG and (B) nonimmune IgG, and with ferritin conjugated goat anti-rabbit IgG. Coated vesicles appear as fuzzy electron dense balls. Note: Ferritin particles are bound only on the coats treated with anticlathrin IgG (A) but not on those treated with nonimmune IgG (B). (A) original magnification X 73,000; (B) original magnification X 110,000.

immunodiffusion, microcomplement fixation, immunoelectrophoresis and radioimmuno assay (Garbern and Wu, 1981; Lin et al., 1982). The direct proof came from the demonstration of clathrin on reassociated coats by immunoferritin technique (Lin et al., 1982). Figures 6A and 6B are electron micrographs of the reassociated coats treated with anti-clathrin IgG (Fig. 6A) or nonimmune IgG (Fig. 6B) and ferritin-labeled goat anti-rabbit IgG. Ferritin particles are bound specifically to the coats treated with immune IgG, but not bound to coats treated with nonimmune IgG.

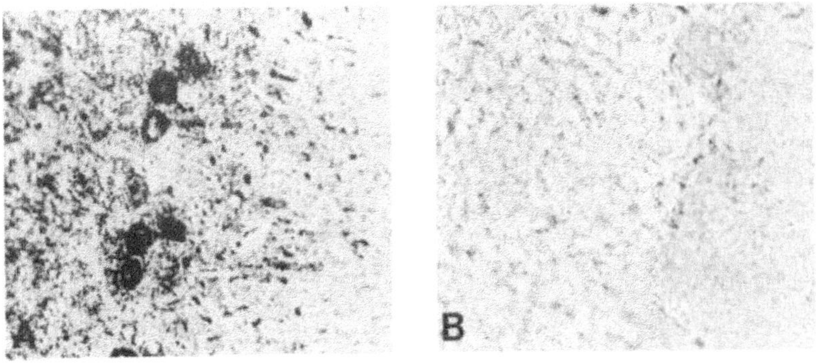

Fig. 7. A: Tissue section from the rat cerebellum treated with antiserum against clathrin. The reaction products are found in the Purkinje cells and in the molecular and granular layers (X380).

B: A control section of the rat cerebellum treated with preimmune serum shows no reaction product (X714).

Immunocytochemical Localization of Clathrin in the Cerebellum

When 50 μm sections of rat cerebellum were treated with anti-clathrin IgG and peroxidase-labeled goat anti-rabbit IgG, the peroxidase reaction products were seen in most Purkinje cell bodies and their dendrites and in granule cells and neurofibers (Fig. 7A). No staining was seen in the control sections which were treated with IgG from pre-immune serum instead of anti-clathrin IgG (Fig. 7B). When the thin sections were examined by electron microscopy the most striking finding in the granule cells was the staining of the budding vesicle from the Golgi apparatus (Fig. 8, arrows), and its associated cisternae (Fig. 8B, arrows), and the coated vesicles in the vicinity of Golgi region (Figs. 8A, arrowheads, and 8B). The clathrin-positive reaction products were also apparent on the membrane of multi-vesicular bodies (Fig. 8C, large arrows), on some endoplastic reticulum and on the amorphous ground substance (Fig. 8C, short arrows). However, no staining was apparent in the cisternae of the Golgi apparatus or endoplasmic reticulum. In the molecular layer, certain presynaptic terminals

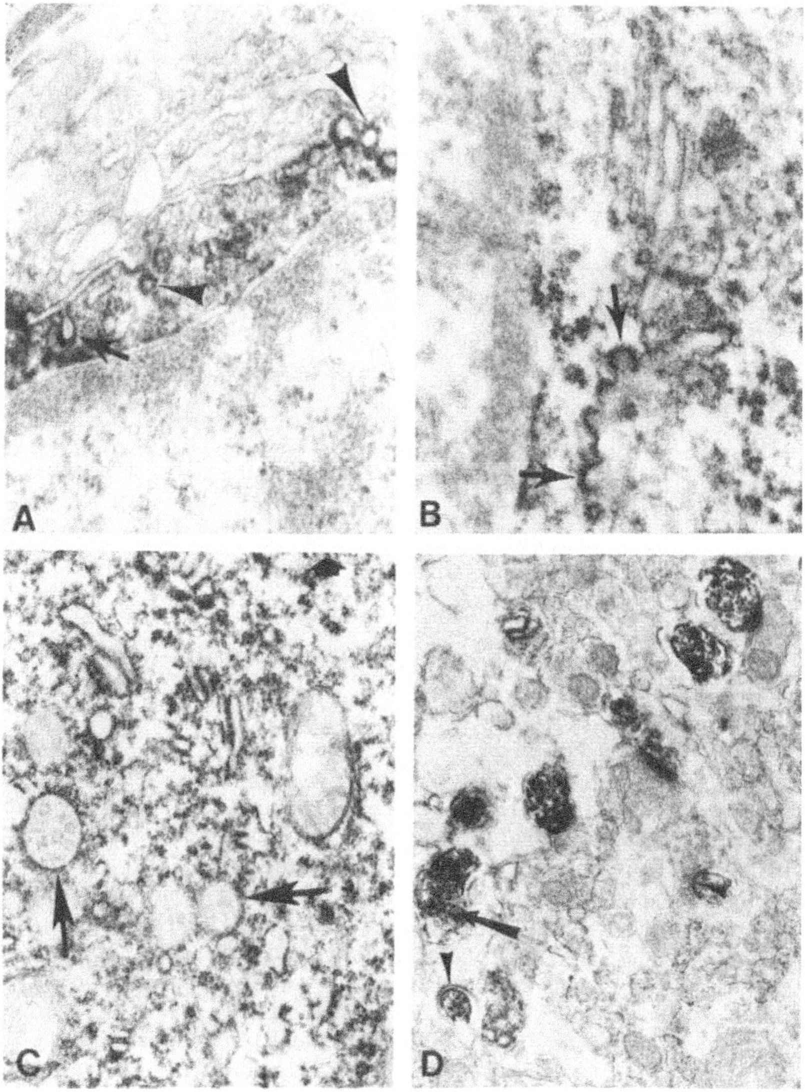

Fig. 8: Ultrastructural localization of clathrin in the rat cerebellum granular and molecular layers.

A: One of the granule cells in the rat cerebellum shows several clathrin-positive vesicles (arrows) budding from the Golgi apparatus. The cisternae of the Golgi apparatus are unstained. Arrow heads: stained-coated vesicle (X41,200).

B: Another granule cell shows a Golgi apparatus associated membranous structure with reaction product coated on the external membranes of several budding vesicles (arrows) (X32,000).

C: Several clathrin-positive coated vesicles are shown in one of the Purkinje cell bodies. The surfaces of a few multivesicular bodies (long arrows), some of the smooth ER membrane and the amorphous ground substance (short arrows) are also stained (X13,700).

D: In the molecular layer of the rat cerebellum there are several presynaptic terminals including the presynaptic vesicles and presynaptic membrane (arrows) and some postsynaptic membranes (arrow heads) (X19,200) labeled heavily with clathrin-positive products.

and postsynaptic density with heavy staining were also evident (Fig. 8D).

Immunocytochemical Localization of Clathrin in Kidney

When the sections of rat kidney were treated with anti-clathrin and peroxidase-labeled anti-rabbit IgG, peroxidase reaction products were seen most prominently in the proximal convoluted tubular epithelia (Lin *et al*; 1982). The staining was heavier in the apical portion than in the basal portion of the tubular epithelia. When the sections were examined by electron microscopy, clathrin-positive reaction products were found on the coated pits, coated vesicles, centriole, free polysomes as well as some ribosomes associated with rough endoplasmic reticulum (RER) (Fig. 9A: arrows and arrowheads). However, no reaction product was found within the cisternae of nuclear envelope, RER and Golgi apparatus. The smooth endoplasmic reticulum was entirely unstained. Similarly, most of the mitochondria showed no reaction product. Occasionally some autophagic vacuoles may contain vesicles which were stained (Fig. 9B and inset). These observations suggest that clathrin is synthesized on bound and free ribosomes and discharged into the cytosol where it becomes associated with a variety of membranes and ground substances and assembles on coated pits, coated vesicles, and Golgi associated vesicles. Clathrin may be finally degraded in autophagic vacuoles.

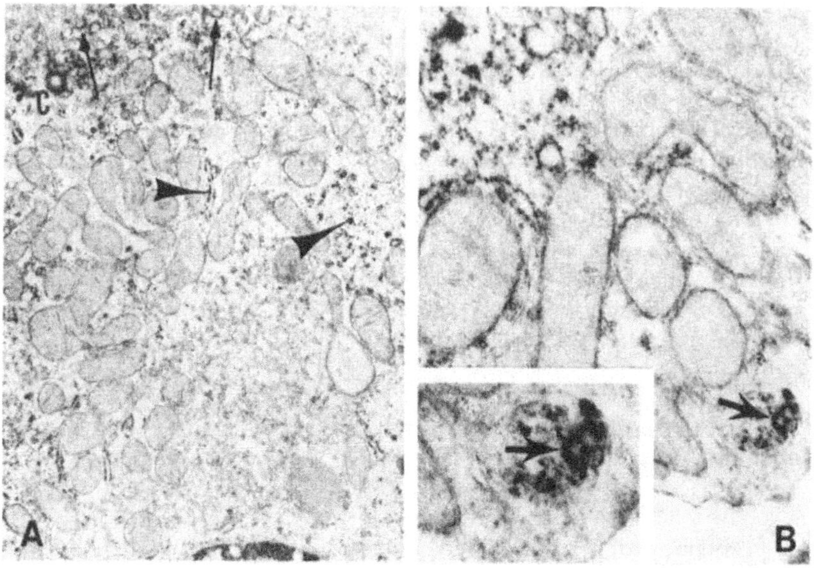

Fig. 9: Ultrastructural localization of clathrin in rat renal sections treated with antiserum against clathrin.

A: A proximal convoluted tubular epithelium from a rat kidney shows reaction products attached on the coated pits and coated vesicles (arrows). The reaction products are also present on some bound ribosomes and free polysomes (arrow head), but not in the cisternae of the rough ER. The smooth ER is unstained. One centriole (C) is also stained with reaction products (X10,000).

B: Ultrastructural localization of clathrin in rat renal tubular epithelia. Several stained-coated vesicles (arrow) and some free reaction products are seen in an autophagic vacuole near the basal cell membrane of one tubular epithelium (X18,000). Inset: A higher magnification of the autophagic vacuole. Several stained coated vesicles (arrow) and some free reaction products are shown in the vacuole (X33,000).

ACKNOWLEDGEMENT

The works described in this review were supported in part by grants NS13224 and EY03909 from National Institutes of Health.

REFERENCES

Anderson, R.G.W., Goldstein, J.L. and Brown, M.S. (1976) Localization of low density lipoprotein receptors on plasma membrane of normal human fibroblasts and their absence in cells from a familial hypercholesterolemia homozygote. Proc. Nat. Acad. Sci. USA 73, 2434-2438.

Anderson, R.G.W., Brown, M.S. and Goldstein, J.L. (1977) Role of the coated endocytic vesicle in the uptake of receptor-bound low density lipoprotein in human fibroblasts. Cell 10, 351-364

Baxter, C. F. (1970) The Nature of γ-aminobutyric acid. In Handbook of Neurochemistry, Vol.3 (Lajtha, A. ed.) Plenum Press, New York: pp. 289-353.

Braestrup, C. and Squires, R.F. (1978) Pharmacological characterization of benzodiazepine receptors in the brain. Eur. J. Pharmac. 48, 263-270.

Brandon, C., Lam, D.M.K. and Wu, J.-Y. (1979) The γ-aminobutyric acid system in rabbit retina: localization by immunocytochemistry and autoradiography. Proc. Nat. Acad. Sci. USA 76, 3557-3561.

Brandon, C., Lam, D.M.K., Su, Y.Y.T. and Wu, J.-Y. (1980) Immunocytochemical localization of GABA neurons in the rabbit and frog retina. Brain Res. Bull. 5, 21-29, (Suppl. 2).

Chan-Palay, V., Palay, S. L. and Wu, J.-Y. (1979) Gamma-aminobutyric acid pathways in the cerebellum studied by retrograde and anterograde transport of glutamic acid decarboxylase antibody after in vivo injections. Anat. Embryol. 157, 1-14.

Chan-Palay, V., Nilaver, G., Palay, S.L., Beinfeld, M.C., Zimmerman, E.E. and Wu, J.-Y. (1981) Chemical heterogeneity in cerebellar purkinje cells: existence and coexistence of glutamic acid decarboxylase-like and motilin-like immunoreactivities. Proc. Nat. Acad. Sci. USA 78, 7787-7791.

Choi, D.W., Farb, D.H. and Fischback, G.D. (1977) Chlordiazepoxide selectively augments GABA action in spinal cord cell cultures. Nature 269, 342-344.

Cook, L. and Sepinwall, J. (1975) Behavioral analysis of the effects and mechanisms of action of benzodiazepines. In: Mechanism of Action of Benzodiazepines (Costa, E. and Greengard, P., eds.) Raven Press, New York: pp. 1-28.

Costa, E. and Guidotti, A. (1979) Molecular mechanisms in the receptor action of benzodiazepines. Ann. Rev. Pharmacol. Toxicol. 19, 531-545.

Dray, A. and Straughan, D.W. (1976) Benzodiazepines: GABA and glycine receptors on single neurons in the rat medulla. J. Pharm. Pharmacol. 28, 314-318.

Ericsson, J. (1965) Transport and digestion of hemoglogin in the proximal tubule II electron microscopy. Lab. Invest. 14, 16-39.

Florey, E. (1957) Further evidence for the transmitter function of factor I. Naturwissenschaften 44: 424-425.

Fonnum, F. (1975) Metabolic compartmentation and neurotransmission (Berl, S., Clarke, D.D., and Schneider, D., eds.) Plenum Press, New York: pp. 99-122.

Friend, D.S. and Farquhar, M.G. (1967) Functions of coated vesicles during protein absorption in the rat vas deferens. J. Cell Biol. 35, 357-376.

Fuxe, K., Agnati, L.F., Bolme, P., Hokfelt, T., Lidbrink, P., Ljungdahl, A., Perez Dela Mora, M. and Ogren, S.O. (1975) The possible involvement of GABA mechanisms in the action of benzodiazepine on central catecholamine neurons. In: Mechanism of Action of Benzodiazepine (Costa, E. and Greengard, P., eds.) Raven Press, New York: pp. 45-61.

Gallager, D. W.(1978) Benzodiazepines: potentiation of a GABA inhibitory response in the dorsal-raphe nucleus. Eur. J. Pharmacol. 49, 133-143.

Garbern, J.-Y and Wu, J.-Y. (1981) Purification and characterization of clathrin from bovine brain J. Neurochem. 36, 602-612.

Goldowitz, D., Vincent, S.R., Wu, J.-Y. and Hokfelt, T. (1982) Immunohistochemical demonstration of plasticity in GABA neurons of the adult rat dentate gyrus. Brain Res. 238, 413-420.

Gottesfeld, Z., Brandon, C., Jacobowitz, D.M. and Wu, J.-Y (1980) The GABA system in the mammalian habenula. Brain Res. Bull. 5, 1-6, (Suppl. 2).

Gottesfeld, Z., Brandon, C. and Wu, J.-Y (1981) Immunocytochemistry of glutamate decarboxylase in the deafferented habenula. Brain Res. 208, 181-186.

Guidotti, A. (1978) Synaptic mechanisms in the action of benzodiazepine. In: Psychopharmacology; A Generation of Progress (Lipton, M. A., DiMascio, A. and Killam, K.F., eds.) Raven Press, New York: pp. 1349-1357.

Guidotti, A., Toffano, G. and Costa, E. (1978) An endogenous protein modulates the affinity of GABA and benzodiazepine receptors in rat brain. Nature 275, 553-555.

Guidotti, G., Toffano, M., Baraldi, O., Schwartz, J.P. and Costa, E. (1979) A molecular mechanism for the facilitation of GABA receptor function by Benzodiazepine. In:

GABA-Neurotransmitter Pharmacochemical, Biochemical and Pharmacological Aspects (Krogsgaard-Larsen, P., Scheel-Kruger, J. and Kofod, H., eds.) Academic Press, New York: pp. 406-415.

Hendrickson, A.E., Hunt, S. and Wu, J.-Y. (1981) Immunocytochemical localization of glutamic acid decarboxylase in monkey striate cortex. Nature 292, 605-607.

Heuser, J.E. and Reese, T.S. (1973) Evidence for recycling of synaptic vesicle membrane during transmitter release at the frog neuromuscular junction. J. Cell Biol. 57, 315-344.

Heuser, J.E. and Reese, T. (1977) Structure of the synapse. In: Handbook of Physiology: The Nervous System (Kandel, E., ed.) Bethesda, American Physiological Society pp. 261-294.

Holtzman, E., Novikoff, A.B. and Villaverde, H. (1967) Lysosomes and GERL in normal and chromatolytic neurons of the rat ganglion nodosum. J. Cell Biol. 33, 419-435.

Hunt, S.P., Kelly, J.S., Emson, P.C., Kimmel, J.R., Miller, R.J. and Wu, J.-Y. (1981) An immunohistochemical study of neuronal populations containing neuropeptides or GABA within the superficial layers of the rat dorsal horn. Neuroscience 6, 1883-1898.

Karobath, M. and Sperk, G. (1979) Stimulation of benzodiazepine receptor binding by γ-aminobutyric acid. Proc. Natl. Acad. Sci. USA 76, 1004-1006.

Köhler, G. and Milstein, C. (1975) Continuous cultures of fused cells secreting antibody of predefined specificity. Nature 256, 495-497.

Köhler, G. and Milstein, G. (1976) Derivation of specific antibody-producing tissue culture and tumor lines by cell fusion. Eur. J. Immunol. 6, 511-519.

Kozhechkin, S.N. and Ostovskaya, R.V. (1977) Are benzodiazepines GABA antagonists. Nature 269, 72-73.

Kravitz, E.A. and Potter, D.D. (1965) A further study of the distribution of γ-aminobutyric acid between excitatory and inhibitory axons of the lobster. J. Neurochem. 12, 323-328.

Kravitz, E.A. (1967) Acetylcholine, γ-aminobutyric acid and glutamic acid: physiological and chemical studies related to their roles as neurotransmitter agents. In: The Neurosciences (Quarton, G.C., Melnechuk, T. and Schmitt, F. O. eds.), Rockefeller Univ. Press, New York, pp. 433-444.

Krnjevic, K. (1974) Chemical nature of synaptic transmission in vertebrates. Physiol. Rev. 54, 418-540.

Kuriyama, K., Haber, B., Sisken, T. and Roberts, E. (1966) The γ-aminobutyric acid system in rabbit cerebellum. Proc. Nat. Acad. Sci. USA 55, 846-849.

Lam, D.M.K., Su, Y.Y.T., Swain, L., Marc, R.E., Brandon, C. and Wu, J.-Y. (1979) Immunocytochemical localization of glutamic acid decarboxylase in goldfish retina. Nature 278, 565-567.

Lin, C.T., Garbern, J.-Y. and Wu, J.-Y. (1982) Light and electron microscopic immunocytochemical localization of clathrin in rat cerebellum and kidney. J. Histochem. Cytochem. 30, 853-863.

MacDonald, R. and Barker, J.L. (1978) Benzodiazepines specifically modulate GABA-mediated postsynaptic inhibition in cultured mammalian neurones. Nature 271, 563-564.

Matsuda, T., Wu, J.-Y. and Roberts, E. (1973) Immunochemical studies on glutamic acid decarboxylase from mouse brain. J. Neurochem. 21, 159-166.

Maunsbach, A. (1966) Absorption of ferritin by rat kidney proximal tubule cells. Electron microscopic observations of the initial uptake phase in cells of microperfused single proximal tubules. J. Ultrastr. Res. 16, 1-12.

McLaughlin, B.J., Wood, J.G., Saito, K., Barber, R., Vaughn, J.E., Roberts, E. and Wu, J.-Y. (1974) The fine structural localization of glutamate decarboxylase in synaptic terminals of rodent cerebellum. Brain Res. 76, 377-391.

McLaughlin, B.J., Barber, R., Saito, K., Roberts, E. and Wu, J.-Y. (1975a) Immunocytochemical localization of glutamate decarboxylase in rat spinal cord. J. Comp. Neur. 164, 305-322.

McLaughlin, B.J., Wood, J.G., Saito, K., Roberts, E. and Wu, J.-Y. (1975b) The fine structural localization of glutamate decarboxylase in developing axonal processes and presynaptic terminals of rodent cerebellum. Brain Res. 85, 355-371.

McNamara, J.O., Peper, A.M. and Patrone, V. (1980) Repeated seizure induce long-term increase in hippocampal benzodiazepine receptors. Proc. Nat. Acad. Sci. USA 77, 3029-3032.

Miller, F. (1960) Hemoglobin absorption by the cells of the proximal convoluted tubule in mouse kidney. J. Biophys. Biochem. Cytol. 8 689-718.

Miller, F. and Palade, G. (1964) Lytic activities in renal protein absorption droplets: An electron microscopical cyotchemical study. J. Cell Biol. 23, 519-552.

Möhler, H. and Okada, T. (1977) Benzodiazepine receptor: demonstration in the central nervous system. Science 198, 849-851.

Möhler, H., Richards, J.G. and Wu, J.-Y. (1981) Autoradiographic localization of benzodiazepine receptors in immunocytochemically identified GABAergic synapses. Proc. Nat. Acad. Sci, USA 78, 1935-1938.

Olson, R.W. (1981) GABA-benzodiazepine-barbiturate receptor interactions. J. Neurochem. 37, 1-13.

Pearse, B.M.F. (1975) Coated vesicles from pig brain: purification and biochemical characterization. J. Mol. Biol. 97, 93-98.

Pearse, B.M.F. (1976) Clathrin: a unique protein associated with intracellular transfer of membrane by coated vesicle. Proc. Nat. Acad. Sci. USA 73, 1255-1259.

Saito, K., Barber, R., Wu, J.-Y., Matsuda, T., Roberts, E. and Vaughn, J. E. (1974a) Immunohistochemical localization of glutamic acid decarboxylase in rat cerebellum. Proc. Nat. Acad. Sci. USA 71, 269-273.

Saito, K., Wu, J.-Y. and Roberts, E. (1974b) Immunochemical comparisons of vertebrate glutamate acid decarboxylase. Brain Res. 65, 277-285.

Snyder, S.H. and Enna, S.J. (1975) The role of central glycine receptors in the pharmacologic actions of benzodiazepines. In: Mechanism of Action of Benzodiazepines (Costa, E. and Greengard, P., eds.) Raven Press, New York: pp. 81-91.

Squires, R.F. and Breastrup, C. (1977) Benzodiazepine receptors in rat brain. Nature 266, 732-734.

Stein, L., Wise, C.D. and Belluzzi, J.D. (1975) Effect of benzodiazepine on central serotonergic mechanisms In: Mechanism of Action of Benzodiazepines (Costa, E. and Greengard, P., eds.) Raven Press, New York: pp. 29-44.

Su, Y.Y.T., Wu, J.-Y. and Lam, D.M.K. (1979) Purification of L-glutamic acid decarboxylase from catfish brain. J. Neurochem. 33, 169-179.

Supavilai, P. and Karobath, M. (1980) Interaction of SQ 20009 and GABA-like drugs as modulators of benzodiazepine receptor binding. Eur. J. Pharmacol. 62, 229-233.

Tallman, J.F., Thomas, J.W. and Gallager, D.W. (1978) GABA-ergic modulation of benzodiazepine binding site sensitivity. Nature 274, 383-385.

Tsuchiya, T. and Fukushina, K. (1978) Effects of benzodiazepines and pentobarbitone on the GABA-ergic recurrent in hibition of hippocampal neurons. Eur. J. Pharmacol. 48, 421-424.

Vincent, S.R., Hokfelt, T. and Wu, J.-Y. (1982a) GABA neuron systems in hypothalamus and the pituitary gland: immunohistochemical demonstration using antibodies against glutamate decarboxylase neuroendocrinology, 34, 117-125.

Vincent, S. R., Hokfelt, T., Wu, J.-Y., Elde, R. P., Morgan, L.M. and Kimmel, J.R. (1982b) Immunohistochemical studies of the GABA system in the pancreas. Neuroendocrinology (in press).

Wastek, G.J., Speth, R.C., Reisine, T.D. and Yamamura, H.I. (1978) The effect of γ-aminobutyric acid on ^{3}H-flunitrazepam binding in the rat brain. Eur. J. Pharmacol. 50, 445-447.

Wong, E., Schousboe, A., Saito, K., Wu, J.-Y. and Roberts, E. (1975) Glutamate decarboxylase and GABA-transaminase from six mouse strains. Brain Res. 68, 133-139.

Wu, J.-Y. (1976) Purification and properties of L-glutamate decarboxylase (GAD) and GABA-aminotransferase (GABA-T). In: GABA In Nervous System Function (Roberts, E., Chase, T. and Tower, D., eds.) Raven Press, New York, pp. 7-55.

Wu, J.-Y. (1982a) Purification and characterization of cysteic/cysteine sulfinic acids decarboxylse and L-glutamate decarboxylase in bovine brain. Proc. Nat. Acad. Sci. USA 79, 4270-4274.

Wu, J.-Y., (1982b) Preparation of glutamic acid decarboxylase as immunogen for immunocytochemistry. In: Neuroimmunocytochemistry (IBRO Handbook Series: Methods in the Neurosciences) (Cuello, A. C., ed.) John Wiley & sons Ltd., Sussex, in press.

Wu, J.-Y., Matsuda, T. and Roberts, E. (1973) Purification and characterization of glutamate decarboxylase from mouse brain. J. Biol. Chem. 248, 3029-3034.

Wu, J.-Y., Brandon, C., Su, Y.Y.T. and Lam, D.M.K. (1981a) Immunocytochemical and autoradiographic localization of GABA system in the vertebrate retina. Molec. Cell. Biochem. 39, 229-238.

Wu, J.-Y., Su, Y.Y.T., Lam, D.M.K., Schousboe, A. and Chude, O. (1981b) Assay methods, purification and characterization of L-glutamate decarboxylase and GABA-transaminase. Research Methods in Neurochemistry, Vol. 5, 129-177.

Wu, J.-Y., Lin, C.-T., Brandon, C., Chan, D.-S., Mohler, H. and Richards, J.G. (1982a) Regulation and immunocytochemical characterization of GAD. In: Cytochemical Methods in Neuroanatomy (Palay, S. and Palay V. eds.) Alan R. Liss Inc., New York, pp. 279-296.

Wu, J.-Y., Su, Y.Y.T. and Lin, C.T. (1982b) Characterization of GABA and benzodiazapine systems in vertebrate nervous systems. In: Advances in Neuropharmacology (Lee, C.Y., ed.) Acadmia Sinica, Taipei, pp. 109-121.

ROLES OF SYNAPTIC MEMBRANOUS PHOSPHOLIPIDS IN THE MODULATION OF CEREBRAL GABA AND BENZODIAZEPINE RECEPTOR BINDINGS.

Kinya Kuriyama and Yukio Yoneda

Department of Pharmacology, Kyoto Prefectural University of Medicine, Kyoto 602, Japan

INTRODUCTION

It has been well established that γ-aminobutyric acid (GABA) plays an inhibitory neurotransmitter role in the mammalian central nervous system (CNS) (Kuriyama et al., 1966; Iversen and Bloom, 1972; Yoneda and Kuriyama, 1978) as well as in the invertebrate nervous system (Kravitz et al., 1965). The specific binding of [^{3}H] GABA to synaptic membrane preparations obtained from the rat brain has been thought to reflect the association of this neuroactive amino acid with its physiologically relevant synaptic receptors (Zukin et al., 1974; Enna and Snyder, 1977). Recently, it has been also demonstrated that muscimol, which is isolated from the mushroom *Amanita muscaria*, has a potent agonistic action on synaptic GABA receptors (Krogsgaard-Larsen et al., 1976), and [^{3}H] muscimol is usable as a specific radioligand for biochemical (Beaumont et al., 1978; Snodgrass, 1978) as well as autoradiographic (Chan-Palay, 1979) studies on GABA receptors in the mammalian CNS. In addition, muscimol is known to be a more potent agonist than GABA itself in terms of the induction of conductance changes at the crustacean inhibitory synapses (Hori et al., 1978).

On the other hand, it has been shown that there exists high affinity, stereospecific and saturable binding sites for benzodiazepines (BZPs) in the mammalian CNS (Möhler and Okada, 1977; Squires and Braestrup, 1977). The high correlation between pharmacological potency and affinity of a series of BZPs to this receptor has suggested that this site may be involved in the exhibition of pharmacological actions of BZPs (Möhler et al., 1978). Several lines of evidence have also indicated that at least a part of the actions of BZPs results from a specific interaction with cerebral GABA systems (Costa et al., 1975; Gallager, 1978). Furthermore, it has been reported that in vitro addition of GABA facilitates the specific binding of BZP to synaptic membrane fractions obtained from the mammalian brain (Brilely and Langer, 1978; Karobath and Sperk, 1979; Tallman et al., 1978). These findings have led to the concept that BZP binding site (receptor) may be located at postsynaptic membrane with a significant functional coupling to the synaptic GABA receptor which resides in the vicinity of the binding site for this psychotropic drug.

In this chapter, we have attempted to demonstrate the specific interaction between GABA and BZP receptors with a special reference to possible modulatory roles of membrane phospholipids in the bindings of both ligands.

MODULATION OF GABA RECEPTOR BINDING

Treatment of synaptic membrane fractions from the brain with a nonionic detergent, Triton X-100, is known to induce a significant increase in GABA receptor binding (Enna and Snyder, 1977; Beaumont et al., 1978; Yoneda and Kuriyama, 1980 a). This enhancement of GABA receptor binding has been considered to be attributable to the removal of endogenous inhibitory substance(s) from the membrane, such as phosphatidylethanolamine (a phospholipid) (Johnston and Kennedy, 1978), GABA itself (Greenlee et al., 1978), acidic protein (GABA-modulin)(Toffano et al., 1978) and a low molecular substance (GRIF)(Yoneda and Kuriyama,1980 b). Although it seems possible that these endogenous modulators may have a functional role in the regulation of synaptic transmission at GABAergic neurons as observed in kainic acid-induced denervation supersensitivity phenomenon (Kuriyama et al., 1980), biochemical characterization and identification of these modulators as well as their physiological significances remain to be solved by future studies.

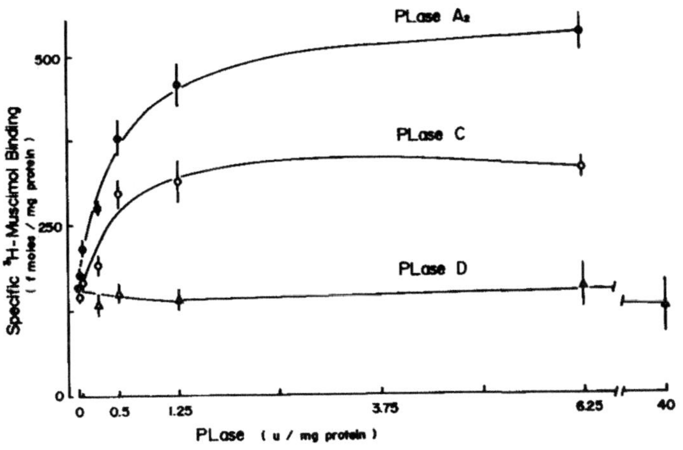

Fig. 1. Effect of treatments with various phospholipases (PLases)on [^{3}H] muscimol binding to synaptic membrane. Crude synaptic membrane fractions obtained as described previously (Zukin et al., 1974) were preincubated at 37°C for 60 min with the indicated amount of PLase A_2 (from Naja naja venom), PLase C (from B. cereus) or PLase D (from cabbage) in 50 mM Tris HCl buffer (pH 7.4) containing 1 mM $CaCl_2$. After terminating the preincubation by centrifuging at 48,000 × g for 20 min, resultant precipitates were subjected to the binding assay for [^{3}H] muscimol (10.1 Ci/mmole, New England Nuclear) using the filtration method as described previously (Kuriyama et al., 1980). Each value represents the mean ± S.E.M. obtained from 3 separate experiments done in triplicate.

PHOSPHOLIPIDS AND GABA RECEPTOR

The treatment of cerebral synaptic membrane with phospholipase (PLase) A_2 and PLase C induced a significant stimulatory effect on [^{3}H] muscimol binding, while PLase D treatment had no such a significant effect on the binding (Fig. 1). Although the increment by PLase C was in good agreement with previous findings (Giambalvo and Rosenberg, 1976; Lloyd and Davidson, 1979), it has been also shown (Giambalvo and Rosenberg, 1976; Lloyd and Davidson, 1979) that PLase A_2 has an inhibitory action and/or no profound effect on the GABA receptor binding, which is apparently contradictory to the present results. The enhancement of [^{3}H]

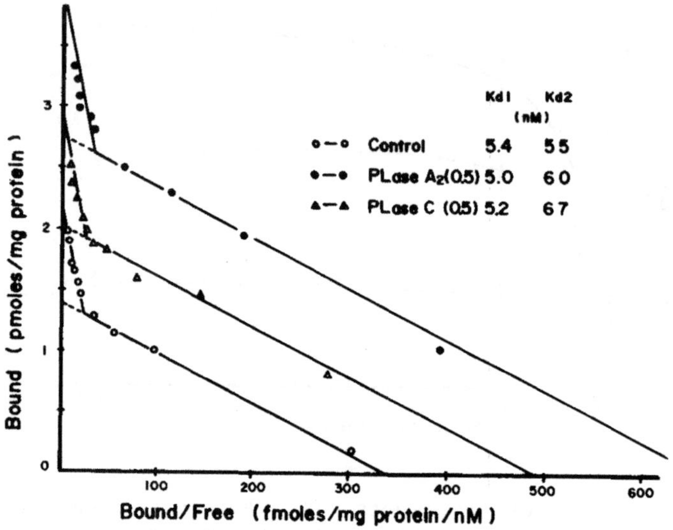

Fig. 2. Scatchard analysis of PLases-induced increase in [^{3}H] muscimol binding. Following the termination of pre-incubation with 0.5u/mg protein of PLase A_2 or PLase C by centrifuging at 48,000 × g for 20 min, the synaptic membrane suspension was incubated with 10 nM [^{3}H] muscimol in the absence and presence of varying concentrations of non-radioactive muscimol. Each value represents the mean from 3 - 5 separate experiments. Control membrane, Kd_1 = 5.4 nM, $Bmax_1$= 1.39 pmoles/mg protein, Kd_2 = 55 nM, $Bmax_2$ = 2.17 pmoles/mg protein; PLase A_2 membrane, Kd_1 = 5.0 nM, $Bmax_1$ = 2.74 pmoles/mg protein, Kd_2 = 60 nM, $Bmax_2$ = 3.92 pmoles/mg protein ; PLase C membrane, Kd_1 = 5.2 nM, $Bmax_1$ = 2.02 pmoles/mg protein, Kd_2 = 67 nM, $Bmax_2$ = 2.92 pmoles/mg protein, respectively.

muscimol binding by PLase A_2 was, however, found to be dependent on the presence of Ca^{2+} which was shown to be an essential factor for the exhibition of the maximal enzyme activity. Scatchard analysis revealed that the specific binding of [^{3}H] muscimol to synaptic membrane consists of two components with high and low affinities for the radioligand, respectively (Fig. 2). Treatments of the membrane with PLase A_2 and PLase C induced an increase in the density

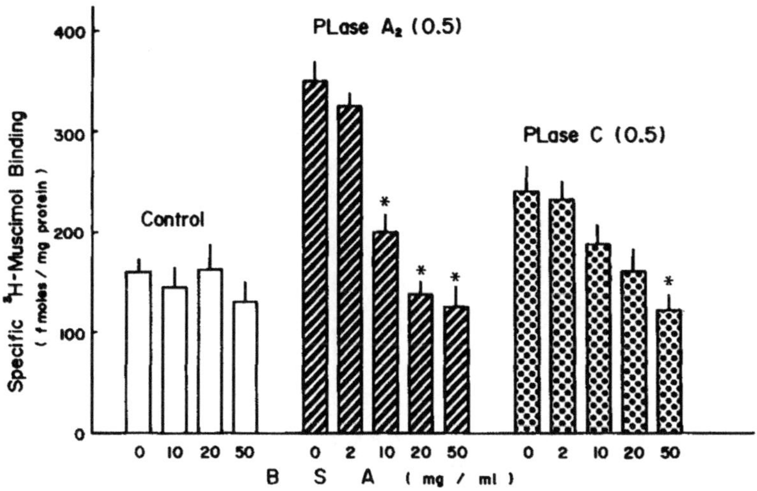

Fig. 3. Effect of preincubation with bovine serum albumin (BSA) on PLases-induced increase in [^{3}H] muscimol binding. Synaptic membrane suspensions were preincubated at 37°C for 60 min with 0.5 u/mg protein of PLase A_2 or PLase C in 50 mM Tris HCl buffer (pH 7.4) containing 1mM $CaCl_2$ with and without indicated amount of BSA. Each value represents the mean ± S.E.M. obtained from 3 - 4 separate experiments. * $P < 0.05$, compared with each control value.

of both components without affecting their affinities (Fig. 2). These results suggest that treatments with both enzymes may exhibit stimulatory action on the binding through a similar mechanism.

Considering the fact that PLase C as well as PLase A_2 produces free fatty acids as a common reaction product (Wolfe, 1982), it seems possible that these free fatty acids may be responsible for the activation of GABA receptor binding induced by these PLase treatments. To examine such a possibility, we have used bovine serum albumin (BSA) which is known to have a potential binding capacity toward free fatty acids (Goodman, 1958). It was found that preincubation of the membrane with BSA not only suppressed the PLase A_2-induced increase in the binding, but also attenuated the PLase C-induced elevation of the binding without affecting the

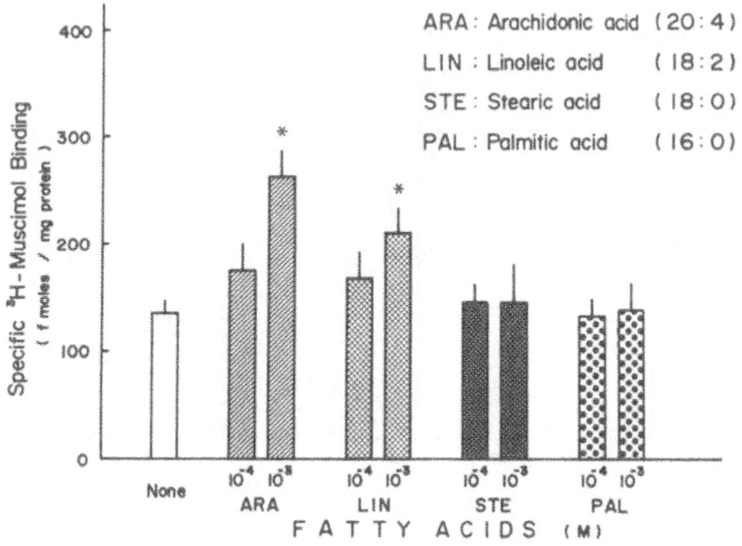

Fig. 4. Effect of preincubation with various free fatty acids on [^{3}H] muscimol binding. Synaptic membrane fractions were preincubated at 37°C for 60 min in 50 mM Tris HCl buffer (pH 7.4) containing 1 mM $CaCl_2$ in the presence and absence of various free fatty acids indicated in the figure. The preincubation was terminated by subsequent centrifugation at 48,000 × g for 20 min, and the pellets thus obtained were subjected to the binding assay for [^{3}H] muscimol. Each value represents the mean ± S.E.M. obtained from 3 - 4 separate experiments, * $P < 0.05$, compared with no-addition value.

binding to non-treated membrane (Fig. 3). No significant preventive effect was observed, however, when BSA was added to the incubation mixture following the treatments with PLase A_2 and PLase C. These results clearly show that the PLase-induced increase in [^{3}H] muscimol binding results from the production by these enzymes of free fatty acids which are susceptible to scavenging action of BSA rather than from the destruction of membrane phospholipids *per se*. It is also suggested that the enhancement of [^{3}H] muscimol binding caused by PLases treatments may be irreversible in the sense of antagonism by BSA.

In vitro addition of various free fatty acids was found to exhibit no significant alteration in [^{3}H] muscimol binding to non-treated as well as PLase A_2-treated membranes in a concentration range from 10^{-5} M to 10^{-3} M. In contrast, the enhancement of [^{3}H] muscimol binding was induced by the pretreatment of intact synaptic membranes with unsaturated free fatty acids such as arachidonic acid (20:4) and linoleic acid (18:2), while that with saturated free fatty acids such as stearic acid (18:0) and palmitic acid (16:0) elicited no significant action on the binding (Fig 4). These findings suggest that the [^{3}H] muscimol binding may be elevated possibly as a result of the conversion of unsaturated free fatty acids to some actually active compound(s) during the pretreatment of synaptic membrane. Scatchard analysis has revealed that the pretreatment with arachidonic acid enhances the [^{3}H] muscimol binding through the increase in the density of both binding sites with high and low affinities without significantly affecting their affinities(Fig. 5). Comparing these results with those found in PLases treatments (Fig. 2), it seems likely that there exists a common stimulatory mechanism between PLase-induced and fatty acid-induced increases in [^{3}H] muscimol binding.

PHOSPHOLIPIDS AND BENZODIAZEPINE RECEPTOR

To examine possible regulatory mechanisms on BZP receptor in the CNS as observed in the case of GABA receptor, effect of pretreatments of cerebral synaptic membrane with various enzymes on specific [^{3}H] diazepam binding was also investigated. As shown in Table 1, proteolytic enzymes such as trypsin and pronase exhibited a significant reduction of [^{3}H] diazepam binding, while neuraminidase, β-galactosidase, lipoxygenase and PLase D had no profound action on the binding. In contrast, treatments with PLase A_2 and PLase C induced a significant increase in [^{3}H] diazepam binding as well as [^{3}H] flunitrazepam binding to cerebral synaptic membranes. The accentuating effects of the PLase treatment, however, were not detected in the solubilized BZP receptor preparations which are devoid of membrane lipids (Ueno and Kuriyama, 1981). These results suggest that membrane phospholipids may play a significant modulatory role in the association of BZPs with its receptor in addition to eliciting a regulatory function on the permeability of these high-

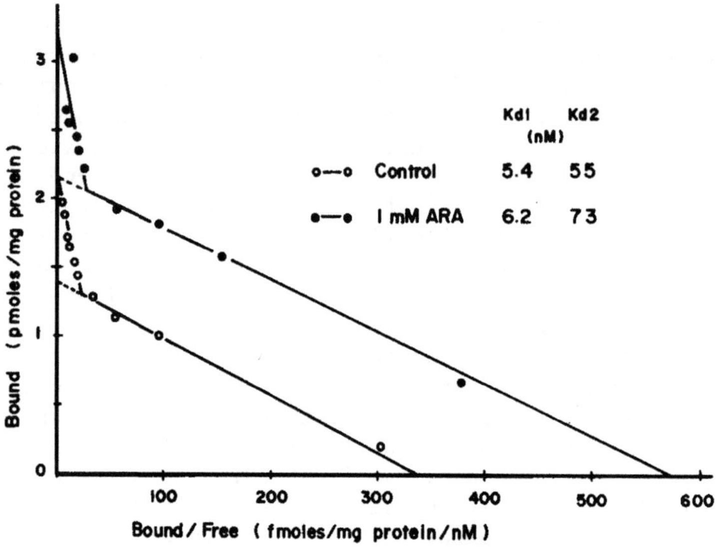

Fig. 5. Scatchard analysis of arachidonic acid (ARA)-induced increase in [^{3}H] muscimol binding. Following the termination of preincubation at 37°C for 60 min with 1 mM arachidonic acid by centrifuging at 48,000 × g for 20 min, the synaptic membranes were incubated at 2°C for 30 min with 10 nM [^{3}H] muscimol in the presence and absence of non-radioactive muscimol. Each value represents the mean from 3 - 5 separate experiments. Control membrane, Kd_1 = 5.4 nM, $Bmax_1$= 1.39 pmoles/mg protein, Kd_2 = 55 nM, $Bmax_2$ = 2.17 pmoles/mg protein; ARA membrane, Kd_1 = 6.12 nM, $Bmax_1$ = 2.15 pmoles/mg protein, Kd_2 = 79 nM, $Bmax_2$ = 3.18 pmoles/mg protein, respectively.

ly lipid soluble drugs to various biomembranes.

COUPLING OF GABA AND BZP RECEPTORS

It has been demonstrated (Costa et al., 1979) that the [^{3}H] diazepam binding is significantly enhanced in a dose dependent manner by the *in vitro* addition of anions which are electrophysiologically known to penetrate the activated postsynaptic membrane of cat motorneurons (Eccles, 1964), whereas non-permeable anions such as fluoride and oxalate

Table 1. Effect of various enzymatic treatments of synaptic membranes on [^{3}H] diazepam binding

Treatment	(unit/mg of membranous protein)	[^{3}H]diazepam bound (% of control)
Trypsin	(50)	51
Pronase	(50)	69
Neuraminidase	(0.12)	113
β-Galactosidase	(0.25)	108
Lipoxygenase	(10)	114
Phospholipase C*	(0.5)	150
Phospholipase A_2*	(0.5)	128
Phospholipase D*	(0.5)	117

* Incubated with 1 mM $CaCl_2$. Following the preincubation of the membrane with each enzyme at 37°C for 30 min, the preparations were centrifuged at 48,000 × g for 20 min. The resultant pellets were resuspended with Tris-HCl buffer and used for the binding assay. Each value represents the mean from 3 to 7 separate experiments. The concentration of [^{3}H] diazepam used was 0.5 M. (Ueno and Kuriyama, 1981)

have no significant effect on the binding. These results suggest the presence of a close relationship between BZP receptor and chloride ion channel (Cl^- ionophore). Since it has been well documented that GABA exerts its physiological action through the activation of chloride ion channel after the specific association with its postsynaptic receptor, it is reasonable to assume that GABA and BZP receptors are closely associated and may be jointly coupled with a chloride ion channel mechanism in postsynaptic membrane (Costa et al., 1979). In fact, several reports are available which indicate that a single macromolecular receptor protein complex possesses the binding ability to GABA as well as BZP (Asano and Ogasawara, 1980; Gavish and Snyder, 1981; Stephenson et al., 1982). We have also demonstrated (Ito and Kuriyama, 1982) that BZP binding site is co-chromatographed with GABA receptor having a molecular weight of 270,000 daltons on gel filtration by Sephdex G-200 (Fig. 6). In addition, it has been found (Ito et al., 1982) that the apparent molecular weight of BZP binding site is estimated to be 51,500 daltons when solubilized GABA receptor fraction (ASP fraction) was photolabeled using [^{3}H] flunitrazepam in the presence of UV light (Fig. 7). These results strongly suggest

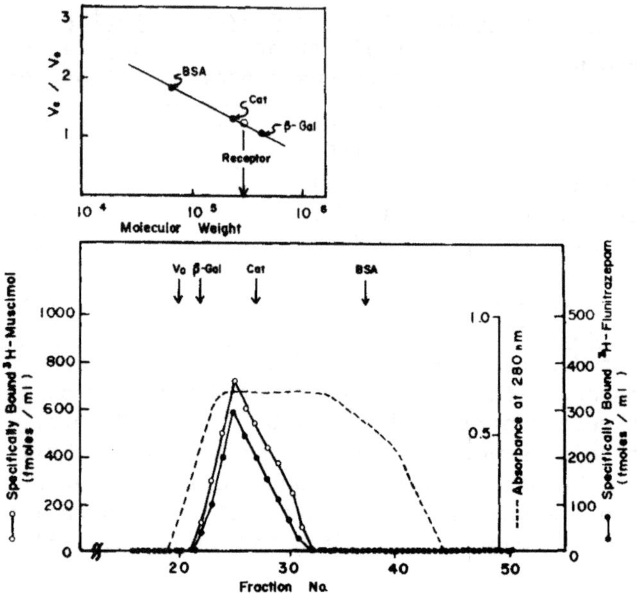

Fig. 6. Sephadex G-200 gel filtration of solubilized (ASP) fraction. Two ml of solubilized (ASP) fraction (15 mg of protein) was applied to a Sephadex G-200 column (70 × 1.5 cm) equilibrated with 50 mM Tris-citrate buffer (pH 7.1) containing 0.1 % Nonidet P-40. Three ml each of aliquot was collected at flow rate of 0.1 ml/min. The aliquots were assayed for specific [^{3}H] muscimol (o — o) and [^{3}H] flunitrazepam bindings (● — ●). The elution position of the protein was determined by absorbance at 280 nm (····). Marker proteins were chromatographed under the same experimental conditions. Insert: the elution volume (Ve) to void volume (Vo) of the various proteins was plotted against the log of the molecular weight(●). The molecular weight of the receptor was located on the regression analysis(o). (Ito and Kuriyama, 1982)

that GABA and BZP receptors may reside on the same macromolecular protein, and may be functionally similar.

The dose-dependent stimulatory effect of GABA on [^{3}H] diazepam binding to cerebral synaptic membranes treated with

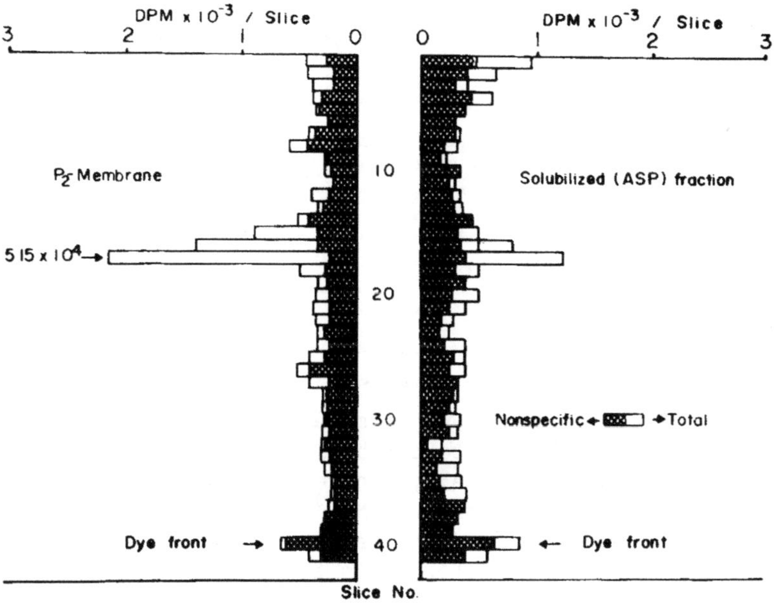

Fig. 7. SDS-polyacrylamide gel electrophoresis of photo-labeled synaptic P_2-membrane and solubilized (ASP) fractions. Electrophoresis was conducted with 10 % SDS-polyacrylamide gels. (Ito et al., 1982)

PLase C and PLase A_2 is shown in Fig. 8. Although the treatments with both PLases significantly increased the specific binding of [^{3}H] diazepam measured in the absence of GABA, the magnitude of the stimulation by added GABA was significantly attenuated following PLase A_2 and PLase C treatments. These results suggest that the coupling of GABA and BZP receptors may also be modulated by synaptic membranous phospholipids.

DISCUSSION

There is no doubt that synaptic membranous phospholipids have a functional significance in the association of GABA, a neurotransmitter, with its physiologically relevant postsynaptic receptor which is considered to be functionally coupled to the receptor for BZPs. Although molecular mecha-

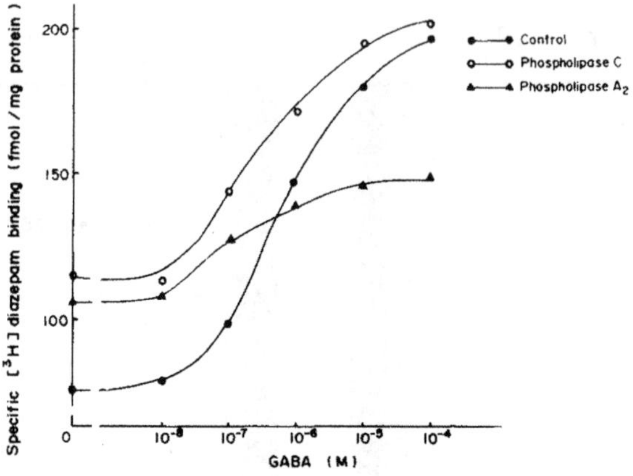

Fig. 8. The dose-dependent stimulatory effect of GABA on [^{3}H] diazepam binding to cerebral synaptic membranes treated with PLase C and PLase A_2. Each enzymatically treated membrane was incubated with 0.5 nM [^{3}H] diazepam. Each point represents the mean from at least 2 separate and closely checked experiments performed in triplicate. (Ueno and Kuriyama, 1981).

nisms underlying these phenomena are not clarified at present, it is highly probable that the formation of unsaturated free fatty acids is indeed responsible for the augmentation of GABA receptor binding caused by the PLases treatments. It is well known that PLase A_2 hydrolyzes phospholipid to free fatty acid and lysophospholipid, whereas PLase C yields a 1, 2-diacylglycerol which is in turn deacylated to a free fatty acid by membrane-associated diglyceride lipase. Arachidonic acid (20:4) is one of the most probable products yielded by these reactions, since naturally occuring phospholipids are known to have this fatty acid as a constituent in the 2-position of the glycerol moiety.

It has been shown that arachidonic acid is successively converted to prostaglandins or thromboxanes via the action of membrane-bound cyclo-oxygenase and transformed to other novel products such as 12-hydroxy-5, 8, 14-<u>cis</u>, 10-<u>trans</u>-eicosatetraenoic acid (12-HETE), 5-hydroxy-6-<u>trans</u>, 8, 11, 14,-<u>cis</u>-eicosatetraenoic acid (5-HETE) and leukotrienes by the initiating action of lipoxygenase in the mammalian CNS (Wolfe, 1982). Since it has been found that the enhancement of [^{3}H] muscimol binding induced by neither PLase A_2 nor arachidonic acid is suppressed by the co-incubation of the membrane with a cyclo-oxygenase inhibitor "indomethacin" (10^{-5} - 10^{-3}M) (preliminary experiments), it seems unlikely that the products formed by the catalytic action of cyclo-oxygenase such as prostaglandins and thromboxanes have a stimulatory effect on the binding of GABA to its synaptic receptor. The exact functional significances as well as molecular mechanisms underlying the enhancement of GABA receptor binding observed in the present study, however, remain to be elucidated. It is conceivable that free arachidonic acid may be released from membrane phospholipids during neuronal excitation and one of its lipoxygenase products may augment the association of GABA released from presynaptic nerve terminals with its postsynaptic receptor, which results in the facilitation of GABAergic transmisson.

The importance of synaptic membranous phospholipids at GABA receptor is further augmented by considering the fact that cerebral BZP receptor binding as well as the coupling of GABA and BZP receptors is also modulated by these phospholipids which are susceptible to PLase treatments. It is well known that BZP has a high lipid solubility and the increase in temperature of incubation significantly diminishes BZP receptor binding (Braestrup and Squires, 1977). Such facts, coupled with the present results, suggests tnat the observed modulating roles of phospholipids on BZP receptor binding may be exhibited, at least in part, via an alteration in biophysical properties such as the fluidity of synaptic membrane. A similar concept may also be applied to the modulating roles of phospholipids on the coupling of GABA and BZP receptors. Studies to examine such a possibility need to be carried out.

CONCLUSION

Synaptic membrane phospholipids were found to play multiple modulating roles on cerebral GABA-BZP receptor complex: modulating roles on GABA receptor binding , BZP receptor binding and the coupling of GABA and BZP receptors, respectively.

The modulatory effect on GABA receptor binding seems attributable to the production of an unsaturated free acid, probably arachidonic acid, from membranous phospholipids through the actions of PLase A_2 and/or of PLase C and diglyceride lipase, all of which are supposed to be present in cerebral synaptic membrane. It is likely that arachidonic acid exerts its stimulatory effect on the GABA receptor binding after being transformed to some actually active compounds. In contrast, the modulating roles of membranous phospholipids on BZP receptor binding as well as on the coupling of GABA and BZP receptors seem to be related to an alteration in biophysical properties of the synaptic membrane such as membrane fluidity. The exact molecular and biochemical mechanisms underlying these phenomena, however, remain to be elucidated by future studies.

ACKNOWLEDGMENTS

This work was supported, in part, by Grant-in-Aid for Scientific Research (Nos. 56480104, 57870019 and 57770174, 1982) from The Ministry of Education, Science and Culture, Japan.

REFERENCES

Asano T. and Ogasawara N. (1980) Solubilization of γ-aminobutyric acid receptor from rat brain. Life Sci. 26, 1131-1137.

Beaumont K., Chilton W., Yamamura H.I., and Enna S.J. (1978) Muscimol binding in rat brain: Association with synaptic GABA receptors. Brain Res. 148, 153-162.

Braestrup C. and Squires R.F. (1977) Specific benzodiazepine receptors in rat brain characterized by high affinity [^{3}H] diazepam binding. Proc. Natl. Acad. Sci. USA 74, 3805-3809.

Brilely M.S. and Langer S.Z. (1978) Influence of GABA receptor agonists and antagonists on the binding of [^{3}H] diazepam to the benzodiazepine receptor. Eur. J. Pharmacol. 52, 129-132.

Chan-Palay V. (1979) Autoradiographic localization of γ-aminobutyric acid receptors in the rat central nervous system by using [^{3}H] muscimol. Proc. Natl. Acad. Sci. USA 75, 1024-1028.

Costa E., Guidotti A., Mao C.C., and Suria A. (1975) New concepts on the mechanism of action of benzodiazepines. Life Sci. 17, 167-186.

Costa T., Rodbard D., and Pert C.B. (1979) Is the benzo-

diazepine receptor coupled to a chloride anion channel ? Nature 277, 315-317.
Eccles J.C. (1964) in The Physiology of Synapses (Eccles J. C., ed), Academic Press, New York.
Enna S.J. and Snyder S.H. (1976) Influences of ions, enzymes and detergents on γ-aminobutyric acid receptor binding in synaptic membranes of rat brain. Mol. Pharmacol. 13, 442-453.
Gallager D.W. (1978) Benzodiazepines: potentiation of a GABA inhibitory response on the dorsal raphe nucleus. Eur. J. Pharmacol. 49, 133-143.
Gavish M. and Snyder S.H. (1981) γ-Aminobutyric acid and benzodiazepine receptors: Copurification and characterization. Proc. Natl. Acad. Sci. USA 78, 1939-1942.
Giambalvo C.T. and Rosenberg P. (1976) The effect of phospholipases and proteases on the binding of γ-aminobutyric acid to junctional complexes of rat cerebellum. Biochim. Biophys. Acta 436, 741-756.
Goodman D.S. (1958) The interaction of human serum albumin with long-chain fatty acid anions. J. Am. Chem. Soc. 80, 3892-3898.
Greenlee D.V., Van Ness P.G., and Olsen R.W. (1978) Endogenous inhibitor of GABA binding in mammalian brain. Life Sci. 22, 1953-1962.
Hori N., Ikeda K., and Roberts E. (1978) Muscimol, GABA, picrotoxin: effects on membrane conductance of a crustacean neuron. Brain Res. 141, 364-370.
Ito Y., Kuriyama K., Ueno E., Nishimura C., and Yoneda Y. (1982) Solubilization and partial purification of cerebral GABA receptors, in Problems in GABA Research (Okada Y. and Roberts E., eds), pp316-327, Excerpta Medica, Amsterdam.
Ito Y. and Kuriyama K. (1982) Some properties of solubilized GABA receptor. Brain Res. 236, 351-363.
Iversen L.L. and Bloom F.E. (1972) Studies on the uptake of [^{3}H]GABA and [^{3}H]glycine in slices and homogenates of rat brain and spinal cord by electron microscopic autoradiography. Brain Res. 41, 131-143.
Johnston G.A.R. and Kennedy S.M.E. (1978) GABA receptor and phospholipids, in Amino Acids as Chemical Transmitters (Fonnum F., ed), pp507-516, Plenum Press, New York.
Karobath M. and Sperk G. (1979) Stimulation of benzodiazepine receptor binding by γ-aminobutyric acid. Proc. Natl. Acad. Sci. USA 76, 1004-1006.
Kravitz E.A., Morinoff P.B., and Hall Z.W. (1965) A comparison of the enzymes and substrates of gamma-aminobutyric acid

metabolism in lobster excitatory and inhibitory axons. Proc. Natl. Acad. Sci. USA 54, 778-782.
Krogsgaard-Larsen P., Johnston G.A.R., Curtis D.R., Game C.J.A., and McCullock R.H. (1976) Structure and biological activity of a series of conformationally restricted analogues of GABA. J. Neurochem. 25, 803-809.
Kuriyama K., Haber B., Sisken B., and Roberts E. (1966) The γ-aminobutyric acid system in rabbit cerebellum. Proc. Natl. Acad. Sci. USA 55, 846-852.
Kuriyama K., Kurihara E., Ito Y., and Yoneda Y. (1980) Increase in striatal [^{3}H]muscimol binding following intrastriatal injection of kainic acid: A denervation supersensitivity phenomenon. J. Neurochem. 35, 343-348.
Lloyd K.G. and Davidson L. (1979) [^{3}H]GABA binding in brains from Huntington's chorea patients: Altered regulation by phospholipids? Science 205, 1147-1149.
Möhler H. and Okada T. (1977) Benzodiazepine receptors: demonstration in the central nervous system. Science 198, 849-851.
Möhler H., Okada T., Heity P.H., and Ulrich J. (1978) Biochemical identification of the site of action of benzodiazepines in human brain by [^{3}H]diazepam binding. Life Sci. 22, 985-996.
Snodgrass S.R. (1978) Use of [^{3}H]muscimol for GABA receptor studies. Nature 273, 392-394.
Squires R.F. and Braestrup C. (1977) Benzodiazepine receptors in rat brain. Nature 266, 732-734.
Stephenson F.A., Watkins A.E., and Olsen R.W. (1982) Physiochemical characterization of detergent-solubilized γ-aminobutyric acid and benzodiazepine receptor proteins from bovine brain. Eur. J. Biochem. 123, 291-298.
Tallman J.F., Thomas J.W., and Gallager D.W. (1978) GABAergic modulation of benzodiazepine binding site sensitivity. Nature 274, 383-385.
Toffano G., Guidotti A., and Costa E. (1978) Purification of an endogenous protein inhibitor on the high affinity binding of γ-aminobutyric acid to synaptic membranes of rat brain. Proc. Natl. Acad. Sci. USA 75, 4024-4028.
Ueno E. and Kuriyama K. (1981) Phospholipids and benzodiazepine recognition sites of brain synaptic membranes. Neuropharmacol. 20, 1169-1176.
Wolfe L.S. (1982) Eicosanoids: prostaglandins, thromboxanes, leukotrienes, and other derivatives of carbon-20 unsaturated fatty acids. J. Neurochem. 38, 1-14.
Yoneda Y. and Kuriyama K. (1978) A comparison of microdistributions of taurine and cysteine sulphinate decarboxylase

activity with those of GABA and L-glutamate decarboxylase activity in rat spinal cord and thalamus. J. Neurochem. 30, 821-825.

Yoneda Y. and Kuriyama K. (1980 a) Some characteristics of [^{3}H] muscimol binding to synaptic membrane from rat brain. Brain Res. 197, 554-560.

Yoneda Y. and Kuriyama K. (1980 b) Presence of a low molecular weight endogenous inhibitor on ^{3}H-muscimol binding in synaptic membranes. Nature 285, 670-673.

Zukin S.R., Young A.B., and Snyder S.H. (1974) Gamma-aminobutyric acid binding to receptor sites in the rat central nervous system. Proc. Natl. Acad. Sci. USA 71, 4802-4807.

CHARACTERIZATION AND ISOLATION OF DOPAMINE-BINDING PROTEINS IN RAT CNS BY GEL ELECTROPHORESIS & AFFINITY CHROMATOGRAPHY

Louise L. Hsu[1], Kayoko Moroi[1] and Angelos E. Halaris[2]

[1]Department of Psychiatry and Behavioral Sciences, The University of Texas Medical Branch, Galveston, Texas USA 77550
[2]Department of Psychiatry, UCLA School of Medicine and Veterans Administration Hospital, Brentwood, Los Angeles, California, USA 90073

During the last decade, neurotransmitter receptors in the central nervous system have received considerable attention from investigators in search of unknown factors that may be involved in the etiology of various neurological and psychiatric disorders. In particular, dopamine (DA) receptors have been shown to be related to tardive dyskinesia (Baldessarini and Tarsey, 1980; Creese and Snyder, 1980 and Jeste and Wyatt, 1981), Parkinson's disease (Lee et al.,1981 and Rinne, et al., 1981), Huntington's Chorea (Login et al., 1982) and schizophrenia (Crow et al., 1978; Lee et al., 1978; Lee and Seeman, 1980; and Mackay et al., 1980). Numerous reports have focused on the pharmacological and biochemical characterization and identification of multiple forms of DA receptors in various regions of the brain (Hsu and Halaris, 1977; Creese and Sibley, 1979; Creese et al., 1979; Clement-Cormier and George, 1979; Kebabian and Calne, 1979;Sokoloff, et al., 1980 a and b; and Nishikori, et al.; 1980, Titeler and Seeman, 1980, Creese et al., 1981). At least four subtypes of DA receptors have been characterized in mammalian brain and differentiated by their relationship to adenylate cyclase and their affinity for DA agonists and antagonists (Sokoloff et al., 1980b). The pre- and post-synaptic localization of these DA receptors has been determined by the effects of kainate and 6-hydroxy-dopamine lesions. However, it remains to be proven if these DA receptors represent the same protein molecules with different binding sites. Also, whether or not DA agonists and antagonists bind to the same site of each receptor protein awaits elucidation. With these goals in mind, we have undertaken to characterize and purify these DA receptors in various brain regions and subcellular fractions using disc gel electrophoresis and DA-affinity chromatography techniques. Here we report on the evidence that DA agonists and antagonists bind to identical as well as different protein molecules in brain and that the subtypes of DA receptors represent different protein molecules.

MATERIALS AND METHODS

Adult male Sprague-Dawley rats (Madison, Wisconsin), 150-200g in weight, were used. Animals were kept in an air-conditioned room at 22°C and 55% relative humidity with a 7 a.m. to 7 p.m. light-dark cycle. The rats were housed six per cage and had ad libitum access to Purina rat chow and water.

DA-affinity gel beads were synthesized according to the method of Cuatrecasas (1970). Sigma 4B-200 Sepharose with a side arm of 1,6-hexanediamine (w-amino-hexyl-sepharose) was treated with p-nitrobenzoyl azide (Eastman Kodak) for 1 hour at room temperature in 0.2 N sodium borate buffer pH 9.3 and 40% dimethylformamide (v/v) to yield p-nitrobenzamido-hexyl-Sepharose which was subsequently reduced to p-aminobenzamido-hexyl-Sepharose by sodium dithionite in 0.5 M $NaHCO_3$, pH 8.5. This derivative was diazotized by 0.1 M sodium nitrite at 4°C for 7 minutes. The yellowish diazonium-Sepharose derivative was coupled at once with DA in 0.2 N sodium borate buffer, pH 10, at 4°C overnight to yield orange colored DA-Sepharose affinity beads. Polyacrylamide disc gels were prepared by the method of Davis (1964).

Subcellular fractions from whole rat brain or brain regions were obtained according to the method of Whittaker and Barker (1972) as modified by Halaris and Freedman (1977). Brain tissues were homogenized in 9 volumes of 0.32 M sucrose and centrifuged at 1,000 g for 10 min. The supernatant fraction was further centrifuged at 12K g for 20 min. to yield the crude mitochondrial fraction. The supernatant fraction was further centrifuged at 200K g for 30 min. to yield the cytosol fraction, S_3. The pellet (crude mitochondria) was suspended in an appropriate amount of 0.32 M sucrose, layered over a sucrose gradient (0.8 M sucrose over 1.2 M sucrose), and centrifuged at 100K g for 30 min. The synaptosomal fraction (P_{2B}) was collected between the interface of 0.8 M and 1.2 M sucrose gradients. The P_{2B} fractions from whole rat brain were diluted with an equal volume of distilled water and centrifuged at 200K g for 30 min. The resultant P_{2B} pellet was suspended in cold distilled water (2 ml/g orig. tissue), sonicated for 90 sec, and centrifuged at 12K g for 20 min to yield the synaptic membrane fraction P_{2D} and the supernatant fraction, S_{p2B}. S_{p2B} was further centrifuged at 200K g for 30 min to yield the intrasynaptic soluble fraction, S.

For some experiments, rat brain regions rich in DA innervation, including corpus striatum, hypothalamus, amygdala, olfactory tubercles and septum, were dissected freehand over ice (Hsu and Mandell, 1975). Cerebellum and hippocampus, which lack DA terminals, were also included for comparison. The synaptosomal or nerve ending fraction (P_{2B}) from each brain region was obtained according to the procedures described above. Aliquots of P_{2B} (0.1-0.3 mg protein) from each region were incubated with 20 nM (^{3}H)-DA in a total volume of 0.2 ml potassium-phosphate buffer (KPB), Ph 7.5, containing 1 µM pargyline in the absence or presence of 10 µM d- or l-butaclamol, 0.1 mM d-LSD, haloperidol (HAL) or chlorpromazine (CPZ), at 37°C for 20 minutes. The incubation mixtures were subsequently applied to polyacrylamide disc gels and electrophoresed at 4 mA/gel for 1.5 hour in the cold with bromophenol blue as the dye marker. After electrophoresis, some gels were stained with Coomassie brilliant Blue G250 for visualization of protein patterns and other gels were sliced into 12 segments, each segment being meshed and dissolved in 1 ml of Soluene and 10 ml of Aquasol for measurement of radioactivity in a Tri-carb Scintillation Spectrophotometer.

In other experiments, the synaptic membrane fraction (P_{2D}), cytosol fraction (S_3) and intrasynaptic soluble fraction (S) from whole rat brain were prepared according to the method described above. P_{2D} fraction was treated with 1% Triton X-100 at 4°C overnight to yield the solubilized synaptic membrane fraction, P_{2DT}. Aliquots of P_{2DT}, S_3 and S (0.1-0.3 mg protein) were incubated with 10 nM each of (^{14}C)-DA and (^{3}H)-HAL simultaneously in 0.5 M KPB, pH 7.5 containing 1 mM $NaHSO_3$ in the absence or presence of 0.04 mM unlabeled DA, HAL, cis- or trans- flupenthixol of phencyclidine at 37°C for 30 minutes. At the end of incubation, the reaction mixtures were applied to disc polyacrylamide gels (7.25%) and electrophoresed for 1.5 hour in the cold. The gels were processed the same way as described above for measurements of radioactivity of both (^{14}C) and (^{3}H), simultaneously.

For isolation and further characterization of DA-binding proteins in the synaptic membranes, P_{2BT} fraction obtained from 6 rat brains was incubated with DA-Sepharose affinity beads at 37°C for 40 min. At the end of the incubation, the beads were transferred to a Sephadex column and washed with 0.2 M KPB, pH 7.5, until no more protein was detected in the washings. The DA-binding proteins were eluted with 0.1 mM of HAL, APM and CPZ, consecutively. After each drug elution, the beads were washed with the same buffer. Each drug eluate was collected, dialyzed against 0.02 M KPB overnight and concentrated to a minimal volume by sucrose crystals. Aliquots of each concentrated drug eluate were incubated with 10 nM of either (^{3}H)-DA or (^{3}H)-HAL in the presence of 1μM pargyline at 37°C for 0.5 hour. At the end of incubation, reaction mixtures were applied to disc polyacrylamide gels for electrophoretic analysis. Some gels were stained with Coomassie blue 250 G for visualization of protein bands and other gels were sliced into 18 segments for measurements of radioactivity.

RESULTS

1. Disc gel electrophoretic patterns of (^{3}H)-DA binding proteins in the synaptosomal fractions of various rat brain regions.

After incubation with (^{3}H)-DA, the intact P_{2B} fraction of corpus striatum showed a large binding peak at the origin of the gel, and three slight binding peaks at gel segments 5, 8 and 10, respectively (Fig. 1). Both d- and l- butaclamol (10 μM) blocked all these DA-binding peaks. The P_{2B} fraction of hypothalamus exhibited a sharp binding peak at gel segment 2 and two small peaks at gel segments 5 and 8, respectively. Again, these peaks were blocked by d- and l- butaclamol. The synaptosomal fraction of amygdala showed a similar binding pattern to that of corpus striatum at the origin of the gel but did not show any binding peaks in the middle part of the gel. Both d- and l- butaclamol blocked this binding peak near the origin. Olfactory tubercle showed two small binding peaks at gel segments 3 and 8, respectively, which were also blocked by d- and l- butaclamol. Surprisingly, septum did not show any DA-binding peaks in

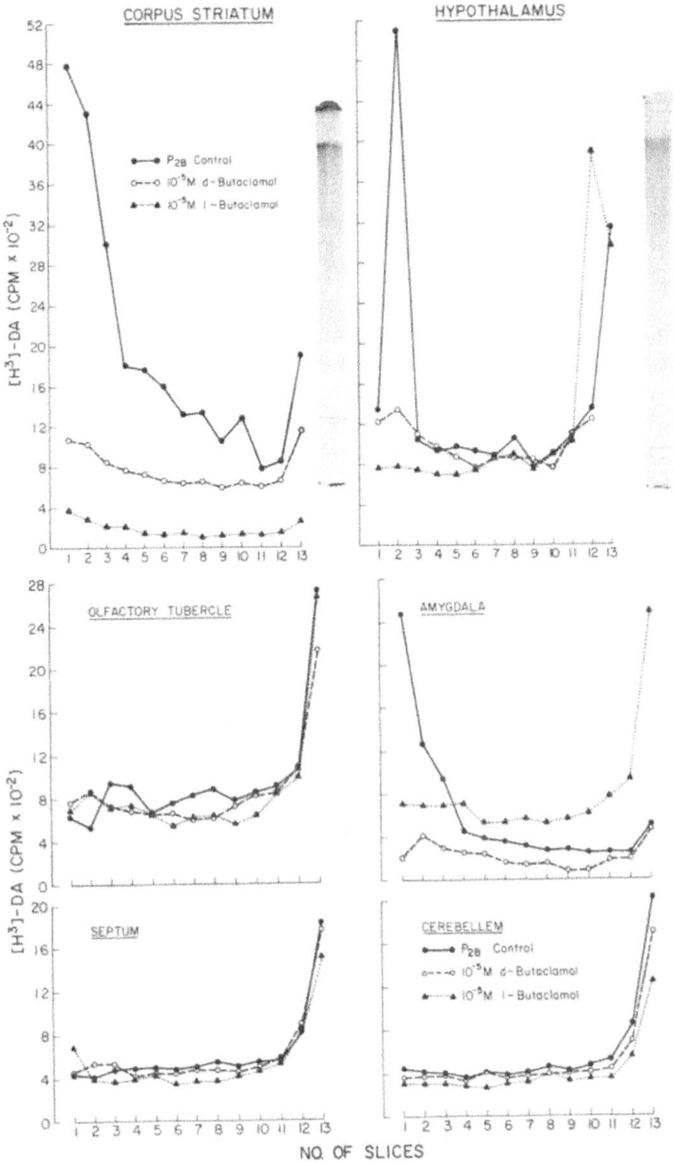

FIG. 1. Disc gel electrophoretic protein patterns and (^{3}H)-DA binding profiles of P_{2B} from rat brain regions. Aliquots of P_{2B} of each brain region were incubated with 10 nM (^{3}H)-DA with or without drugs at 37°C for 20 min. & applied to gels for electrophoretic analysis.

the gel at all. As expected, cerebellum did not exhibit any DA-binding peaks in the gel.

2. Further characterization of the gel electrophoretic patterns of (^{3}H)-DA binding proteins in P_{2B} fractions of various brain regions by various drugs.

The properties of the above described DA-binding peaks were further characterized by the effects of cold DA, HAL, CPZ, APM and d-LSD. As shown in Fig. 2, at 0.1 mM, all of these drugs blocked the DA-binding peaks in P_{2B} fraction of hypothalamus and P_{2BT} (1% Triton X-100 treated) fraction of corpus striatum. As expected, P_{2B} fraction of hippocampus did not show any DA-binding peaks either in the absence or presence of various drugs. Triton X-100 seemed to have markedly inhibited the DA-binding in P_{2B} fraction of corpus striatum (Figs. 1 & 2).

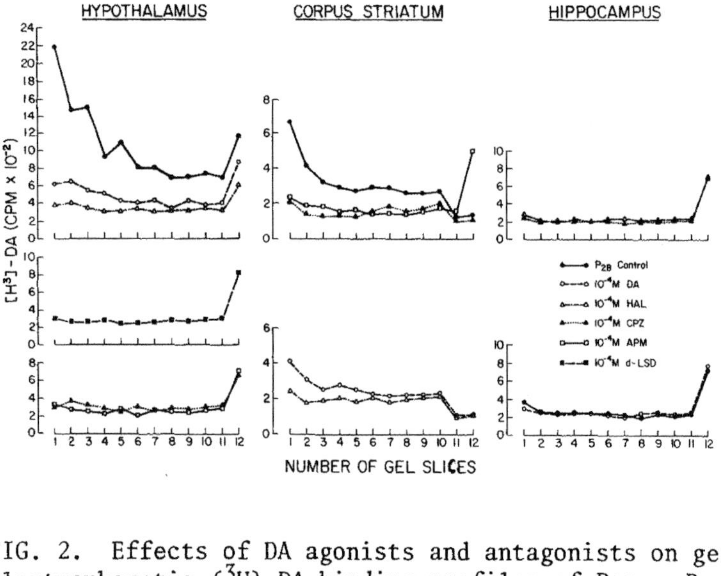

FIG. 2. Effects of DA agonists and antagonists on gel electrophoretic (^{3}H)-DA-binding profiles of P_{2B} or P_{2BT} fraction from rat brain regions. P_{2B} fractions of hypothalamus and hippocampus, and P_{2BT} (Triton X-100 treated) fraction of corpus striatum were used in this experiment. Methods were the same as described in legend for FIG. 1 except that different drugs were used.

3. Disc gel electrophoretic patterns of (^{14}C)-DA and (^{3}H)-HAL binding proteins in P_{2DT}, S, and S_3 fractions of whole rat brains and drug effects.

The finding that the intact P_{2B} fractions of corpus striatum, hypothalamus and amygdala showed DA-binding peaks in the polyacrylamide gels suggested that some kind of soluble DA-binding proteins either inside or outside the synaptosomes were able to enter into the gels during electrophoresis. This also raised the question whether there might be soluble DA-binding proteins in the intrasynaptic soluble fraction, S, as well as in the cytosol fraction, S_3. In a pilot study of the possible DA-binding protein(s) in the S_3 and S fraction of rat brain using Sephadex-G50 gel filtration methods, we observed that a (^{3}H)-DA-binding peak coincided with the protein peak in both fractions. Therefore, we proceeded to further characterize the soluble binding protein(s) in S_3 and S, and compared their electrophoretic patterns with that obtained for the Triton X-100 solubilized synaptic membrane fraction, P_{2DT}. In order to determine whether or not HAL binds to the same proteins as DA does in these three fractions, a double labeling technique was employed using both (^{14}C)-DA and (^{3}H)-HAL as ligands in the same incubation mixtures.

As shown in Fig. 3, P_{2DT} fraction exhibited a slight (^{14}C)-DA binding peak at the gel origin, a medium size peak at gel segment 5 and a large sharp peak at gel segment 10. On the other hand, there was a small and broad (^{3}H)-HAL binding hump at gel segments 1-4 and possibly two (^{3}H)-HAL binding peaks at gel segments 5-11. Unlabeled DA inhibited the (^{14}C)-DA binding at both gel segments 5 and 10 but enhanced the binding at gel segments 8 and 9; it slightly enhanced the (^{3}H)-HAL binding at the gel origin and gel segments 8-9 but inhibited the binding at gel segments 6 and 7. Unlabeled HAL broadened the (^{14}C)-DA binding at gel segments 4-6 and enhanced the (^{14}C)-DA binding peak at gel segment 10; it inhibited the (^{3}H)-HAL binding at gel segment 8 but did not affect the (^{3}H)-HAL binding at other segments to any significant extent. trans-Flupenthixol slightly increased the (^{14}C)-DA binding at the origin, and the binding at gel segment 6 shifting the second binding peak to segment 6 but did not affect the (^{14}C)-DA binding peak at segment 10; it enhanced the (^{3}H)-HAL binding at the gel origin and segment 4 but decreased the (^{3}H)-HAL binding at gel segments

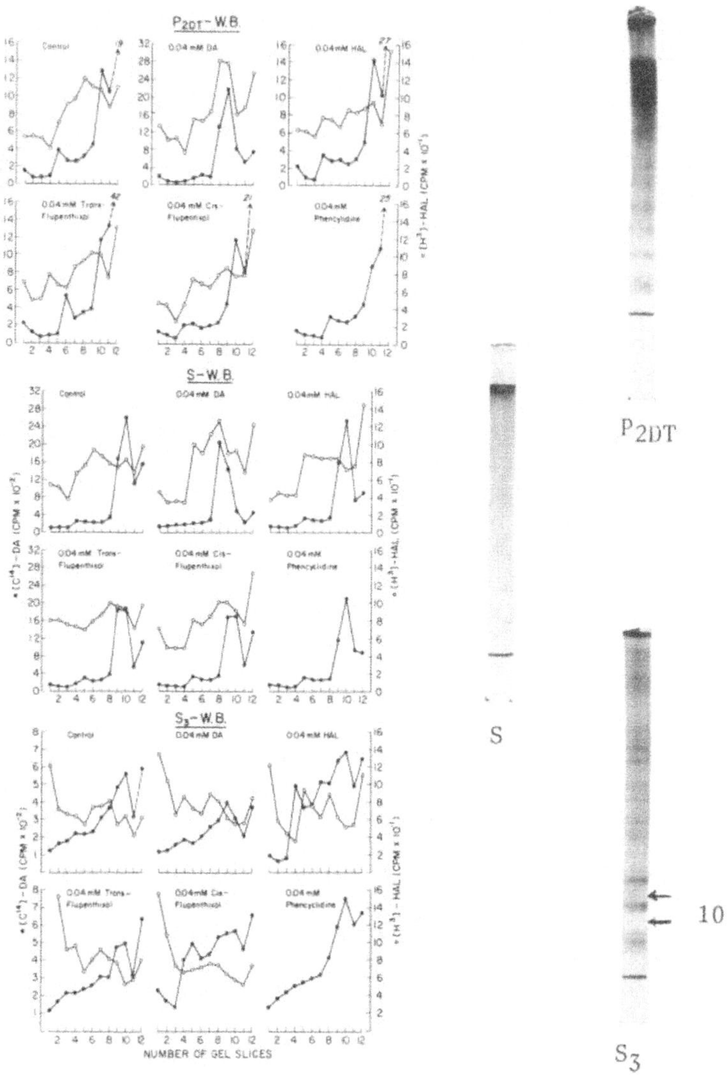

FIG. 3. Disc gel electrophoretic protein patterns and (^{14}C)-DA and (^{3}H)-HAL binding profiles of different subcellular fractions of rat brain. For details, see Materials and Methods.

5 through 10 altering the (^{3}H)-HAL binding profile to three medium size peaks. cis-Flupenthixol did not affect the (^{14}C)-DA binding patterns to any noticeable degree but it significantly reduced the (^{3}H)-HAL binding at gel segments 6-10. Phencyclidine did not affect the (^{14}C)-DA binding at segment 5 but it reduced the (^{14}C)-DA binding at segment 10 by 30%.

The S fraction also showed a small (^{14}C)-DA binding peak at gel segment 4, and a very distinct and sharp binding peak at gel segment 10. On the other hand, S showed a slight (^{3}H)-HAL binding peak at the gel origin and two to three medium sized binding peaks at gel segments 4-10. Unlabeled DA drastically reduced the (^{14}C)-DA binding peak at gel segment 10 shifting the peak to segment 8, but it enhanced the (^{3}H)-HAL binding peaks at gel segments 4-10 causing the largest peak to appear at gel segment 8 instead of segment 6. Unlabeled HAL seemed to have slightly enhanced the (^{14}C)-DA binding at gel segments 4-5 but did not affect the binding peak at gel segment 10. It did not alter the binding pattern of (^{3}H)-HAL to any significant extent either. Interestingly and maybe significantly, both cis- and trans- flupenthixol reduced the (^{14}C)-DA binding peak at gel segment 10 by about 30% but did not affect the binding at gel segment 9 and thus gave rise to a flattened (^{14}C)-DA binding peak at gel segments 9 and 10. trans-Flupenthixol greatly enhanced the (^{3}H)-HAL binding at the gel origin and the gel segments 2-3 but slightly reduced the (^{3}H)-HAL binding at gel segment 6 shifting the binding peak to gel segment 8. cis-Flupenthixol did not affect the (^{3}H)-HAL binding at gel origin to any noticeable extent nor did it alter the (^{3}H)-HAL binding at gel segments 5-10. Phencyclidine inhibited the (^{14}C)-DA binding at gel segments 9 and 10 significantly but did not affect the small DA binding peak at gel segment 5.

The cytosol fraction S_3 showed a small binding peak at gel segment 10 for (^{14}C)-DA whereas it showed a large (^{3}H)-HAL binding peak at the origin and 2 to 3 medium size binding peaks at gel segments 6-10. Unlabeled DA reduced the binding peak for (^{14}C)-DA at gel segment 10 but did not alter the binding at segment 4. The unlabeled DA enhanced the large binding peak for (^{3}H)-HAL and altered the binding patterns at gel segments 4-10. Unlabeled HAL, on the other hand, enhanced both binding peaks for (^{14}C)-DA at gel segments 4 and 10, contrary to our prediction, and it also altered the binding patterns of (^{3}H)-HAL at the origin as well as at gel segments 4-10, resulting in two sharp peaks at gel segments 5 and 8. trans-Flupenthixol slightly inhibited the (^{14}C)-DA binding at gel segment 10 but did not affect the (^{3}H)-HAL binding to any significant extent. cis-Flupenthixol, however, greatly enhanced the (^{14}C)-DA binding at the origin and gel segments 4-5, and broadened the binding peak at segments 8-10. It greatly enhanced the (^{3}H)-HAL binding at the origin, but reduced the HAL-binding at gel segments 4-10, emerging the binding into a small flattened peak. Unexpectedly, phencyclidine enhanced and sharpened the (^{14}C)-DA binding peak at gel segments 8-10.

4. Isolation and characterization of DA binding protein(s) in the synaptosomal fraction of rat brain.

To further investigate the DA-binding protein(s) in the synaptosomal fraction (P_{2B}) of the rat brain, we partially purified these binding proteins by affinity chromatography and characterized them by disc gel electrophoresis.

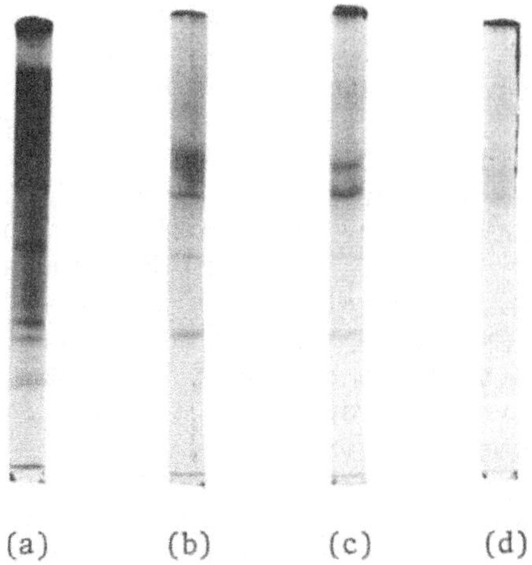

FIG. 4. Disc gel electrophoretic protein patterns of DA binding proteins in P_{2BT} fraction of the rat brain eluted from DA-Sepharose affinity column. (a) P_{2BT} fraction, (b) HAL eluate, (c) APM eluate and (d) CPZ eluate. For details, see Materials and Methods.

Results in Fig. 4 indicated that, in addition to the protein band at the origin, there were at least 5 protein bands in the HAL eluate, 4 corresponding protein bands in the APM eluate and only 3 protein bands in the CPZ eluate with various intensities. Furthermore, as shown in Fig. 5, the HAL eluate gave rise to two major (^{3}H)-DA binding peaks, one at the gel origin, the other at gel segments 3-5, and two slight peaks at gel segment 8 and 12, respectively, whereas the same eluate demonstrated a broad (^{3}H)-HAL binding peak at gel segments 4-13 with fluctuations. The APM eluate showed three (^{3}H)-DA binding peaks at gel segments 4-11 and a broad (^{3}H)-HAL binding peak at the same gel segments. Similarly, the CPZ eluate also yielded several (^{3}H)-DA binding peaks at gel segments 4-13 but it showed a very small and broad (^{3}H)-HAL binding peak at gel segments 4-13.

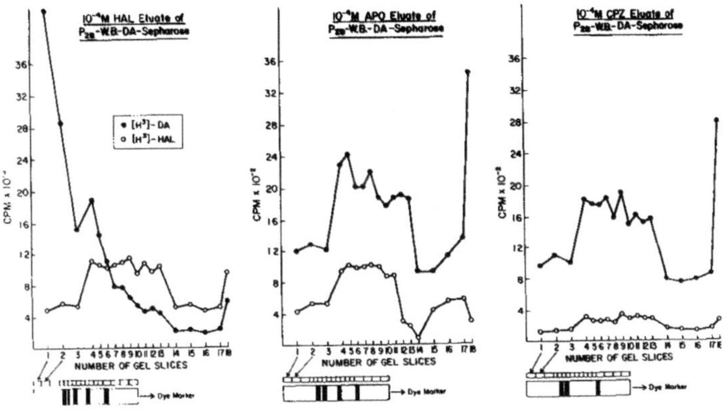

FIG. 5. Disc gel electrophoretic DA and HAL binding profiles of DA binding proteins in P_{2B} fraction eluted from DA-affinity column by various drugs. Bottom: Diagram of gel electrophoretic protein patterns for corresponding drug eluates. For details, see Materials and Methods.

DISCUSSION

The results from (^{3}H)-DA binding studies with synaptosomal fractions of various brain regions indicated that the DA-binding proteins in P_{2B} of one region might be different from those in another region. For example, in hypothalamus, we observed high binding capacity with slow moving proteins whereas in corpus striatum, we observed markedly less binding in the same slow moving region but moderately high binding in fast moving protein bands (Fig. 1). As expected, very little binding was observed in the hippocampal synaptosomes due to lack of DA terminals in this region. HAL, APM and CPZ were more potent in blocking the (^{3}H)-DA binding than unlabeled DA in all the regions examined. This may indicate that HAL, APM and CPZ compete with (^{3}H)-DA for the binding site(s) but unlabeled DA simply exerts a dilution effect.

Our data from double labeling studies of DA- and HAL-binding proteins in subcellular fractions (Fig. 3) suggest that there are multiple (^{14}C)-DA and (^{3}H)-HAL binding proteins in both the membranous fraction, P_{2DT}, and the soluble fractions, S_3 and S. The major (^{14}C)-DA binding protein(s) were observed at gel segments 8-10 in all the three subcellular fractions investigated while the minor binding was observed at gel segments 4-6. On the other hand, the major (^{3}H)-HAL binding protein(s) in S and P_{2DT} were observed at gel segments 5-10 and the minor binding proteins were observed near the origin; whereas, the major (^{3}H)-HAL binding protein(s) in S_3 were observed at the origin with the minor binding protein(s) observed at gel segments 4-6.

The (^{14}C)-DA and (^{3}H)-HAL binding peaks in P_{2DT} and S fractions overlapped at gel segments 9 and 10 but not so much in S_3. This suggests that in P_{2DT} and S there may be common binding protein(s) for (^{14}C)-DA and (^{3}H)-HAL appearing at gel segment 9 and 10 with different affinities. Furthermore, there may be binding proteins in these two fractions specific only for (^{3}H)-HAL but not for (^{14}C)-DA between the gel origin and segment 8. There is relatively less DA-binding protein(s) with fast mobility in S_3 at gel segment 10 than in P_{2DT} or S, and relatively more (^{3}H)-HAL binding protein(s) with slow mobility in S_3 at the origin than in P_{2DT} or S fraction. This indicates that the solubilized DA-binding proteins from P_{2DT} fraction are similar in nature to those in S fraction but different from those in S_3 fraction. To further characterize the HAL binding proteins in S_3 at gel origin, low concentration of polyacrylamide gels would have to be used.

The multiple protein bands in P_{2BT} fraction eluted from DA-affinity column by HAL, APM and CPZ strongly suggest that there are multiple DA-binding proteins in the synaptosomal fraction of rat brain. Observations that HAL, APM and CPZ eluted at least three identical DA-binding protein bands from the affinity column further support the notion that DA agonists and antagonists bind to the same protein molecules in P_{2B} fraction but with different affinities. Moreover, the electrophoretic patterns in Fig. 4 also suggest that the top DA binding protein band in HAL eluate might be specific for HAL or had highest affinity toward HAL, because it did not appear in either the APM or the CPZ eluate.

In summary, the gel electrophoretic protein patterns as well as ligand binding patterns of subcellular fractions and various drug eluates from DA-affinity columns provide physical evidence that there are multiple forms of DA-binding proteins in the synaptic membrane and soluble fractions of rat brain.

The identity of each DA-binding protein to various types of DA receptors remains to be determined. Experiments are underway in our laboratory to achieve this goal. Recently, we have been able to isolate one single DA-binding protein with fast mobility in the HAL eluate from the synaptic membrane fraction. (Hsu and Moroi, 1982, details will be published elsewhere). Preliminary SDS gel electrophoretic analyses of this single DA-binding protein showed at least two distinct protein bands with R_f values of 0.5 and 0.65, respectively. The apparent molecular weights of these subunits were estimated to be 70K and 45K daltons, respectively. A larger quantity of this DA binding protein is being collected for studies on adenylate cyclase activity, ligand specificity and other binding properties. Furthermore, we have succeeded in obtaining monoclonal antibodies specific to the DA-binding proteins (partially purified) in rat brain and these antibodies will be used for further purification of each DA-binding protein in the drug eluates from DA-Sepharose affinity columns. They will also be used for immunohistochemical localization of various types of DA receptors in the brain.

ACKNOWLEDGEMENTS

This work was supported in part by a grant from Robert A. Welch Foundation and a grant from the Scottish Rite

Foundation. We thank Nilda Munoz and Robert Chang for performing the assays and Terry Bellais for the typing.

REFERENCES

Baldessarini, R.J. and Tarsy, D. (1980) Pathophysiologic basis of tardive dyskinesia in long-term effects of neuroleptics. Adv. Biochem. Psychopharmacol. 24, 451-455.

Clement-Cormier, Y.C. and George, R.J. (1979) Multiple dopamine binding sites: subcellular localization and biochemical characterization. J. Neurochem. 32, 1061-1069.

Creese, I. and Sibley, D.R. (1979) Radioligand binding studies: evidence for multiple dopamine receptors. Commun. Psychopharmacol. 3, 385-396.

Creese, I., Usdin, T., and Snyder, S.H. (1979) Guanine nucleotides distinguish two dopamine receptors. Nature (London) 278, 577-578.

Creese, I. and Snyder, S. (1980) Chronic neuroleptic treatment and dopamine receptor regulation. Long-term effects of neuroleptics. Adv. Biochem. Psychopharmacol. 24, 89-94.

Creese, I., Sibley, D.R., Leff, S. and Hamblin, M. (1981) Dopamine receptors: subtypes, localizations and regulation. Fed. Proc. 40, 147-152.

Crow, T.J., Johnstone, E.C., Longden, A.J. and Owen, F. (1978) Dopaminergic mechanisms in schizophrenia: The antipsychotic effects and the disease process. Life Sci. 23, 563-568.

Cuatrecasas, P. (1970) Protein purification by affinity chromatography. J. Biol. Chem. 245, 3059-3065.

Davis, B.J. (1964) Disc electrophoresis-II. Method and application to human serum proteins. Ann. New York Acad. Sci. 121, 404-427.

Halaris, A.E. and Freedman, D.S. (1977) Vesicular and juxtavesicular serotonin: effects of lysergic acid ditheylamide and reserpine. J. Pharmacol. Exp. Ther. 203(3), 575-586.

Hsu, L.L. and Halaris, A.E. (1977) Characterization and isolation of dopamine binding protein(s) in rat CNS by affinity chromatography and gel electrophoresis. Proc. Int. Soc. Neurochem. 6, 624.

Hsu, L.L. and Mandell, A.J. (1975) Enzymatic formation of tetrahydro-β- carboline from tryptamine and 5-methyltetrahydrofolic acid in rat brain fractions: regional

and subcellular distribution. J. Neurochem. 24, 631-636.
Hsu, L.L. and Moroi, K. (1982) Dopamine binding proteins in synaptical membranes of rat brain. Trans. Amer. Soc. Neurochem. 13(1), 249.
Jeste, D.V. and Wyatt, R.J. (1981) Dogma disputed: Is tardive dyskinesia due to postsynaptic dopamine receptor supersensitivity? J. Clin. Phychiatry 42(12), 455-457.
Kebabian, J.W. and Calne, D.B. (1979) Multiple receptors for dopamine. Nature(London) 277, 93-96.
Lee, T. and Seeman, P. (1980) Elevation of brain neuroleptic /dopamine receptors in schizophrenia. Amer. J. Psychiatry 137, 191-197.
Lee, T., Seeman, P., Tourtellotte, W.W., Farley, I. and Hornykiewicz, O. (1978) Binding of (^{3}H)-neuroleptics and (^{3}H)-apomorphine in schizophrenic brains. Nature (London) 274, 897-900.
Lee, T., Seeman, P., Hornykiewicz, O., Bilbao, J., Deck, J. and Tourtellotte, W.W. (1981) Parkinson's disease: low density and presynaptic location of D_3 dopamine receptors. Brain Res. 212, 494-498.
Login, I.S., Coronin, J.J. and MacLeod, R.M. (1982) Neuroleptic malignant syndrome caused by dopamine depleting drugs. Neurology 32, 218-219.
Mackay, A.V.P., Bird, E.D., Iversen, L.L., Spokes, E.G., Creese, I. and Snyder, S.H. (1980) Dopaminergic abnormalities in post-mortem schizophrenic brain. Adv. Biochem. Psychopharmacol. 24, 325-333.
Nishikori, K., Noshiro, O., Sano, K. and Maeno, H. (1980) Characterization, solubilization and separation of two distinct dopamine receptors in canine caudate nucleus. J. Biol. Chem. 255(22), 10909-10915.
Rinne, U.K., Lonnberg, P. and Koskinen, V. (1981) Dopamine receptors in the Parkinsonian brain. J. Neural Transmis. 51, 97-106.
Sokoloff, P., Martres, M.-P. and Schwartz, J.-C. (1980a) (^{3}H)-Apomorphine labels both dopamine postsynaptic receptors and auto-receptors. Nature 288, 283-286.
Sokoloff, P., Martres, M.-P. and Schwartz, J.-C. (1980b) Three classes of dopamine receptor (D-2, D-3, D-4), identified by binding studies with (^{3}H)-apomorphine and (^{3}H)-domperidone. Naunyn-Schmiedelberg's Arch. Pharmacol. 315, 89-102.
Titeler, M. and Seeman, P. (1980) Radioreceptor labeling of pre- and postsynaptic dopamine receptors. Long-term effects of neuroleptics. Adv. Biochem. Psychopharmacol.

24, 159-165.

Whittaker, V.P. and Barker, L.A. (1972) The subcellular fractionation of brain tissue with special reference to the preparation of synaptosomes and their component organelles. In: Methods of Neurochemistry (R. Fried, ed.) Vol. 2, Marcel Dekker, New York.

EFFECTS OF ASCORBATE ON POST-SYNAPTIC SEROTONIN BINDING SITES IN RAT BRAINS

Jean Chen Shih, Samar F. Muakkassah-Kelly,
Jeffrey W. Andresen and Paul Hochstein
Institute for Toxicology, School of Pharmacy, University of Southern California, Los Angeles, CA 90033

Serotonin (5-HT) is a neurotransmitter which occupies a uniquely important place in neurobiology because of its relationship to certain animals (Jacobs, 1976) and human behavior such as sleep (Henriksen et al., 1974), sexual activity, aggression (Valzelli, 1974) and mental depression (Murphy et al., 1978). Like other neurotransmitters, serotonin is released into the synaptic junction and exerts its effect on specific receptor proteins on the post-synaptic membranes. Based on electrophysiological studies (McCall and Aghajanian, 1979) and on the differences in the potencies of various drugs on 5-HT receptors, it has been suggested that two types of 5-HT receptors exist in rat frontal cortex (Peroutka and Snyder, 1979). The 5-HT agonist, ^{3}H-5-HT, labels the 5-HT_1 receptor which is linked to adenylate cyclase and is regulated by guanine nucleotides (Peroutka et al., 1979). On the other hand, the 5-HT antagonist, spiperone, labels the 5-HT_2 receptor which is not regulated by guanine nucleotides but is related to 5-HT behavioral syndrome (Peroutka et al., 1981). ^{3}H-d-LSD labels both of them. The molecular nature of the two types of 5-HT receptors is not understood at the present time.

This report describes the dual effects of ascorbate on serotonin and spiperone binding in rat cortical membranes. Our results indicate that ascorbate-induced lipid peroxidation decreased both ^{3}H-serotonin and ^{3}H-spiperone binding similarly, suggesting that the lipid environment may be similar for these two types of 5-HT binding sites. In contrast, the direct reducing effect of ascorbate decreased only ^{3}H-5-HT binding, but not ^{3}H-spiperone binding, suggesting a fundamental biochemical difference between the two apparent 5-HT binding sites.

DECREASED ^{3}H-5-HT AND ^{3}H-SPIPERONE BINDING CONSEQUENT TO ASCORBATE-INDUCED LIPID PEROXIDATION

As shown in fig. 1 and 2 when ascorbate (0.025-6.0 mM) was incubated with rat cortical membranes, MDA formation reached a maximum at 0.5 mM ascorbate. Conversely, the specific binding of ^{3}H-5-HT (fig. 1) and of ^{3}H-spiperone binding (fig. 2) to washed ascorbate-pretreated membranes decreased to a minimum at 0.5 mM ascorbate.

At higher ascorbate concentrations (0.5 mM), MDA formation was inhibited. Interestingly, the specific binding of (^{3}H)5-HT and (^{3}H)spiperone correspondingly increased. Inhibition of lipid peroxidation at high ascorbate concentrations has been suggested to result from a disruption of the ratio between ferric and ferrous ions which are essential for the autooxidation process (Wills, 1969).

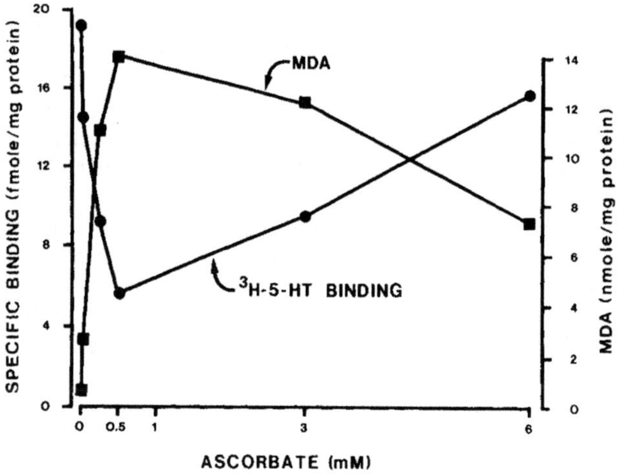

Fig. 1 The effect of ascorbate concentrations on lipid peroxidation and specific (^{3}H)5-HT binding. Lipid peroxidation was produced by incubating the membranes with the

various concentrations of ascorbate at 37°C for 30 min, aliquots were analyzed for MDA production (squares)(Wilbur et al, 1949). The remaining membrane suspension was spun down, washed with 0.05M Tris-HCl pH 7.4 and then resuspended in 5-HT binding assay buffer. The specific binding of 1 nM (^{3}H)5-HT (circles) was determined as described before. Values are averages of triplicate determinations from experiments repeated several times. Detailed procedures for this experiment were published previously (Muakkassah-Kelly et al, 1982).

ASCORBATE-INDUCED LIPID PEROXIDATION REDUCED THE MAXIMUM NUMBER OF BINDING SITES (B MAX) OF BOTH ^{3}H-5-HT AND ^{3}H-SPIPERONE BINDING

Scatchard analysis shows that the maximum number of

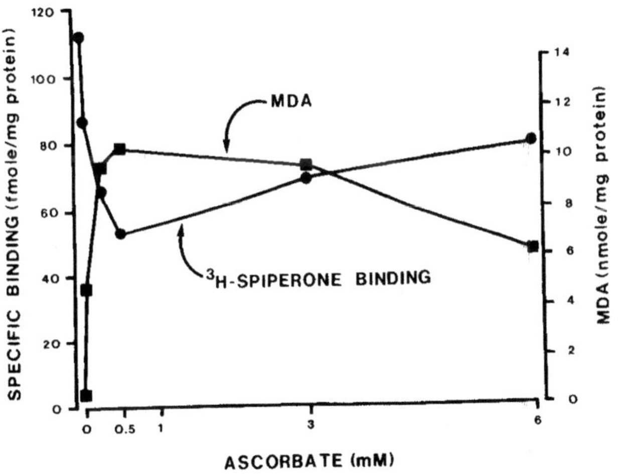

Fig. 2 The effect of ascorbate concentration on lipid peroxidation and specific (^{3}H)spiperone binding. The conditions were the same as in the legend of Fig. 1. MDA formation and specific (^{3}H)spiperone binding (1 nM) are designated as squares and circles respectively. Values are averages of triplicate determinations from experiments repeated several times.

binding sites (B max) of both ^{3}H-5-HT (fig. 3) and ^{3}H-spiperone binding (fig. 4) decreased by 52 and 58% where 9.22 and 11.21 nmoles MDA/mg were produced respectively. No significant changes in the dissociation constants (Kd) were observed.

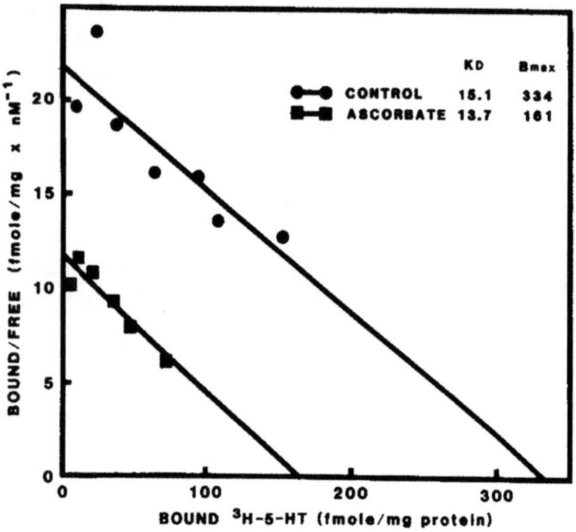

Fig. 3 Scatchard plots of ^{3}H-5-HT specific binding in control membranes subjected to ascorbate induced lipid peroxidation. Freshly prepared cortical membranes (0.8-0.9 mg protein/ml) were incubated in phosphate buffer, pH 7.4, without or with 0.25 mM ascorbate. The reaction was carried out at 37^{0}C for 30 min. MDA production was determined in aliquots. The remaining membrane suspensions were washed twice by centrifugation with 0.05 M Tris-HCl pH 7.4, and the final pellet resuspended in the same buffer containing 5.7 mM ascorbate, 4 mM $CaCl_2$ and 10 μM pargyline. The specific binding of ^{3}H-5-HT was then determined. The amount of MDA produced was 0.23 and 11.21 nmoles/mg protein for control (circles) and ascorbate treated membranes (squares), respectively. The line was drawn by linear regression. The estimated Kd and Bmax in control membranes were 15.4 nM and 334 fmoles/mg protein and in ascorbate treated membranes 13.7 nM and 161 fmoles/mg protein, respectively.

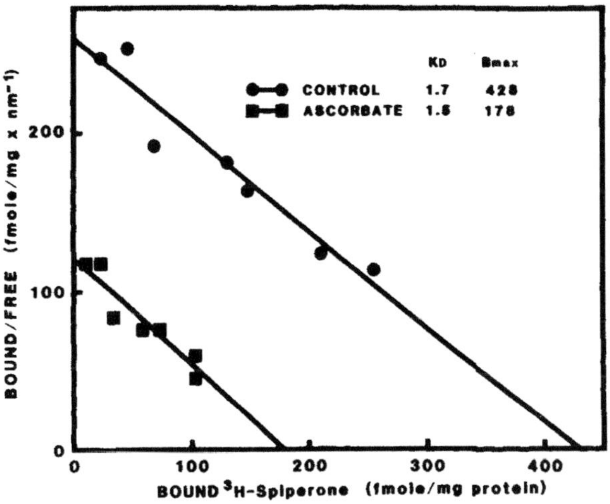

Fig. 4 Scatchard plots of ^{3}H-spiperone specific binding in control membranes and in membranes subjected to ascorbate induced lipid peroxidation. Cortical membranes were treated as described in the legend of fig. 3. The amount of MDA produced was 0.16 and 9.22 nmoles/mg protein in control (circles) and ascorbate treated membranes (squares), respectively. Estimated Kd and Bmax in control membranes were 1.7 nM and 428 fmoles/mg protein and in ascorbate treated membranes 1.5 nM and 178 fmoles/mg protein, respectively.

ANTAGONISM BY EDTA AND BUTYLATED HYDROXYTOLUENE (BHT) OF THE DECREASE IN ^{3}H-5-HT AND ^{3}H-SPIPERONE BINDING DUE TO ASCORBATE-INDUCED PEROXIDATION

Table 1 shows that MDA formation in the presence of 0.25 mM ascorbate is prevented by lipid peroxidation inhibitors EDTA and BHT with a simultaneous blockage of the inactivation of both ^{3}H-5-HT and ^{3}H-spiperone binding. These results provided further evidence that the effect of ascorbate on both types of 5-HT binding sites are indeed mediated through lipid peroxidation.

TABLE 1

Treatment	MDA production nmoles/mg	Specific binding of ^{3}H-spiperone % of control	Specific binding of ^{3}H-serotonin % of control
A. Control	0.34 ± 0.06	100	100
Ascorbate	8.71 ± 0.76	51 ± 2	73
EDTA	0.20 ± 0.04	101 ± 4	99
EDTA + ascorbate	0.22 ± 0.05	109 ± 4	110
B. Control	0.43 ± 0.11	100	100
Ascorbate	10.2 ± 0.44	56 ± 2	57
BHT	0.24 ± 0.08	107 ± 3	103
BHT + ascorbate	1.83 ± 0.15	82 ± 7	89

The effects of EDTA and BHT on the inhibition of ^{3}H-serotonin and ^{3}H-spiperone binding by ascorbate-induced peroxidation. Cortical membranes were treated as described in the legend of Fig. 3 with the additions listed in this table. The final concentrations of ascorbate, EDTA and butylated hydroxytoluene (BHT) were 250, 100 and 10 μM respectively. BHT was dissolved and added in 10 μl acetone. All treatments in group B received an equal volume of acetone. Values are mean±S.E.M. of three separate experiments each done in triplicate. For experiments involving serotonin, values are the average of 2 separate experiments. Control values for the specific binding of 1 nM ^{3}H-spiperone and 2 nM ^{3}H-serotonin were 133.3 ± 11.7 and 28 fmoles/mg protein, respectively.

EFFECT OF ASCORBATE AND OTHER REDUCING AGENTS ON ^{3}H-5-HT BINDING IN THE ABSENCE OF LIPID PEROXIDATION

Since most commonly used binding assays contain 5.7 mM ascorbate, we investigated its effect on lipid peroxidation and on ^{3}H-5-HT binding. Similar to the results shown in fig. 1, significant amounts of MDA were produced when cortical membranes were incubated at 37^{0}C for 12 min (incubation time for the binding assay). However, in the presence of 4 mM $CaCl_2$, a standard constituent routinely used in 5-HT binding assays no significant amounts of MDA were detected. The profound effect of $CaCl_2$ on lipid peroxidation and 5-HT binding is reported elsewhere (submitted to J. Neurosci). Interestingly, under the standard assay conditions (5.7 mM ascorbate and 4 mM $CaCl_2$), $CaCl_2$ inhibited MDA production by 88%, whereas the inactivation of ^{3}H-5-HT binding was inhibited only by 57% suggesting an additional effect of ascorbate which is independent of lipid peroxidation. This additional effect of ascorbate was also detected when lipid peroxidation was completely inhibited by 0.5 mM EDTA and was reversible after washing the membranes.

Table 2 shows that the reducing agents glutathione(GSH), dithiothreitol (DTT), and sodium metabisulfite significantly diminished the specific binding of ^{3}H-5-HT to cortical membranes. The reducing agents were effective in the μM range and the inhibition of the binding was concentration dependent. Oxidized glutathione (GSSG) did not affect ^{3}H-5-HT binding when compared to an equimolar concentration of GSH which caused a 51% decrease. The decrease in ^{3}H-5-HT binding due to ascorbate or glutathione in the absence of lipid peroxidation was reversible (data not shown). In non-washed membranes, the specific binding of ^{3}H-5-HT was decreased by 85 and 32% in the presence of 100 μM GSH or 5.7 mM ascorbate respectively. When the membranes were centrifuged and the pellet washed and resuspended in fresh assay buffer, no difference was detected between the control and pretreated membranes.

TABLE 2

Additions		^{3}H-spiperone specific binding % of control	(n)	^{3}H-5-HT specific binding % of control	(n)
None		100	(13)	100	(21)
Ascorbate	(5.7 mM)	96 ± 3	(3)	72 ± 4	(6)
GSH	(10 μM)	-- [a]		49 ± 3	(7)
	(100 μM)	--		22 ± 4	(5)
	(500 μM)	97 ± 7	(3)	25 ± 5	(3)
DTT	(10 μM)	--		61 ± 5	(3)
	(100 μM)	--		33 ± 2	(6)
	(500 μM)	99 ± 4	(3)	30 ± 3	(3)
Na meta-	(10 μM)	--		53 ± 2	(5)
bisulfite	(50 μM)	94 ± 4	(6)	21 ± 4	(4)
GSSG	(10 μM)	--		90 ± 5	(3)
	(100 μM)	96 ± 6	(4)	--	

[a] not determined

Effect of ascorbate and other reducing agents on specific ^{3}H-spiperone and ^{3}H-5-HT binding in the absence of lipid peroxidation. Cortical membranes were suspended at 4-6 mg protein/ml in 50 mM Tris-HCl buffer pH 7.4, containing 10 μM pargyline, 4 mM $CaCl_2$ and 0.5 mM EDTA with the additions indicated above. After 10 min incubation at 37^{0}C samples were transferred to ice and the specific binding of 2 nM ^{3}H-5-HT and 1 nM ^{3}H-spiperone determined as described in the legend of Table 1. Values are expressed as % of appropriate controls. The specific binding in control incubations was 184 ± 21 and 137 ± 41 fmoles/mg protein for ^{3}H-5-HT and H-spiperone, respectively. Values are the mean ± SEM of (n) independent experiments.

LACK OF AN EFFECT OF ASCORBATE OR OTHER REDUCING AGENTS ON ^{3}H-SPIPERONE BINDING IN THE ABSENCE OF LIPID PEROXIDATION

As shown earlier in this paper, both ^{3}H-spiperone and ^{3}H-5-HT binding decreased in cortical membranes that were exposed to ascorbate-induced lipid peroxidation. In contrast to that of ^{3}H-5-HT, ^{3}H-spiperone was not affected by either ascorbate or by any of the other reducing agents examined (GSH, DTT and sodium metabisulfite) in the absence of lipid peroxidation (Table 2). The above results suggest fundamental differences in the biochemical nature of ^{3}H-5-HT and ^{3}H-spiperone binding sites.

EFFECT OF ASCORBATE ON ^{3}H-5-HT AND ^{3}H-SPIPERONE BINDING IN THE ABSENCE OF LIPID PEROXIDATION

Scatchard plots of the specific binding of ^{3}H-5-HT and ^{3}H-spiperone to cortical membranes in the absence or presence of ascorbate are shown in Figures 5 and 6. In these experiments, 1 mM EDTA was included in the assay buffer. Under these conditions no MDA was detected in both control and treated membranes. In the absence of ascorbate, ^{3}H-5-HT binding exhibited a Kd of 7.6 nM and a Bmax of 2710 fmoles/mg protein. (A representative experiment is shown in figure 5.) In the presence of ascorbate, Kd = 3.3 nM and Bmax = 298 fmoles/mg protein. Similar results were also obtained when 10 μM GSH was used instead of ascorbate (Fig. 5). This experiment has been repeated five times.

In contrast to those of ^{3}H-5-HT, ^{3}H-spiperone binding parameters were not significantly altered by ascorbate. The Kd's were 2.1 ± 0.6 and 1.8 ± 0.44 nM and the Bmax were 616 ± 275 and 605 ± 184 fmoles/mg protein in the absence or presence of ascorbate, respectively, in three independent experiments. A representative experiment is shown in Figure 6. The above results demonstrate that in the absence of lipid peroxidation, ascorbate and other reducing agents have different effects on ^{3}H-5-HT and ^{3}H-spiperone binding sites.

The direct interaction of ascorbate with ^{3}H-5-HT binding sites may be attributed to its chemical reducing properties. Ascorbate has also the potential to reduce disul-

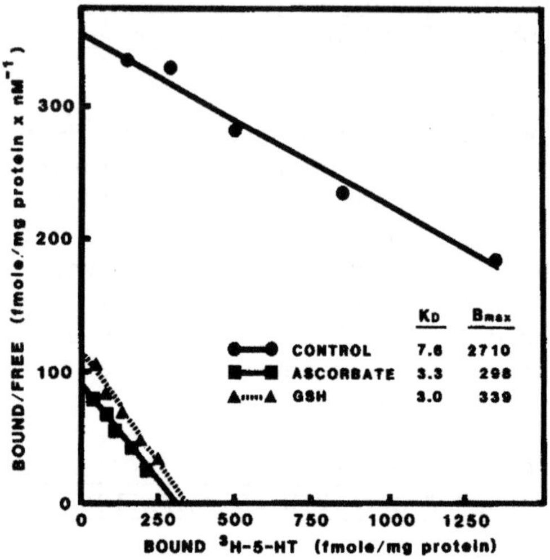

Fig. 5 . Scatchard plots of ^{3}H-5-HT specific binding with or without ascorbate in the absence of lipid peroxidation. Cortical membranes were suspended at 0.4-0.6 mg protein/ml in assay buffer containing 50 mM Tris-HCl, 10 μM pargyline, 5 mM $CaCl_2$ and 1 mM EDTA either without (circles) with 10 μM GSH (triangles), or with (squares) 5.7 mM ascorbate. After incubation for 12 min at 37^{0}C, the membrane suspension was transferred to ice and the specific binding of ^{3}H-5-HT determined as described in methods. Values are the average of triplicate determinations from one experiment replicated five times.

fide bonds as evidenced by its ability to reduce the disulfide agent 5,5'-dithiobis(2-nitrobenzoic acid) (Newton, 1962). Furthermore, our results indicate that this effect of ascorbate is to decrease the binding of only the agonist ^{3}H-5-HT but not the antagonist ^{3}H-spiperone. Similar results have been reported for other neurotransmitters. For example, Kaylaap and Neff (1980) and Kaylaap et al. (1981) reported that ascorbate decreased ^{3}H-dopamine binding to striatal membranes but not that of the dopamine antagonist ^{3}H-spiperone. A direct chemical effect resulting from a change in the oxidation state of the binding sites was proposed. In one of these studies, the ascorbate in-

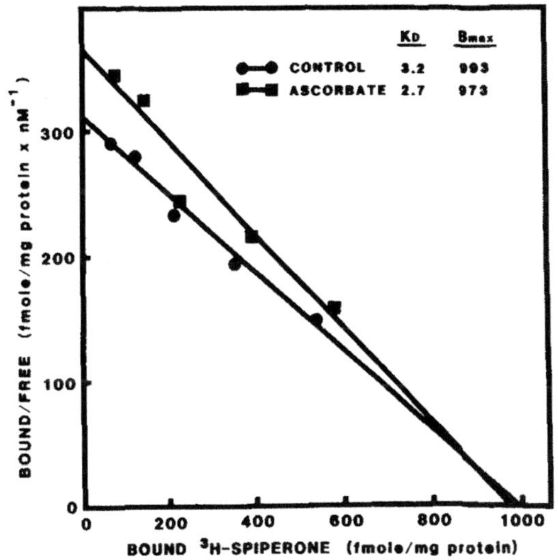

Fig. 6 Scatchard plots of ^{3}H-spiperone specific binding with or without ascorbate in the absence of lipid peroxidation. Cortical membranes were treated as described in the legend of fig. 5. Values for the specific binding of ^{3}H-spiperone without (circles) or with 5.7 mM ascorbate (squares) represent one single experiment replicated three times.

duced decrease in ^{3}H-dopamine binding occurred in the presence of 5 mM EDTA. Under these conditions, presumably lipid peroxidation is not involved.

This direct effect of ascorbate on dopamine receptors was confirmed by Heikkila et al. (1982), who also found that at high ascorbate concentration (6 mM) where negligible amounts of MDA were detected, ^{3}H-dopamine binding in striatal membranes was much less than that in the absence of ascorbate, whereas, ^{3}H-spiperone binding was not altered. The disulfide reducing agent DDT has also been reported to decrease ^{3}H-dopamine but not ^{3}H-spiperone binding in striatal membranes (Suen et al., 1980). No Scatchard plot analysis was done in the above studies. Lucas et al. (1977)

reported a decrease in β-adrenergic receptor affinity in the presence of DTT. The mechanism of the direct effect of ascorbate in decreasing the Bmax and the Kd of ^{3}H-5-HT binding (Fig. 5) remains speculative. It is possible that in the presence of ascorbate or other reducing agents, membrane proteins involved in ^{3}H-5-HT binding may undergo conformational changes,thereby increasing the affinity and decreasing the number of binding sites.

In summary, this report demonstrates that ascorbate has a dual and differential effect on serotonin binding sites. First, ascorbate-induced lipid peroxidation, as measured by malonyldialdehyde(MDA) production, caused irreversible decreases in Bmax of both ^{3}H-5-HT and ^{3}H-spiperone binding. Second, independent of lipid peroxidation, ascorbate causes decreases in both the Kd and Bmax of ^{3}H-5-HT but not of ^{3}H-spiperone binding, indicating a difference in the biochemical nature of the two serotonin binding sites.

ACKNOWLEDGEMENT

Supported by grant MH 37020 from NIMH, Department of Health and Human Services. J.W.A. is supported by a fellowship from American Lung Association. S.F.M-K. is supported by a grant from the Eli Lilly Foundation. Spiperone is kindly provided by Janssen Pharmaceutica.

REFERENCES

HEIKKILA R. E., CABBAT F. S. and MANZINO L. (1982) Inhibitory effects of ascorbic acid on the binding of(^{3}H) dopamine antagonists to neostriatal membrane preparations: Relationship to lipid peroxidation. J. Neurochem. 38, 1000-1006.

HENRIKSEN S., DEMENT, W. and BARCHAS J. (1974) The role of serotonin in the regulation of a phasic event of rapid eye movement sleep: The Ponto-geniculo-occipital wave. In:Serotonin-New Vistas. COSTA E., GESSA G. L., SANDLER M., eds., Raven Press, New York, pp. 169-180.

JACOBS B. L. (1976) An animal behavior model for studying central serotonergic synapses. Life Sci. 19, 777-786.

KAYAALP S.O. and NEFF N. H. (1980) Differentiation by ascorbic acid of dopamine agonist and antagonist binding sites in striatum. Life Science 26, 1837-1841.

KAYAALP S.O., RUBENSTEIN J. S. and NEFF N. H. (1981)

Inhibition of dopamine D-1 and D-2 binding sites in neuronal tissue by ascorbate. Neuropharmacol., 20, 409-410.

LUCAS M., HANOUNE J and BOCKAERT J (1977) Chemical modification of the beta-adrenergic receptors coupled with adenylate cyclase by disulfide bridge-reducing agents. Mol. Pharmacol. 14, 227-236.

McCALL R. B. and AGHAJANIAN G. K. (1979) Serotonergic facilitation of facial motorneuron excitation. Brain Res. 169, 11-27.

MUAKKASSAH-KELLY S. F., ANDRESEN J. W., SHIH J. C. and HOCHSTEIN P. (1982) Decreased (^{3}H) serotonin and (^{3}H) spiperone binding consequent to lipid peroxidation in rat cortical membranes. Biochem. Biophys. Res. Commun. 104, 1003-1010.

MURPHY D. L., CAMPBELL I. C. and COSTA J. L. (1978) The brain serotonergic system in the affective disorders. Prog. Neuro-psychopharmacol. 2, 1-31.

NEWTON J. W., (1962) A disulfide photoreduction system in chromatophores of Rhodospirillum rubrum. J. Biol. Chem. 237, 3282-3286.

PEROUTKA S. J. and SNYDER S. H. (1979) Multiple serotonin receptors: Differential binding of (^{3}H)5-hydroxytryptamine, (^{3}H)lysergic acid diethylamide and (^{3}H)spiroperidol. Mol. Pharmacol. 16, 687-699.

PEROUTKA S. J., LEVOVITZ R. M. and SNYDER S. H. (1979) Serotonin receptor binding sites affected differentially by guanine nucleotides. Mol. Pharmacol. 16, 700-708.

PEROUTKA S. J., LEBOVITZ R. M. and SNYDER S. H. (1981) Two distinct central serotonin receptors with different physiological functions. Science 212, 827-829.

SUEN E. T., STEFANINI E and CLEMENT-CORMIER Y. C. (1980) Evidence for essential thiol groups and disulfide bonds in agonist and antagonist binding to the dopamine receptor. Biochem. Biophys. Res. Comm. 96, 953-960.

VALZELLI L. (1974) 5-Hydroxytrptamine in aggressiveness. In: Serotonin-New Vistas COSTA E., GESSA G. L., SANDLER M., eds., Raven Press, New York, pp. 255-264.

WILBUR K. M., BERNHEIM and SHAPIRO O. W. (1949) The thiobarbituric acid reagent as a test for the oxidation of unsaturated fatty acids by various agents. Arch. Biochem. Biophys. 24, 305-313.

WILLS E. D. (1969) Lipid peroxide formation in microsomes. Biochem. J. 113, 315-324.

MEMBRANE PROPERTIES OF ISOLATED HORIZONTAL CELLS IN THE TELEOST RETINA

Dominic Man-Kit Lam and George S. Ayoub

Cullen Eye Institute and Program in Neuroscience
Baylor College of Medicine, Houston, TX 77030 U.S.A.
Kevin Hsu Research Foundation, Hong Kong

Neurons in the intact vertebrate retina are interconnected through a complex network of electrical and chemical synaptic connections which make it difficult if not impossible to study the intrinsic biophysical and biochemical properties of individual cells. During the past decade, enzymatic methods have been developed to dissociate adult retinas into single, viable and readily identifiable cells (Lam, 1972, 1976; Drujan and Svaetichin, 1972). This approach has been used for a number of biochemical (Lam, 1972, 1975, 1976; Drujan and Svaetichin, 1972; Sarthy and Lam, 1978, 1979a, b; Van Buskirk and Dowling, 1981, 1982), histochemical (Sarthy et al., 1981) and physiological (Drujan and Svaetichin, 1972; Kaneko et al., 1976; Bader et al., 1978; Johnston and Lam, 1981; Tachibana, 1981; Lasater and Dowling, 1982) studies. For the past eight years, we have studied the biochemical and biophysical properties of one type of isolated neuron in the vertebrate retina: the horizontal cell.

Horizontal cells are retinal interneurons which mediate lateral interactions in the outer retina. In certain teleost fish retinas, the horizontal cells are particularly large and were therefore the first class of retinal neurons whose electrophysiological properties were studied by intracellular recordings (Svaetichin and MacNichol, 1958). More recently, in cyprinid and catfish retinas, a number of studies have pointed to γ-aminobutyric acid (GABA) as the neurotransmitter used by a type of horizontal cell (H1) which receives input predominantly from red-sensitive cones (Lam and Steinman, 1971; Lam, 1975; Marc et al., 1978; Murakami et al., 1978, Lam et al., 1978, 1979; Wu and Dowling, 1980). Since isolated GABA-ergic horizontal cells can be readily obtained and easily identified even in isolation, we have recently used these cells as a model system to examine the intrinsic membrane properties as well as the cellular mechanisms regulating GABA uptake, synthesis and release from an identified CNS neuron. This article is an overview of some of our work to-date on the biophysical, neurochemical and pharmacological properties of isolated GABA-ergic horizontal cells in goldfish and catfish retinas.

METHODS

Texas channel catfish (*Ictalurus punctatus*) and common goldfish (*Carassius auratus*) were used for these studies. The procedure for dissociating teleost retinas into single cells has been described in detail elsewhere (Lam, 1972, 1975, 1976; Johnston and Lam, 1981). For electrophysiological studies, 0.2 ml of the isolated cell suspension was placed on a depression slide and the cells were allowed to sediment for 5 minutes. Cone horizontal cells were impaled under Nomarski interference optics with 30-50 MΩ KCl-filled micropipettes. Studies on these cells were then made using the single-electrode clamp system (SEC; Johnston, 1981).

The biosynthesis of GABA by cone horizontal cells isolated from the catfish retina was studied using a technique described earlier (Lam, 1975). In this retina, the characteristic morphologies of the isolated cone and rod horizontal cells allow a ready identification of these cells under microscopic observation (Fig. 1). For instance, the cone horizontal cell has the largest soma (about 50 μm in diameter) among all retinal cells and also possesses a long axon terminal arising from the soma. In contrast, the rod horizontal cell is more stellate, has a much smaller soma with many longer processes radiating from the soma.

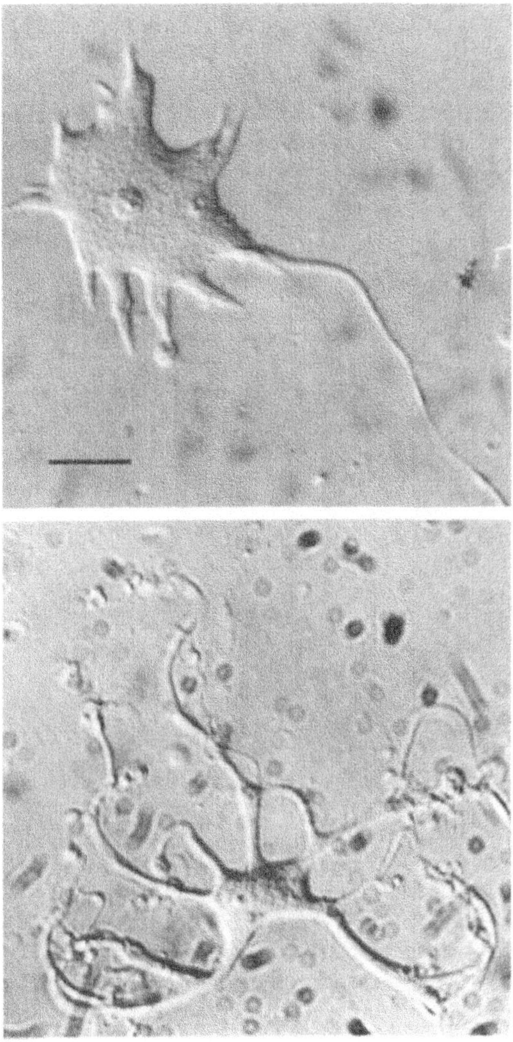

Fig. 1. Morphologies of an isolated cone (external) horizontal cell (top) and isolated rod (intermediate) horizontal cell (bottom) from the catfish retina. The cone horizontal cell has a large soma with short processes and a single long axon terminal that extends for several hundred μm, while the rod horizontal cell is much more stellate and has a smaller soma with many processes 50 to 80 μm long, allowing ready identification. Scale 20 μm.

Futhermore, fractions consisting predominantly of horizontal cells could be separated from the mixed population of isolated retinal neurons by allowing the retinal cells to sediment at unit gravity for 1 to 3 hours (Fig. 2; Lam, 1972, 1975; Sarthy and Lam , 1978). Following velocity sedimentation, individual cone or rod horizontal cells were selected and drawn into a glass micropipet with

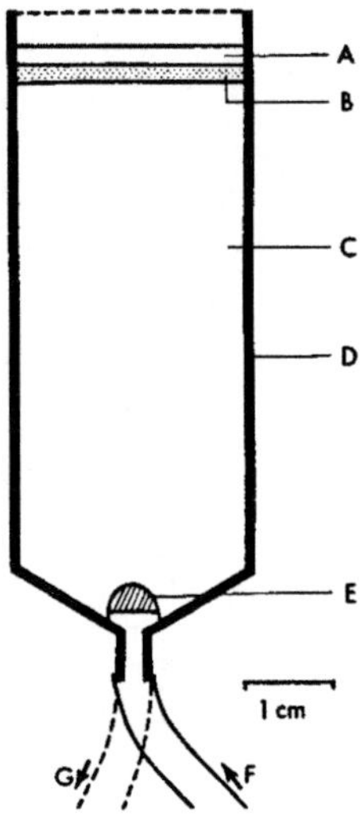

Fig. 2. Cross section of the cylindrical chamber for separating different types of retinal cells by velocity sedimentation. The cells (B) are allowed to sediment at unit gravity to different regions of the gradient on the basis of their sizes and densities. After 2 hours at 4°C, fractions enriched with horizontal cells could be readily obtained and pooled for further studies. (A) layer of saline solution (1 ml); (B) layer of about 4×10^5 retinal cells in 0.3% Ficoll dissolved in a calcium-free saline solution (1 ml); (C) linear density gradient of 0.8 to 3% Ficoll or bovine serum albumin (30 ml); (D) plastic 50 ml syringe; (E) flow deflector; (F) from linear gradient marker; (G) to fraction collector. Reproduced with permission from Lam (1972).

a tip diameter slightly larger than the cells under microscopic observation. After 10 to 30 cells were collected into each micropipet, GABA synthesis or the specific activity of the GABA-synthesizing enzyme, L-glutamic acid decarboxylase, in these cells was measured by radiochemical procedures described previously (Lam, 1975, 1976). GABA uptake into isolated goldfish horizontal cells was examined by incubating fractions enriched for these cells (>80%) at room temperature with isotonic saline solutions containing 1 μM of ^{3}H-GABA (New England Nuclear) for 1 to 60 minutes. Each sample was then transferred to a filtration manifold where the cells were collected on Nuclepore filters (pore diameter: 3 or 5 μm) and washed with 10 volumes of saline solution for 10 seconds. Each filter was placed in a scintillation vial and dried at 80°C. Ten ml of scintillation fluid (ACS) was added to each vial and the amount of radioactivity measured with liquid scintillation counting. The effects of various ions and chemicals on GABA uptake were measured by making appropriate changes in the incubating media. In these experiments an incubation time of 10 minutes was employed to ensure that the initial linear uptake rate was measured.

For studies on the release of the accumulated ^{3}H-GABA, cells obtained after velocity sedimentation were incubated at room temperature with saline solutions containing 1 μM ^{3}H-GABA (1 μCi/ml). After 10 minutes, further ^{3}H-GABA uptake was inhibited by the addition of 200 μM nipecotic acid, which selectively blocks 99% of GABA uptake into these cells. Specific transmitter candidates (such as glutamate, aspartate and dopamine), drugs (such as quisqualate and kainate) and different saline solutions (such as K^+ -rich, and others) were then added to determine their abilities to enhance or diminish release of the newly accumulated ^{3}H-GABA. After 10 minutes, cold saline solution was added and the cells were rapidly collected, washed and the radioactivity measured as described above.

RESULTS

Horizontal cells were isolated from catfish and goldfish retinas. We chose goldfish retinas because most of our work in this area had originally been done in this retina (Lam and Steinman, 1971; Marc et al., 1978; Lam et al., 1979. There are, however, unique advantages to the use of catfish retinas for certain studies. First, this retina contains only two types of photoreceptors: rods and red-sensitive cones (max. absorbance: 620 nm; Naka, 1976, 1977). Hence, all the cone (or external) horizontal cells receive photoreceptor input only from red-sensitive cones. Secondly, cone horizontal cells are relatively large in size for teleost retinas (40-80 μM), allowing unequivocal identification, even in isolation (Fig. 1). It is these features in particular which have facilitated many of our physiological and biochemical studies on isolated horizontal cells.

Biophysical properties.

Neurons in the intact retina are interconnected by a complex network of chemical and electrical synapses. In particular, the electrical interactions between retinal cells make it difficult to determine the active and passive properties of the individual neurons. The isolated cone horizontal cells from the catfish retina, being quite large and readily identifiable are therefore particularly advantageous for such biophysical studies. We have used a single-electrode current-clamp technique to examine the biophysical as well as pharmacological properties of the isolated horizontal cells (Johnston and Lam, 1981). Figure 3a shows one of the cells impaled by a microelectrode. Constant current pulses of 400 msec duration were applied and the resulting voltage transients recorded (Fig. 3b). A plot of the steady-state voltage-current (V-I) relationship obtained from these transients is depicted in Fig. 3c. Over a wide voltage range, we typically measured a decrease in input resistance with large depolarizations or hyperpolarizations.

An accurate measurement of the membrane time constant was necessary for determining the specific membrane capacity. Fig. 3d shows the logarithm of the first derivative of two low-amplitude voltage transients (from Fig. 3b), one depolarizing (10 mV) and the other hyperpolarizing (15 mV), plotted against time. Our calculations reveal single exponentials with time constants of 17.3 and 18.2 msec, respectively. Additionally, the similarity of these time constants supports our assumption that the membrane resistivity is essentially ohmic in this narrow voltage range. The specific membrane properties may then be calculated from information provided in Fig. 3. Our results indicate that these cells have a specific membrane resistivity ranging from 5,000 to 20,000 Ω-cm^2 and a specific

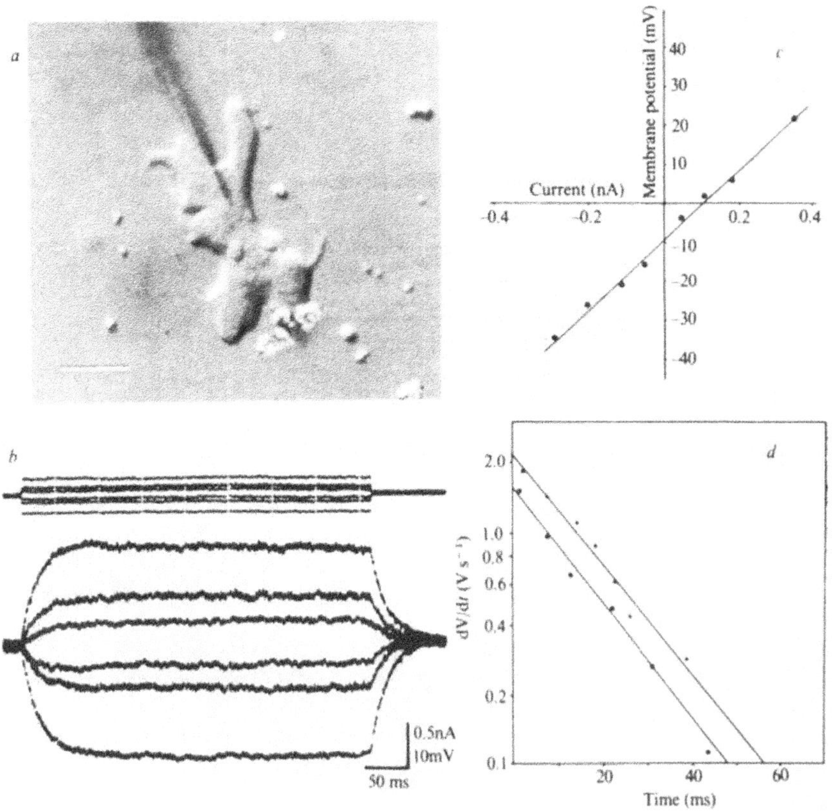

Fig. 3. Morphology and passive membrane properties of an isolated cone horizontal cell. *a*, Photomicrograph of a cone horizontal cell prior to impalement, with a microelectrode above. This cell's surface area was calculated to be 16,000 μm^2. *b*, 400-ms constant current pulses (top traces) and the resultant voltage transients (bottom traces) recorded intracellularly from the cell shown in *a*. *c*, Steady-state V-I plot determined from data shown in *b*. An input resistance of 89.6 MΩ was calculated from the slope. *d*, The log of the first derivative of two voltage transients taken from *b* and plotted against time. The top line employs data from a 10 mV depolarizing transient, while the bottom line is derived from a 15 mV hyperpolarizing transient. These fit single exponentials, implying the cell was isopotential, and the average time constant was 17.7 ms. Reproduced with permission from Johnston and Lam (1981).

membrane capacity of about 1 $\mu F/cm^2$ (Johnston and Lam, 1981).

In the majority of the isolated cells, depolarizing current pulses revealed a regenerative voltage response, as defined by an increasing positive slope in the membrane potential transient (Fig. 4). The addition of tetrodotoxin

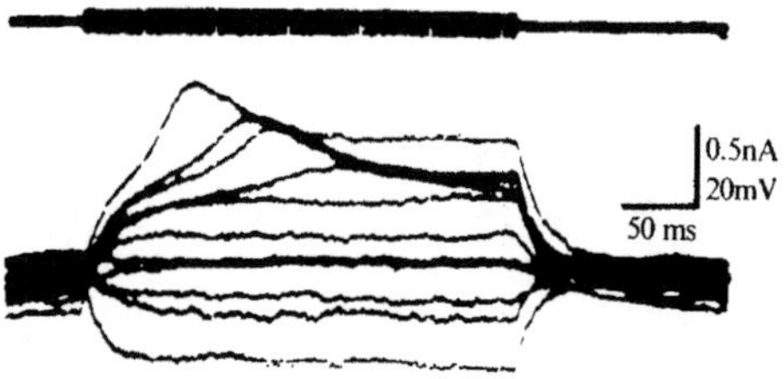

Fig. 4. Regenerative voltage responses from a catfish cone horizontal cell. At a membrane potential of -65mV, regenerative responses could be obtained in response to 300 msec constant current pulses (top traces). The passive membrane properties of this cell were: R_N = 250 MΩ, A = 3,740 μm^2, R_m = 9,350 Ω-cm^2, τ_m = 9.6 msec and C_m = 1.0 $\mu F/cm^2$.

R_N is the input resistance, A is the estimated surface area of the cell, R_m is the specific membrane resistivity. τ_m is the average membrane time constant and C_m is the specific membrane capacity. Reproduced with permission from Johnston and Lam (1981).

(TTX, 10 μg/ml) to the bathing medium had no apparent effect on the regenerative responses of these cells. Therefore, we tested whether Ca^{++} might be involved by applying $CoCl_2$, $MnCl_2$ and $NiCl_2$ to the medium, which are known to block inward Ca^{++} currents in various systems (Hagiwara, 1975; Akaike, 1978). A rapid and complete blockade of all regenerative voltage responses was effected by application of any one of these divalent cations at concentrations less than 10 mM (Johnston and Lam, 1981). These results show that the regenerative voltage responses recorded from cone horizontal cells are TTX-insensitive and probably Ca^{++}-dependent.

GABA biosynthesis.

Our autoradiographic studies in the goldfish retina have shown that, while H1 horizontal cells possess a high-affinity uptake system for GABA, H2, H3 and rod horizontal cells do not (Lam and Steinman, 1971; Marc et al., 1978). Immunocytochemical techniques revealed that the somas and axons of H1 horizontal cells contain high levels of L-glutamic acid decarboxylase (GAD), the enzyme responsible for GABA biosynthesis (Lam et al., 1979). Also, only GABA, and no other examined transmitter candidates, are synthesized by isolated axons of cone horizontal cells when they are incubated with radioactively-labeled precursors of these transmitters (Lam, 1975, 1976). These studies indicate that H1 horizontal cells possess GABA uptake and biosynthetic mechanisms.

All the cone horizontal cells of the catfish retina receive input predominantly from red-sensitive cones, and hence are physiologically similar to type H1 horizontal cells of the goldfish retina. In addition, all catfish cone horizontal cells have been shown by autoradiography to possess an uptake mechanism for GABA (Lam et al., 1978). Moreover, since isolated rod and cone horizontal cells in the catfish retina can be readily distinguished from each other (Fig. 2), we have used techniques reported earlier (Lam, 1975; 1976) to study transmitter synthesis in these cells. Cone horizontal cells of the catfish retina are similar to those from the goldfish in their capability for GABA synthesis and in their high levels of L-glutamic acid decarboxylase activities. Thus, type H1 horizontal cells, which receive input predominantly from red-sensitive cones, possess a high-affinity system for GABA uptake, significant levels of L-glutamic acid decarboxylase and the ability to synthesize GABA from its precursor L-glutamic acid in both catfish and goldfish retinas.

GABA uptake.

Autoradiographic studies in the goldfish retina show that exogenously supplied GABA is taken up by one type of horizontal cell (H1) and at least one type of amacrine cell (probably type Ab) (Lam and Steinman, 1971; Marc et al., 1978). Kinetic analyses indicate that GABA uptake in this retina is mediated by a single Na^+- and temperature-dependent high-affinity system with an apparent K_m of 18.2 μM and V_{max} of 26.1 pmoles/min/mg protein (Lam et al., 1980). The accumulation of ^3H-GABA by H1 horizontal cells in the intact goldfish retina has also been shown to be enhanced by light stimulation. Since the amount of ^{3}H-GABA accumulated into H1 cells is determined by the combination of GABA uptake, metabolism, storage and release, it is extremely difficult, if not impossible, to determine in the intact retina the cellular basis for this stimulation-dependent GABA accumulation by H1 cells. A direct approach to this problem is to dissociate the retina and separate the retinal cells by velocity sedimentation at unit gravity to obtain a population enriched with horizontal cells (Lam, 1975). These cells can be incubated or perfused with media containing ^{3}H-GABA, drugs or of different ionic compositions and collected on translucent Nuclepore filters (Fig. 5). Using this method, we found

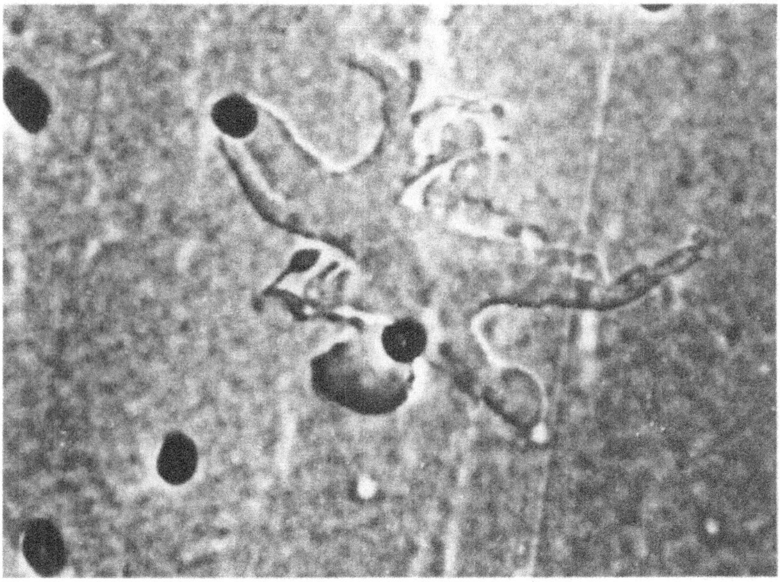

Fig. 5. The morphology of the soma of an isolated cone horizontal cell from the goldfish retina. For our studies on GABA uptake and release, these cells were placed on translucent Nuclepore filters with pore diameters of 3 or 5 μm. In this figure, the pore diameter is 5 μm.

that the somas and axons of isolated horizontal cells in the goldfish retina possess a Na^+- and temperature-dependent ^{3}H-GABA accumulation mechanism (Ayoub and Lam, 1980), with a steady rate of accumulation for at least 30 minutes of incubation at room temperature (Fig. 6).

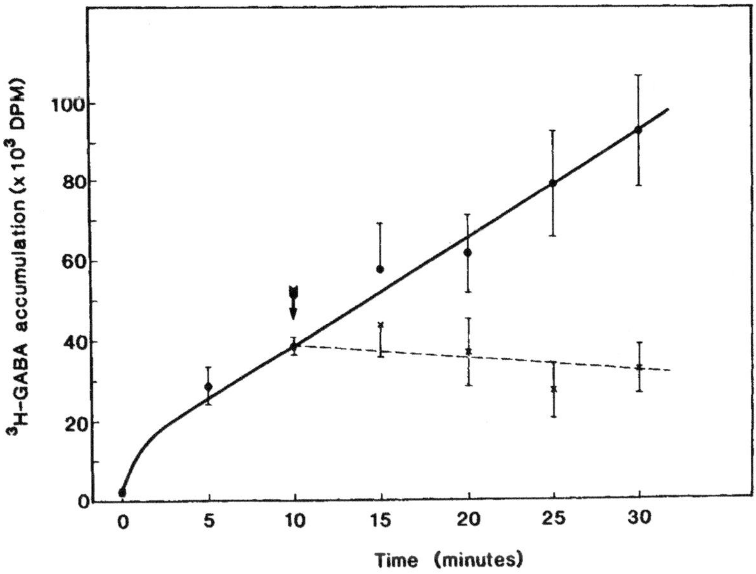

Fig. 6. ^{3}H-GABA accumulation by isolated H1 horizontal cells is linear for the first 30 minutes of incubation (solid line) but is almost completely abolished by the addition of 200 μM nipecotic acid (arrow and dashed line). Quantitative studies of ^{3}H-GABA release from these cells by different media and chemicals can therefore be achieved after using nipecotic acid to inhibit further ^{3}H-GABA uptake. The values are means ± standard errors of at least 6 experiments.

Additionally, the initial rates of GABA uptake into isolated horizontal cells are influenced by the external K^+ concentration (K^+_o) such that high K^+_o decreases this uptake (Fig. 7). In this regard it is of interest that Tachibana (1981) has recently demonstrated electrophysiologically that isolated goldfish horizontal cells are

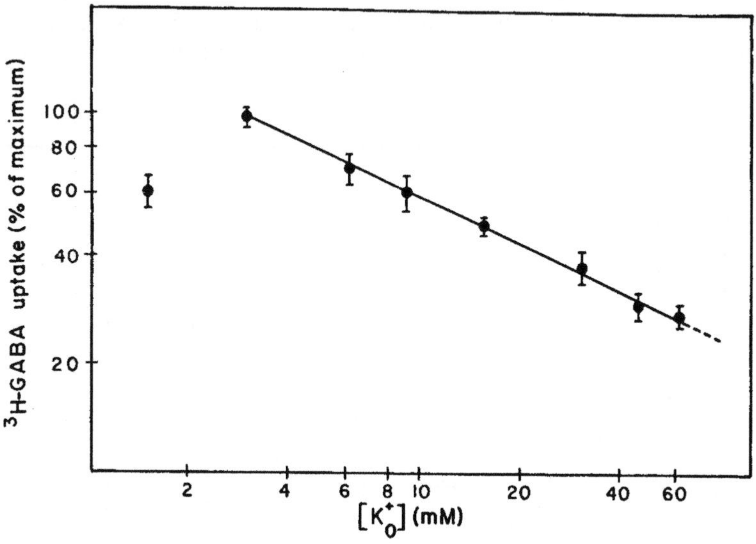

Fig. 7. The effects of external K^+ concentrations on the initial rates of ^{3}H-GABA uptake by isolated H1 horizontal cells. Isotonicity of the external media is maintained by substituting K^+ for Na^+. One μM valinomycin was added to all media to ensure that the membrane potential of the cells is predominantly dependent upon the transmembrane K^+ equilibrium potential, and modeled in a Nernstian fashion. The values represent means ± standard errors of at least 6 experiments.

hyperpolarized (~-80mV) in normal saline solutions and are depolarized by high K_o^+ in a quantitative fashion. Our results are therefore consistent with the hypothesis that GABA transport into these cells is at least partially regulated by the transmembrane potential such that, within the physiological range, uptake is increased by hyperpolarization and decreased by depolarization (Ayoub and Lam, 1980). Also, our findings are in general agreement with the models proposed by Martin (1976) and Marc et al. (1978), as described in the Discussion.

GABA release.

The ^{3}H-GABA taken up by H1 horizontal cells and Ab amacrine cells from goldfish and catfish retinas can be released by appropriate light- or K^+-stimulation (Marc et al., 1978; Lam et al., 1980; Lasater and Lam, in preparation). Since there are at least two types of GABA-accumulating neurons, it becomes difficult to demonstrate unequivocally the release of ^{3}H-GABA by horizontal cells in the intact retina. Because the release of GABA from these cells is an important criterion for GABA being their neurotransmitter, we have once again utilized the isolated horizontal cell preparation in order to examine the release of recently accumulated ^{3}H-GABA in response to saline solutions of various ionic compositions, as well as containing different transmitter candidates. We have made the following observations using the procedured described above.

First, it has been established that horizontal cells are hyperpolarized in the absence of photoreceptor input (Dowling and Ripps, 1973; Kaneko and Shimazaki, 1975). Therefore, isolated H1 cells are expected to be in a hyperpolarized state. In fact, Tachibana (1981) has demonstrated that these cells are hyperpolarized (~-80 mV) in normal saline solutions and are depolarized by increasing external K^+ in a Nerstian fashion. We have found that ^{3}H-GABA release from horizontal cells is also stimulated by increasing the concentration of K^+ in the incubation medium (Fig. 8). Since the external Na^+ concentration was held constant by substituting KCl with choline chloride, our result indicates that ^{3}H-GABA can be released from H1 horizontal cells by depolarization with high external K^+. Furthermore, in light of Tachibana's result (1981), our data suggest that within the physiological range, the amount of ^{3}H-GABA released is almost linearly proportional to the membrane potential (Fig. 8).

Secondly, since L-glutamic acid and L-aspartic acid have been postulated as neurotransmitter candidates for cones in the teleost retina (Murakami et al., 1972; Wu and Dowling, 1978; Ishida and Fain, 1981; Marc and Lam, 1981; Lasater and Dowling, 1982), we have also examined the effects of these substances on GABA release. As shown in Fig. 9, both L-glutamate and L-aspartate stimulate ^{3}H-GABA release in a stereo-specific and dose-dependent manner. In our isolated cell preparation, L-glutamate is generally more potent in eliciting GABA release than L-aspartate. At saturating concentrations, L-glutamate is about 3 times as effective as L-aspartate. However, the most interesting observation is probably that in this preparation, L-glutamate is effective at very low concentrations, stimulating detectable GABA release at concentrations less than 0.5 μM and eliciting one half the maximal release at about 3 μM. Additionally, the putative L-glutamate agonists, quisqualic acid and kainic acid, at concentrations of 1 to 10 μM are equally effective in stimulating ^{3}H-GABA release (Ayoub and Lam, in preparation).

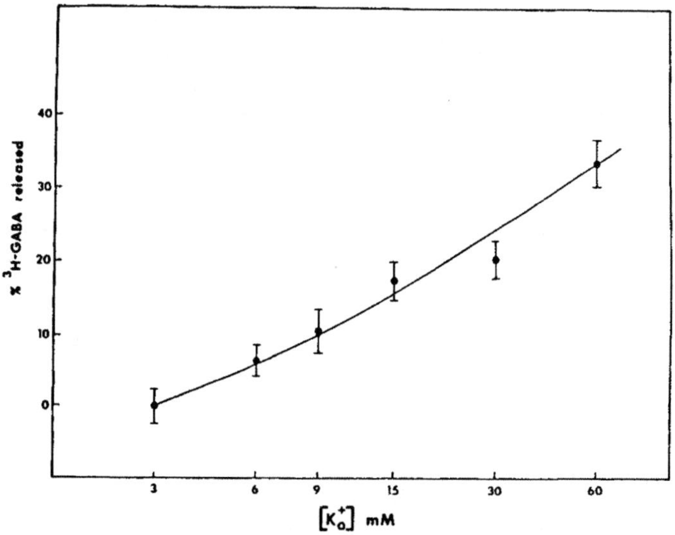

Fig. 8. ^{3}H-GABA accumulated by isolated goldfish H1 horizontal cells is released by increasing the K^+ concentration in the incubation medium. The cells are loaded with ^{3}H-GABA by incubation in normal saline solutions containing 1 μM ^{3}H-GABA for 10 minutes. ^{3}H-GABA release in this series of experiments was measured while retaining the external Na^+ concentration at 60 mM. Isotonicity of the K^+-rich (3 to 60 mM K^+) media was maintained by adding appropriate amounts of KCl and choline chloride. The values represent means ± standard errors of at least 6 experiments.

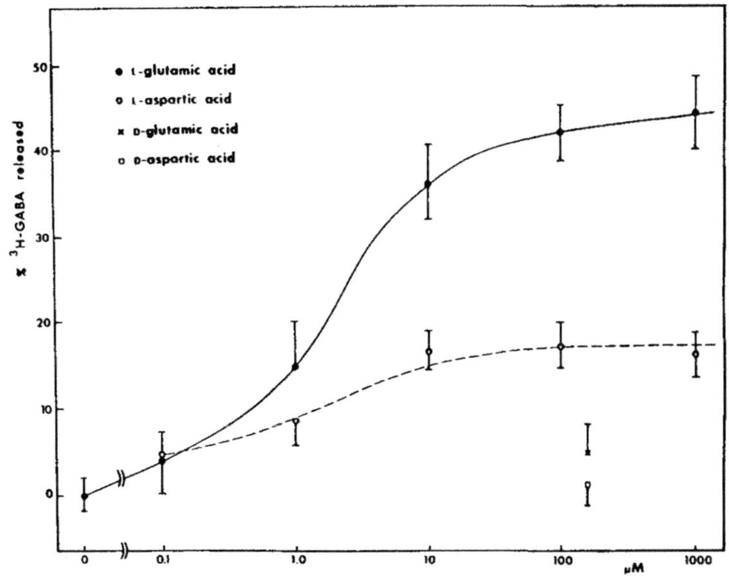

Fig. 9. The effects of transmitter candidates on ^{3}H-GABA release from isolated H1 horizontal cells. The abscissa represents the concentrations in μM of the different acidic amino acids exogenously applied. The values are means ± standard errors of at least 6 experiments.

DISCUSSION

GABA as the neurotransmitter for H1 horizontal cells. In the teleost retina, there is substantial evidence that horizontal cells receiving input predominantly from red-sensitive cones use GABA as their neurotransmitter (Lam et al., 1980). Studies performed on the intact goldfish retina, indicate that these cells possess a high-affinity mechanism for GABA uptake (Lam and Steinman, 1971; Marc et al., 1978), contain L-glutamic acid decarboxylase (Lam et al., 1979) and release GABA in response to depolarizing stimuli (Marc et al., 1978; Lam et al., 1980). Electrophysiological techniques have revealed that the feedback from H1 horizontal cells to cones may be GABA-ergic in teleost retinas (Murakami et al., 1978; Lam et al., 1978; Wu and Dowling, 1980; J. Toyoda, personal communication).

We have substantiated these findings with our work on isolated horizontal cells from goldfish and catfish retinas. Our work has shown that H1 horizontal cells isolated from these retinas contain L-glutamic acid decarboxylase, are capable of synthesizing GABA, have a high-affinity uptake mechanism for exogenous GABA, and release accumulated GABA in response to known depolarizing stimuli such as L-glutamate or high external K^+ (Lam, 1975, 1976; Ayoub and Lam, 1980, 1982). In addition to confirming and extending the results obtained from intact retinas, our experiments on isolated horizontal cells enable us to also examine the intrinsic biophysical properties of these cells in the absence of complex neural interactions in the retina, as well as the cellular mechanisms regulating the uptake and release of GABA in an identified CNS neuron. Several interesting findings have emerged from these studies.

Biophysical properties of isolated horizontal cells.

First, horizontal cells from the intact retina are known to respond to light stimulation by sustained graded potentials (Svaetichin and MacNichol, 1958; Werblin and Dowling, 1969; Kaneko, 1970; Naka, 1972). Additionally, isolated horizontal cells of the teleost retina have recently been shown to demonstrate regenerative voltage responses which are most probably Ca^{++}-dependent (Johnston and Lam, 1981; Tachibana, 1981; Lasater and Dowling, 1982). The extensive electrical coupling between horizontal cells *in situ* most likely prevent these regenerative events from occurring in the intact retina under physiological conditions. However, our data indicates that there exist Ca^{++} channels on horizontal cells which can also be activated by depolarization in the intact retina. In light of this, it is of interest to note that Byzov and Trifonov (1981), in their analyses of the ionic mechanisms underlying the nonlinearity of the horizontal cell membrane, demonstrated that Ca^{++} channels probably exist on these cells and suggested that depolarization of the horizontal cell membrane increases its Ca^{++} conductance. Hence, an examination of the physiological roles played by these channels in horizontal cell function is of great interest. Also, the observation made by us and others that enzymatic dissociation of the retina does not appear to destroy the voltage-dependent Ca^{++} channels in horizontal cells suggests that this preparation may be useful for studying the biophysics of Ca^{++} channels in a CNS neuron.

Cellular mechanisms regulating GABA transport.

Secondly, in addition to a Na^+- and temperature-dependence known to be required for most high-affinity transport systems, we have shown that the initial rate of GABA uptake into isolated GABA-ergic horizontal cells is influenced by the membrane potential of the cells, in a fashion consistent with hyperpolarization increasing GABA uptake and depolarization decreasing this uptake (Ayoub and Lam, 1980). Our studies on whole goldfish retinas revealed that GABA accumulation into H1 horizontal cells is enhanced by light stimulation and decreased by incubation in darkness. Since H1 horizontal cells are hyperpolarized by light (Luminosity-type; Kaneko, 1970), our findingon GABA uptake in isolated cells lends support to the hypothesis that light-dependent stimulation of GABA accumulation in H1 cells can be explained by an increase in the uptake of GABA when these cells are hyperpolarized. Certain other factors also contribute to the net amount of GABA accumulated by these cells. For instance, a depolarization by darkness, K^+ or L-glutamate can also stimulate release of the accumulated GABA (Fig. 5 and 7; Marc et al., 1978; Ayoub and Lam, 1982). Since the ^{3}H-GABA taken up by H1 cells is not significantly metabolized in 30 minutes (Lam and Steinman, 1971), the accumulated ^{3}H-GABA therefore comprises the difference between the amount taken up and the amount released. Thus, by simultaneously enhancing GABA uptake and decreasing GABA release, the net effect of light-induced hyperpolarization in H1 cells would be to increase the amount of ^{3}H-GABA accumulated by these cells.

The functional significance for the GABA-ergic neurons with a high-affinity mechanism for GABA uptake has not been elucidated. Since GABA-transaminase, the degradative enzyme for GABA, has not been localized to any extracellular space or synaptic cleft examined (Varon et al., 1965; Salganicoff and DeRobertis, 1965), the termination of GABA-mediated transmission may be facilitated by the uptake of this neurotransmitter. In the teleost retina, where only the GABA-ergic neurons themselves appear to possess this uptake system, this role of transmission termination by a clearance mechanism may be especially important. There is no low-affinity GABA uptake system in these retinas and the Müller (glial) cells do not accumulate GABA *in situ* (Lam et al., 1980).

One possible function for these potential-dependent GABA uptake and release processes may be to increase the signal-to-noise ratio during transmitter-mediated neurotransmissions. However, it is of importance to recognize that the influence of membrane potential on the uptake of GABA or any other amino acid transmitter is by no means a novel idea or observation (see, for instance, Hammerstad and Cutler, 1972; Bennett et al., 1973; Levy et al., 1973; Martin, 1976; Johnston, 1977; Marc et al., 1978; Marc and Lam, 1981). In fact, using a modified single energy-barrier model (Jack et al., 1975), Martin (1976) has proposed an equation to predict the relationship between the transmembrane GABA gradient and membrane potential:

$$GABA_i/GABA_o = (Na^+_o/Na^+_i)^n (K^+_i/K^+_o)^k \exp((k-n)F(\psi_i-\psi_o)/RT)$$

where the subscripts i and o indicate intracellular and extracellular, respectively; n and k are the stoichiometric coefficients for Na^+ and K^+ (respectively) in the transport reaction $GABA_o + nNa^+ + kK^+_o \quad GABA_i + nNa^+_i + kK^+_o$ and $(\psi_i - \psi_o)$ is the membrane potential. Although the values of n and k are unknown, in view of the evidence that the initial transport of GABA requires Na^+_o but perhaps not K^+_o (Martin, 1976), it is probable that n>k under most physiological conditions. In these situations, it is evident that $GABA_i/GABA_o$ is strongly dependent upon the membrane potential, such that the ratio is increased by hyperpolarization and decreased by depolarization, a prediction quite consistent with our findings.

Depolarization-mediated GABA release.

Isolated horizontal cells have been shown to be depolarized by high external K^+ (Kaneko et al., 1976; Tachibana, 1981; Lasater and Dowling, 1982). In particular, Tachibana (1981) demonstrated that over a wide physiological range, the membrane potentials of isolated goldfish horizontal cells could be accounted for to a large extent by the transmembrane K^+ equilibrium potential. Therefore, our observation that ^{3}H-GABA efflux from isolated H1 horizontal cells is enhanced by increasing the external K^+ concentration (Fig. 8), indicates that cells can release GABA in response to depolarizing stimuli. Moreover, if the membrane potentials of our isolated goldfish horizontal cells are dependent upon the external K^+ concentrations in the same manner as Tachibana's preparation (1981), then our result suggests that the efflux of ^{3}H-GABA from H1 cells is to the first approximation linearly proportional to the membrane potential of the cells (Fig. 8).

Transmitter-mediated GABA release.

L-glutamic and L-aspartic acids have been postulated as transmitter candidates of vertebrate photoreceptors for many years. In the intact retina, these amino acids depolarize horizontal cells and block their light-evoked responses (Murakami et al., 1972; Cervetto and MacNichol, 1972; Dowling and Ripps, 1973; Wu and Dowling, 1978). Utilizing autoradiographic and kinetic analyses, Marc and Lam (1981) have recently shown, in the goldfish retina, that photoreceptors are the only neural elements which possess a high-affinity system for aspartate and glutamate uptake. Additionally, although rods accumulate L-glutamate about 30 times more effectively than L-aspartate, red- and green-sensitive cones take up glutamate and aspartate with similar effectiveness, albeit they accumulate aspartate much more effectively than rods. Blue-sensitive cones show essentially no high-affinity aspartate or glutamate uptake (Marc and Lam, 1981). These results raise the possibilities that (1) there exist high-affinity transport systems in goldfish photoreceptors that can distinguish between glutamate and aspartate, (2) pharmacological studies to identify the photoreceptor transmitters in the intact retina may be biased by powerful uptake systems and (3) rods and different types of cones may use different transmitters.

Intracellular recordings from isolated horizontal cells, and a testing of the effects of the various transmitter candidates and their analogues on these cells is another means by which the photoreceptor transmitters may be identified. Lasater and Dowling (1981) employed this procedure to show that in the carp retina, L-glutamate and its analogues, quisqualate and kainate, caused an initial graded depolarization accompanied apparently by a resistance increase. L-Aspartate, however, had little effect on the membrane potential of most of the horizontal cells recorded. Earlier studies, however, using the intact carp retina found L-aspartate to be much more effective than L-glutamate in depolarizing cone horizontal cells (Wu and Dowling, 1978). These findings raise the possibility that (1) the post-synaptic receptors for L-aspartate normally on horizontal cells are localized primarily to the fine processes where most cone to horizontal cell synapses are found, and may be lost or damaged by the enzymatic cell dissociation procedure, or (2) the apparent potency of L-aspartate over L-glutamate in depolarizing cone horizontal cells in the intact retina may be due to a more effective uptake system for L-glutamate than L-aspartate (Marc and Lam, 1981; Ishida and Fain, 1981; Lasater and Dowling, 1982).

Our further investigation of this problem employed the depolarization-mediated GABA release from isolated goldfish H1 horizontal cells. This was used as an assay system to examine the efficacies of various transmitter candidates and their analogues in releasing GABA from these cells. The most striking observation is that a significant amount of recently incorporated ^{3}H-GABA (>10%) can be released from isolated H1 cells by very low concentrations of L-glutamic acid (<1 μM) (Fig. 9). This concentration is about 10^3 times less than the minimal concentrations required for either L-glutamate or L-aspartate to elicit a response from horizontal cells in isolated retina (Murakami et al., 1972; Wu and Dowling, 1978). The uptake of glutamate and aspartate is therefore implicated in playing an important role in pharmacological studies which utilize intact retinas. We have also found L-aspartate to be effective, but considerably less potent than L-glutamate, in eliciting the release of GABA from isolated H1 cells (Fig. 9). D-glutamate and D-aspartate have no significant effect, however, suggesting that these amino acids have a stereo-specific action. Furthermore, we found that the L-glutamate agonists, quisqualic acid and kainic acid, are also very potent at eliciting GABA release from isolated H1 cells, in agreement with the studies of Lasater and Dowling (1982). At μM concentrations, the order of potency is generally quisqualate > kainate≃L-glutamate>L-aspartate.

General Conclusions.

Our studies and those of others have shown that a powerful tool for investigating certain problems in neurobiology and vision research may be that of the isolated horizontal cell preparation. Our biophysical studies indicate that the active and passive membrane properties of cone horizontal cells in the catfish retina, except for the lack of TTX-sensitive sodium channels, appear to be similar to those reported for various neurons from the central and peripheral nervous systems of both vertebrates and invertebrates (Cole, 1968; Hagiwa, 1974; Akaike et al., 1978; Brown et al., 1981). Of considerable significance, our biochemical data lead to the observation that type H1 horizontal cells of the teleost retina have most of the GABA-specific properties to be classified as a conventional GABA-ergic neuron. Finally, these GABA-ergic cells may be an appropriate model system to study the cellular mechanisms regulating GABA transport and release in an isolated, identified CNS neuron.

ACKNOWLEDGEMENTS

We thank Professors Daniel Johnston, Robert Marc and Sam Wu for valuable assistance and Ms. Patricia Cloud for typing the manuscript. This work was supported by grants from the Retina Research Foundation (Houston) and the U.S. National Institutes of Health (EY02423). D.M.K. Lam is recepient of a Research Career Development Award from NIH and an Olga K. Wiess Scholar Award from Research to Prevent Blindness, Inc.

This article is based on the work that one of us (D.M.K.L.) presented at the International Symposium on Neural Membranes, sponsored by the National Yang Ming Medical College and the National Science Council and held in Taipei, Taiwan on August 2-5, 1982. D.M.K. Lam wishes to thank the Organization Committee of the Symposium and Dr. Wang Chi Wu, Vice Chairman of the National Science Council for their kind invitation to attend this symposium.

REFERENCES

Akaike, N. (1978) The calcium current of Helix neuron. J. Gen. Physiol. 71, 509-531.

Ayoub, G.S. and D.M.K. Lam (1980) GABA uptake by isolated cells of the goldfish retina. Soc. Neurosci. Abst. 10, 54.

Ayoub, G.S. and D.M.K. Lam (1982) Uptake and release of ^{3}H-GABA by isolated goldfish retinal neurons. Invest. Ophthalmol. Vis. Sci., suppl. 22, 227.

Bader, C.R., P.R. MacLeish and E.A. Schwartz (1978) Responses to light of solitary rod photoreceptors isolated from tiger salamander retina. Proc. Natl. Acad. Sci. 75, 3507-3511.

Bennett, J.P., Jr., W.J. Logan and S.H. Synder (1973) Amino acids as central nervous transmitters: the influence of ions, amino acid analogues, and ontogeny on transport systems for L-glutamic and L-aspartic acids and glycine into central nervous synaptosomes of the rat. J. Neurochem 21, 1533-1550.

Brown, T.H., D.H. Perkel, J.C. Norris and J.H. Peacock (1981) Electronic structure and specific membrane properties of mouse dorsal root ganglion neurons. J. Neurophysiol. 45, 1-15.

Byzov, Al and Y.S. Trifonov (1981) Ionic mechanisms underlying the nonlinearity of horizontal cell membrane. Vis. Res. 21, 1573-1578.

Cervetto, L. and E. MacNichol (1972) Inactivation of horizontal cells in turtle retina by glutamate and aspartate. Science 178, 767-768.

Cole, K.S. (1968) Membranes, Ions and Impulses. Univ. of Calif. Press, Berkeley.

Dowling, J.E. and H. Ripps (1973) Effect of magnesium on horizontal cell activity in the skate retina. Nature 242, 101-103.

Drujan, B.D. and G. Svaetichin (1972) Characterization of different classes of isolated retinal cells. Vision Res. 12, 1777-1784.

Hagiwara, S. (1975) Ca-dependent action potential. In: Membrane. A Series of Advances. Vol. 3, ed. G. Eiseman (Dekker, New York), 359-381.

Hammerstad, J.P. and R.W.P. Culter (1972) Efflux of amino acid neurotransmitters from brain slices: role of membrane transport, Eur. J. Pharm. 20, 118-121.

Ishida, A.T. and G.L. Fain (1981) D-Aspartate potentiates the effects of L-glutamate on horizontal cells in goldfish retina. Proc. Nat. Acad. Sci. 78, 5890-5894.

Jack, J.J.B., D. Noble and R.W. Tsien (1975) Nonlinear properties of excitable membranes in: Electric Current Flow in Excitable Cells. Claredon Press, Oxford. pp. 225-260.

Johnston, D. (1981) Passive cable properties of hippocampal CA3 pyramidal neurons. Cell. Molec. Neurobiol. 1, 41-55.

Johnston, D. and D.M.K. Lam (1981) Regenerative and passive membrane properties of isolated horizontal cells from a teleost retina. Nature 292, 451-454.

Johnston, G.A.R. (1977) Effects of calcium on the potassium-stimulated release of radioactive β-alanine and γ-aminobutyric acid from slices of rat cerebral cortex and spinal cord. Brain Res. 121, 179-181.

Kaneko, A. (1970) Physiological and morphological identification of horizontal, bipolar and amacrine cells in goldfish retina. J. Physiol. 207, 623-633.

Kaneko, A. and H. Shimazaki (1975) Effects of external ions on the synaptic transmission from photoreceptors to horizontal cells in the carp retina. J. Physiol. 252, 509-522.

Kaneko, A., D.M.K. Lam and T.N. Wiesel (1976) Isolated horizontal cells of a elasmobranch retinae. Brain Res. 105, 567-572.

Lam, D.M.K. (1972) Biosynthesis of acetylcholine in turtle photoreceptors. Proc. Nat. Acad. Sci. 60, 1787-1981.

Lam, D.M.K. (1975) Biosynthesis of γ-aminobutyric acid by isolated axons of cone horizontal cells in the goldfish retina. Nature 254, 345-347.

Lam, D.M.K. (1976) Synaptic chemistry of identified cells in the vertebrate retina. Cold Spring Harbor Symp. Quant. Biol. XL, 571-579.

Lam, D.M.K. and L. Steinman (1971) The uptake of (γ-^{3}H) aminobutyric acid in the goldfish retina. Proc. Nat. Acad. Sci. 68, 2777-2781.

Lam, D.M.K., E.M. Lasater, K-I. Naka (1978) γ-Aminobutyric acid: a neurotransmitter candidate for cone horizontal cells of the catfish retina. Proc. Nat. Acad. Sci. 75, 6310-6313.

Lam, D.M.K., Y.Y.T. Su, L. Swain, R.E. Marc, C. Brandon, and J.-Y. Wu (1979) Immunocytochemical localisation of L-glutamic acid decarboxylase in the goldfish retina. Nature 278, 565-567.

Lam, D.M.K., Y.Y.T. Su, C.A. Chin, C. Brandon, J.-Y. Wu, R.E. Marc and E.M. Lasater (1980) GABA-ergic horizontal cells in the teleost retina. Brain Res. Bull. 5, 137-140.

Lasater, E.M. and J.E. Dowling (1982) Carp horizontal cells in culture respond selectively to L-glutamate and its agonists. Proc. Nat. Acad. Sci. 79, 939-940.

Levy, W.B., D.A. Redburn and C.W. Cotman (1973) Stimulus - coupled secretion of γ-aminobutyric acid from rat brain synaptosomes. Science 181, 676-678.

Marc, R.E., W.K. Stell, D. Bok and D.M.K. Lam (1978) GABA-ergic pathways in the goldfish retina. J. Comp. Neurol. 182, 221-245.

Marc, R.E. and D.M.K. Lam (1981) Uptake and aspartic and glutamic acid by photoreceptors in goldfish retina. Proc. Nat. Acad. Sci. 78, 7185-7189.

Martin, D.L. (1976) Carrier-mediated transport and removal of GABA from synaptic regions in: GABA in Nervous System Function E. Roberts, T.N. Chase and D.B. Tower, ed. Raven Press, New York, pp. 347-386.

Murakami, M., K. Ohtsu and T. Ohtsuka (1972) Effects of chemicals on receptors ad horizontal cells in the retina. J. Physiol. 227, 899-913.

Murakami, M., Y. Shimoda and K. Nakatani (1978) Effects of GABA on neuronal activities in the distal retina of the carp. Sensory Proc. 2, 334-338.

Naka, K.I. (1972) The horizontal cells. Vision Res. 12, 573-588.

Naka, K.I. (1976) Neuronal circuitry in the catfish retina. Invest. Opthalmol. 15, 926-935.

Naka, K.I. (1977) Functional organization of catfish retina. J. Neurophys. 40, 26-43.

Salganicoff, L. and E. DeRobertis (1965) Subcellular distribution of the enzymes of the glutamic acid, glutamine and gamma-aminobutyric acid cycles in rat brain. J. Neurochem. 12, 287-309.

Sarthy, P.V. and D.M.K. Lam (1978) Biochemical studies of isolated glial (Muller) cells from the turtle retina. J. Cell Biol. 78, 675-684.

Sarthy, P.V. and D.M.K. Lam (1979a) The uptake and release of (^{3}H)dopamine in the goldfish retina. J. Neurochem. 32, 1269-1277.

Sarthy, P.V. and D.M.K. Lam (1979b) Endogenous levels of neurotransmitter candidates in photoreceptor cells of the turtle retina. J. Neurochem. 32, 455-461.

Svaetichin, G. and E.F. MacNichol, Jr. (1958) Retinal mechanisms for chromatic and achromatic vision. Ann. New York Acad. Sci. 74, 385-404.

Tachibana, M. (1981) Membrane properties of solitary horizontal cells isolated from goldfish retina. J. Physiol. 321, 141-161.

VanBuskirk, R. and J.E. Dowling (1981) Isolated horizontal cells from carp retina demonstrate dopamine-dependent accumulation of cyclic AMP. Proc. Natl. Acad. Sci. 78, 7825-7829.

Varon, S., H. Weinstein, C.F. Baxter and E. Roberts (1965) Uptake and metabolism of exogenous γ-aminobutyric by subcellular particles in a sodium-containing medium. Biochem. Pharmacol. 14, 1755-1764.

Werblin, F.S. and J.E. Dowling (1969) Organization of the retina of mudpuppy, Necturus maculosus. II. intracellular recording. J. Neurophysiol. 32, 339-355.

Wu, S.M. and J.E. Dowling (1978) L-aspartate: evidence for a role in cone photoreceptor synaptic transmission in the carp retina. Proc. Nat. Acad. Sci. 75, 5205-5209.

Wu, S.M. and J.E. Dowling (1980) Effects of GABA and glycine on the distal cells of the cyprinid retina. Brain Res. 199, 401-414.

Part III: Membrane Structure and Functions

THE KINDLING EFFECT OF ETHANOL ON NEURONAL MEMBRANES

Albert Y. Sun

Sinclair Comparative Medicine Research Farm and Department of Biochemistry, University of Missouri, Columbia, MO 65201

SUMMARY

Many membrane transport processes in brain are altered due to hydrophobic interaction of alcohol with membrane. Since the activity of synaptosomal (Na^{+}+K^{+})-ATPase is very much dependent on the microenvironment of the synaptic plasma membranes (SPM), we have investigated the effect of acute and chronic ethanol on this enzyme in the brain. After acute intragastric intubation of ethanol (30% w/v) to adult Sprague Dawley rats, there was no significant increase in synaptosomal (Na^{+}+K^{+})-ATPase activity at dose levels of 2 and 4 g/kg, but the activity increased more than 30% at dose levels of 6 and 8 g/kg as compared with controls ($p<0.01$). The increase in enzymic activity was first observed at 2 hr after administration and persisted for 16 hr before returning to basal levels after 24 hr. Besides changes in membrane transport activity, the physicochemical properties of SPM are also altered after chronic ethanol treatment. The biphasic response of TNS-fluorescence to different concentrations of ethanol disappeared and the membrane became more susceptible to lipid peroxidation. Furthermore, a lower transition temperature was observed with

SPM isolated from alcoholic rats as compared with controls. The kindling development of membrane excitability during chronic ethanol administration is thus correlated to a change in physicochemical properties of the SPM due to a disturbance of the dynamic turnover of membrane phospholipids. It is apparent that this type of excitable membrane can adapt and compensate for the stress imposed on the membranes due to presence of ethanol.

INTRODUCTION

Alcohol drinking in man not only produces an immediate pattern of acute effects, but also gives rise to tolerance and physical dependence upon repeated exposure and intoxication. Physical dependence is presumably due to development of biochemical or biophysical abnormalities during chronic alcohol administration. Cessation of alcohol drinking after the development of physical dependence has been associated with manifestation of the alcohol withdrawal syndrome. The biochemical mechanism of alcohol tolerance and physical dependence characteristic of alcohol addiction is not well understood.

Animal studies have shown that physical dependence on alcohol can be developed in a relatively short time (Goldstein, 1976), and this development is likely to be an adaptive process due to the drug action. Upon chronic ingestion of ethanol, many organs may develop adaptive changes in order to sustain their functions. Many biochemical mechanisms are proposed for the adaptation in brain; e.g, development of redundant pathways (Martin, 1968), synthesis of excessive receptors (Collier, 1964), and result of enzyme derepression (Goldstein and Goldstein, 1968). However, there is no conclusive evidence supporting any of the hypotheses. It is obvious that understanding the adaptive mechanism in brain is difficult due to the highly complex organization of this organ.

ALCOHOL-MEMBRANE INTERACTION

There is strong evidence that ethanol depresses the nervous function by exerting a direct action on the excitable membranes (Wallgren, 1970). Thus, the cellular mechanism of action of ethanol is related to the biophysical properties of the membrane (Sun and Samorajski, 1970; Kalant, 1971; Seeman, 1972, 1975; Sun, 1980). It is suggested that ethanol molecules are inserted into the lipid bilayer of the memoranes, thereby altering the lateral mobility of membrane lipids or the hydrophobic portion of membrane proteins to produce steric changes. The effects of ethanol on neuronal memorane processes such as active transport, sodium pump-associated ATPase activity and evoked potential have been investigated. Due to the alcohol-membrane interaction, an adaptive mecnanism may exist in brain in order to offset the continuous stimulative or depressive action of the compound.

A) The biphasic phenomenon

The effect of ethanol on membrane excitability was shown to be biphasic in nature (Kalant, 1971). Recently, we also observed such a biphasic effect of ethanol *in vitro* on some membrane transport activities (Burke et al., 1977; Sun and Seaman, 1980). At low concentrations (0.1-0.5%), ethanol enhanced mitochondrial Ca^{2+}-accumulation and plasma membrane $(Na^{+}+K^{+})$-ATPase activity (Fig 1). Activity was inhibited at concentrations higher than 1%. It is known that most memorane transport processes require a high degree of fluidity. In the presence of low concentration of ethanol, these molecules may become embedded in the hydrophobic regions of the membrane causing an increase in membrane fluidity (Seeman, 1975; Chin and Goldstein, 1977a,b). By using 2-toluidinonaphthalene-sulfonate (TNS) as a fluorescent probe, we were able to demonstrate a decrease in polarization of the TNS-membrane complex with increasing ethanol concentration

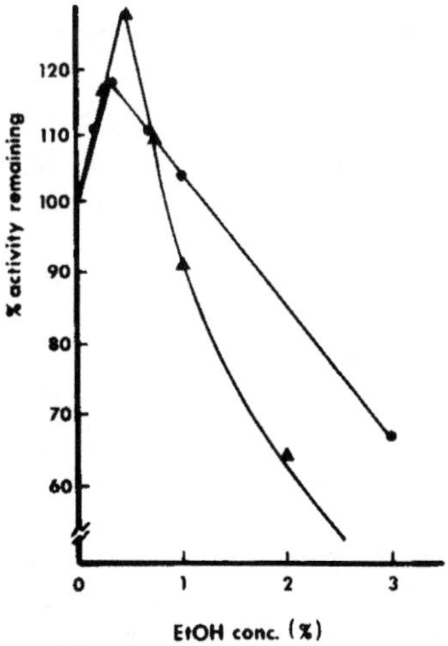

Fig 1. Effect of ethanol on synaptosomal (Na^+ + K^+)-ATPase (●—●) and mitochondrial Ca^{++}-uptake (▲—▲) activities. Data adapted from Burke et al. (1977) and Sun and Seaman (1980).

(Sun and Seaman, 1980). Based on the Perrin's equation for rotational depolarization of non-spherical fluorescent probes, a decrease in polarization is correlated with a decrease in the microviscosity or an increase in the membrane fluidity (Shinitzky et al., 1971; Shinitzky and Inbar, 1974). It has been well demonstrated that many membrane-bound enzymes are affected by their fluid environment (Raison, 1973; Papahadjopoulos and Kimelberg, 1973; Fourcans and Jain, 1974; Singer, 1974; Fox, 1975), and an increase in membrane fluidity is correlated to an enhanced membrane transport activity (Pang et al., 1979). The biphasic effect of ethanol on (Na^++K^+)-ATPase activity of neuronal membranes may have an important physiological implication for the excitability and behavioral manifestation in humans after chronic ingestion.

The plasma membrane enzyme (Na^+ + K^+)-ATPase is known to be embedded in the lipid bilayer of the membrane. For normal action, this enzyme is free to undergo conformational change or lateral translational movement in order to manifest its

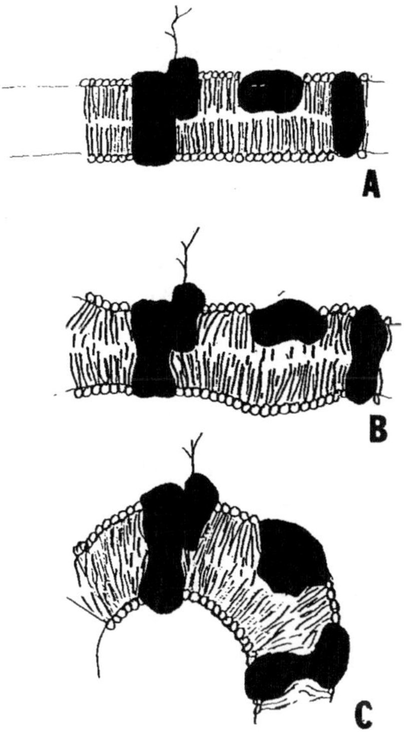

Fig 2. A model depicting that different amounts of ethanol imbedded in membranes may alter membrane physico-chemical properties.

trans-membrane transport function. The phospholipid acyl groups and cholesterol in the membrane are important in giving the enzyme a proper environment for its function as shown in Fig 2A. Ethanol is shown to increase the mobility of spin label (Chin and Goldstein, 1977a; 1977b) or fluorescent probes (Sun and Seaman, 1980; Harris and Schroeder, 1981), and it can lower the phase transition temperature of membranes (Levental and Tabakoff, 1980). The fluidizing effect of low concentration of ethanol can further facilitate the trans-membrane movement of the protein, thus resulting in an increase in $(Na^+ + K^+)$-ATPase activity as shown in Fig 2B. However, at higher concentrations (1-5%), ethanol may actually disrupt the fine arrangement of lipid-lipid or lipid-protein interaction in the membrane, thus leading to an inhibition of the transmembrane processes as depicted in Fig 2C.

B) Acute ethanol administration and ($Na^+ + K^+$)-ATPase

We have demonstrated that the activity of synaptosomal (Na^++K^+)-ATPase increased after a single large dose (6-8 g/kg) of ethanol administered to rats by intragastric intubation. The increase in (Na^++K^+)-ATPase activity was dose-dependent (Fig 3); however, the increase in activity did not appear until 2 hr after intubation. This effect persisted for at least 16 hr before activity returned to the normal level (Fig 4). Israel and Kuriyama (1971) also demonstrated a 36% increase in mouse brain mitochondrial ATPase activity at 3 hr after ethanol (4 g/kg) administration. The exact mechanism for the delayed increase in enzymic activity after an acute dose of ethanol is not yet clearly understood.

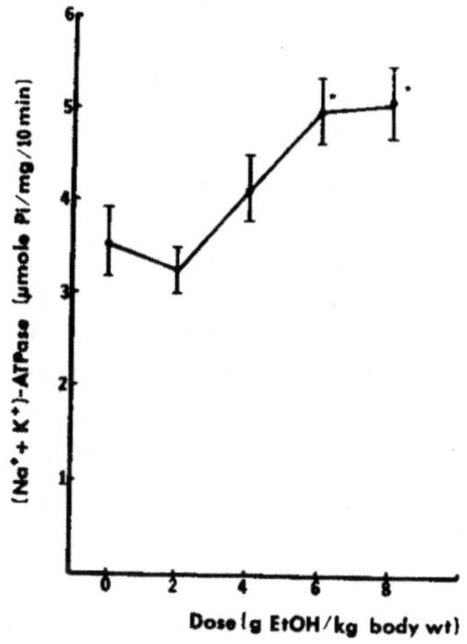

Fig 3. Effect of acute ethanol administration on ($Na^+ + K^+$)-ATPase activity in rat brain synaptosomes. Ethanol was administered via intragastric intubation. (*indicates statistical significance at p<0.01; from Sun et al., 1979.)

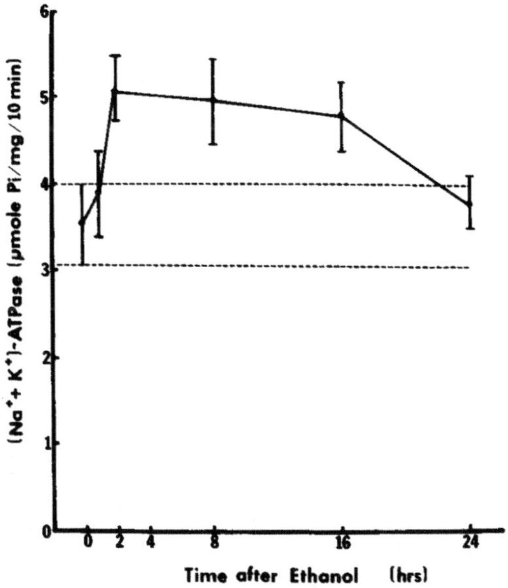

Fig 4. The time course of $(Na^+ + K^+)$-ATpase activity after acute ethanol administration (8 g/kg). (From Sun et al., 1979.)

C) Chronic ethanol administration and $(Na^+ + K^+)$-ATPase

Chronic ethanol administration is shown to induce an increase in $(Na^+ + K^+)$-ATPase activity in brain homogenates as well as in the synaptic plasma membranes (Fig 5). A similar increase in enzymic activity has been reported by other laboratories (Israel et al., 1970; Knox et al., 1972; Roach et al., 1973). In addition, a further increase in enzyme activity was correlated to excitation due to withdrawal of ethanol (Wallgren et al., 1975). Since $(Na^+ + K^+)$-ATPase is known to play an important role in membrane excitability, the ethanol-induced increase in activity may be associated with the hyperactivity frequently observed after withdrawal of ethanol (Roach et al., 1973; Sun et al., 1977).

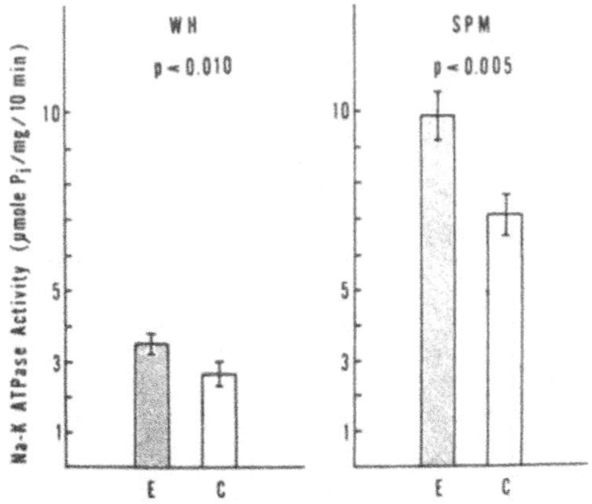

Fig 5. Effect of chronic ethanol administration (shaded bars) on (Na^{+}+K^{+})-ATPase activity in the brain whole homogenates (WH) and synaptic plasma membranes (SPM). Animals were given 8 g/kg/day for 21 days.

D) The kindling phenomenon

Periodic stimulation of the brain has been shown to give rise to increasingly intense electrical and behavioral seizure activity in a number of animal species (Goddard et al., 1969). Under this condition, each stimulation may exert an additive effect which apparently can lead to a higher level of membrane excitability through the adaptive kindling mechanism. Consequently, neuronal membrane is able to develop tolerance and physical dependence by the kindling induction of hyperactive membrane-dependent enzymes. This mechanism may explain the increase in synaptosomal (Na^{+}+K^{+})-ATPase activity after chronic ethanol administration. Due to the depressant action of ethanol, (Na^{+}+K^{+})-ATPase activity may be inhibited initially after administration of an acute dose (6-8 gm/kg) but soon becomes elevated after 2 hr. The increase in enzyme activity after chronic administration may thus represent an adaptive change resulting from direct interaction of ethanol on the membrane. After continuous challenging of the membrane by a high dose of ethanol, the kindling development of hyperactivity may occur. As shown in Fig 6, with each additional amount of ethanol ingested (shown by arrow), the increased membrane $(Na^{+}+K^{+})$-ATPase activity reinforces itself in order to withstand the stress imposed by the presence of ethanol instead of returning to basal level such as occurs after an acute dose (shown by dotted line).

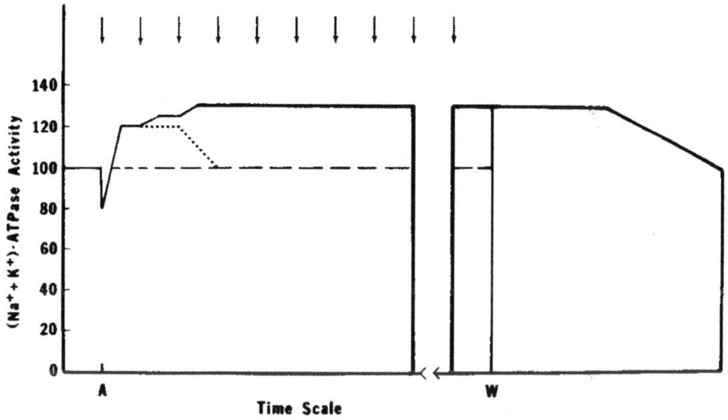

Fig 6. Adaptive kindling of $(Na^{+}+K^{+})$-ATPase activity in rat brain after chronic ethanol administration and withdrawal. Arrows denote time of ethanol administration (8 g/kg/day). Dotted line shows time course of activity following a single acute dose of ethanol (8 g/kg).

In the chronic state, $(Na^{+}+K^{+})$-ATPase activity is found elevated even at 32-48 hr after the last ethanol dose, long after blood ethanol has disappeared. Therefore, the adaptive change in the chronic stage is probably due to a more permanent induction of the enzyme. This adaptive increase in $(Na^{+}+K^{+})$-ATPase activity would enable the animal to function normally in the presence of ethanol. Continued elevation in enzymic activity after termination of ethanol administration is significant because this increase is associated with hyper-excitability of the neuronal membranes. Other membrane-dependent biochemical processes, such as Ca^{2+}-transport, may also be affected similarly. We have shown that NE-uptake activity was elevated by 20% even at 24 hours after the last dose of ethanol (Sun et al., 1977). The kindling development of membrane hyperactivities is therefore an adaptive mechanism related to tolerance and dependence, and may form the biochemical basis of behavioral abnormalities related to ethanol withdrawal.

INTRINSIC CHANGES IN SYNAPTIC PLASMA MEMBRANES AFTER CHRONIC ETHANOL INGESTION

In addition to the increase in membrane-dependent transport activities, prolonged ethanol administration is also correlated to changes in intrinsic properties of the neuronal plasma membranes. Chin et al. (1977) indicated a change in resistance to the fluidizing effect of ethanol after mice were exposed to long-term ethanol treatment. Changes in double bond index and cholesterol content of the membrane have also been reported after chronic ethanol administration (Littleton and John, 1977; Nandini-Kishore et al., 1979; Johnson et al., 1979; Starich and Reitz, 1979; Goldstein and Chin, 1981). Recently, we have used biochemical and biophysical methods to examine the mechanism of alcohol-membrane interaction. These results are pertinent to development of the kindling hypothesis of alcohol addiction.

A) Fluorescence studies

2-Toluidinonaphthalene-sulfonate (TNS) was used to probe the effect of ethanol on micro-environment of membrane matrix. TNS was chosen due to its amphipathic properties and because this probe can be reversibly associated with biological membranes. This compound emits a rather weak fluorescence when excited at 335 nm in aqueous solution but fluorescence becomes intensified when this compound is incorporated into the membrane bilayer (Sun and Seaman, 1980). We found that low concentrations of ethanol (0.1-0.5%) elicited an enhancement of the TNS-membrane complex formation and yielded greater fluorescence intensity. At high concentrations, however, ethanol may "squeeze" some of the TNS molecules out of the membrane matrix, thus resulting in a lower fluorescence intensity (Fig 7).

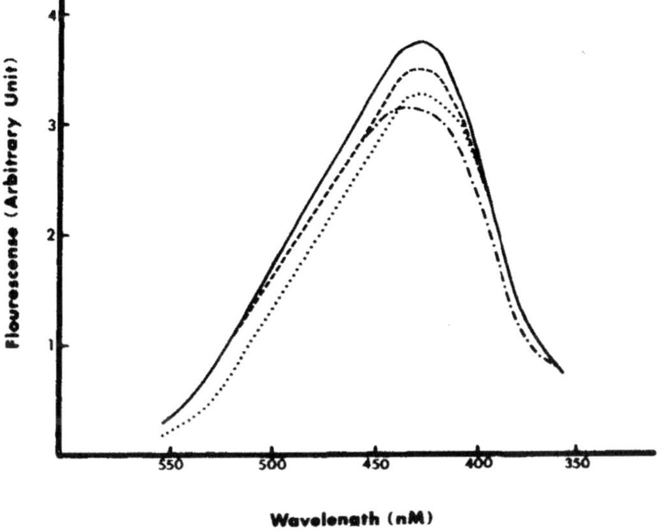

Fig 7. Effect of ethanol on the fluorescence intensity of TNS-membrane complex. The emission spectra were scanned 5 min after the addition of 0 (----), 0.2% (——), 2% (••••) and 3% (•—•—) ethanol solution in the presence of 2.7 x 10^{-5} M TNS and SPM (25 ug protein/ml).

An experiment was designed to examine the fluorescence response of SPM towards ethanol after chronic alcohol administration. A group of Sprague-Dawley rats weighing about 250 grams were subjected to alcohol treatment by giving 8 g/kg/day of 30% ethanol solution intragastrically in 2 fractional doses daily for 21 days. Animals were sacrificed about 24 hr after the last ethanol intubation. Control animals received the same volume of sucrose solution (30%, w/v). A third group of rats were offered 15% ethanol as the sole drink *ad lib* for 3 months. As shown in Fig 8, the control animals showed a typical biphasic response to various concentrations of ethanol (squares). However, the biphasic response of TNS-membrane complex formation to ethanol disappeared with the membranes (SPM) from ethanol-tolerant animals (circles). There was no apparent change in physical properties in SPM from the third group as revealed by the fluorescence studies.

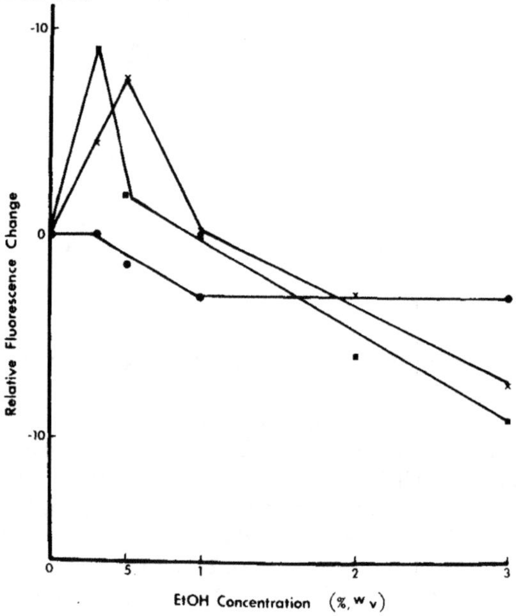

Fig 8. Effect of chronic ethanol ingestion on TNS-fluorescence of rat brain SPM exposed to ethanol *in vitro*. Squares, control animals receiving sucrose; x, animals receiving 15% ethanol *ad lib* for 3 months; circles, animals intubated with 8 g/kg/day for 21 days.

B) Susceptibility of membrane to lipid peroxidation

In the presence of Fe^{++}, mitochondrial membranes isolated from alcoholic rats were shown to be more vulnerable to lipid peroxidation than controls (Sun et al., 1981) (Fig 9). No significant difference was observed with rats offered 32% alcohol *ad lib*. It is possible that after chronic high dose of ethanol administration, some of the membrane acyl groups are exposed, making them more susceptible to free radical attack and lipid peroxidation activity.

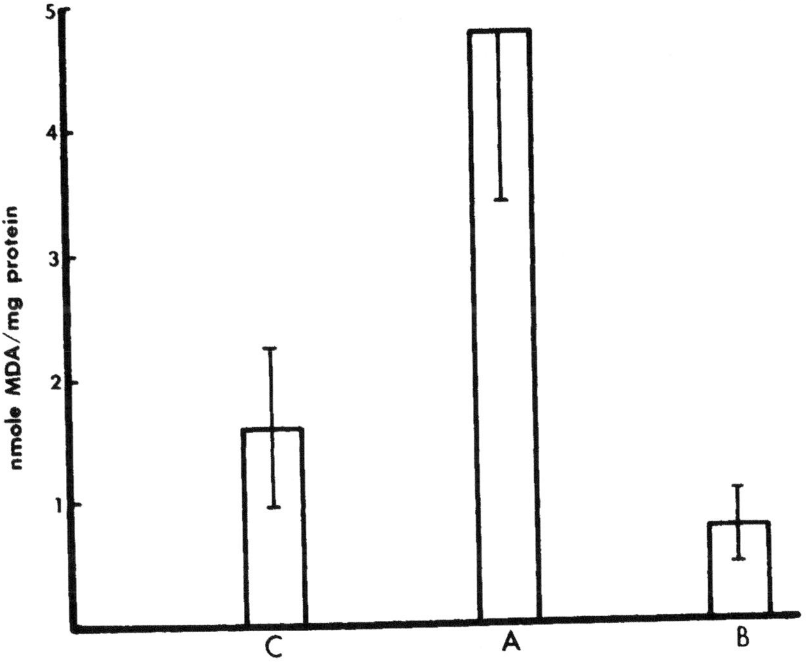

Fig 9. Lipid peroxidation of rat hepatic mitochondrial membranes after chronic ethanol ingestion. C = controls; A = rats intubated with 6-12 g/kg ethanol daily for 14 days; B = rats given 32% ethanol *ad lib* for 28 days. Values are the mean ± S.E.M. for 4-5 animals.

C) Transition temperature

Since membrane fluidity is affected by a change in the temperature, it is possible to use this phenomenon to assess the change in physical properties of SPM (Sun and Seaman, 1980). Synaptosomal plasma membranes were isolated from rats administered ethanol (30%) at 8 g/kg/day by intragastric intubation twice daily for 21 days and from control animals which were handled similarly, except that ethanol was replaced with the same volume of 30% sucrose solution. The (Na^+ + K^+)-ATPase activity of SPM was assayed at different temperatures in order to construct the Arrhenius plots. As shown in Fig 10, a discontinuity representing a transition from gel to liquid crystalline state of the membrane was observed with the membranes isolated from either control or alcoholic rats. There was no significant changes in the energy of activation required for the catalysis as shown by the slope of the

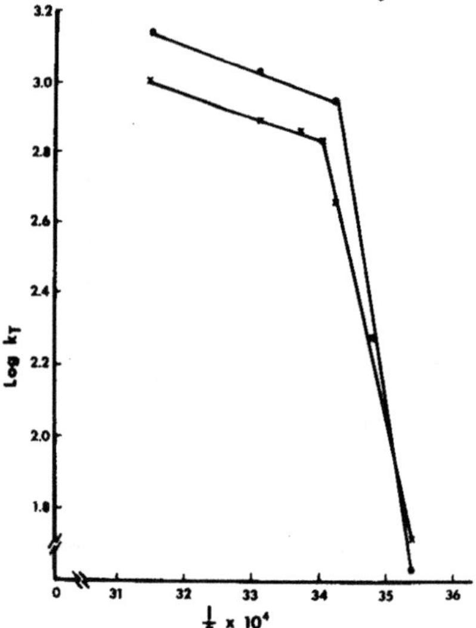

Fig 10. Arrhenius analysis of the synaptosomal (Na^++K^+)-ATPase from control (x——x) and chronically ethanol-treated (•——•) rats. K_T represents the initial rate of reaction. T represents the absolute temperature of the assay medium.

curves. However, the phase transition temperature was lowered from 21° to 18° in membrane isolated from chronic ethanol group. Similar changes have been demonstrated with mice fed a liquid diet containing 6% ethanol (Levental and Tabakoff, 1980). Thus, the results indicate a change in membrane properties due to chronic ethanol ingestion. Since the unsaturated fatty acids of the membrane phospholipids and/or cholesterol content may both influence the fluidity of the biological membranes, the decrease in transition temperature with membranes of ethanol tolerant animals suggests that some intrinsic properties of the membrane are changed towards a more fluid state.

D) Membrane lipid changes

Although the increase in membrane transport activity may be due to ethanol acting as a fluidizing agent, there is some evidence that chronic ethanol administration may alter brain membrane phospholipids and acyl group composition. We have examined the phospholipid acyl groups of myelin and SPM after chronic administration of ethanol to guinea pigs (Sun and Sun, 1979). These guinea pigs were given a liquid ethanol diet containing 5% ethanol for 3 weeks. There was no apparent change in the acyl group composition of myelin lipids; however, there was a decrease in the proportion of monoenes in both choline- and ethanolamine-phosphoglycerides. The decrease in monoenes was marked by an increase in the saturated fatty acids in choline phosphoglycerides, whereas an increase in polyenes was observed in the ethanolamine phosphoglycerides (Table 1).

Thus, ethanol may be associated with changes in acyl group composition of specific types of phosphoglycerides, probably due to the asymmetric localization of the lipid molecules. Changes in acyl group composition of membrane phospholipids were reported after chronic inhalation of ethanol vapor (Littleton and John, 1977). The lack of

Table 1. Acyl group composition of choline and ethanolamine phosphoglycerides from guinea pig synaptic plasma membrane fraction.

Acyl groups	Choline phosphoglycerides Control	Choline phosphoglycerides Ethanol	Ethanolamine phosphoglycerides Control	Ethanolamine phosphoglycerides Ethanol[a]
		%, wt.		
16:0	49.5 ± 1.1	52.6 ± 0.6*	9.5 ± 0.1	8.1 ± 0.5
18:0	14.3 ± 0.4	13.3 ± 0.2	27.5 ± 0.2	25.0 ± 1.0*
18:1	28.1 ± 0.5	25.9 ± 0.6*	17.1 ± 1.19	12.4 ± 1.2*
18:2	1.4 ± 0.2	1.6 ± 0.5	2.4 ± 0.1	2.6 ± 0.4
20:1	1.0 ± 0.1	0.9 ± 0.0	1.8 ± 0.1	1.6 ± 0.0
20:3	---	---	0.5 ± 0.1	0.5 ± 0.0
20:4	2.9 ± 0.1	2.9 ± 0.4	15.8 ± 0.3	16.1 ± 0.3
22:4	0.5 ± 0.0	0.7 ± 0.1	7.3 ± 0.1	9.0 ± 0.3**
22:5	0.3 ± 0.1	0.3 ± 0.1	3.4 ± 0.4	4.3 ± 0.0
22:6	2.0 ± 0.2	2.2 ± 0.1	14.5 ± 2.55	19.7 ± 1.5*
Saturated	63.7 ± 0.88	65.9 ± 0.80	37.0 ± 0.27	33.1 ± 1.50*
Monoenes	29.2 ± 0.59	26.8 ± 0.65*	18.9 ± 1.15	14.0 ± 1.10
Polyenes	7.1 ± 0.38	8.6 ± 1.15	44.0 ± 1.38	51.9 ± 2.40*

Results are mean ± S.E.M. (n=3) from 6 controls and 6 guinea pigs treated with ethanol. [a]Only two samples were analyzed by GLC. Values from ethanol group are significantly different from controls as shown by analysis of variance, *$p<0.05$, **$p<0.005$. (Taken from Sun and Sun, 1979) change in myelin lipids with respect to this mode of ethanol administration is not entirely surprising, since myelin is a more inert type of membrane and changes in myelin components due to dietary manipulation require a much longer induction time. However, acyl group changes in SPM seemed to agree well with the increase in membrane acyltransferase activity, as well as other transport activities after chronic ethanol administration (Sun et al., 1977). The activity of arachidonoyl transfer to 1-acyl-GPC in a group of rats (n=14) administered 30% ethanol by intragastric intubation for 21 days was 24% higher than control. The increased turnover of phospholipid acyl groups as mediated by the deacylation-reacylation mechanism after chronic ethanol administration may be part of the adaptive phenomenon in order to overcome the inhibitory effect of ethanol on membrane functions.

A change in the membrane cholesterol level was also observed after chronic alcohol administration (Chin et al., 1978; Goldstein and Chin, 1981; Johnson et al., 1979). We found that the cholesterol to protein ratio was lower in the group given ethanol as sole drink, but the intubated group was not different from controls (Sun et al., 1979). Besides phospholipids and cholesterol, there are other lipid components which may alter membrane properties. Experiments towards identifying some of these lipid-metabolites are in progress.

CONCLUDING REMARKS

The neuronal membrane transport system, especially the enzymes for $Na^{+}+K^{+}$ and Ca^{++} transport, is known to require specific phospholipids for optimal activities. In this chapter, we have described effects of acute and chronic ethanol administration on some of the membrane transport activities. These changes may be due to a change in the microviscosity and fluidity of the membranes, which in turn may result from a disturbance of the dynamic equilibrium of the membrane phospholipids. Our results indicated an increase in polyunsaturated acyl groups of ethanolamine phosphoglyceride of SPM. Changes in other intrinsic properties of the membranes were observed after chronic ethanol ingestion. For example, the neuronal membranes become more resistant to the fluidizing effect of ethanol (Chin and Goldstein, 1977a,b). They are also more vulnerable to lipid peroxidation (Sun et al., 1980). They can accomodate more TNS molecules. A change in membrane fluidity could explain the disappearance of the biphasic fluorescence response in the presence of ethanol (Fig 8).

Most membrane proteins require a rather specific microenvironment of lipids in the membrane for optimal functioning. Thus, the lipid phase of membranes is intimately related to membrane functions such as ion permeability and active transport of small molecules across the membrane. In the biological system, phase transition changes may be influenced by interaction of divalent cations or hydrophobic drugs with the membrane. The intrinsic alteration of membrane phase transition after chronic alcohol ingestion may also modify membrane functions through regulating calcium metabolism. This would lead to a modification of the calcium dependent processes such as neurotransmitter release and cyclic AMP mediated receptor response (Hunt and Majchrowicz, 1979). Similarly, the supra-additivity between ethanol and other hydrophobic drugs such as barbituates may also be related, in part, to an alteration of the membrane fluidity (Deitrich and Petersen, 1979).

Alcohol consumption could enhance lipid peroxidation in both _in vitro_ and _in vivo_ studies (Di Luzio and Hartman, 1967; Litov et al., 1978). The increase in membrane vulnerability to lipid peroxidation may lead to in increased cytotoxicity (Di Luzio, 1973) or other pathological and aging phenomena (Tappel et al., 1973; Harman, 1973).

The stimulation of membrane excitability by repeated administration of high dose of ethanol is explained by the adaptive kindling phenomenon. In this paper, we have shown that ethanol plays a primary role in kindling development of $(Na^{+}+K^{+})$-ATPase. However, other membrane dependent processes may be affected similarly. The adaptive change is also the basis for explaining alcohol tolerance and physical dependence and manifestation of withdrawal syndrome upon termination of alcohol. Understanding of this adaptive kindling mechanism may be important for developing new methods towards prevention and treatment of alcoholism.

Acknowledgement: The author is indebted to Drs. G. Sun, L. Foudin and J. Wu for their helpful criticism of this paper. The author also expresses his thanks to Mrs. D. Torres for help in preparing the manuscript. This work has been supported in part by a grant from National Institute of Alcohol Abuse and Alcoholism (AA 02054).

REFERENCES

Burke, J.P., Tumbleson, M.E., Seaman, R.N. and Sun, A.Y. (1977) Alcohol-membrane interaction: Calcium uptake in mitochondria. Res. Commun. Chem. Path. Pharm. 18: 569-572.

Chin, J.H. and Goldstein, D.B. (1977a) Drug tolerance in biomembranes: A spin label study of the effects of ethanol. Science 196: 684-685.

Chin, J.H. and Goldstein, D.B. (1977b) Electron paramagnetic resonance studies of ethanol on membrane fluidity. Adv. Exp. Med. Biol. 85A:111-122.

Chin, J.H., Parsons, L.M. and Goldstein, D.B. (1978) Increased cholesterol content of erythrocyte and brain membranes in ethanol-tolerant mice. Biochim. Biophys. Acta 513: 358-363.

Collier, H.O.J. (1964) Tolerance, physical dependence and receptors. A theory of the genesis of tolerance and physical dependence through drug-induced changes in the number of receptors. Adv. Drug Res. 3: 171-188.

Deitrich, R.A. and Petersen, D.R. (1979) Interaction of ethanol with other drugs. In: Biochemistry and Pharmacology of Ethanol (E. Majchrowicz and E.P. Noble, eds.) Vol. 2, pp. 283-302, Plenum Press, New York.

Di Luzio, N.R. (1973) Antioxidants, lipid peroxidation and chemical-induced liver injury. Fed. Proc. 32: 1875-1881.

Di Luzio, N.R. and Hartman, A.D. (1967) Role of lipid peroxidation in the pathogenesis of the ethanol-induced fatty liver. Fed. Proc. 26: 1436.

Fourcans, B. and Jain, M.K. (1974) Role of phospholipids in transport and enzymic reactions. Adv. Lipid Res. 12: 147-226.

Fox, C.F. (1975) Transitions in model systems and membranes. In: MTP International Review of Science. Biochemistry of Cell Walls and Membranes Biochemistry Series 1 (C.F. Fox, ed.) Vol 2, pp. 279-306, University Park Press, Baltimore, MD.

Goddard, G.V., McIntyre, D.C. and Leech, C.K. (1969) A permanent change in brain function resulting from daily electric stimulation. Experimental Neurol. 25: 295-330.

Goldstein, D.B. (1976) Pharmacological aspects of physical dependence on ethanol. Life Sci. 18: 553-562.

Goldstein, D.B. and Goldstein, A. (1968) Possible role of enzyme inhibition and repression in drug tolerance and addiction. Biochem. Pharmacol. 8: 48.

Goldstein, D.B. and Chin, J.H. (1981) Interaction of ethanol with biological membranes. Fed. Proc. 40: 2073-2076.

Harman, D. (1973) Free radical theory of aging. Triangle 12: 153-158.

Harris, R.A. and Schroeder, F. (1981) Ethanol and the physical properties of brain membranes fluorescence studies. Mol. Pharmacol. 20: 128-137.

Hunt, W.A. and Majchrowicz, E. (1979) Alterations in neurotransmitter function after acute and chronic treatment with ethanol. In: Biochemistry and Pharmacology of Ethanol (E. Majchrowicz and E.P. Noble, eds.) Vol. 2, pp. 167-186, Plenum Press, New York.

Israel, Y. and Kuriyama, (1971) Effects of in vivo ethanol administration on adenosine triphosphatase activity of subcellular fractions of mouse brain and liver. Life Sci. 10: 591-599.

Israel, Y., Kalant, H., LeBlanc, A.E. (1970) Changes in cation transport and (Na^+,K^+)-activated adenosine triphosphatase produced by chronic administration of ethanol. J. Pharmacol. Exp. Ther. 174: 330-336.

Johnson, D.A., Lee, N.M., Cooke, R. and Loh, H.H. (1979) Ethanol-induced fluidization of brain lipid bilayers: Required presence of cholesterol in membranes for the expression of tolerance. Mol. Pharmacol. 15: 739-746.

Kalant, H. (1971) Absorption diffusion, distribution and elimination of ethanol: Effects on biological membranes. In: The Biology of Alcoholism (B. Kissin and H. Begleiter, eds.) Vol. 1, pp. 1-62, Plenum Press, New York.

Knox, W.H., Perrin, R.G., and Sen, A.K. (1972) Effect of chronic administration of ethanol on (Na^+,K^+)-activated ATPase in six areas of the cat brain. J. Neurochem. 19: 2881-2884.

Levental, M. and Tabakoff, B. (1980) Sodium-potassium-activated adenosine triphosphatase activity as a measure of neuronal membrane characteristics in ethanol-tolerant mice. J. Pharmacol. Exp. Ther. 212: 315-319.

Litov, R.E., Irving, D.H., Downey, J.E. and Tappel, A.L. (1978) Lipid peroxidation: A mechanism involved in acute ethanol toxicity as demonstrated by in vivo pentane production in the rats. Lipids 13: 305-307.

Littleton, J.M. and John, G. (1977) Synaptosomal membrane lipids of mice during continuous exposure to ethanol. J. Pharm. Pharmacol. 29: 579-580.

Martin, W.R. (1968) A homeostatic and redundancy theory of tolerance to and dependence on narcotic analgesics. Research Publications

Association for Research in Nervous and Mental Disease 46: 206-223.

Nandini-Kishore, S.G., Mattox, S.M., Martin, C.E. and Thompson, Jr., G.A. (1979) Membrane changes during growth of tetrahymena in the presence of ethanol. Biochim. Biophys. Acta 551: 315-327.

Pang, K.Y., Chang, T.L. and Miller (1979) On the coupling between anesthetic induced membrane fluidization and cation permeability in lipid vesicles. Mol. Pharmacol. 15: 729-738.

Papahadjopoulos, D. and Kimelberg, H.K. (1973) Phospholipid vesicles (liposomes) as models for biological membranes: Their properties and interactions with cholesterol and proteins. In: Progress in Surface Sciences (S.G. Davison, ed.) pp. 141-232, Pergamon Press, Oxford.

Raison, J.K. (1973) The influence of temperature-induced phase changes on the kinetics of respiratory and other membrane-associated enzyme systems. Bioenergetics 4: 285-309.

Roach, M.K., Khan, M.M., Coffman, R. (1973) Brain (Na^+,K^+)-activated adenosine triphosphatiase activity and neurotransmitter uptake in alcohol-dependent rats. Brain. Res. 63: 323-329.

Seeman, P. (1972) The membrane actions of anesthetics and tranquilizers. Pharmacol. Res. 24: 583-655.

Seeman, P. (1975) The actions of nervous system drugs on cell membranes. In: Cell Membranes (G. Weissman and R. Clairborne, eds.), pp. 239-248, H.P. Publishing Co., New York.

Shinitzky, M. and Inbar, M. (1974) Differences in microviscosity induced by different cholesterol levels in the surface membrane lipid layers on normal lymphocytes and malignant lymphoma cells. J. Mol. Biol. 85: 603-615.

Shinitzky, M. Dianoux, A.C., Gitler, C. and Weber, G. (1971) Microviscosity and order in the hydrocarbon region of micelles and membranes determined with fluorescence probes. I. Synthetic micelles. Biochemistry 10: 2106-2113.

Singer, S.J. (1974) The molecular organization of membranes. Ann. Rev. Biochem. 43: 805-833.
Starich, G.H. and Reitz, R.C. (1979) Effects of chronic ethanol ingestion in the fatty acids and phospholipids of spinal cord. Toxicol. Lett. 4: 1-5.
Sun, A.Y. (1980) Biochemical and biophysical approaches in the study of ethanol-membrane interaction. In: Biochemistry and Pharmacology of Ethanol (E. Majchrowicz and E.P. Noble, eds.) Vol. 2, pp. 81-100, Plenum Press, New York.
Sun, A.Y. and Samorajski, T. (1970) Effects of ethanol on the activity of adenosine triphosphatase and acetylcholinesterase in synaptosomes isolated from guines pig brain. J. Neurochem. 17: 1365-1372.
Sun, A.Y. and Seaman, R.N. (1980) Physicochemical approaches to the alcohol-membrane interaction in brain. Neurochem. Res. 5: 537-545.
Sun, G.Y. and Sun, A.Y. (1979) Effect of chronic ethanol administration on phospholipid acyl groups of synaptic plasma membrane fraction isolated from guinea pig brain. Res. Commun. chem. Pathol. Pharamacol. 24: 405-408.
Sun, A.Y., Sun, G.Y. and Middleton, C.C. (1977) Alcohol-membrane interaction in the brain: Effect of chronic ethanol administration. Curr. in Alcoholism: Clin. Expt. Res. 1: 81-94.
Sun, A.Y., Seaman, S.L. and Seaman, R.N. (1979) Alcohol-membrane interaction in the brain: Biological adaptation. Curr. in Alcoholism 5: 63-69.
Sun, G.Y., Danopoulos, V. and Sun, A.Y. (1980) Effect of chronic ethanol administration on myelin lipids. Curr. in Alcoholism 7: 83-81.
Sun, A.Y., Foudin, L. and Middleton, C.C. (1981) Vulnerability of liver membrane towards lipid peroxidation after chronic ethanol administration. In: Alcoholism: A Perspective (F.S. Messiha and G.S. Tyner, eds.) pp. 458-466, P.J.D. Publications Ltd., Westbury, New York.
Tappel, A., Flettcher, B. and Deamer, D. (1973) Effect of antioxidants and nutrients on lipid

peroxidation fluorescent products and aging parameters in the mouse. J. Geront. 28: 415-424.

Wallgren, H. (1970) Absorption, diffusion, distribution and elimination of ethanol, effect on biological membranes. In: Alcohols and Derivatives (J. Tremolieres, ed.) Vol. 1, pp. 161-188, Pergamon Press, Oxford.

Wallgren, H., Nikander, P., and Virtanen, P. (1975) Ethanol-induced changes in cation-stimulated adenosine triphosphatase activity and lipid-proteolipid labeling of brain microsomes. Adv. Exp. Med. Biol. 59: 23-36.

RESPONSE OF BRAIN CELL CULTURE TO ETHANOL: MORPHOLOGICAL AND BIOCHEMICAL STUDIES

T.C. Jou and A.Y. Sun

Neuroscience Institute, National Yang-Ming Medical College, Taipei, Taiwan, ROC and Sinclair Comparative Medicine Research Farm University of Missouri, Columbia, MO 65201

ABSTRACT

Since ethanol may affect membrane structure and function by hydrophobic interaction, we have used the rat brain tissue cells (BTC) as a model for studying the morphological and biochemical changes in the presence of ethanol. BTC were derived from 3 day-old rat and maintained in our laboratory for 30 months after more than 20 passages. They were predominately glial cells composed of a monolayer of flat polygonal astroglial cells and oligodendrocytes. Active cell movement and division could be seen by time-lapse cinematography. After several days of cultivation, the normal nutrient medium (NNM, DM-160) was replaced by one composed of NNM and 2% ethanol, and the cell culture was allowed to grow for 7 more days. Nearly all the cells underwent morphological changes soon after the addition of ethanol medium. The astroglial cells shrank into smaller and mostly bipolar spindle form with few multipolar cells. Oligodendrocytes were more sensitive to ethanol and most of them died after exposure to ethanol. Ethanol inhibited leucine transport and its incorporation into protein in a dose-dependent fashion. However, protein synthesis was greatly enhanced after a three hr or longer period of exposure to ethanol. Leucine transport across cell membrane was also enhanced after three hrs exposure to ethanol. The results indıcate that ethanol stimulates changes in cell membrane morphology and inhibits transport and protein synthesis initially, but cells resume the active state after longer exposure to ethanol througn biological adaptation.

INTRODUCTION

It is well known that some exogenous agents such as detergents and general anesthetics, including alcohols, can exert profound effects on the biophysical and biochemical properties of memorane (Seeman, 1972; Sun and Seaman, 1980). Ethanol is one of the most extensively investigated of these agents because its effects on humans at the organismal level are well known; yet the mechanism of its action is not well understood. We have previously demonstrated that many memorane transport processes in brain are altered due to the hydrophobic interaction of alcohoı with membranes. The transport processes included in the _in vitro_ studies are $(Na^+ + K^+)$-ATPase (Sun and Samorajski, 1970), norepinephrine (NE) uptake (Sun et al., 1979) and Ca^{2+}-transport (Burke et al., 1977). Thus, after prolonged ethanol intake, the altered membrane-dependent processes may further lead to a modification of membrane structure and functions. An increase in $(Na^+ + K^+)$-ATPase activity has been shown in rats, cats and dogs after chronic ingestion of ethanol (Sun, 1980), and the increase is thought to be due to an adaptive mechanism compensating for the stress imposed on the membrane upon prolonged exposure to ethanol (Sun, 1983, this volume). We believe that other membrane-dependent biochemical processes will be similarly affected and that after chronic ethanol treatment, an adaptational increase in their activities may occur in order to offset the continous challenge by ethanol (Sun and Sun, 1976; Sun et al., 1977). However, it is not clear whether the increase in enzymic or transport activities after chronic ethanol ingestion is due to an increase in net synthesis of the protein components involved, or activation of the existing enzymes by protein conformational changes, or a change in the lipid microenvironment around the enzyme.

In the past, inherent difficulties existed in attempts to ascertain these possibilities, mainly due to the complexity and heterogeneity of the intact animal. Brain cell culture, on the other hand, offers an ideal situation to complement the study of cellular functions, because the cells can be examined under controlled conditions in their intact state. We have thus used brain cell culture to investigate the effects of either single or prolonged exposure to alcohol on cells. Transport activities and protein synthesis are emphasized.

MATERIALS AND METHODS

I. Primary brain tissue cell culture and morphological studies

Brain tissue cells were derived from cerebral corteces of 3 day-old rats and were cultured *in vitro* for approximately 30 months through 20 passages. Cells were grown in a medium consisting of DM-160 (Kyokuto, Japan) medium with 10% of fetal calf serum (v/v) and supplemented with penicillin G (100 u/ml) and streptomycin (100 ug/ml). Experiments were conducted on confluent monolayered culture of brain tissue cells composed mainly of the bed layer flat cells (astrocytes) and oligodendrocytes which were growing on top of the bed layer. At the renewal of medium, different concentrations of ethanol were added to the medium, and the morphological features were examined by phase-contrast microscopy. Time-lapse microphotographs were also taken with an automatic Bolex cinecamera attached to a Nikon inverted phase-contrast microscope. Selected frames were used for this report.

II. Biochemical studies

A. Transport activity

Cells were washed twice with Krebs-Ringer solution and equilibrated in the same solution for a few min at 37°C. [^{14}C]-labeled leucine was used for transport studies. About 1 uM ^{14}C-leucine (1 uCi, New England Nuclear, Boston, MA) was added to the medium with or without ethanol. At the end of incubation, the incubation medium was decanted and cells were washed twice with 5 ml of ice-cold non-labeled Krebs-Ringer solution. The cells were detached from the flask by means of a rubber policeman. Portions of the cell suspension were filtered through millipore filters, washed and transferred to a scintillation vial with 10 ml of Triton X-100: toluene (1:2) scintillation cocktail and radioactivity was assayed by liquid scintillation spectrometry. The radioactive leucine transported inside the cell was calculated as dpm/mg protein.

B. Protein synthesis

Cells were washed twice with Krebs-Ringer solution and equilibrated in the same solution for 10 min at 37°C. [^{14}C]-Leucine (1 uCi) was added to the medium to a final concentration of 10^{-6} M. The incubation was terminated by decanting off the incubation medium, and the cells were washed twice with 5 ml of ice-cold Krebs-Ringer solution containing 1% unlabeled l-leucine. The cells were removed from the flask by means of a rubber policeman, and the protein was precipitated by 5% trichloroacetic acid (TCA). The protein precipitate was washed again with hot 5% TCA and the radioactivity incorporated into the nascent polypeptides was determined by liquid scintillation spectrometry.

RESULTS

A. Morphological observations

The brain cells used for this study were derived from the cerebral corteces of 3 day-old newborn rats and then cultured for approximately 30 months. Even after more than 20 passages, cells were still actively proliferating and showed the normal stratification of oligodendrocytes and astrocytes. Fig 1 illustrates the growth pattern of the primary neural cell population depicting two distinct cell types. The oligodendrocytes are small cells with dark cell body and fine cellular processes resting on top of a bed layer of astrocytes. When 2% ethanol were added to the medium, the cell processes appeared to retract from their extended positions, and with longer exposure time, some of the oligodendrocytes became rounded-up and were

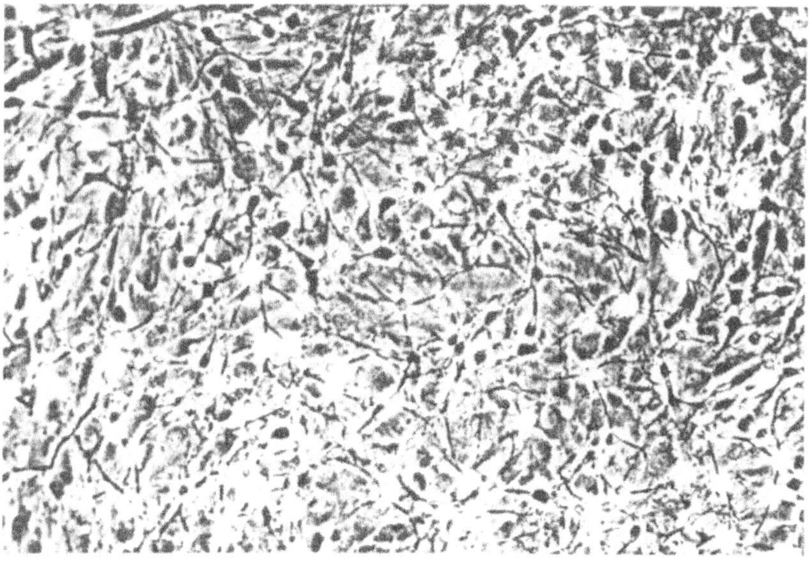

Fig 1. The normal growth pattern of the primary neural cell culture. The cell population consists of two distinct cell types, the bed layer of astrocytes and the small oligodendrocytes with dark cell bodies and fine cellular processes resting on top of bed layer cells (x 174).

detacned from the bottom layer and eventually died as shown under the time-lapse cinematography. Fig 2 shows that after 3 hr of ethanol treatment, some oligodendroglial cells were still attached to the surface. However, the intercellular spaces are clearly visible as a result of cellular retraction. Fig 3 shows that after 24 hr of exposure to ethanol, only a few of the oligodendroglial cells have survived the ethanol treatment. As compared to the astrocytes, the oligodendroglial cells are more sensitive to the ethanol treatment. Although most of the oligodendrocytes died after 7 days of the ethanol treatment, the astroglial cells were alive and became simple dipole type of structure with a few multipolar cells.

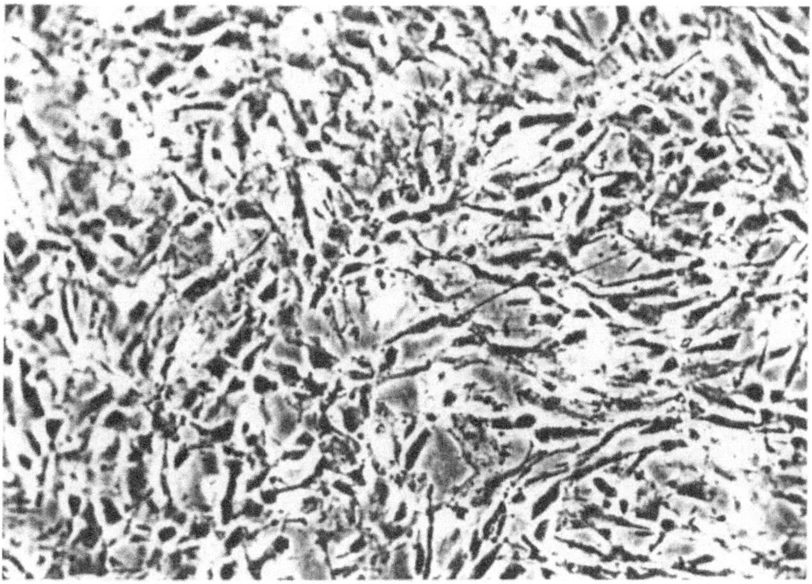

Fig 2. Morphological changes of primary neural cell culture after 2 hr exposure to 2% ethanol. Note that the cell bodies have shrunk and intercellular spaces have appeared (x 174).

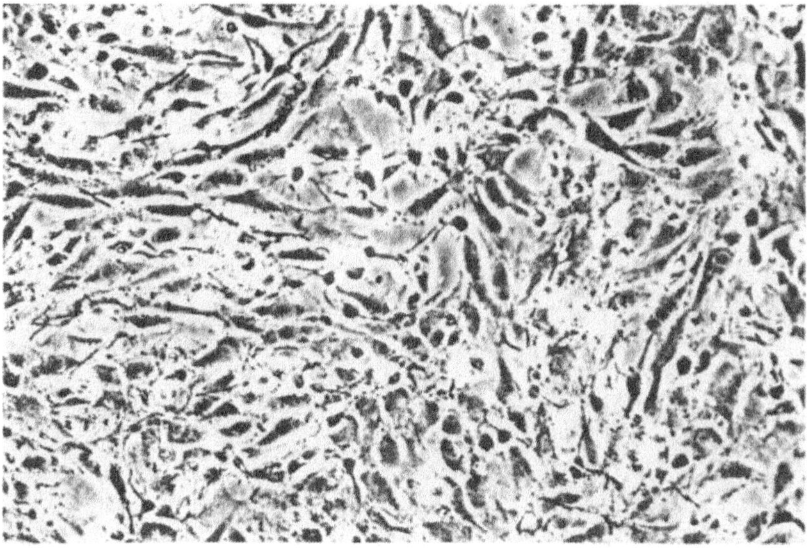

Fig 3. Morphological changes of primary neural cell culture after 24 hr exposure to 2% ethanol. Many of the oligodendrocytes died and floated around (x 174).

B. Effect of ethanol treatment on protein synthesis.

The brain cell culture incorporated [^{14}C]-leucine into cellular proteins in a linear fashion up to 60 min (Fig 4). In the presence of 2% ethanol, protein synthesis was inhibited by 50%. In fact, the inhibitory effect of ethanol increased with increasing concentration as shown in Fig 5. However, upon longer exposure of the cells to ethanol, an adaptive mechanism appeared to take place, and an increase in protein synthesis occurred. As shown in Fig 6, cells were exposed to ethanol for 3 to 20 hr and protein synthesis experiments were performed after removal of ethanol from the medium. After 3 hrs exposure to ethanol, more than 30% increase in protein synthesis was found, and this increase in activity persisted even after 20 hr of the ethanol treatment. The increase in protein synthesis after the exposure of cells to ethanol may be due to an increase in amino acid transport activity. We have therefore studied the effect of ethanol on [^{14}C]-leucine transport in brain cells.

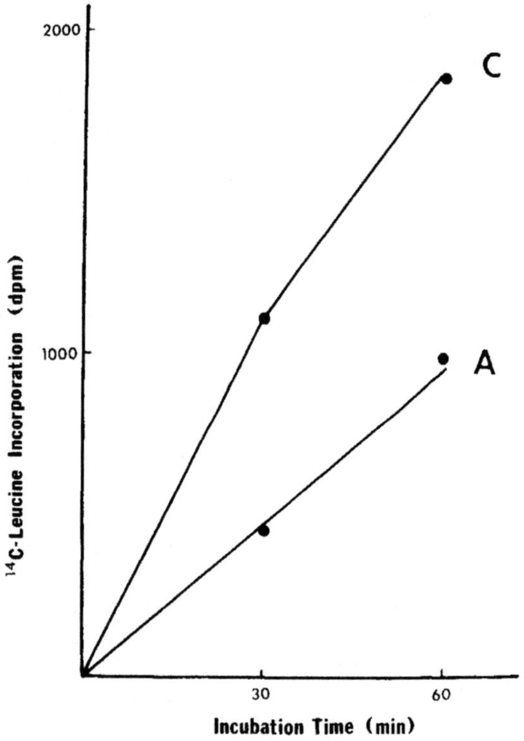

Fig 4. Time course of protein synthesis of the primary neural cell culture in the presence of 2% ethanol (A) as compared with control (C). See text for experimental details. No ethanol was added in the control experiment.

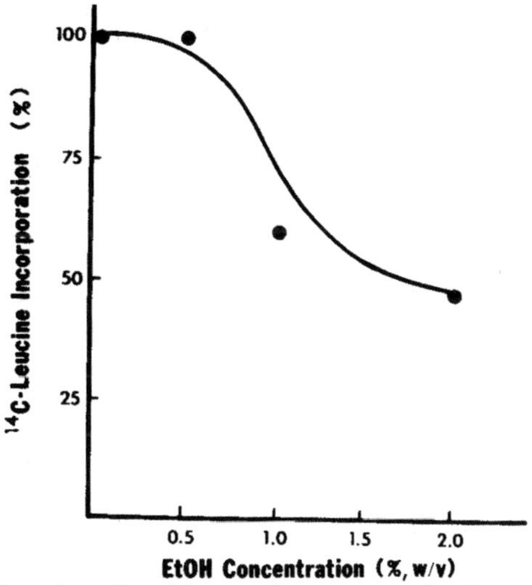

Fig 5. Effect of ethanol on protein synthesis of the primary neural cell culture. Different concentrations of ethanol were added to the medium, and protein synthesis was stopped after 1 hr of incubation at 37°C.

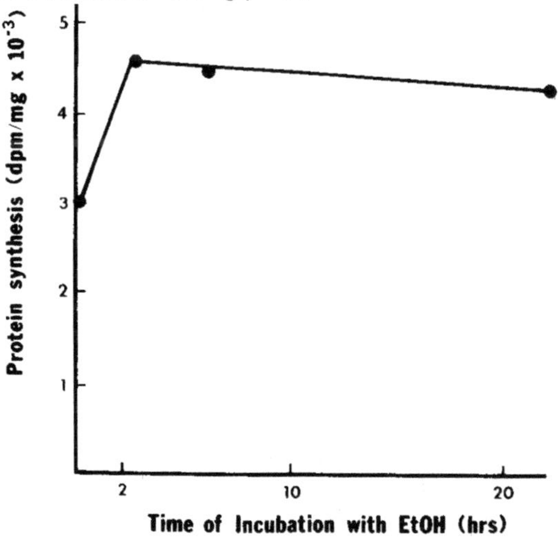

Fig 6. The adaptational development of protein synthetic activity of primary neural cells. After the cells were exposed to 2% ethanol for a certain period of time, the culture medium was changed to Krebs-Ringer solution, and ^{14}C-leucine incorporation into nascent polypeptide was measured as described in the text. Data are the means of three separate experiments.

B. Effect of ethanol treatment on [^{14}C]-leucine transport.

The effect of ethanol on [^{14}C]-leucine transport across cell membranes is less dramatic as compared to its effect on protein synthesis. In the presence of 2% ethanol, only 25% of the leucine transport activity was inhibited as shown in Fig 7. Nevertheless, there was an increase (25%) in transport activity after the cells were exposed to ethanol for 3 hrs as compared with controls (Fig 8). The transport activity of cells remained higher than controls even after the cells were exposed to ethanol for 5 and 20 hrs.

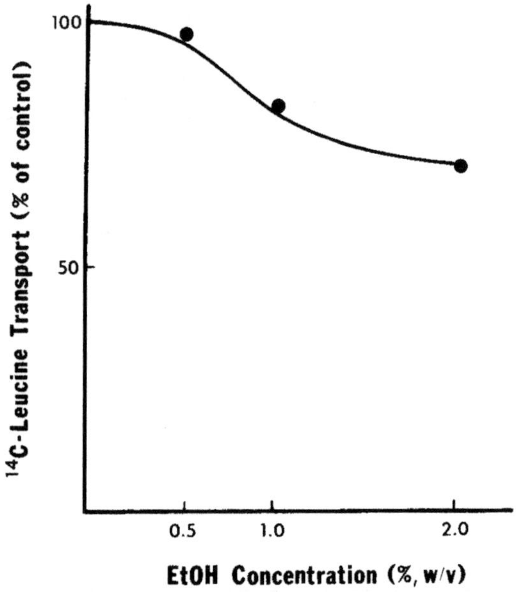

Fig 7. Effect of ethanol on ^{14}C-leucine transport into the primary neural cells in the presence of different concentrations of ethanol. Data represent means of three separate experiments.

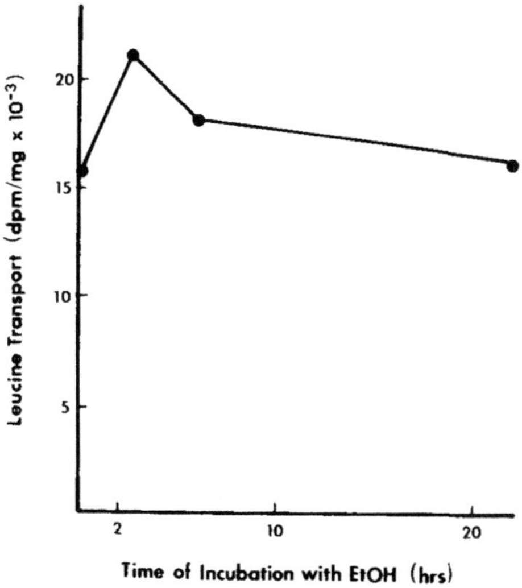

Fig 8. The adaptational development of ^{14}C-leucine transport activity of primary neural cells. After the cells were exposed to 2% ethanol for 3-20 hrs, the culture medium was changed to Krebs-Ringer solution, and ^{14}C-leucine transport inside the primary neural cells was assayed as described in the text. Data are the means of three separate experiments.

DISCUSSION

Animal studies have indicated that high concentrations of ethanol embedded in the membrane may interrupt the lipid-lipid or lipid-protein interaction, thus resulting in general inactivation of membrane enzymes (Sun and Seaman, 1980). It is possible that this kind of inactivation leads to the adaptive development of physical dependence on alcohol. Since the biological adaptation is a very complicated phenomenon, it is difficult to study this kind of mechanism by using animal organs, especially with regard to studying the effects of ethanol on central nervous system. Primary brain cell culture offers a unique system for correlating the cause-effect relationship since the growth environment and ethanol concentrations can be closely controlled.

The effect of ethanol on the growth and morphological features of primary neural cell culture has not been studied extensively. Our observations indicated that there is a dramatic change in cellular process formation. Oligodendrocytes appear more sensitive to ethanol than astroglial cells. After the culture was exposed to ethanol for 7 days, some oligodendrocytes died and floated around the medium, and the surviving cells became much simpler in shape and cellular processes as compared with controls (Fig 3). Similar results have been demonstrated by McCarthy and deVellis (1980) who indicated that the cell number of the primary glial cell culture as greatly reduced when exposed to ethanol for a short period of time. It is possible that some cells, especially oligodendrocytes, may not survive in the presence of high concentrations of ethanol because of the disruption of transport activities.

Using cultured neuroblastoma cells, Syapin et al. (1976) have demonstrated that acute exposure to ethanol results in an inhibition in the activity of Mg^{++}-ATPase and $(Na^{+}+K^{+})$-ATPase, but upon prolonged maintenance of these cells in ethanol, an enhanced activity in the two enzymes was observed. These observations in general seem to give support to the results of studies of brain tissue and subcellular fractions (see Sun, 1983, this volume), although the former investigators have attributed the increase in the ATPases activity to the conformational changes in the components of the ATPase system. In this study, we have demonstrated that chronic exposure to ethanol induced an increase in protein synthesis. Thus the increase in $(Na^{+}+K^{+})$-ATPase both in rat brain and cell culture after chronic ethanol treatment may well be due to an adaptational induction of synthesis of enzyme protein involved. This hypothesis, however, requires quantitative measurement of the enzyme protein for a confirmation.

Increase in active transport activity appears to be the general phenomenon in whole animals and cell culture studies. Besides $(Na^{+}+K^{+})$-ATPase supported Na-transport, Massarelli et al. (1977) also demonstrated an increased uptake of choline after chronic exposure of ethanol by neural cell culture. We have also demonstrated increased transport of $[^{14}C]$-leucine by glial cell culture. The increase in these transport activities may be partially due to the increase in protein synthesis, but a change in the microenvironment of the carrier protein cannot be excluded. With this cell culture system, we may be able to further examine the changes in chemical and physical properties of membrane after chronic exposure to ethanol.

REFERENCES

Burke, J.P., Tumbleson, M.E., Seaman, R.N. and Sun, A.Y. (1977) Alcohol-membrane interaction: calcium uptake in mitochondria. Res. Commun. Chem. Path. Pharmacol. 18, 569-572.

Massarelli, R., Syapin, P.J. and Noble, E.P. (1976) Increased uptake of choline by neural cell cultures chronically exposed to ethanol. Life Sci. 18, 397-404.

McCarthy, K.D. and de Vellis, J. (1980) Preparation of separate astroglial and oligodendroglial cell cultures from rat cerebral tissue. J. Cell Biol. 85, 890-902.

Seeman, P. (1972) The membrane actions of anesthetics and tranquilizers. Pharmacol. Rev. 24, 583-655.

Sun, A.Y. (1980) Biochemical and biophysical approaches in the study of ethanol-membrane interaction. In: Biochemistry and Pharmacology of Ethanol (E. Majchrowicz and E.P. Noble, eds.) Vol. 2, pp. 81-100. Plenum Press, New York.

Sun, A.Y. (1983) The kindling effect of ethanol on neural membranes. This volume.

Sun, A.Y. and Samorajski, T. (1970) Effects of ethanol on the activity of adenosine triphosphatase and acetylcholinesterase in synaptoomes isolated from guinea pig brain. J. Neurochem. 17, 1365-1372.

Sun, A.Y. and Seaman, R.N. (1980) Physicochemical approaches to the alcohol-membrane interaction in brain. Neurochem. Res. 5, 537-545.

Sun, A.Y. and Sun, G.Y. (1976) Functional roles of phospholipids of synaptosomal membrane. In: Function and Metabolism of Phospholipids in CNS and PNS (G. Porcellati, L. Amaducci and C. Galli, eds.) pp. 169-197. Plenum Press, New York.

Sun, A.Y., Seaman, S.L. and Seaman, R.N. (1979) Alcohol-membrane interaction in the brain: Biological adaptation. Curr. Alcoholism 5, 63-69.

Sun, A.Y., Sun, G.Y. and Middleton, C.C. (1977) Alcohol-membrane interaction in the brain: Effect of chronic ethanol administration. Curr. Alcoholism 1, 81-94.

Syapin, P.J., Stefanovic, V., Mandel, P. and Noble, E.P. (1976) The chronic and acute effects of ethanol on adenosine triphosphatase activity in cultured astroblast and neuroblastoma cells. J. Neurosci. Res. 2, 147-155.

STRUCTURE AND FUNCTION OF SYNAPTOSOMAL AND MITOCHONDRIAL MEMBRANES: ELUCIDATION USING NEUROTOXIC METALS AND NEUROMODULATORY AGENTS.

J.C.K. Lai*, P.C.L. Wong** and L. Lim+
*Dementia Research. Serv., Cornell Med. Coll. Burke Rehab. Cntr., 785 Mamaroneck Ave., White Plains, NY 10605, USA; **Biochem. Dept., Hong Kong Univ; +Neurochem. Dept., Inst. of Neurol., London Univ., UK

The structure and function of neural membranes at the cellular and the organelle levels show profound changes during development and aging. These age-related changes are influenced by a variety of insults, be they physiological (e.g., hormonal status, dietary manipulations) or pathophysiological (e.g., intoxication with different drugs or neurotoxic chemicals) (Sun, 1972; Sun and Samorajski, 1975; Sun and Sun, 1976; Su and Sun, 1978; Balazs et al., 1979; Wong and McGeer, 1981; Leung et al., 1981, 1982a; Lai et al., 1982b). Our interests have been to develop a dual approach to studying the structure and function of neural membranes (Bourne et al., 1975; Lai et al., 1976; Owen et al., 1977; Lai et al., 1978, 1980a,b,c,d, 1981a,b; Leung et al., 1982a,b; Lai et al., 1982a,b). On the one hand, we are exploring the possibilities of using neuromodulatory agents (e.g., metals or psychoactive drugs) as tools to elucidate the molecular properties of certain functionally defined entities located in neural membranes (e.g., carrier-mediated transport systems and membrane-bound enzymes). On the other hand, since the agents we have employed have known neurotoxic or neuromodulatory properties, our approach could also help throw some light on the neurochemical mechanisms underlying the pathophysiological effects elicited with these agents.

In this paper, we review some of our studies to test the "theory of selective vulnerability of the nervous system."

A. STRUCTURE AND FUNCTION OF SYNAPTOSOMES: EFFECTS OF NEUROTOXIC METALS

Ionic homeostasis across the neuronal membrane is central to the regulation of neurotransmission and the neurotransmitter uptake system may be an important part of this mechanism (Bogdanski, 1976; Lai et al., 1980c,d). The ionic gradient hypothesis has been proposed to account for how the neurotransmitter uptake system may be energetically linked to the ionic fluxes (Bogdanski, 1976). While this hypothesis is attractive in that it postulates that the energy derived from the sodium electrochemical gradient across the neuronal membrane, maintained by the activity of the membrane-bound Na-K-ATPase, is coupled to the neurotransmitter uptake system(s) (Bogdanski, 1976), the energetic aspect of this hypothesis is, nevertheless, controversial in that other sources of energy (e.g., glycolytic, mitochondrial) may act as alternatives to that generated by the sodium electrochemical gradient. Thus, if one could devise suitable experimental approaches to singly dissect these structure and function aspects of the neuronal membrane, the controversy regarding the energetic aspect of the ionic gradient hypothesis may be resolved (Lai et al., 1980c). We will discuss the evidence in support of the view that neurotoxic metals are useful and suitable tools for elucidating precisely these structure and function aspects of the neuronal membrane.

Metals modulate brain function and behavior, but the underlying mechanisms are poorly understood (Lai et al., 1980c). We have studied cadmium (Cd), manganese (Mn) and aluminum (Al) as these metals are neurotoxic. Al is implicated in Alzheimer's disease, dialysis encephalopathy and several other dementing diseases (Crapper et al., 1978). In Mn encephalopathy, Parkinsonian extrapyramidal symptoms appear after an initial psychotic phase (Cotzias et al., 1971 and 1974). In Cd poisoning, neurological symptoms (e.g., anosmia, vestibular disease) are recognized in addition to the better defined hepatic and renal dysfunctions (Weiss, 1978). Thus, a better understanding of the mechanisms underlying the neurotoxic effects of these metals may throw some light on the pathophysiology of several neurological diseases.

In the following sections, we discuss our studies on the effects of Cd, Mn and Al on carrier-mediated transport systems and ATPases in synaptosomal membranes.

A.1 Effects of Cd, Mn and Al on Forebrain Synaptosomal Uptake Systems and ATPases

Cd is a potent inhibitor of Na-K-ATPase. The level of Cd giving rise to 50% inhibition of Na-K-ATPase activity (i.e. the IC_{50}) is 5 μM (Table 1). Cd, at 10 μM, completely inhibits its activity (Lai et al., 1980c). However, these low levels (5-10 μM) of Cd do not inhibit the uptake of dopamine, noradrenaline, serotonin, choline, GABA and glutamate (Lai et al., 1978, 1980c; Wong et al., 1981; Lai et al., 1981b, 1982a). As these uptake systems are only inhibited at Cd levels above 200 μM (Lai et al., 1980c; Wong et al., 1981; Lai et al., 1981b, 1982a), the inhibition of these systems by Cd cannot be directly related to its inhibition of the Na-K-ATPase (Tables 1 and 2).

Inhibition of the uptake of dopamine, noradrenaline and choline by Mn (Lai et al., 1978, 1980c, 1981b) can be a direct function of its inhibition of Na-K-ATPase as the IC_{50} values for the inhibition of these uptake systems and the Na-K-ATPase by Mn are between 1 and 2 mM (Table 1). However, the IC_{50} values for the inhibition of the uptake of serotonin, glutamate and GABA by Mn are much higher (Tables 1 and 2) (Wong et al., 1981; Lai et al., 1982a).

Unlike Cd and Mn, Al is a poor inhibitor (IC_{50} for Al = 8.3 mM) of Na-K-ATPase (Lai et al., 1980c). On the other hand, Al inhibits different uptake systems at μM concentrations. Consequently, the inhibitory effects of Al on the uptake systems cannot be attributed to its inhibition of Na-K-ATPase (Lai et al., 1980c; Wong et al., 1981; Lai et al., 1981b, 1982a).

The inhibitory effects of Cd, Mn and Al on synaptosomal Mg-ATPase are less pronounced than those on the Na-K-ATPase (Lai et al., 1980c). The rank order of inhibition of Mg-ATPase, similar to that of Na-K-ATPase, is: Cd > Mn >> Al (Table 1). Unlike their effects on the ATPases only one of these metals, namely Cd, is a potent inhibitor (IC_{50} for Cd = 70 μM) of the methylation of phospholipids in the synaptosomal membranes (Wong and Lim, 1981). However, Al slightly stimulates the methylation of synaptosomal membrane phospholipids whereas Mn (up to 1 mM) produces no effect (Wong and Lim, 1981).

That the inhibitory actions of Cd and Al on the neurotransmitter uptake systems are not directly related to their inhibition of Na-K-ATPase contradicts the ionic gradient hypothesis (Bogdanski, 1976) and contrasts with the results of Prakash et al. (1973). However, when the results of Hexam (1974) are compared with those of Silbergeld (1977), the same conclusion as ours can also be arrived at since Hexam (1974) observed that brain microsomal Na-K-ATPase activity is completely inhibited by lead at 100 μM whereas Silbergeld (1977) noted only minor inhibition of synaptosomal choline and dopamine uptake by lead at 10 to 250 μM.

A.2 Effects of Cd, Mn and Al on Uptake Systems in Synaptosomes Isolated from the Developing Forebrain

To determine if the structural changes during development would have any functional relevance, we have compared the effects of Cd, Mn and Al on the uptake of choline, GABA and glutamate by synaptosomes from the developing forebrain.

As shown in Fig. 1A, Cd inhibits choline uptake in a concentration-dependent manner (IC_{50} for Cd between 200 and 400 μM; Table 2). The inhibitory actions of Cd on up-

Table 1. Summary of IC_{50} values.

Uptake/Enzyme System	Ca Status*	IC_{50} Cd	IC_{50} Mn	IC_{50} Al
Dopamine	-	400 μM	2.5 mM	400 μM
	+	400 μM	2.1 mM	240 μM
Noradrenaline	-	250 μM	1.5 mM	430 μM
	+	540 μM	1.5 mM	330 μM
Serotonin	-	610 μM	3.4 mM	370 μM
	+	1.5 mM	4.0 mM	290 μM
Choline	-	363 μM	1.5 mM	224 μM
	+	363 μM	1.2 mM	123 μM
Na-K-ATPase		5 μM	955 μM	8.3 mM
Mg-ATPase		316 μM	5.5 mM	21.9 mM

Lai et al. (1978, 1980c, 1981b, 1982a). * ± 1 mM Ca.

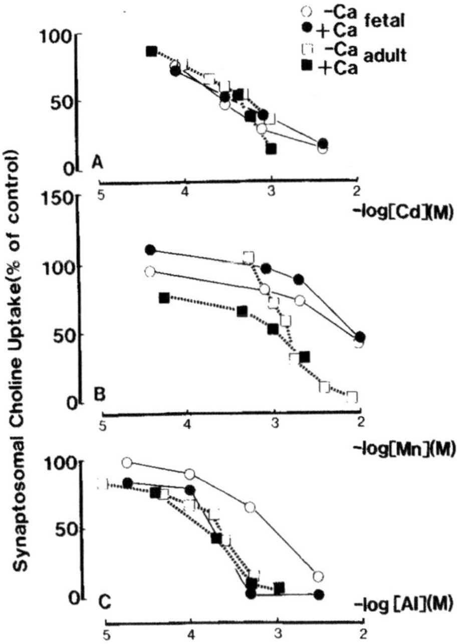

Fig. 1 Effects of Cd, Mn and Al on Choline Uptake by Adult and Fetal Forebrain Synaptosomes.

For adult synaptosomes: all values were the Mean of 3 experiments and SDs (not shown) were < 10% of the Mean; control values of uptake were 2.0 pmol/min/mg protein without Ca and 1.9 pmol/min/mg protein with 1 mM Ca; synaptosomes were isolated as described previously (Lai et al., 1980c).

For fetal synaptosomes: values were the Mean; of 3 separate determinations and SDS (not shown) were < 10% of the Mean; control values (Mean of 3 experiments) of uptake were 1.5 pmol/min/mg protein with 1 mM Ca and 1.7 pmol/min/mg protein without Ca; fetal synaptosomes were prepared using the rats (21 or 22 days of gestation) pooled from 1 or 2 litters as previously described (Lai et al., 1980c) except that the Ficoll-sucrose gradient consisted of 3 ml of isolation medium on 6 ml of 5% (w/w) Ficoll-sucrose medium on 10 ml 7.5% (w/w) Ficoll-sucrose medium and the banded layers at the interphase between the isolation medium and the 5% (w/w) Ficoll-sucrose and that between the 5% (w/w) Ficoll-sucrose and the 7.5% (w/w) Ficoll-sucrose were pooled, washed and used for the uptake studies (Lai et al., 1980c).

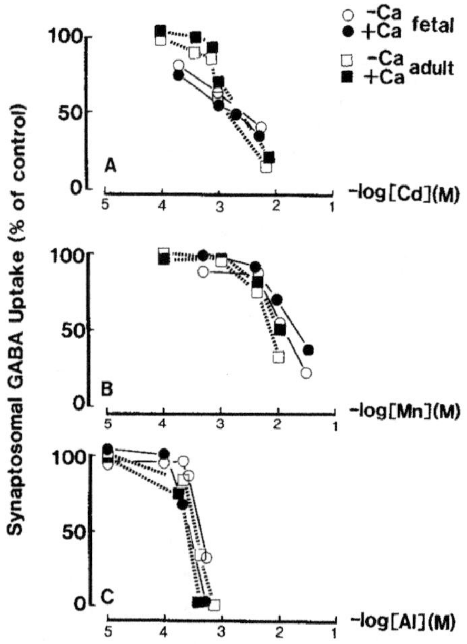

Fig. 2 Effects of Cd, Mn and Al on GABA Uptake by Adult and Fetal Forebrain Synaptosomes.

For adult synaptosomes: all values were the Mean of 3 separate determinations and SDs (not shown) were < 10% of the Mean; control values of uptake were 190 pmol/min/mg protein with or without 1 mM Ca (Mean of 4 experiments); other procedures were as described previously (Wong et al., 1981).

For fetal synaptosomes: all values were the Mean of 3 separate determinations and SDs (not shown) were < 10% of the Mean; control values (Mean of 3 experiments) of uptake were 121 pmol/min/mg protein with 1 mM Ca and 118 pmol/min/mg protein without; uptake measurements were performed as described by Wong et al. (1981); other procedures were as described in the legend to Fig. 1.

take by fetal and adult synaptosomes are similar (Fig. 1A). However, uptake by fetal synaptosomes (IC_{50} for Mn = 6-9 mM) is less susceptible to inhibition by Mn in comparison with uptake by adult synaptosomes (IC_{50} for Mn = 1-2 mM; Fig. 1B and Table 2). Ca enhances the inhibitory effects of Mn on uptake by adult synaptosomes but antagonizes those on fetal synaptosomes (Fig. 1B and Table 2). Ca also greatly enhances the inhibition of uptake by fetal synaptosomes by Al (IC_{50} = 851 μM without Ca and 180 μM with Ca) (Fig. 1C). The enhancement of the inhibitory effect of Al on uptake by adult synaptosomes is less marked (IC_{50} for Al = 224 μM without Ca and 123 μM with Ca) (Fig. 1C and Table 2).

The GABA uptake by fetal synaptosomes is more susceptible to inhibition by low levels of Cd (< 1 mM) than the uptake by adult synaptosomes (Fig. 2A). Despite the similarity of the IC_{50} values for Cd (1-3 mM; Table 2), Ca antagonizes the inhibitory effects of Cd on the uptake by adult synaptosomes but slightly enhances these effects on the uptake by fetal synaptosomes (Fig. 2A and Table 2). However, Ca antagonizes the inhibitory actions of Mn on uptake by both adult and fetal synaptosomes (Fig. 2B and Table 2), but the uptake by fetal synaptosomes (IC_{50} for Mn = 21.9 mM with Ca and 11.7 mM without; Table 2) is less sensitive to inhibition by Mn than the uptake by adult synaptosomes (IC_{50} for Mn = 10.0 mM with Ca and 7.4 mM without; Table 2). GABA uptake by fetal synaptosomes (IC_{50} for Al = 417 μM with Ca and 269 μM without) is less sensitive to inhibition by Al than GABA uptake by adult synaptosomes (IC_{50} for Al = 316 μM with Ca and 149 μM without; Table 2). Ca slightly enhances the inhibitory effects of Al on GABA uptake by fetal or adult synaptosomes (Fig. 2C and Table 2).

Although Ca slightly antagonizes the inhibitory effects of Cd on glutamate uptake by fetal or adult synaptosomes (Fig. 3A and Table 2), the inhibitory effects of Cd on uptake by fetal synaptosomes (IC_{50} for Cd = 3.4 mM with Ca and 2.5 mM without; Table 3) are similar to those

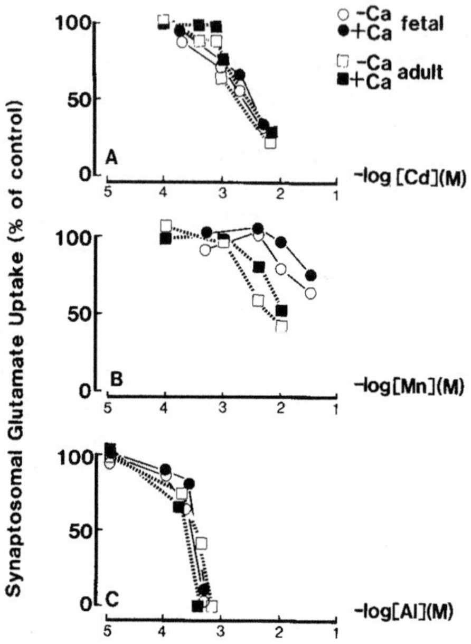

Fig. 3 Effects of Cd, Mn and Al on Glutamate Uptake by Adult and Fetal Forebrain Synaptosomes.

For adult synaptosomes: all values were the Mean of 3 separate determinations and SDs (not shown) were < 10% of the Mean; control values (Mean of 4 experiments) of uptake were 1230 pmol/min/mg protein with 1 mM Ca and 1380 pmol/min/mg protein without; other procedures were as described previously (Wong et al., 1981). For fetal synaptosomes: all values were the Mean of 3 separate determinations and SDs (not shown) were < 10% of the Mean; control values (Mean of 3 experiments) of uptake were 673 pmol/min/mg protein with 1 mM Ca and 704 pmol/min/mg protein without; other procedures were as described in the legend to Fig. 1.

on uptake by adult synaptosomes (IC_{50} for Cd = 3.0 mM with Ca and 2.0 mM without; Table 2). Ca also antagonizes the inhibitory actions of Mn on uptake by fetal and adult synaptosomes although the uptake by fetal synaptosomes is less sensitive to inhibition by Mn (IC_{50} for Mn > 30 mM;

Table 2. Comparison of IC_{50} values for Cd, Mn and Al in the inhibition of uptake by synaptosomes from the fetal and adult forebrain.

Uptake System	Synaptosomal Preparation	Ca Status	IC_{50} Cd	IC_{50} Mn	IC_{50} Al
Choline	fetal	-	275 µM	6.3 mM	851 µM
		+	355 µM	8.9 mM	180 µM
	adult	-	363 µM	1.5 mM	224 µM
		+	363 µM	1.2 mM	123 µM
GABA	fetal	-	2.7 mM	11.7 mM	417 µM
		+	1.9 mM	20.9 mM	269 µM
	adult	-	1.4 mM	7.4 mM	316 µM
		+	2.1 mM	10.0 mM	149 µM
Glutamate	fetal	-	2.5 mM	>30.0 mM	300 µM
		+	3.4 mM	>30.0 mM	363 µM
	adult	-	2.0 mM	7.1 mM	299 µM
		+	3.0 mM	10.0 mM	224 µM

Ca status was ± 1 mM Ca.

Table 2) than the uptake by adult synaptosomes (IC_{50} for Mn = 10.0 mM with Ca and 7.1 mM without; Table 2). Ca slightly antagonizes the inhibitory effect of Al on uptake by fetal synaptosomes (IC_{50} for Al = 363 µM with Ca and 300 µM without; Table 2) but enhances the same effect on uptake by adult synaptosomes (IC_{50} for Al = 224 µM with Ca and 299 µM without; Table 2).

The effects of Cd, Mn and Al on GABA and glutamate uptake by synaptosomes from the 15-day-old forebrain were also studied (Table 3) to determine if these effects differ from those observed in fetal and adult synaptosomes (Table 2). At 2 mM, Cd inhibits GABA and glutamate uptake by 15-day-old forebrain synaptosomes by more than 50% (Table 3): these effects of Cd in the 15-day-old forebrain synaptosomes are similar to those in the adult and fetal synaptosomes (Figs. 2A and 3A).

Mn, at 10 mM, inhibits GABA uptake by 15-day-old forebrain synaptosomes by about 50% (Table 3). This effect is comparable to that of Mn on uptake by adult forebrain synaptosomes (Fig. 2B) but is more pronounced than that of Mn on uptake by fetal forebrain synaptosomes (Fig. 2B). The same concentration of Mn is more effective in inhibiting glutamate uptake by 15-day-old forebrain synaptosomes than uptake by fetal forebrain synaptosomes (Table 3 and Fig. 3B) but is less effective in inhibiting uptake by adult forebrain synaptosomes (Table 3 and Fig. 3B).

Ca, at 1 mM, slightly enhances the inhibitory actions of Al on GABA and glutamate uptake by 15-day-old forebrain synaptosomes (Table 3). Al, at 0.2 mM, is more effective in inhibiting GABA uptake by 15-day-old forebrain synaptosomes (Table 3) than uptake by fetal or adult forebrain synaptosomes (Fig. 2C). In the absence of Ca, 0.2 mM Al inhibits glutamate uptake by 15-day-old forebrain synaptosomes (Table 3) to a greater extent than it does uptake by fetal or adult forebrain synaptosomes (Fig. 3C). The presence of 1 mM Ca further accentuates these differences (Table 3 and Fig. 3C).

The results discussed above (Figs. 1-3 and Tables 2 and 3) suggest that, during development, the structure and function of the synaptosomal membranes may show differential susceptibility to the effects of neurotoxic metals such as Cd, Mn and Al.

A.3 Effects of Cd, Mn and Al on Forebrain Synaptosomal Dopamine Uptake: the Kinetic Aspects

To obtain some insights into the more molecular aspects of the interactions of metals with the membrane-bound carrier-mediated processes, the kinetics of the inhibition of dopamine uptake by Cd, Mn and Al have been studied (Lai et al., 1980a,b). The results of these experiments demonstrate that the three metals interact differently with the high-affinity dopamine uptake system. At levels around its IC_{50}, Cd is an uncompetitive inhibitor (Lai et al., 1980b) of dopamine uptake. Unlike Cd, Mn, at levels both below and just around its IC_{50}, inhibits uptake competitively indicating that Mn interacts directly with the dopamine uptake site (Lai et al., 1980a).

However, the interactions of Al with dopamine uptake are far more complex. Depending on the level, Al inhibits uptake either competitively (at 50 µM), or uncompetitively (at 200 µM), or noncompetitively (at 500 µM), suggesting that Al interacts with at least several sites on the dopamine uptake system.

A.4 Comparison of the Effects of Cd, Mn and Al on Dopamine Uptake by Striatal and Forebrain Synaptosomes.

Another of our approaches (Lai et al., 1980d, 1981b) is based on the argument that the structure and function of neuronal membranes show brain regional specialization or differentiation. In the light of this approach, we have compared the effects of Cd, Mn and Al on dopamine uptake by striatal and forebrain synaptosomes (Lai et al., 1981b).

The effects of these metals on uptake by striatal synaptosomes differ from those on uptake by forebrain synaptosomes in several important respects. (i) Ca, at 1 mM, antagonizes the inhibitory effects of all three metals on uptake by striatal synaptosomes but not the same effects

Table 3. Effects of Cd, Mn and Al on GABA and glutamate uptake by synaptosomes from the 15-day-old forebrain.

Metal (Conc.)	Ca Status	Uptake Rates (% of Control*) GABA	Glutamate
Cd^{2+} (2 mM)	-	42 ± 1	46 ± 1
	+	36 ± 4	41 ± 1
Mn^{2+} (10 mM)	-	50 ± 5	61 ± 7
	+	46 ± 6	72 ± 4
Al^{3+} (0.2 mM)	-	65 ± 2	70 ± 3
	+	56 ± 2	54 ± 4

All values are Mean ± SD of 3 determinations. *The control values are: GABA uptake, 304 ± 2 pmol/min/mg protein without Ca and 280 ± 20 pmol/min/mg protein with 1 mM Ca; glutamate uptake, 219 ± 10 pmol/min/mg protein without Ca and 202 ± 15 pmol/min/mg protein with 1 mM Ca.

on uptake by forebrain synaptosomes (Lai et al., 1981b and Table 4). (ii) In the absence of Ca, the IC_{50} of Cd (250 µM) and Mn (1.7 mM) on the inhibition of uptake by striatal synaptosomes are less than the corresponding values (IC_{50} for Cd = 400 µM, for Mn = 2.5 mM) for uptake by forebrain synaptosomes (Table 4). (iii) Ca, at 1 mM, antagonizes both the stimulatory as well as the inhibitory effects of Al on uptake by striatal synaptosomes but only antagonizes the stimulatory effects of Al on uptake by forebrain synaptosomes (Lai et al., 1981b). However, the same concentration of Ca actually enhances the inhibitory actions of Al on uptake by forebrain synaptosomes only (Lai et al., 1981b).

B. STRUCTURE AND FUNCTION OF SYNAPTOSOMES AND MITOCHONDRIA: COMPARISON OF THE EFFECTS OF PSYCHOACTIVE DRUGS

A number of psychoactive drugs act on the central monoaminergic system through the selective inhibition of either MAO, or synaptosomal amine uptake, or both (Roth, 1976). As these effects are observed in two different

Table 4. Comparison of IC_{50} values for Cd, Mn and Al in the inhibition of dopamine uptake by forebrain and striatal synaptosomes.

Synaptosomal Preparation	Ca Status	IC_{50} Cd	IC_{50} Mn	IC_{50} Al
Forebrain	-	400 µM	2.5 mM	400 µM
	+	400 µM	2.1 mM	240 µM
Striatum	-	250 µM	1.7 mM	400 µM
	+	400 µM	6.3 mM	400 µM

Ca status was ± 1 mM Ca. From Lai et al. (1981b).

categories of neural membranes (i.e., the synaptosomal and the mitochondrial membranes), a comparative study of the effects of these psychoactive drugs will not only give us more insight into the structure and function of these membrane systems but also yield us more information about the mechanisms of action of these drugs.

B.1 Effects of Clorgyline and Deprenyl on Monoamine Oxidase in Synaptic and Non-synaptic Mitochondria

One useful way of classifying the multiple forms of MAO into basically two forms was first introduced by Johnston (1968). Type A MAO preferentially oxidizes serotonin and is selectively inhibited by clorgyline whereas type B MAO oxidizes benzylamine or phenylethylamine and is selectively inhibited by l-deprenyl (Houslay et al., 1976). Although the precise nature of these two forms of MAO has not been fully elucidated (Sandler and Youdim, 1972; Houslay et al., 1976), there is some evidence that they may be different polypeptides (Cawthon and Breakefield, 1979). Since brain mitochondria are heterogeneous in their enzyme contents (Lai et al., 1977; Lai and Clark, 1979) and methods have been devised whereby metabolically active and distinct populations of brain mitochondria may be isolated (Lai et al., 1977; Lai and Clark, 1979), the activities of MAO are studied in non-synaptic and synaptic mitochondria to elucidate the functional significance of the heterogeneity of MAO (Bourne et al., 1975; Lai et al., 1976; Owen et al., 1977).

Based on the oxidation of serotonin and phenylethylamine, the MAO activities in the two populations of synaptic mitochondria (SM and SM2) are much higher than the activities in the non-synaptic mitochondria (M) (Bourne et al., 1975; Lai et al., 1976; Owen et al., 1977). The MAO-A to MAO-B ratio in fraction SM of synaptic mitochondria is higher than the ratios in the other mitochondrial fractions (Owen et al., 1977). This suggests that the latter (SM) is more metabolically adapted to oxidizing neurotransmitter amines such as serotonin and noradrenaline. The MAO-A to MAO-B ratio can also be determined by measuring the ratio of the % inhibition of MAO by clorgyline to the % inhibition of MAO by deprenyl. By the latter, the ratio in the SM fraction of synaptic mitochondria is also found to be higher (Lai et al., 1976; Owen et al., 1977). These results lend further support to the view that the multiple forms of MAO show heterogeneous distribution in different populations of brain mitochondria (Bourne et al., 1975; Lai et al., 1976; Owen et al., 1977).

As the original observation was made with forebrain mitochondria, it may simply be a reflection of the heterogeneous distribution of MAO activities at the brain regional level (Owen et al., 1977). We recently addressed this issue more directly (Lai et al., 1981a; T.K.C. Leung, J.C.K. Lai and L. Lim, in preparation). Two populations of synaptic (SM and SM2) and one of non-synaptic (M) mitochondria from each of three different regions (cerebral cortex, striatum and pons and medulla) have been isolated by the Lai and Clark (1979) method. The heterogeneity of MAO in these mitochondrial populations has been investigated with different substrates (serotonin, benzylamine and dopamine) and specific inhibitors (clorgyline and deprenyl). The results of these experiments support our previous conclusion that the multiple forms of brain MAO show heterogeneous subcellular distribution (Bourne et al., 1975; Lai et al., 1976; Owen et al., 1977); in addition, they suggest that the distribution of the multiple forms of MAO in synaptic and non-synaptic mitochondria shows distinct, region-specific patterns (Lai et al., 1981a; T.K.C. Leung, J.C.K. Lai and L. Lim, in preparation). Other approaches to address the functional relevance of the multiple forms of MAO also reveal that the microenvironment within the mitochondrial membranes exerts definite influences on the functional expression of the multiple forms of MAO *in situ* (Huang and Faulkner, 1980; Leung et al., 1982a).

B.2. Effects of Clorgyline and Deprenyl on Synaptosomal Uptake of Amines

The inhibition of intraterminal MAO activity may alter the synaptosomal amine flux and consequently change the amine uptake. As uM concentrations of levels inhibit dopamine uptake whereas nialamide, another MAO inhibitor, has no such effect (Lai et al., 1980d), experiments have been devised to address the following questions: (i) Do clorgyline and deprenyl inhibit serotonin and noradrenaline uptake in addition to dopamine uptake? (ii) If so, can the inhibitory effects of clorgyline and deprenyl on the uptake of these neurotransmitter amines be predicted on the bases of their selective effects towards the mitochondrial MAO-A and MAO-B? (iii) Do the inhibitory actions of clorgyline and deprenyl on amine uptake show any brain regional differences?

Both clorgyline and deprenyl inhibit amine uptake by forebrain synaptosomes (Lai et al., 1980d). The rank order of inhibition of uptake by clorgyline is: serotonin (IC_{50} = 10 μM) > dopamine (IC_{50} = 56 μM) > noradrenaline (IC_{50} = 66 μM) and that by l-deprenyl is: noradrenaline (IC_{50} = 26 μM) > serotonin (IC_{50} = 460 μM) > dopamine (IC_{50} = 530 μM) (Lai et al., 1980d).

The effects of clorgyline and l-deprenyl on amine uptake by synaptosomes isolated from the striatum, a region particularly enriched in dopaminergic nerve-endings, contrast with those in forebrain synaptosomes (Lai et al., 1980d).

That clorgyline is a more potent inhibitor than l-deprenyl of synaptosomal dopamine and serotonin uptake (Lai et al., 1980d) is consistent with the fact that clorgyline (IC_{50} < 0.1 μM) is a more effective inhibitor than l-deprenyl (IC_{50} > 10 μM) of MAO in synaptic and non-synaptic mitochondria (Bourne et al., 1975; Lai et al., 1976; Owen et al., 1977; Lai et al., 1981a; T.K.C. Leung, J.C.K. Lai and L. Lim, in preparation). Furthermore, as predicted from serotonin being a MAO-A substrate, serotonin uptake is selectively inhibited by clorgyline (Lai et al., 1980d). Paradoxically, noradrenaline uptake is equally sensitive to inhibition by clorgyline and l-deprenyl (Lai et al., 1980d) though noradrenaline is a MAO-A substrate.

C. CONCLUDING REMARKS

Our results indicate that metals such as Cd, Mn and Al exert differential and different effects on the various carrier-mediated uptake systems in the synaptosomal membranes. These effects are modulated - either enhanced or antagonized - by 1 mM Ca and are not directly the consequences of the inhibitory actions of these metals on synaptosomal Na-K-ATPase activities.

In addition to being selective inhibitors of MAO, clorgyline and l-deprenyl are inhibitors of synaptosomal amine uptake. However, that clorgyline is at least two orders of magnitude more potent in inhibiting MAO than amine uptake indicates that the central actions of clorgyline are likely to involve the selective inhibition of MAO. On the other hand, l-deprenyl is equally effective in inhibiting MAO and amine uptake suggesting that l-deprenyl may act centrally via the inhibition of both MAO and amine uptake.

Our studies suggest that neuromodulatory agents could be fruitfully employed for elucidating the structure and function of neural membranes. At the same time by combining the various approaches which take into consideration (i) organelle specialization and differentiation, (ii) kinetic analysis, (iii) brain regional heterogeneity of anatomical structures and (iv) developmental perspective, one has the means to dissect, at the membrane level, the more molecular aspects of the biochemical mechanisms underlying the central action of these neuromodulatory agents.

Acknowledgements. The interest and encouragement of Professor A.N. Davison and the support of the Worshipful Company of Pewterers and the Brain Research Trust are gratefully acknowledged. P.C.L. Wong was the recipient of a China Medical Board/University of Hong Kong Fellowship. While this paper was in preparation, J.C.K. Lai was supported by NIH Grant # AAO 3883, the Will Rogers Institute and the Winifred Masterson Burke Relief Foundation; J.C.K. Lai acknowledges the encouragement of Professor J.P. Blass and the assistance of Mrs. A. Baker with the art work.

REFERENCES

Balazs R., Lewis P.D., and Patel A.J. (1979) Nutritional deficiencies and brain development, In Human Growth (Falkner F. and Tanner J.M., eds), pp. 415-479. Plenum Press, New York.

Bogdanski D.F. (1976) Mechanisms of transport for the uptake and release of biogenic amines in nerve endings, In Transport Phenomena in the Nervous System (Levi G., Battistin L., and Lajtha A., eds), pp. 291-305. Plenum Press, New York.

Bourne R., Lai J.C.K., and Owen F. (1975) Monoamine oxidase in distinct populations of rat brain mitochondria. Brit. J. Pharmacol. 55, 298P.

Cawthon R.M. and Breakefield X.O. (1979) Differences in A and B forms of monoamine oxidase revealed by limited proteolysis and peptide mapping. Nature 281, 692-694.

Cotzias G.C., Papavasiliou P.S., Ginos J., Steck A., and Duby S. (1971) Metabolic modification of Parkinson's disease and of chronic manganese poisoning. Annu. Rev. Med. 22, 305-326.

Cotzias G.C., Papavasiliou P.S., Mena I., Tang L.C., and Miller S.T. (1974) Manganese and catecholamines, In Advances in Neurology, Vol. 5: Second Canadian-American Conference on Parkinson's Disease (McDowell F.H. and Barbeau A, eds), pp. 235-243. Raven Press, New York.

Crapper D.R., Karlik S., and De Boni U. (1978) Aluminum and other metals in senile (Alzheimer) dementia, In Aging, Vol. 7 (Katzman R., Terry R.D. and Bick K.L., eds), pp. 471-485. Raven Press, New York.

Hexam T.D. (1974) Studies on the reaction catalysed by transport (Na^+, K^+) adenosine triphosphatase: I. Effects of divalent metals. Biochem. Pharmacol. 23, 3441-3447.

Houslay M.D., Tipton K.F., and Youdim M.B.H. (1976) Multiple forms of monoamine oxidase: fact and artefact. Life Sci. 19, 467-478.

Huang R.H. and Faulkner R. (1980) Lipid-protein interactions in the multiple forms of monoamine oxidases: lipases as probes using purified intact rat brain mitochondria. Mol. Pharmacol. 18, 267-273.

Johnston J.P. (1968) Some observations upon a new inhibitor of monoamine oxidase in brain tissue. Biochem. Pharmacol. 17, 1285-1297.

Lai J.C.K., Owen F., Bourne R., and Williams R. (1976) Monoamine oxidase - heterogeneity in rat brain mitochondria. 10th International Congress of Biochem., Hamburg, Germany, Abstracts Vol., p. 554.

Lai J.C.K., Walsh J.M., Dennis S.C., and Clark J.B. (1977) Synaptic and non-synaptic mitochondria from rat brain: isolation and characterization. J. Neurochem. 28, 625-631.

Lai J.C.K., Guest J.F., Lim L., and Davison A.N. (1978) The effects of transition-metal ions on rat brain synaptosomal amine-uptake systems. Biochem. Soc. Trans. 6, 1010-1012.

Lai J.C.K. and Clark J.B. (1979) Preparation of synaptic and non-synaptic mitochondria from mammalian brain. Methods Enzymol. 55, Part F, 51-60.

Lai J.C.K., Lim L., and Davison A.N. (1980a) The kinetics of the inhibition of rat brain synaptosomal dopamine uptake by manganous ions. Biochem. Soc. Trans. 8, 67-68.

Lai J.C.K., Lim L., and Davison A.N. (1980b) Inhibition of rat brain synaptosomal dopamine uptake by Cd^{2+}: a kinetic study. Biochem. Soc. Trans. 8, 68.

Lai J.C.K., Guest J.F., Leung T.K.C., Lim L., and Davison A.N. (1980c) The effects of cadmium, manganese and aluminum on sodium-potassium-activated and magnesium-activated adenosine triphosphatase activity and choline uptake in rat brain synaptosomes. Biochem. Pharmacol. 29, 141-146.

Lai J.C.K., Leung T.K.C., Guest J.F., Lim L., and Davison, A.N. (1980d) The monoamine oxidase inhibitors clorgyline and l-deprenyl also affect the uptake of dopamine, noradrenaline and serotonin by rat brain synaptosomal preparations. Biochem. Pharmacol. 29, 2763-2767.

Lai J.C.K., Leung T.K.C. and Lim L. (1981a) Monoamine oxidase in synaptic and non-synaptic mitochondria from brain regions. 8th Meeting of the International Society for Neurochem., Abstracts Vol., p. 291.

Lai J.C.K., Lim L., and Davison A.N. (1981b) Differences in the inhibitory effect of Cd^{2+}, Mn^{2+} and Al^{3+} on the uptake of dopamine by synaptosomes from forebrain and from striatum of the rat. Biochem. Pharmacol. 30, 3123-3125.

Lai J.C.K., Lim L., and Davison A.N. (1982a) Effects of Cd^{2+}, Mn^{2+} and Al^{3+} on rat brain synaptosomal uptake of noradrenaline and serotonin. J. Inorganic Biochem. 17, 215-225.

Lai J.C.K., Leung T.K.C., and Lim L. (1982b) Differences in the neurotoxic effects of manganese during development and aging: some observations on brain regional neurotransmitter and non-neurotransmitter metabolism in a developmental rat model of chronic manganese encephalopathy. Neurotoxicology (in press).

Leung T.K.C., Lai J.C.K., Marr W., and Lim L. (1980) The activities of the A and B forms of monoamine oxidase in liver, hypothalamus and cerebral cortex of the female rat: effects of administration of ethinyl oestradiol and the progestogens norethisterone acetate and d-norgestrel. Biochem Soc. Trans. 8, 615-616.

Leung T.K.C., Lai J.C.K., and Lim L. (1981) The regional distribution of monoamine oxidase activities towards different substrates: effects in rat brain of chronic administration of manganese chloride and of aging. J. Neurochem. 36, 2037-2043.

Leung T.K.C., Lai J.C.K., and Lim L. (1982a) Type A and type B monoamine oxidase activities in rat brain and liver mitochondria: a comparison of their properties using the non-ionic detergent Triton X-100. Comp. Biochem. Physiol. 71C, 219-222.

Leung T.K.C., Lai J.C.K. and Lim L. (1982b) The effects of chronic manganese feeding on the activity of monoamine oxidase in various organs of the developing rats. Comp. Biochem. Physiol. 71C, 223-228.

Owen F., Bourne R.C., Lai J.C.K. and Williams R. (1977) The heterogeneity of monoamine oxidase in distinct populations of rat brain mitochondria. Biochem. Pharmacol. 26, 289-292.

Prakash N.J., Fontana J., and Henkin R.I. (1973) Effect of transition metal ions on ($Na^+ + K^+$) ATPase activity and the uptake of norepinephrine and choline by brain synaptosomes. Life Sci. 12, 249-259.

Roth J.A. (1976) Multiple forms of monoamine oxidase and their interaction with tricyclic psychomimetic drugs. Gen. Pharmacol. 7. 381-386.

Sandler M. and Youdim M.B.H. (1972) Multiple forms of monoamine oxidase: functional significance. Pharmacol. Rev. 24, 331-349.

Shih J.C. (1979) Monoamine oxidase in aging human brain, In Monoamine Oxidase: Structure, Function, and Altered Function (Singer T.P., von Korff R.W., and Murphy D.L., eds), pp. 413-421. Academic Press, New York.

Silbergeld E.K. (1977) Interactions of lead and calcium on the synaptosomal uptake of dopamine and choline. Life Sci. 20, 309-318.

Su K.L. and Sun G.Y. (1978) Acyl group composition of metabolically active lipids in brain: variances among subcellular fractions and during post-natal development. J. Neurochem. 31, 1043-1047.

Sun A.Y. and Samorajski T. (1975) The effects of age and alcohol on ($Na^+ + K^+$)-ATPase activity of whole homogenate and synaptosomes prepared from mouse and human brain. J. Neurochem. 24, 161-164.

Sun A.Y. and Sun G.Y. (1976) Functional roles of phospholipids of synaptosomal membrane, In Function and Metabolism of Phospholipids in the Central and Peripheral Nervous Systems (Porcellati G., Amaducci L., and Galli C., eds), pp. 169-197. Plenum Press, New York.

Sun G.Y. (1972) Effects of a fatty acid deficiency on lipids of whole brain, microsomes, and myelin in the rat. J. Lipid Res. 13, 56-62.

Weiss B. (1978) The behavioral toxicity of metals. Fed. Proc. 37, 22-27.

Wong P.C.L. and Lim L. (1981) The effects of aluminum, manganese and cadmium chloride on the methylation of phospholipids in the rat brain synaptosomal membrane. Biochem. Pharmacol. 30, 1704-1705.

Wong P.C.L., Lai J.C.K., Lim, L., and Davison A.N. (1981) Selective inhibition of L-glutamate and gamma-amino-butyrate transport in nerve ending particles by aluminum, manganese and cadmium chloride. J. Inorganic Biochem. 14,253-260.

Wong P.T.-H. and McGeer E.G. (1981) Postnatal changes of gabaergic and glutamatergic parameters. Develop. Brain Res. 1, 519-529.

PROPERTIES OF A BRAIN MEMBRANE AMINOENKEPHALINASE: INHIBITION STUDIES

Koon-Sea Hui, Maria Hui, and Abel Lajtha

Center for Neurochemistry, Rockland Research Institute
Ward's Island, New York, New York 10035

ABSTRACT

Enkephalins are inactivated by a number of enzymes, such as amino-, endo-, and carboxypeptidases. We recently purified to apparent homogeneity an aminopeptidase from rat brain membranes that may be of functional significance in cerebral enkephalin metabolism. The enzyme which split enkephalin (aminoenkephalinase) was extracted with 1% Triton X-100 and purified by chromatography, successively on DEAE-Sepharose, Bio-Gel HTP, and Sephadex G-200 column. The purified enzyme showed one band in disc gel electrophoresis with an apparent molecular weight of 250,000. The substrate specificities, pH profile, and subcellular localization distinguished it from other enzymes capable of splitting enkephalins (leucine aminopeptidase, aminopeptidase A, aminopeptidase B, aminopeptidase M, angiotensin-converting enzyme, and the cytosolic aminoenkephalinase). The enzyme activity was assayed by using reverse-phase high-performance liquid chromatography with Met-enkephalin as substrate. The disappearance of the peptide (or the appearance of tyrosine) was determined spectrophotometrically at 280 nm. The enzyme released tyrosine from the enkephalin and left the metabolites intact. Met-enkephalin and Met-enkephalin-Lys^6 were the most rapidly split substrates for the enzyme among those tested. Extension of the enkephalin at the NH_2- or the COOH-terminus made the peptide more resistant to the enzyme, and these peptides became potent inhibitors of the aminoenkephalinase. Arg^0-Met-enkephalin analogs tested at IC_{50}=0.56 uM. Its inhibitory potency is similar to that of bestatin (IC_{50}=0.8 uM), amastatin (IC_{50}=0.17 uM), and ACTH $_{1-24}$ (IC_{50}=0.55 um). Arg^0-Met-enkephalin is also converted to Met-enkephalin by the membrane aminoenkephalinase. The properties of the enzyme, in view of the physiological occurrence of a number of enkephalin analogs, may be important in regulation of the enkephalin levels and bioavailability under physiological conditions.

INTRODUCTION

Since Met- and Leu-enkephalin were discovered, vigorous research has been focused on their biosynthesis and metabolism. Even though β-endorphin shares the same amino acid sequence as Met-enkephalin (Tyr-Gly-Gly-Phe-Met), overwhelming data show that it is not a precursor for Met-enkephalin. Instead, the candidate for the precursor of Met-enkephalin has a molecular weight of 45,000, which comprises six copies of Met-enkephalin and one copy of Leu-enkephalin (Kojima et al., 1982; Gubler et al., 1982; Noda et al., 1982). In the proenkephalin, four copies of the Met-enkephalin and the Leu-enkephalin are flanked by a pair of basic amino acids at the N-terminal and the C-terminal. The octapeptide (Tyr-Gly-Gly-Phe-Met-Arg-Gly-Leu) is also bracketed by a pair of basic amino acids at each terminus. The heptapeptide (Tyr-Gly-Gly-Phe-Met-Arg-Phe) is located at the C-terminal of the precursor; its N-terminal, however, is also linked to a pair of basic amino acids. Since Met-enkephalin-Arg^6, Met-enkephalin-Lys^6, Met-enkephalin, Leu-enkephalin, Leu-enkephalin-Lys^6 and Met-enkephalin -Arg^6-Phe^7 were found in adrenal medulla, brain, and mysenteric plexus, it appears that there is a general pathway for enkephalin biosynthesis (Kojima et al., 1982). Enkephalins and the enkephalin-containing peptides (ECPs) are co-released by secretogogues (Kilpatrick et al., 1980). The relative proportions of these ECPs released from the gland are the same as those found in the intact storage vesicles (Kilpatrick et al.. 1980). Met-enkephalin-Arg^6-Phe^7, which binds on a different subpopulation of opiate receptors (Audigier et al., 1980), has been found to be a potent inhibitor of contraction of the electrically stimulated guinea-pig ileum (Inturrisi et al., 1980).

Several membrane-associated peptidases of the brain have been shown to degrade enkephalins in vitro. They are aminopeptidase(s) (Knight & Klee, 1978), angiotensin-converting enzyme (Erdos et al., 1978), enkephalin dipeptidyl carboxypeptidase (Malfroy et al., 1978), and enkephalin dipeptidyl aminopeptidase (Gorenstein & Snyder, 1979). Their role in physiological inactivation of enkephalins needs further clarification. Enkephalins are hydrolyzed most rapidly by the release of the N-terminal tyrosine, which terminates their opioid effects (Schwartz et al., 1981). Aminopeptidase was found to be the primary degradative enzyme for the inactivation of exogenous enkephalin in guinea-pig ileum (Geary et al., 1982).

It was suggested that enkephalin bound to the opiate receptors is available for degradation by aminopeptidase (Knight & Klee, 1978). However, the physiological role of the membrane aminopeptidase is still unclear.

We have recently isolated an aminopeptidase from rat brain membrane (Hui et al., 1982b), which rapidly removes the N-terminal Tyr from Met-enkephalin to yield Gly-Gly-Phe-Met. β-Endorphin, which is apparently resistant to the aminopeptidase, competitively inhibited the hydrolysis of Met-enkephalin by this membrane aminopeptidase with a K_i value of 11.5 uM. Arg^0-β_h-endorphin was found to be 10 times more potent than β_h-endorphin (Hui et al., 1981a,b). In addition, various neuropeptides, including ACTH $_{1-24}$, ACTH $_{1-10}$, somatostatin, and substance P, also inhibit this reaction, but ACTH $_{18-39}$, corticotrophin-like intermediate lobe peptide (CLIP), and bradykinin do not affect Met-enkephalin breakdown by the enzyme. We proposed that neuropeptides may regulate neuropeptide metabolism in the brain (Graf et al., 1982). We report here our study on the inhibition of Met-enkephalin degradation by Met-enkephalin-containing peptides, to shed light on their relationship and physiological function.

MATERIALS AND METHODS

O-Sulphated Leu^5-enkephalin was purchased from Cambridge Research Biochemicals (Cambridgeshire, England). Arg-Tyr-Gly-Gly-Phe-Met (LPH_{60-65}), Lys-Arg-Tyr-Gly-Gly-Phe-Met (LPH_{59-65}), Asp-Lys-Arg-Tyr-Gly-Gly-Phe-Met (LPH_{58-65}), Lys-Asp-Lys-Arg-Tyr-Gly-Gly-Phe-Met

(LPH_{57-65}), and Pro-Lys-Asp-Lys-Arg-Tyr-Gly-Gly-Phe-Met (LPH_{56-65}) were kindly synthesized and supplied by Dr. N. Ling, the Salk Institute. Met- and Leu-enkephalin were obtained from Boehringer Mannheim (Indianapolis, IN). Other enkephalin analogues were purchased from Peninsula.

[Ethylene-Bis (oxyethylenenitrile)]-tetracetic acid (EGTA) was purchased from Baker Chemical Co. (Philipsburg, NJ); dithiothreitol, from Sigma; HPLC-grade acetonitrile, from Fisher. Amastatin ((2S,3R)-3-amino-2-hydroxyl-5-methylhexanoyl-L-Val-L-Val-L-Asp) and bestatin ((2S,3R)-3-amino-2-hydroxyl-4-phenyl-butanoyl-L-Leu) were generously supplied by Dr. H. Umezawa, Microbial Chemistry Research Foundation, Japan. Epibestatin ((2R,3R)-3-amino-2-hydroxyl-4-phenylbutanoyl-L-Leu) and epiamastatin ((2S,3R)-3-amino-2-hydroxyl-5-methylhexanoyl-L-Val-L-Val-L-Asp) were purchased from Sigma, and puromycin dihydrochloride was obtained from Nutritional Biochemicals Corporation (Cleveland, Ohio).

Purification of Membrane-bound Aminopeptidase

The details of the preparation of the aminopeptidase were published elsewhere (Hui et al., 1982b). In brief, the enzyme was prepared by homogenization of 5 rat brains (10 g) with 9 volumes of 0.32 M sucrose in a glass homogenizer with a motor drive. The homogenate was centrifuged at 800 x _g_ for 10 min, and the debris was discarded. After centrifugation at 30,000 x _g_ for 20 min, the supernatant was discarded and the pellet was resuspended in 50 mM Tris-HCl buffer, pH 7.5, containing 1% (W/V) Triton X-100, and incubated at 37°C for 45 min. The solubilized enzyme, obtained after centrifugation at 30,000 x _g_ for 10 min, was applied to a DEAE-Sepharose CL-6B column (Pharmacia; 1.5 x 25 cm) that had been equilibrated with 25 mM Tris-HCl buffer, pH 7.5, containing 2 mM EGTA, and 0.2 mM dithiothreitol (Buffer A). The column was washed with two bed volumes of Buffer A, and the enzyme was eluted with 500 ml of salt gradient of 0-0.4 M NaCl in Buffer A at a flow rate of 0.4 ml/min. Fractions in 10 ml-volumes were collected. The aminopeptidase activity was assayed by reverse-phase high-performance liquid chromatography (Waters HPLC Model 204) by measuring the disappearance of Met-enkephalin (Hui et al., 1980, 1981a, 1981b).

The enzyme peak from ion-exchange chromatography was applied directly to a Bio-Gel HTP (hydroxylapatite) column (Bio-Rad, 0.9 x 12 cm) equilibrated with Buffer A. The column was washed with two bed volumes of the same buffer, and the retained proteins were eluted at a rate of 0.4 ml/min with 500 ml of Buffer A containing 0-200 mM sodium phosphate gradient, pH 7.5.

The pooled enzyme fractions from the Bio-Gel HTP chromatography were concentrated to aproximately 2 ml with an Amicon PM 10-membrane filter and applied to a Sephadex G-200 column (Pharmacia; 1.5 x 85 cm) equilibrated with Buffer A. The protein was eluted with the same buffer at a flow rate of 0.05 ml/min. The eluate was collected in 2-ml fractions, and the fractions (30-34) were pooled and kept at 4°C until use.

Aminopeptidase Activity on Different Peptides

Peptide (33 uM) was incubated with the enzyme (20 ng protein) in a final volume of 150 ul of Buffer A at 37°C for 30 min. The reaction was terminated by adding 20 ul of 20% trichloroacetic acid (TCA). The mixture was centrifuged, and 50 ul of the supernatant was submitted to HPLC analysis on a Waters high-performance liquid chromatograph equipped with a Radial-Pak C_8 (5 x 100 mm) column with particle size of 10 um. The sample was eluted isocratically at ambient temperature with a mixture of acetonitrile and 0.1 M phosphate buffer, pH 3.0. The flow rate was 2 ml per min, resulting in a back-pressure of 500-1,000 lb/in^2. The peptides were measured spectrophotometrically at 280 nm and/or 205 nm. The solvent system (acetonitrile:0.1 M phosphate buffer, pH 3.0) for Met-enkephalin, (O)-Met-enkephalin, O-Sulphated Leu-enkephalins, Met-enkephalin-Arg^6, and Met-enkephalin-Arg^6-Phe^7 was 20:80; for Met-enkephalin-Lys^6, Met-enkephalin-Arg^6-Arg^7, Leu-enkephalin-Arg^6, Leu-enkephalin-NH_2, and Leu-enkephalin-Lys^6 it was 27.5:72.5; and for Tyr-Gly-Gly-Phe and Gly-Gly-Phe-Met it was 10:90. In these systems, the retention time for Met-enkephalin was 1.5 min; for Leu-enkephalin, 2.1 min; for Met-enkephalin-Arg^6, 1.38 min; for Met-enkephalin-Arg^6-Phe^7, 4.2 min; for Met-enkephalin-Lys^6, 1.45 min; for Met-enkephalin-Arg^6-Arg^7, 1.77 min; for Leu-enkephalin-Arg^6, 2.32 min; for Leu-enkephalin-Lys^6, 2 min; for Tyr-Gly-Gly-Phe, 3 min; for

Gly-Gly-Phe-Met, 3.5 min; for (O) Met-enkephalin, 1 min; for Leu-enkephalin-NH_2, 2.86 min; and for O-sulphated Leu-enkepnalin, 1.4 min.

LPH_{56-65}, LPH_{57-65}, LPH_{58-65}, LPH_{59-65}, and LPH_{60-65} were separated on a different column (Waters Bondapak C_{18} column, 3.9 x 300 mm, particle size 10 um), with an isocratic solvent system acetonitrile:phosphate buffer (27.5:72.5) at a flow rate of 2 ml/min. In this system, the retention time for Met-enkephalin (LPH_{61-65}) was 2.23 min; for LPH_{56-65}, 2.08 min; for LPH_{57-65}, 1.57 min; for LPH_{58-65}, 1.55 min; for LPH_{59-65}, 2.36 min; for LPH_{60-65}, 2.5 min.

Inhibition Studies

11.25 nmol of Met-enkephalin, 20 ng aminopeptidase, and the inhibitor were incubated in a final volume of 200 ul of Buffer A at 37°C for 30 min. The reaction was terminated by adding 20 ul of 20% TCA. As internal standard, 11.25 nmol of Leu-enkephalin was added. The sample was centrifuged, and 30 ul of the supernatant was submitted to HPLC analysis. Breakdown of the Met-enkephalin was determined spectrophotometrically by measuring the disappearance of the peptides, or the formation of the Tyr released from the N-terminal of the peptide. Tyr was detected by HPLC on a LiChrosorb RP-18 column (250 x 4.6 mm, particle size 7 um, Unimetrics, Westwood, NJ). The sample was eluted isocratically with acetonitrile and phosphate buffer (4:96) at a flow rate of 1.5 ml/min. The retention time for Tyr was 4 min, which resolved from Tyr-Gly (6.9 min) and Tyr-Gly-Gly (5.7 min). Met-enkephalin could not be eluted from the column with this solvent system.

Formation of Met-enkephalin from LPH_{60-65}

5 nmol of LPH_{60-65} was incubated with 20 ng of the membrane aminopeptidase with 200 ul of buffer A at 37°C and incubated for varying periods of time. The reaction was stopped by adding 20 ul of 20% TCA. 30 ul of the sample was submitted to HPLC analysis. The sample was separated on a Waters Radial-Pak C_8 (5 x 100 mm) column with particle size of 10 um. The sample was eluted isocratically with acetonitrile: phosphate buffer (20:80).

The retention time was 2.5 min for Gly-Gly-Phe-Met, 4.5 min for Met-enkephalin, and 5.6 min for LPH_{60-65}.

RESULTS

Effect of Aminopeptidase and Cholinesterase Inhibitors

The cholinesterase inhibitors neostigmine and erserine did not inhibit the degradation of enkephalin by the membrane aminopeptidase (IC_{50}>50 uM). IC_{50} values are 50% inhibition concentrations of the inhibitor. Amastatin was the most potent inhibitor among the aminopeptidase inhibitors we studied. Its IC_{50} for the membrane aminopeptidase was 0.17 uM. Amastatin is a strong inhibitor for leucine aminopeptidase and aminopeptidase A, and bestatin is specific for leucine aminopeptidase and aminopeptidase B. Bestatin inhibited the enzyme with the IC_{50} of 0.9 uM. The stereoisomers of these two microbial inhibitors were less potent: the IC_{50} was 1.35 uM for epiamastatin and 14 uM for epibestatin. Puromycin was less inhibitory than amastatin or bestatin, with an IC_{50} of 1.6 uM for our enzyme.

Effect of Enkephalin Analogues

The enkephalin analogues we studied can be classified into three categories: _A_, peptides with no more than 5 amino acid residues; _B_, enkephalin analogues with C-terminal extensions; and _C_, enkephalin analogues with N-terminal extensions (Table I). The peptides in category _A_ did not inhibit the membrane aminoenkephalinase (IC_{50}>50 uM) even though some of them are comparatively good substrates. Enkephalin analogues with modification of the Tyr or the C-terminal amino residues were not inhibitors. The peptides in the _B_ category are the immediate precursors of enkephalin in different tissues. They are favorable substrates for the enzyme and they are inhibitors for the enzymes. Tyr-Gly-Gly-Phe-Met-Arg-Arg is the most potent inhibitor (IC_{50}=1.4 uM), and Tyr-Gly-Gly-Phe-Met-Lys is the weakest inhibitor (IC_{50}=18 uM). Their inhibitory potencies are similar to the bigger peptide analogues Arg^{0}-β-endorphin (IC_{50}=1 uM) and β-endorphin (IC_{50}=14 uM) (Hui et al., 1982). All the peptides in

Table I Effect of Enkephalin Analogues on Met-enkephalin Degradation

Peptide	Relative Specific Activity (Met-enkephalin=100)	IC_{50}* (uM)
A		
Tyr-Gly	0	>50
Tyr-Gly-Gly	0	>50
Tyr-Gly-Gly-Phe	78	>50
Gly-Gly-Phe-Met	0	>50
Tyr-Gly-Gly-Phe-Leu	77	>50
Tyr-Gly-Gly-Phe-(O)Met	85	>50
(SO_3H) Tyr-Gly-Gly-Phe-Leu	n.d.**	>50
Tyr-Gly-Gly-Phe-Leu-NH^2	n.d.	>50
B		
Tyr-Gly-Gly-Phe-Met-Arg	77	4.5
Tyr-Gly-Gly-Phe-Met-Lys	98	18
Tyr-Gly-Gly-Phe-Leu-Arg	44	3.6
Tyr-Gly-Gly-Phe-Leu-Lys	49	11.3
Tyr-Gly-Gly-Phe-Met-Arg-Arg	10	1.4
Tyr-Gly-Gly-Phe-Met-Arg-Phe	32	3.2
C		
Arg-Tyr-Gly-Gly-Phe-Met	43	0.56
Lys-Arg-Tyr-Gly-Gly-Phe-Met	0	1.4
Asp-Lys-Arg-Tyr-Gly-Gly-Phe-Met	0	2.1
Lys-Asp-Lys-Arg-Tyr-Gly-Gly-Phe-Met	0	1.1
Pro-Lys-Asp-Lys-Arg-Tyr-Gly-Gly-Phe-Met	0	8.6

* The concn. of the Met-enkephalin is 56 uM.
** n.d. = not determined.

category C except Arg-Tyr-Gly-Gly-Phe-Met were resistant to digestion by the membrane aminopeptidase. The peptides in this category are fragments of β-lipotropin (LPH). Arg-Tyr-Gly-Gly-Phe-Met was the most potent inhibitor. There seems to be a tendency that the longer the N-terminal extension, the weaker is the inhibitory potency. The IC_{50} was 0.56 uM for the Arg-Tyr-Gly-Gly-Phe-Met, which is similar to those for the microbial inhibitors amastatin and bestatin. Lys-Arg-Tyr-Gly-Gly-Phe-Met was not split by the membrane aminopeptidase.

Table II Substrate Specificity of Rat Brain Aminoenkephalinase

	K_m (uM)	V_{max} (nmol/min per mg)	K_{cat}/K_m* mol/sec X 10^5
Met-enkephalin	280	45	6.7
Arg^0-Met-enkephalin	19	0.5	1.1
Met-enkephalin-Lys^6	170	75	24.7
Met-enkephalin-Arg^6	29	32	46
Leu-enkephalin-Arg^6	39	12	13
Leu enkephalin-Lys^6	54	39	30.1
Met-enkephalin-Arg^6-Arg^7	2.0	2.0	41.7
Met-enkephalin-Arg^6-Phe^7	2.8	2.8	41.7

* The molecular weight of the purified enzyme used for the calculation was 250,000.

The amount of enzyme used for Met-enkephalin, Met-enkephalin-Arg, and Met-enkephalin-Lys was 20 ng; for Leu-enkephalin-Arg, Leu-enkephalin-Lys, Met-enkephalin-Arg-Phe, and Arg-Tyr-Gly-Gly-Phe-Met, 40 ng; and for Met-enkephalin-Arg-Arg, 100 ng. The rate of the hydrolysis was determined by measuring the disappearance of the peptides.

Substrate Specificity

All the enkephalin analogues except Arg-Tyr-Gly-Gly-Phe-Met with extension either at the C- or the N-terminal had higher hydrolytic coefficients than Met-enkephalin, demonstrating that they are more favorable than Met-enkephalin as substrates for the enzyme (Table II). All of the enkephalin analogues have lower Km's than the Met-enkephalin. Met-enkephalin-Arg^6-Arg^7 has the lowest Km (20 uM), and Met-enkephalin has the highest (278 uM).

The hydrolytic coefficient (K_{cat}/K_m) is less than 20% of that for Met-enkephalin. The Lineweaver-Burk plot shows that the breakdown of Arg^0-Met-enkephalin by the membrane aminopeptidase is substrate inhibited (Fig 1); i.e., the enzyme is inhibited by increasing the concentration of substrate.

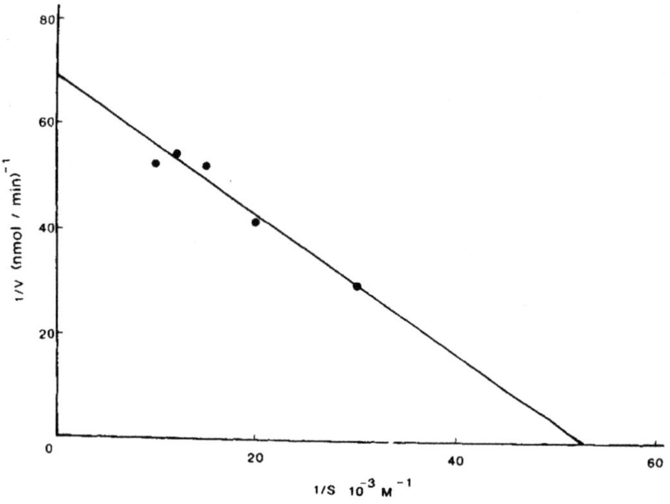

Fig. 1. Lineweaver-Burk plot of the degradation of Arg^0-Met-enkephalin

Inhibitory Effect of Arg^0-Met-enkephalin

The kinetics of the inhibition of the Met-enkephalin hydrolysis by Arg^0-Met-enkephalin (Fig 2) indicate that the hexapeptide is a mixed type of inhibitor for the aminopeptidase, since it changes both the K_m and the V_{max} of the enzyme. This type of inhibition is different from the kinetics of the Arg^0-β-endorphin and β-endorphin in the Met-enkephalin breakdown by the enzyme, which is competitive (Hui et al., 1982a).

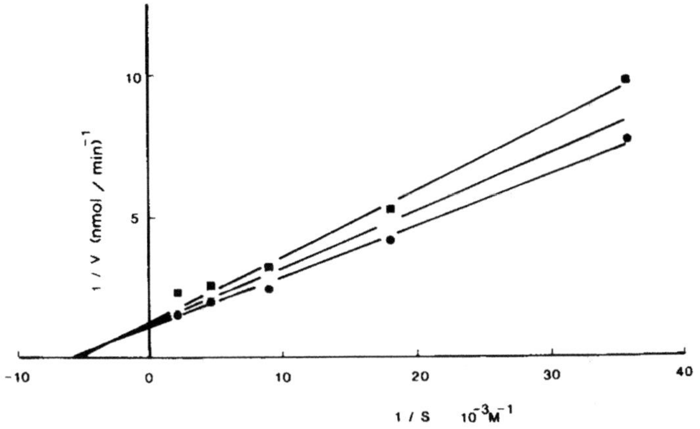

Fig. 2. Lineweaver-Burk plot of hydrolysis of Met-enkephalin (●) and the inhibition of the degradation by 0.25 uM (□) and 0.5 uM Arg^0-Met-enkephalin (■)

Conversion of Arg^0-Met-enkephalin to Met-enkephalin

Aminopeptidase released the N-terminal Arg from Arg^0-Met-enkephalin to form Met-enkephalin (Fig 3). The disappearance (the breakdown) of Arg^0-Met-enkephalin and the generation of the Met-enkephalin was in 1:1 ratio in the first hour of incubation. Only a small amount of Gly-Gly-Phe-Met was formed during this time. As the Arg^0-Met-enkephalin decreased, the Met-enkephalin began to be metabolized, to form des-Tyr-Met-enkephalin (Gly-Gly-Phe-Met). After 240 min of incubation, all the Arg^0-Met-enkephalin was metabolized, and after 360 min of incubation, the only product in a peptide was Gly-Gly-Phe-Met, which is resistant to our enzyme (Hui et al., 1982b).

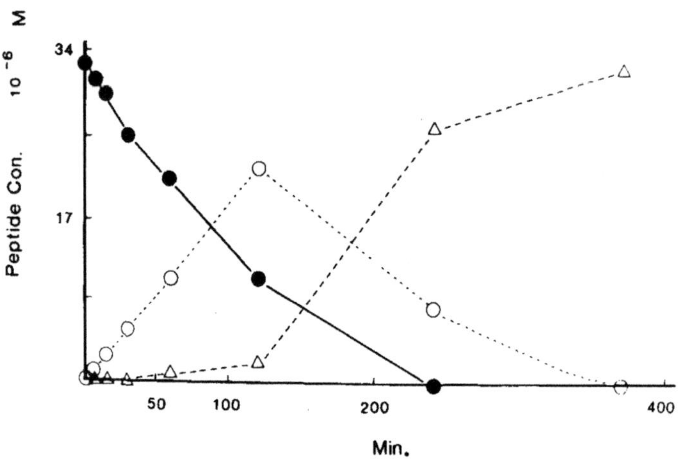

Fig 3. Time-curve of the Arg^0-Met-enkephalin breakdown by aminopeptidase (●) and the formation of Met-enkephalin (○) and Gly-Gly-Phe-Met (△)

DISCUSSION

The membrane aminopeptidase we purified from the rat brain is different from leucine aminopeptidase, aminopeptidase A, aminopeptidase B, aminopeptidase M, and the soluble aminopeptidase for enkephalin degradation (Hui et al., 1982b). It resembles the aminopeptidase MII that had been recently partially characterized by Hersh (1981), except that the enzyme is more sensitive to the puromycin, with IC_{50}=1.6 uM (IC_{50} for MII > 1 mM). Acetylcholinesterase inhibitor, neostigmine, and eserine did not inhibit the aminopeptidase, indicating that the peptidase and the cholinesterase are separate enzymes.

The microbial aminopeptidase inhibitors amastatin ((2S,3R)-3-amino-2-hydroxyl-5-methylhexanoyl-L-Val-L-Val-L-Asp) and bestatin ((2S,3R)-3-amino-2-hydroxyl-4-phenylbutanoyl-L-Leu) were good inhibitors in our system. However, their analogues, epiamastatin and epibestatin, with the hydroxyl group in reverse geometry ((2R,3R)-3-amino-2-hydroxyl), were less so, thus supporting the stereochemical requirements of these functional groups in affecting inhibition. The configuration of the hydroxyl (C_2) and amino (C_3) group of the microbial inhibitors is important as sites for the inhibitor to form a metal complex with the enzyme (Nishizawa et al., 1977). The configuration requirements for the inhibitors of leucine aminopeptidase and aminopeptidase B are more rigorous than membrane aminopeptidase. The epi-compounds are 750 times less potent than their parent compounds in leucine aminopeptidase (Nishizawa et al., 1977). Amastatin and bestatin are only 10 times more potent than their respective epi-compounds in inhibiting an aminopeptidase. This indicates that the requirement for the S configuration at C_2 of the inhibitor is more strict for leucine aminopeptidase or aminopeptidase B than for the membrane aminopeptidase. It also suggests that in the enzyme reaction the hydration of the carbonyl group in leucine aminopeptidase and aminopeptidase B is more stereospecific. The inhibition of the stereoisomers of the membrane aminopeptidase did not parallel exactly that with leucine aminopeptidase or aminopeptidase B. This may be due to a difference in the environment of the active site in these enzymes.

Membrane aminopeptidase does not break down Tyr-Gly, Tyr-Gly-Gly, Gly-Gly-Phe-Met, or the Met-enkephalin analogues with N-terminal extensions, except Arg^0-Met-enkephalin (Table I). Arg^0-Met-enkephalin is a substrate, but not as good as Met-enkephalin. The hydrolytic coefficient for Arg^0-Met-enkephalin is one sixth of that for Met-enkephalin (Table II). This differs from Arg β-naphthylamide, which has a hydrolytic coefficient 3 times that of Met-enkephalin (Hui et al., 1982a). Therefore kinetic constants of aminopeptidase for Arg βNA and Arg peptides are remarkably different. With Arg^0-Met-enkephalin as a substrate, the enzyme is substrate inhibited. This may be due to the peptides binding to the R' site of the aminopeptidase similar to the one for the leucine aminopeptidase as proposed by Bryce & Rabin (1964). The substrate has three reactive sites, the unchanged amino group, the peptide amide or ester oxygen atom, and the water binding sites; and two binding sites, the R group of the N-terminal amino acid residue and the R' that is the peptide chain. These sites act cooperatively in the substrate specificity of the enzyme. Some peptides can not be hydrolyzed by the enzyme because they cannot fulfill the requirements for the binding and reactive sites; thus it is likely that Met-enkephalin analogues with less than 5 amino acid residues cannot inhibit the enkephalin breakdown because their affinity to the reactive site of the enzyme is too weak. O-Sulphated Leu-enkephalin, an endogenous peptide in brain (Unsworth et al., 1982) is inactive as an inhibitor of the enzyme. It is believed that O-sulphated Leu-enkephalin is a metabolite rather than a precursor.

Met-enkephalin analogues with extension at the C-terminal can be degraded by the aminopeptidase but less than Met-enkephalin, even though they have higher hydrolytic coefficients. Arg analogues are more potent than the Leu analogues, and Leu-enkephalin analogues are more potent than Met-enkephalin analogues as inhibitors for the enzyme. Analogues with longer extensions (two amino acids such as Arg-Arg, Arg-Phe) at the C-terminal were better than those with one amino acid. This may be because the longer peptides are more hydrophobic; β-endorphin (31 amino acids) is more potent than the γ-endorphin (17 amino acids) (Hui et al., 1982a). The C-terminal extensions make it more favorable for the binding of the R' site of the enzyme. We found recently that some residues within region 18-31 of the β-endorphin are important for the binding. In addition, other peptides with different amino acid sequences also exert inhibitor effects on the membrane aminopeptidase, such as $ACTH_{1-24}$, $ACTH_{1-10}$, somatostatin, and substance P. However, all these inhibitory peptides have a cluster of hydrophobic and basic residues between sequence 6 and 9 (Hui et al., 1982a).

Our results showed a relationship in that the longer the N-terminal extension with the Met-enkephalin, the less the inhibitory effect. This may be due to the difficulty of the more bulky peptide to get into the binding sites. LPH (91 amino acids) does not affect the breakdown at all (Hui et al., 1982a). Since the membrane aminopeptidase hydrolyzes different neutral and basic amino acyl β-naphthylamides (Hui et al., 1982b) the binding site of the enzyme to the R group of the substrate is not highly specific. Therefore, high specificity of the membrane aminopeptidase in enkephalin breakdown is due to the specificity of its 2 binding and 3 reactive sites.

The degradation of Arg^{0}-Met-enkephalin by our enzyme is of great interest, since several peptides with Arg at the N-terminal are resistant to the enzyme, namely, Arg^{0}-β-endorphin, substance P (Arg-Pro-Lys-Pro-), bradykinin (Arg-Pro-Pro-Gly-Phe-), and CLIP (Arg-Pro-Val-Lys-Val-) (Hui et al., 1982a, Graf et al., 1982). At higher substrate concentrations of Arg^{0}-Met-enkephalin, the metabolic rate progressively falls off. This is not product inhibition, because the high concentration of the product (Met-enkephalin) does not affect the rate. There are seven possible mechanisms involved in substrate inhibition (Webb, 1963). Since the substrate in the [ES] complex forms a multipoint attachment to the enzyme, it is possible that at high substrate concentrations two or more substrate molecules can be bound simultaneously at the active center.

The inhibition of Met-enkephalin breakdown by Arg^{0}-Met-enkephalin is of mixed type. The kinetics of the inhibition is both competitive and non-competitive (Fig 2). Arg^{0}-Met-enkephalin can be a precursor for Met-enkephalin (Fig 3), but its natural occurrence in the tissue has not been established. The biosynthesis of the enkephalin is by a trypsin-like enzyme, which acts at the carboxyl end of a basic amino acid. Met-enkephalin in the proenkephalin molecule is flanked by a pair of basic amino acids at the C- and N-terminal of the peptide. Met-enkephalin-Arg^{6} and Met-enkephalin-Arg^{6}-Arg^{7} are endogenous ECPs. These Met-enkephalin-containing peptides form Met-enkephalin by trimming of the basic amino acids with a carboxylpeptidase-B-like enzyme (Lewis et al., 1980), which was recently purified and characterized (Fricker and Snyder, 1982). It is very possible that Arg^{0}-Met-enkephalin is an endogenous peptide and a precursor for enkephalin (Brownstein and Gainer, 1982).

ECPs coexist and are co-released with enkephalin (Kojima et al., 1982; Kilpatrick et al., 1980). There is a significant amount of ECPs in chromaffin cells; on a molar basis the free enkephalins represent no more than 9%

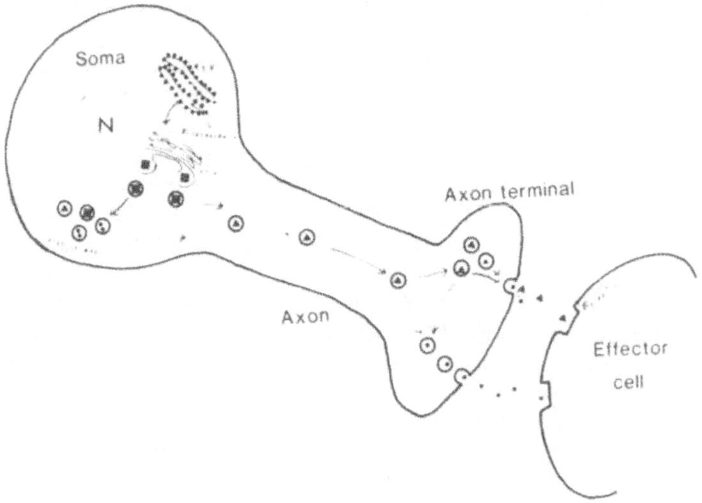

Fig. 4. The hypothetical model of biosynthesis, translocation, processing, release, and binding of enkephalin in a enkephalinergic neuron. (■) Big enkephalin-containing peptides; (R.E.R.) Rough endoplasmic reticulum; (▲) small enkephalin-containing peptides; and (•) enkephalin.

of the total compounds containing enkephalin sequences. On a weight basis, the enkephalins represent less than 0.4% of the total (Lewis et al., 1980). Arg^0-Met-enkephalin and ECPs with enkephalin at the N-terminal are active as inhibitors of contraction of the electrically stimulated guinea-pig ileum (Ling et al., 1978; Inturrisi et al., 1980; Kilpatrick et al., 1980; Law et al., 1977). Met-enkephalin-Arg^6-Phe^7 was found to bind to a subtype of opiate receptor (Audigier et al., 1980). ECPs can control the level of enkephalins by converting themselves to enkephalins in a substrate inhibition manner, and by inhibiting the metabolism of enkephalin (Fig 4).

This control is important for Met-enkephalin transport, release, and binding to the receptors. ECPs are degraded more slowly than Met-enkephalin; for this and other reasons, the possibility has to be considered that enkephalins may be modulators or transmitters at the δ-binding site, while ECPs may act hormonally, that is, at a distance from the site of release, for example, at k-binding sites for Met-enkephalin-Arg^6-Phe^7 (Audigier et al., 1980).

REFERENCES

Audigier Y., Mazarquie H., Rossier J., and Cros J. (1980) Met-enkephalin [Arg^6,Phe^7] interacts with the k-receptors on guinea-pig ileum. Eur. J. Pharmacol. 68, 237-238.

Brownstein M.J. and Gainer H. (1982) Neuropeptides: An Overview, in Handbook of Neurochemistry 2nd Ed. (Lajtha A., ed) Vol 1 pp. 181-186, Plenum Press, New York.

Bryce G.F. and Rabin B.R. (1964) The function of the metal ion in leucine aminopeptidase and the mechanism of action of the enzyme. Biochem. J. 90, 513-518.

Erdos E.G., Johnson A.R., and Boyden N.T. (1978) Hydrolysis of enkephalin by cultured human endothelial cells and by purified peptidyl dipeptidase. Biochem. Pharmacol. 27, 843-848.

Fricker L.D. and Snyder S.H. (1982) Enkephalin convertase: Purification and characterization of a specific enkephalin synthesizing carboxypeptidase localized to adrenal chromaffin granules. Proc. Natl. Acad. Sci. (USA) 79, 3886-3890.

Geary L.E., Wiley K.S., Scott W.L., and Cohen M.L. (1982) Degradation of exogenous enkephalin in the guinea-pig ileum: Relative importance of aminopeptidase, enkephalinase and angiotensin converting enzyme activity. J. Pharmacol. Expt. Ther. 221, 104-111.

Gubler U., Seeburg P., Hoffman B.J., Gage L.P., and Udenfriend S. (1982) Molecular cloning establishes proenkephalin as precursor of enkephalin-containing peptides. Nature 295, 206-208.

Graf L., Hui K.-S., Neidle A., and Lajtha A. (1982) Neuropeptide may regulate neuropeptide metabolism in the brain. Neuropeptide 2, (3-4) 169-173.

Hersh L.B. (1981) Solubilization and characterization of two rat brain membrane-bound aminopeptidases active on Met-enkephalin. Biochemistry 20, 2345-2350.

Hui K.-S., Graf L., and Lajtha A. (1982a) β-endorphin inhibits Met-enkephalin breakdown by a brain aminopeptidase: Structure-activity relationships. Biochem. Biophys. Res. Commun. 105, 1482-1487.

Hui K.-S., Salshutz M., Davis B.A., and Lajtha A. (1980) Separation of alkylaminonaphthylenesulfonyl peptides and amino acids: Methods for measuring melanotropin inhibiting factor breakdown. J. Chromatogr. 192, 341-350.

Hui K.-S., Wang Y.-J., and Lajtha A. (1982b) Purification and characterization of an enkephalin aminopeptidase from rat brain membrane. Biochemistry (in press).

Hui K.-S., Wang Y.-J., Tsai H., Wong K.-H., and Lajtha A. (1981a) The effect of naloxone on enkephalin catabolism. Peptides 21, 89-94.

Hui K.-S., Wang Y.-J., Wong K.-H., Tsai H., and Lajtha A. (1981b) Regional and subcellular distribution of Met- and Leu-enkephalin-degrading activity in rat brain. Neuropeptides 1, 167-173.

Inturrisi C.E., Umans J.G., Wolff D., Stern A.S., Lewis R.V., Stein S., and Udenfriend S. (1980) Analgesic activity of the naturally occurring heptapeptide [Met]enkephalin-Arg^{6}-Phe^{7}. Proc. Natl. Acad. Sci. (USA) 77, 9. 5512-5514.

Kilpatrick D.L., Lewis R.V., Stein S., and Udenfriend S. (1980) Release of enkephalins and enkephalin-containing polypeptides from perfused beef adrenal glands. Proc. Natl. Acad. Sci. (USA) 77, 7473-7475.

Knight M. and Klee W.A. (1978) The relationship between Enkephalin degradation and opiate receptor occupancy. J. Biol. Chem. 253, 3843-3847.

Kojima K., Kilpatrick D.L., Stern A.S., Jones B., and Udenfriend S. (1982) Proenkephalin: A general pathway for enkephalin biosynthesis in animal tissues. Arch. Biochem. Biophys. 215, 638-643.

Law P.Y., Wei E.T., Tseng L.F., Loh H.H., and Way E.L. (1977) Opioid properties of -Lipotropin fragments 60-65, H-Arg-Tyr-Gly-Gly-Phe-Met-OH. Life Sci. 20, 251-260.

Lewis R.V., Stern A.S., Kimura S., Rossier J., Stein S., and Udenfriend S. (1980) An about 50,000-Dalton protein in adrenal medulla: A common precursor of [Met]- and [Leu]enkephalin. Sci. 208, 1459-1461.

Ling N., Minick S., and Guillemin R. (1978) Amino-terminal extension analogs of methionine-enkephalin. Biochem. Biophys. Res. Commun. 83, 2, 565-570.

Nishizawa R., Saino T., Takita T., Suda H., Aoyagi T., and Umezawa H. (1977) Synthesis and Structure-activity relationships of bestatin analogues, inhibitors of aminopeptidase. J. Med. Chem. 20, 510-515.

Noda M., Furutani Y., Takahashi H., Toyosato M., Hirose T., Inayama S., Nakanishi S., and Numa S. (1982) Cloning and sequence analysis of cDNA for bovine adrenal preproenkephalin. Nature 295, 202-206.

Schwartz J.-C., Malfroy B., and De La Baume S. (1981) Biological inactivation of enkephalins and the role of enkephalin-dipeptidyl-carboxypeptidase (Enkephalinase) as neuropeptidase. Life Sci. 29, 1715-1740.

Unsworth C.D., Hughes J., and Morley J.S. (1982) O-sulphated Leu-enkephalin in brain. Nature 295, 519-522.

Webb J.L. (1963) Enzyme and metabolic inhibitors. Vol I Academic Press, New York and London.

CHOLINE ACETYLTRANSFERASE: PURIFICATION, CHARACTERIZATION, LOCALIZATION, AND INDUCTION OF A FATAL AUTOIMMUNE DISEASE

Li-Pen Chao

Department of Neurology, Reed Neurological Research Center

UCLA School of Medicine, Los Angeles, California 90024

Choline acetyltransferase (ChAc; acetyl-CoA: Choline-O-acetyltransferase, EC 2.3.1.6) catalyzes the synthesis of the neurotransmitter acetylcholine (ACh) and is the specific enzymatic maker for cholinergic neurons. Because of its uniqueness and importance, this enzyme has been widely studied, especially in recent years. In our laboratory, this enzyme was purified from bovine caudate nuclei. The purified preparation was characterized by various physicochemical methods in order to examine its molecular properties and homogeneity. We have also localized ChAc in both central and peripheral nervous systems by an immunohistochemical method, using an antiserum raised against the purified ChAc. We found, unexpectedly, that a fraction of the purified ChAc could produce a fatal, apparently autoimmune disease. The purpose of this publication is to present our findings, together with results from other laboratories.

I. PURIFICATION AND CHARACTERIZATION

The purification and characterization of ChAc is discussed in great detail in a recent review (Chao, 1980), and will be re-examined briefly. Our purification procedure is summarized in Table I (Chao and Wolfgram, 1973).

The homogeneity of purified bovine ChAc has been demonstrated by:

(a) The presence of a single band on polyacrylamide gel electrophoresis (PAGE) at different pH values and gel concentrations (Chao and Wolfgram, 1973);

Table I. Summary of the Purification Procedure for ChAc From Bovine Caudate Nuclei

	Procedure	Volume (ml)	Activity (units)[a]	Protein (mg)	Specific Activity (units/mg) x 10^3	Yield	Purification
1.	Acetone-chloroform powder						
2.	Extraction	92.5	17.00	1961	8.7	-	-
3.	Organomercurial-Sepharose column	151	9.90	362.0	27.2	58.0	3.2
4.	DEAE-cellulose column	188	9.80	158.8	61.9	57.8	7.2
5.	Hydroxyapatite column	49	9.30	78.4	118.3	54.6	13.7
6.	Organic solvent	49	8.81	21.0	419.5	51.8	48.4

[a] 1 unit = 1 μmole ACh/min

(b) Demonstration of a single amino-terminal residue of serine and a single carboxyl-terminal residue of alanine in ChAc (Chao, 1975);

(c) Specific chemical cleavage at the methionine or tryptophan residues produced the exact number of peptides predicted from the amino acid analysis of the purified ChAc (Chao, 1975);

(d) The presence of a single precipitin arc in immunodiffusion and immunoelectrophoresis (Eng, et al, 1974); and

(e) Direct visualization of the presence of ChAc in structures which are known to be cholinergic: the ventral horn motor neurons of spinal cord (Eng. et al, 1974; Kan et al, 1978), neuromuscular junctions of rabbit diaphragm (Kan and Chao, 1981) and brain structures thought to be cholinergic (Kan et al, 1978, 1980; Chao, et al. 1982a). We believe that other preparations of "purified" ChAc have not so far been subjected to such rigorous criteria of purity.

Improved purification procedures for ChAc from human brain (Peng et al, 1980) and fly heads (Slemmon et al, 1982) have been reported from other laboratories. In addition, other purification methods for ChAc with varying degrees of purity from catfish (Su et al, 1980), chicken (Peng, et al, 1981), rat (Dietz & Salvaterra, 1980), pig (Eckenstein et al, 1981) and bovine (Cozzari & Hartman, 1980) brains have been published. Some of the newly purified ChAc had high specific activites [SA; (μmole/min)/mg. of protein], leading some authors to use this as a measure of purity. However, it is my belief that SA alone does not provide proof of the purity

(Chao, 1980). Some of my reasons for this conclusion include:

(a) ChAc is stablized by certain contaminating impurities, so that a purer enzyme may be less stable;

(b) Up to a 30-fold change in activity with slight changes of the numerous factors involving the measurement of activity in the incubation medium; and

(c) The final yield is too minute to determine accurately the protein concentration by any available methods. To illustrate this point, I have compiled a Table (II) to show the final yield of protein, SA, and the methods used to determine the protein concentration from some mammalian brain ChAc preparations with "high" SA.

No extinction coefficient (Malthe-Soressen, et al, 1978) was given to calculate the concentration of ChAc for A_{280nm}. In some cases the final yield on protein is either too low or made from one determination by the popular Lowry method. The determination of protein by scanning the gels obtained from PAGE in the presence of sodium dodecyl sulfate (SDS) carries many pitfalls. For instance, the protein had to be calculated by assuming that (a) all the proteins in the final purification entered the gel; (b) the staining of each protein was identical; (c) the peaks were well separated (in fact, the peaks were not fully resolved) (Ryan, 1976); and (d) one inactive band (Rossier, 1976) or several bands (Dietz & Salvaterra, 1980) were identified as the enzyme in a pattern of multiple bands. This author agrees with Ryan's conclusion that "the method must be regarded as approximate at best" (Ryan 1976). Under such circumstances, specific activity alone cannot be used as a guide of purity and is an indication of the frustration encountered by many investigators working with this elusive enzyme.

Table II. Some Mammalian Brain ChAc With High SA

	Tissue	Protein (mg)	SA	Method
Malthe-Soressen, et a. 1978	Bovine	0.14	28.5	$A_{280\ nm}$ and Lowry
Ryan & McClure, 1979	Bovine	Not given	53.76	Scan SDS-PAGE gels
Cozzari & Hertman 1980	Bovine	Not given	120-160	Lowry
Eckenstein, et al. 1981	Pig	0.089	135	Lowry
Rossier, 1976	Rat	0.1	20	Lowry
Ryan & McClure, 1979	Rat	Not given	40.2	Scan SDS-PAGE gels
Dietz & Salvaterra, 1980	Rat	0.0025	73	Scan SDS-PAGE gels

We have studied the instability of our purified ChAc and hypothesized why impurities have a stabilizing effect on this enzyme. Our preliminary results indicate that the purified bovine ChAc exists in two forms; one is ChAc (active) and the other ChAc (inactive). In solution, these forms can be depicted as follows:

$$\underset{\text{(Active)}}{\text{ChAc}} \rightleftharpoons \underset{\text{(Inactive)}}{\text{ChAc}} \qquad \text{(Equation 1)}$$

Once all contaminants are completely removed, some of ChAc (active) will oscillate to ChAc (inactive), as illustrated in Equation 1. This can explain why the purer ChAc is, the less stable this enzyme will be (Chao, 1980). Based on the following experiments, we believe our purified enzyme is in this state:

When our purified ChAc was chromatographed in Bio-gel A column at neutral pH (Chao & Wolfgram, 1973), ChAc dissociated into two peaks with a large loss of enzymatic activity [only one protein peak was obtained at alkaline pH (Chao, 1975)]. The purified ChAc was then fractionated on a column of immobilized Cibacron blue resin with a stepwise elution system (Chao, et al, 1982b). Two major peaks (A and C) and a really minute peak B in between were obtained. Only fraction C was active and fractions A and B were enzymatically inactive. When fraction C was rechromatographed under identical conditions, again two major peaks (A and C) and one minute peak B were obtained, and again peak C was the only active fraction. This demonstrates that active fraction C can oscillate to inactive fractions A and B, as in the original enzyme complex depicted in Equation 1. (Chao, unpublished results).

Antisera were produced against fractions A and C in the same manner used to raise antisera against the purified bovine ChAc (Eng et al, 1974). The antibodies produced against fractions A and C were indistinguishable from the antibodies raised against the purified ChAc (mixture) in immunoelectrophoresis against either concentrated crude extract of bovine caudate nuclei, partially purified, or purified ChAc. When each of three antisera (purified ChAc, fractions A and C) were used for the localization of ChAc in rabbit cerebellum by an immunohistochemical technique (Sternberger, 1979), there were no differences in the staining pattern and the reaction products were localized only in some mossy fibers and their synaptic sites, glomeruli (Kan et al, 1978; 1980).

Unexpectedly, when the pooled minute fraction B was used for the production of antiserum, the injected animals died about 14 days after the first injection, apparently from an autoimmune disease (Chao et al, 1982b). This will be discussed under "Induction of a fatal autoimmune disease."

In conclusion, the biochemical, immunological and immunohistochemical evidence suggest that:

(a) Our purified ChAc consists of ChAc (active) and ChAc (inactive);

(b) Once the ChAc (active) is isolated from the original mixture, some of it oscillates to ChAc (inactive), which is more stable; and

(c) ChAc (active), ChAc (inactive) and the purified ChAc (mixture) are identical immunologically and immunohistochemically.

If the goal is an enzyme preparation with "high" SA, ChAc (inactive) will be continuously removed until there is almost no protein left in solution (Table II).

Our purified ChAc showed one band in PAGE in the absence of SDS at different pH values and gel concentrations (Chao and Wolfgram, 1973). Although eight bands with various intensities are observed in SDS-PAGE (Chao, 1975), the purified ChAc is homogeneous by numerous biochemical (Chao 1975, 1980) and immunological (Eng. et al, 1974) criteria and is localized in structures which are known to be cholinergic (Kan et al, 1978; Kan and Chao 1981). Based on this and other evidence, we conclude that the multiple bands in SDS-PAGE are all derived from the purified ChAc, and this is just one of the unusual properties of ChAc (Chao et al, 1977; Chao, 1978a; 1980).

We propose that bovine ChAc is a hexamer with identical subunits and has a molecular weight of 89,000 daltons by biochemical evidence (Chao, 1975, 1980). Each subunit contains 133 amino acid residues with a molecular weight of 14,700 daltons (Chao, 1975, 1980). When the purified bovine ChAc was chromatographed on a column of Bio-gel A by varying the ionic strength and pH of the buffer system, we detected an enzyme peak corresponding to the monomer (15,000), trimer (47,000), tetramer (60,000) (Chao and Wolfgram, 1974), pentamer (75,000) and hexamer (89,000) (Chao 1975 and unpublished results). Except for the hexamer which is inactive at alkaline pH (Chao, 1975), all are enzymatically active. Although no one has been able to demonstrate only one band on SDS-PAGE for any mammalian ChAc, some investigators (Rossier, 1976; Ryan and McClure 1979) believe that only one band should be detected and lead to some spuriously high SA values. Various reports have concluded that ChAc has two bands in bovine brain (Cozzari and Hartman, 1980), four subunits in catfish brain (Su et al, 1980) and multiple bands in fly head (Slemmon, et al, 1982) and rat brain (Dietz and Salvaterra, 1980) in SDS-PAGE.

Although there are numerous methods to study the forward reaction of this enzyme (Chao & Wolfgram, 1972; Chao, 1978b, 1980), no reliable method exists to monitor the rate of reverse reaction. Recently we reported a method for the measurement of the reverse reaction product, acetyl-CoA, by separating it from substrate ACh with a cation exchange resin (Hsu and Chao, 1982). The reverse reaction of ChAc was linear with incubation time up to 40 minutes, and with enzyme protein concentration up to 5 μg. The pH optimum was 7.0, the same as we reported for the forward reaction (Chao and Wolfgram, 1973). Monovalent chloride and bromide salts activated the reverse ChAc activity, but fluoride and iodide salts inhibited it. Kinetic studies without salt showed that K_{ACh} was 0.62 ± 0.06 mM, $K_{CoA \cdot SH}$ was 12.68 ± 1.21 mM, and V_{max} was 11.6 ± 1.0 nmol acetyl-CoA/mg protein/min. As these data are in disagreement with the values reported for partially purified ChAc from bovine brain (Glover and Potter, 1981) and placental enzyme (Hersh, 1980), further studies are necessary to resolve these differences.

There is an active sulfhydryl (SH) group(s) either in or adjacent to the active center of ChAc (Chao, 1980). Most of the reagents that modify histidine, tyrosine, and other amino acid residues, react more rapidly or at least at a comparable rate with SH of cysteine residue(s) (Glazer, 1970). A reactive arginine residue(s) in a partially purified squid ChAc without the consideration of the active cysteine residue(s) was recently reported (Mautner et al, 1981).

II. IMMUNOHISTOCHEMICAL LOCALIZATION OF ChAc

The distribution of the cholinergic pathways can best be done by localization of ChAc at present. Among all the cholinergic markers, ChAc is the most specific. ACh is also specific, but there is no method available for the histochemical identification of this neurotransmitter. Furthermore, the concentration of ACh in postmortem tissue is known to be labile and difficult to measure accurately. Acetylcholinesterase (AChE) is not limited to cholinergic neurons, and cholinergic receptors identify the cholinoceptive structures. The high affinity choline uptake (HACU) mechanism cannot provide information concerning which neurons utilize

ACh as their neurotransmitter. The localization of ChAc by means of an immunohistochemical method remains the best alternative.

In our laboratory, antisera to the purified bovine ChAc were raised in guinea pigs (Eng et al, 1974). The purity of the enzyme and the monospecificity of the antisera were demonstrated by several widely accepted immunochemical techniques. This antiserum was found to cross-react strongly with ChAc from the extract of rabbit brain (Kan et al, 1978) and human caudate nuclei (Kan et al, 1980). We have used the peroxidase-antiperoxidase (PAP) immunohistochemical method to localize ChAc in rabbit and human brains (Kan et al, 1978, 1980; Chao et al, 1982a).

ChAc was initially localized in the cholinergic ventral horn motor neurons of bovine spinal cord by an immunofluorescent method (Eng et al, 1974). We have used the PAP method for all our other localization studies in rabbit and human brains since then. When a section of rabbit spinal cord was treated with antiserum to ChAc, no cells were stained in the apical area of the dorsal horn while the perikarya of the ventral horn motor neurons, especially in the Nissl substance, were stained (Kan et al, 1978). Using a double staining technique, ChAc was localized at the neuromuscular junctions of rabbit diaphragm with PAP staining, which produced a brown color in combination with the histochemical staining of cholinesterase (ChE), and produced a blue color in motor end plates on the same tissue section (Kan and Chao, 1981). We clearly showed that ChE in blue color was localized at the postsynaptic site, while the ChAc in brown color was visualized at the presynaptic nerve terminals and axions. The innervation of the nerve terminals at the neuromuscular junctions was similar when a silver impregnation was substituted for the immunohistochemical staining of ChAc. Since the innervation of the neuromuscular junctions is known to be cholinergic, the presence of ChAc in the ventral horn motor neurons and in the neuronal element of the muscular junction establishes both the specificity of our localization method and confirmed the purity of the antigen preparation. The ability to localize ChAc in this known cholinergic system is a prerequisite in order to demonstrate the specificity of any antiserum to ChAc unequivocally.

In rabbit cerebellum, ChAc was localized in some of the mossy fibers of the cerebellar folia and in their synaptic sites, the glomeruli (Kan et al, 1978). In the cross-sections of cerebellar folia treated with antiserum, there was no peroxidase activity in the molecular layer, Purkinje cell layer, granular cells or Golgi cells. Staining was restricted to those axons in the medullary layer which extended into the granular layer to branch and occasionally swell. When cerebellar folia were cut tangentially through the granular layer, cross-sections of glomeruli could be observed (Kan et al, 1978). Dark staining in the center of each glomerulus corresponds to a mossy fiber rosette. There were differences in the number of mossy fibers stained in different folia, and it appeared that some mossy fibers, but probably not all, are cholinergic. This finding is consistent with biochemical studies of the distribution of ChAc within various layers of the rabbit cerebellum (Kan et al, 1978 and references therein). It appears that the cholinergic mossy fibers originate outside the cerebellum, since ChAc activity disappears from cerebellar folia after transection of the cerebellar peduncles (Fonnum, 1972) or surgical isolation of the vermis (Kasa & Silver, 1969). Recently, ChAc activity was demonstrated in mossy fibers of mouse cerebellar cultures, but only when dorsal pontine tissue was included. Cultures without dorsal pontine tissue had 20 to 60 times less ChAc activity (Woodward et al, 1982). This is consistent with our finding that some mossy fibers are cholinergic.

When the formalin-fixed and paraffin-embedded human cerebellum sections were treated with antiserum to our purified bovine ChAc, positive brown reaction deposits were observed in some mossy fibers in the medullary and the granular layers, as in rabbit cerebellum (Kan et al, 1980). The stained fibers never seemed to reach the molecular layer, although some were close to the Purkinje cell layer. Staining was also observed in glomeruli, either as a swelling of the mossy fibers or as a rosette-like structure when the cut was made across the glomeruli. Similar staining was observed in normal and in Huntington's disease brains. It has been reported that ChAc is a reasonably stable enzyme in postmortem tissue as determined by its enzymatic activity (Kan et al, 1980 and references therein). Our studies on human cerebellum showed that the antigenicity of this enzyme is also preserved in autopsy tissue which was taken from between 2 and 69 hours after death.

The immunohistochemical localization of ChAc was extended in the study of rabbit forebrain (Chao et al, 1982a). ChAc positive reaction products were found to be widely distributed in this area. The results of this study generally correlate well with findings obtained by biochemical assays or histochemical methods for other cholinergic markers, i.e., ACh, AChE, cholinergic receptor, as well as HACU, and with the cholinergic pathways previously proposed on the basis of such evidence.

Olfactory system: Positively stained fibers and bouton-like structures were observed in every layer except the glomeruli and the olfactory nerve layer. Densely packed ChAc-containing fibers in the form of punctate deposits of the reaction product were localized in the olfactory tract, the molecular layer of the piriform cortex ventral to the rhinal fissure, and the ventral olfactory striae. ChAc-positive fibers were also found in the ganglionic layer and the multiform layer of the piriform cortex, and some of them were radially oriented. Positively stained fibers and bouton-like structures were conspicuously visualized in the island of Calleja, insula magna and other areas of the olfactory tubercle. Microenzymatic assays had also demonstrated ChAc and AChE in all layers of olfactory bulb, with a small amount of both enzymes present in the olfactory nerve layer (Godfrey et al, 1980).

Septal and diagonal band regions: Dorsoventrally projected ChAc-containing axons were found to be more heavily concentrated in the medial septal area close to the sagittal plane than in the lateral region. Bouton-like structures or puncta indicating the presence of possible axon terminals or transversely sectioned axons were also visualized clearly among the stained fibers in the medial septal area. The density of the punctate deposits was greatest in the dorsal area near the corpus callosum but gradually diminished toward the ventral position. Comparatively, more ChAc-positive axons were observed in the diagnoal band region than in the septum. However, only a few bouton-like structures were found in this area. Microenzymatic assay of ChAc (Hoover et al, 1978) had found a more or less even distribution of the activities of ChAc and AChE in the septal region and more of both enzymes in the diagonal band. Muscarinic receptor binding experiments had yielded low grain densities in these regions (Rotter et al, 1979). Since the receptor is a constituent of the cholinoceptive system, this suggests that the ChAc-containing fibers are simply passing through these areas and few form synapses here.

Basal ganglia: Positively stained fibers in the neuropil were widely distributed in the neostriatum. In the caudate nuclei, reaction products were observed in the extensive neuropil. The putamen, on the other hand, exhibited numerous ChAc-positive fascicles oriented rostrocaudally among the stained neuropil and discrete fibers. Dense dark-stained fibers packed in numerous fascicles were also visualized in the globus pallidus. In the area of the claustrum, only a few fibers with brown reaction products were identified. An abundance of ACh, ChAc, and AChE has been found in the basal ganglia by microassays (Chao et al, 1982a and references therein). Lewis and Shute (1978) reported on the basis of biochemical fractionation that the dense neuropil of the neostriatum was rich in ChAc; AChE also was demonstrated there. These findings correlate very well with our immunohistochemical results.

Thalamic, hypothalamic and subthalamic regions: Axons and axon terminals containing ChAc were visualized in the thalamic, hypothalamic and subthalamic regions. Additionally, dorsoventrally oriented fibers with dark ChAc-positive staining in the fasciculus retroflexus were clearly observed emerging from the habenular nucleus. ACh, ChAc, and AChE had also been found in these areas by microassays (Chao et al, 1982a and references therein). Moreover, high grain densities of muscarinic binding sites and the HACU system were also detected in these parts of the brain. When a lesion was made at the habenular nucleus, the activity of ChAc in the interpeduncular nucleus was reduced to less than 20% of the control, but the activity of glutamic acid decarboxylase was unaffected (Kataoka et al, 1973). Based on this and other evidence, a cholinergic pathway from the habenula via the fasiculus retroflexus to the interpeduncular nucleus has been suggested. It correlates well with our immunohistochemical localization of ChAc.

Cerebral cortex: ChAc-positive fibers and bouton-like structures were widely distributed in the neocortex and most distinctly visualized in layers 5 and 6 of the cortex, although they were also observed in superficial areas and near the surface (Chao et al, 1982a). Possibly ChAc-containing axon terminals in the form of bouton-like structures was also evident in the postcentral and parietal areas of cortex. Our findings of cholinergic axons and axon terminals in the cerebral cortex correlates well with the results obtained using other cholinergic markers (Chao et al, 1982a and references therein).

Hippocampal region: In the dorsal hippocampus, we observed ChAc-positive stained fibers and, possibly, axon terminals in all layers, with the exception of the pyramidale layer (Chao et al, 1982a). The pyramidale layer, which was reported to be low in AChE, becomes hypersensitive to ACh after denervation at the medial septum and is cholinoceptive

(Lewis and Shute, 1978). In the dentate gyrus, ChAc was also detected in all layers, but not in the granular layer which was found to contain no AChE (Lewis & Shute, 1978). The distribution of ChAc-positive fibers in the ventral hippocampus was essentially the same as described for the dorsal hippocampus.

Positively stained elements, mostly in punctate form, were localized at relatively deep levels near the white matter in the retrohippocampal formation. However, in the deep portion of the transition zone (between subareas 28a and 28b), an aggregation of reaction products in fascicles was observed (Chao et al, 1982a).

On the basis of histochemical staining of AChE, a septal-hippocampal cholinergic pathway arising from the medial septum and diagonal band nuclei and traveling to the hippocampal formation via the supracallosal (longitudinal) stria, the dorsal fornix, the alveus and the fimbria has been proposed (Storm-Mathisen, 1977). Furthermore, when lesions were made in the medial septum, loss of about 60% of the HACU mechanism, 85-90% of ChAc and 80% of ACh was observed in the hippocampus (Kuhar et al, 1973). Our identification of ChAc in these regions strongly supports the presence of such a cholinergic pathway.

White matter and other areas: ChAc positive elements were found to be present also in the white matter of rabbit forebrain (Chao et al, 1982a). In the corpus callosum, positive stained axons were seen passing through in a rostrocaudal direction. Similar axons packed in fascicles were observed in the internal capsule, while in the external and extreme capsules, the brown fibers were oriented ventrally or ventrolaterally. In addition, ChAc-positive reaction deposits were detected in the longitudinal striae and in the fornix. The deposits in the fornix showed a morphological resemblance to those observed in the alveus and fimbria of the hippocampus (Chao et al, 1982a). In the optic radiation, fibers with brown reaction deposits were also observed.

Shortly after our report on the immunofluorescent localization of ChAc in bovine spinal cord (Eng et al, 1974), McGeer and co-workers (1974) raised antiserum in rabbits to purified human ChAc. Initially, Hattori, et al., of McGeer's group (1976) found that ChAc was localized primarily in the cell bodies in medium sized neurons, but never in the large striated cells. The PAP method yielded almost no staining in the dense neuropil in guinea pig neostriatum. Lewis and Shute (1978) commented that a much higher proportion of cells give a positive reaction than are likely to be cholinergic and that, in contrast, the neuropil, which is known to be rich in ChAc, gave no reaction with the antiserum. Later, Kimura, et al. (1980), from the same group, reported ChAc in the large cells (>25/μm) in the guinea pig caudate nuclei using an antiserum produced against a more highly purified human ChAc. Their observation is supported by the presence of AChE, which was also found in the large cells by histochemical staining. They concluded that only the large cells in guinea pig caudate nucleus are cholinergic and not the medium-sized cells. On the other hand, Kaiya et al. (1979) recently reported that ChAc was localized by a histochemical method in the medium-sized cells in rat caudate nucleus, which is in agreement with the results reported by Hattori, et al. (1976). Kaiya et al. (1980) also found AChE localized in all types of cells in the rat caudate nucleus. They concluded the medium-sized cells are cholinergic and the large cells, which synthesize a large amount of AChE but no ChAc, are cholinoceptive. However, both groups reported no positive reaction of ChAc in the dense neuropil despite the fact that there is much more ChAc in axons and axon terminals than in the cell bodies.

Most recently, Kimura et al. (1981) published an atlas for the localization of ChAc in the central nervous system of the cat, using the PAP method. In contrast to our findings, they localized ChAc primarily in the cell bodies in about forty areas of the cat, from the spinal cord to the olfactory system, excepting the cerebellum. Cozzari and Hartman (1980) used an immunofluorescent technique with antiserum to bovine ChAc to localize ChAc exclusively in the cell bodies. Failure to demonstrate ChAc in axons and axon terminals, where its concentration is highest, raises serious fundamental questions which need to be resolved.

Monoclonal antibodies have been raised both to bovine ChAc (Levey et al, 1981) and ChAc from fly heads (Crawford, et al, 1982). The specificity of these antibodies, as well as their usefulness in localization, remain to be determined.

III. INDUCTION OF A FATAL AUTOIMMUNE DISEASE

In the discussion of purification and characterization, we presented results to show that bovine ChAc is a highly labile enzyme and becomes less stable as it is purified (Chao, 1980). In continuation of this study, the unstable purified bovine ChAc was fractionated on a column of Cibacron blue resin (Chao et al, 1982b). As discussed above, two large peaks (A and C) and one minute peak (B) were obtained accompanied by a loss of enzymatic activity. During antibody production against each fraction, fraction B was unexpectedly found to have the ability to induce an apparently autoimmune disease in guinea pigs, whereas the other two fractions (A and C) were non-pathogenic (Chao et al, 1982b). On the basis of histological, histochemical, immunological, and electromyogram (EMG) examinations, this appears to be a presynaptic autoimmune neuromuscular disease (Chao et al, 1982b).

When fraction B of bovine ChAc, mixed with complete Freund's adjuvant, was injected weekly into guinea pigs, the animals showed progressive weakness, weight loss, reduction of respiratory rate, quiescence during handling and cyanosis followed by death about 14 days from the initial injection (Chao et al, 1982b). Histological and histochemical studies of the experimental animals revealed striking changes with numerous foci of inflammation, particularly conspicuous in certain areas of the diaphragm. In contrast, there were no cellular changes in the muscles of controls which received injections of saline. Furthermore, by a combination of histological staining for cholinesterase (ChE) for the motor end plates, and hematoxylin and eosin (H & E) for the cellular reaction, the sites of these inflammatory changes were identified to be at the neuromuscular junction. With a combination of histochemical staining of ChE and silver impregnation for nerve elements, the sites of the inflammatory infiltrates were further localized around the nerves in the neuromuscular junctions (Chao et al, 1982b). The leg muscles, gastrocnemius, and quadriceps were also affected, but to a lesser degree than diaphragm. Some cellular reactions were occasionally found in the lung, liver, and intestines, but no significant abnormality was found in the central nervous system (brain and spinal cord) of the diseased animals. Based on the clinical, histological and histochemical observations, the animals apparently died from respiratory paralysis.

The pathogenesis of the disease is most likely to be of an autoimmune nature because the experimental lesions were produced by the injection of bovine ChAc in complete Freund's adjuvant (Chao et al, 1982b). Furthermore, when the frozen sections of the diaphragm were incubated with fluorescein-conjugated goat anti-guinea pig IgG, under low magnification, strong fluorescence was observed to be concentrated in certain regions of the diseasedtissues; this was negative in tissues from control animals. This evidence supports the possibility that the syndrome is an autoimmune disease.

Since we have localized ChAc in the axon and axon terminal of the neuromuscular junctions in rabbit diaphragm (Kan & Chao, 1981), the autoimmune responses produced by injection of bovine ChAc (fraction B) were most likely directed against the nerve elements and resulted in denervations as evidenced by the cellular reactions at the neuromuscular junctions and nearby nerves (Chao et al, 1982b). Neurogenic lesions were demonstrated by the occurrence of numerous target and targetoid fibers in muscle. Moreover, the EMGs showed a mild partial denervation, in the form of positive sharp waves and fibrillation discharges in the gastrocnemius muscle of the experimental animals. This evidence was also suggestive of denervation, in agreement with the data from both histological and histochemical studies. This suggests that this presumably autoimmune disease is a presynaptic neuromuscular disease similar to the myasthenic syndrome (Eaton-Lambert Syndrome; ELS) (Brooke, 1977). It may also resemble botulism (Fusfeld, 1970; Kao et al, 1976), and botulinum toxin has been reported to be an inhibitor of ChAc in vivo (Ekstrom et al, 1977).

Recently Dau and Denys (1982) reported that plasmapheresis alone produced modest clinical and EMG improvement in ELS patients. However, the greatest improvement was observed during plasmapheresis combined with immunosuppressive drugs, such as prednisone and azathioprine. This evidence, in combination with the autoimmune disease we produced in guinea pigs, strongly suggests the involvement of a circulating autoantibody which blocks neuromuscular and cholinergic autonomic transmission. This observation resembles the situation in myasthenia gravis (MG), in which a striking and sustained improvement is generally achieved with the addition of prednisone and azathioprine in combination with plasmapheresis. Considerable evidence has been accumulated to show that MG is an autoimmune disease caused by an antibody to cholinergic receptors (Drachman, 1981).

When a soluble fraction of rat sciatic nerves with high ChAc activity, mixed with complete Freund's adjuvant, was injected into female sheep, an experimental allergic neuropathy was demonstrated both electrically in response to endrophonium by Festoff et al. (1977). Some of these results correlate well with our findings, and our study on the autoimmune disease (Chao et al, 1982b) may offer a possible explanation why the soluble fraction of rat sciatic nerve produces a disorder in neuromuscular trasmission.

In conclusion, we have demonstrated the successful purification, characterization and immunohistochemical localization of bovine ChAc. The discovery of the autoimmune disease is an unexpected but a pleasant surprise.

Acknowledgements: The author would like to express his appreciation to Drs. F. Wolfgram, H.E. Hirsch and B. Crandall for their comments and to Dorothy Kuljanski for her typing of this manuscript. This research was supported by U.S. Public Health Service Grant NS-11087.

REFERENCES

Brooke, M.H. (1977) A Clinician's View of Neuromuscular Disease, p. 89, Williams & Wilkins, Baltimore.

Chao, L.-P. (1975) Subunit of choline acetyltransferase. J. Neurochem. 25, 261-266.

Chao, L.-P. (1978a) Choline acetyltransferase: The absence of multiple forms and purification from mouse brain. Neurochem. Res. 3, 549-561.

Chao, L.-P. (1978b) Spectrophotometric determination of choline acetyltransferase in the presence of dithiothreitol. Anal. Biochem. 85, 20-24.

Chao, L.-P. (1980) Choline acetyltransferase: purifiaction and characterization. J. Neurosci. Res. 5, 85-115.

Chao, L.-P., Kan, K.-S, K. and Hung, F.-M. (1982a) Immunohistochemical localization of choline acetyltransferase in rabbit forebrain. Brain Res. 235, 65-82.

Chao, L.-P., Kan, K.-S.K., Angelini, C. and Keesey, J. (1982b) Autoimmune neuromuscular disease induced by a preparation of choline acetyltransferase. Exp. Neurol. 75, 23-35.

Chao, L.-P. and Wolfgram, F. (1972) Spectrophotometric assay for choline acetyltransferase. Anal. Biochem. 46, 114-118.

Chao, L.-P. and Wolfgram, F. (1973) Purification and some properties of choline acetyltransferase from bovine brain. J. Neurochem. 20, 1075-1081.

Chao, L.-P. and Wolfgram, F. (1974) Activation, inhibition, and aggregation of choline acetyltransferase. J. Neurochem. 23, 697-701.

Chao, L.-P., Wolfgram, F. and Eng, L.F. (1977) "Immunohistochemical localization of choline acetyltransferase: real or artefact?" a reply. Neurochem. Res. 2, 323-325.

Cozzari, C. and Hartman, B.K. (1980) Preparation of antibodies specific to choline acetyltransferase from bovine caudate nuclei and immunohistochemical localization of the enzyme. Proc. Natl. Acad. Sci. USA 77, 7453-7457.

Crawford, G., Slemmon, J.R. and Salvaterra, P.M. (1982) Monoclonal antibodies reactive for Drosophila melanogaster choline acetyltransferase. J. Biol. Chem. 257, 3853-3856.

Dau, P.C. and Denys, E.H. (1982) Plasmapheresis and immunosuppressive drug therapy in the Eaton-Lambert Syndrome. Ann. Neurol. 11, 570-575.

Dietz, G.W. and Salvaterra, P.M. (1980) Purification and peptide mapping of rat brain choline acetyltransferase. J. Biol. Chem. 225, 10612-10617.

Drachman, D.B. (1981) The biology of myasthenia gravis. Annu. Rev. Neurosci. 4, 195-225.

Eckenstein, F., Barde, Y.-A. and Thoenen, H. (1981) Production of specific antibodies to choline acetyltransferase purified from pig brain. Neuroscience 6, 993-1000.

Ekstrom, J., Kemplay, S.K., Garrett, J.R. and Duchen, L.W. (1977) Effect of botulinium toxin on the choline acetyltransferase activity in salivary glands of cats. Experientia 33, 1458-1459.

Eng, L.F., Uyeda, C.T., Chao, L.-P. and Wolfgram, F. (1974) Antibody to bovine choline acetyltransferase and immunofluorescent localization of the enzyme in neurons. Nature 250, 243-245.

Festoff, B.W.R., Israel, R.S., Engel, W.K. and Rosenbaum, R.B. (1977) Neuromuscular blockade with anti-axoplasmic antibodies. Neurology 27, 963-970.

Fonnum, F. (1972) Application of microchemical analysis and subcellular fractionation technique to the study of neurotransmitters in discrete areas of mammalian brain. Advan. Biochem. Psychopharmacol. 6, 75-88.

Fusfeld, R.D. (1970) Electromyographic abnormalities in a case of botulism. Bull. L.A. Neurol. Soc. 35, 164-168.

Glazer, A.N. (1970) Specific chemical modification of proteins. Annu. Rev. Biochem. 39, 101-130.

Glover, V.A.S. and Potter, L.T. (1971) Purification and properties of choline acetyltransferase from ox brain striate nuclei. J. Neurochem. 18, 571-580.

Godfrey, D.A., Ross, C.D., Herrman, A.D. and Matschinsky, F.M. (1980) Distribution and derivation of cholinergic elements in the rat olfactory bulb. Neuroscience 5, 273-292.

Hattori, T., Singh, V.K., McGeer, E.G. and McGeer, P.L. (1976) Immunohistochemical localization of choline acetyltransferase containing neostriatal neurons and their relationship with dopaminergic synapses. Brain Res, 102, 164-173.

Hersh, L.B. (1980) Studies on the kinetic mechanism and salt activation of bovine choline acetyltransferase. J. Neurochem. 34, 1077-1081.

Hoover, D.B., Muth, E.A. and Jacobowitz, D.M. (1978) A mapping of the distribution of acetylcholine,choline acetyltransferase and acetylcholinesterase in discrete areas of rat brain. Brain Res. 153, 295-306.

Hsu, L.L. and Chao, L.-P. (1982) Choline acetyltransferase: further studies on the reverse reaction. J. Neurosci. Res. 7, 155-168.

Kaiya, H., Iwata, T., Namba, M., Ago, Y., Mayahara, H. and Ogawa, K. (1979) Ultracytochemical observation of choline acetyltransferase neurons in the rat neostriatum. Exp. Brain Res. 34, 395-400.

Kaiya, H., Kreutzberg, G.W. and Namba, M. (1980) Ultrastructure of acetylcholinesterase synthesizing neurons in the neostriatum. Brain Res. 187, 369-382.

Kan, K.-S.K. and Chao, L.-P. (1981) Localization of choline acetyltransferase at neuromuscular junctions. Muscle & Nerve 4, 91-93.

Kan, K.-S.K., Chao, L.-P. and Eng, L.F. (1978) Immunohistochemical localization of choline acetyltransferase in rabbit spinal cord and cerebellum. Brain Res. 146,221-229.

Kan, K.-S.K., Chao, L.-P. and Forno, L.S. (1980) Immunohistochemical localization of choline acetyltransferase in human cerebellum. Brain Res. 193,165-171.

Kao, I., Drachman, D.B. and Price, D.L. (1976) Botulinium toxin: Mechanism of presynaptic blockade. Science 193, 1256-1258.

Kasa, P. and Silver, A. (1969) The correlation between choline acetyltransferase and acetylcholinesterase activity in different areas of the cerebellum of rat and guinea pig. J. Neurochem. 16, 389-396.

Kataoka, K., Nakamura, Y. and Hassler, R. (1973) Habenulo-interpeduncular tract: a possible cholinergic neuron in rat brain. Brain Res. 62, 254-267.

Kimura, H., McGeer, P.L., Peng, F. and McGeer, E.G. (1980) Choline acetyltransferase-containing neurons in rodent brain demonstrated by immunohistochemistry. Science 208, 1057-1059.

Kimura, H., McGeen, P.L., Peng, J.H., and McGeer, E.G. (1981) The central cholinergic system studied by choline acetyltransferase immunohistochemistry in the cat. J. Comp.

Neurol. 200, 151-201.

Kuhar, M.J., Sethy, V.H., Roth, R.H. and Aghajanian, G.K. (1973) Choline: selective accumulation by central cholinergic neurons. J. Neuronchem. 20, 581-593.

Levey, A.I., Aoki, M., Fitch, F.W. and Wainer, B.H. (1981) The production of monoclonal antibodies reactive with bovine choline acetyltransferase. Brain Res. 218, 383-387.

Lewis, P.R. and Shute, C.C.D. (1978) Cholinergic pathways in CNS, in Handbook of Psychopharmacology vol. 9 (Iversen, L.L., Iversen, S.D. and Synder, S.H. eds) pp. 315-355, Plenum Press, New York.

Malthe-Soressen, D., Lea, T., Fonnum, F., and Eskeland, T. (1978) Molecular characterization of choline acetyltransferase from bovine brain caudate nucleus and some immunological properites of the highly purified enzyme. J. Neurochem. 30, 35-46.

McGeer, P.L., McGeer, E.G. Singh, V.K. and Chase, W.H. (1974) Choline acetyltransferase localization in the central nervous system by immunohistochemistry. Brain Res. 81, 373-379.

Mautner, H.G., Pakula, A.A. and Merrill, R.E. (1981) Evidence for presence of an arginine residue in the coenzyme A binding site of choline acetyltransferase. Proc. Natl. Acad. Sci. USA 78, 7449-7452.

Peng, J.-H., Ma, K. and Sung, S.-C. (1981) Purification of chicken brain choline acetyltransferase. Neurochem. Int. 3, 377-383.

Peng, J.H., McGeer, P.L., Kimura, H., Sung, S.C. and McGeer. E.G. (1980) Purification and immunochemical properties of choline acetyltransferase from human brain. Neurochem Res. 5, 943-962.

Rossier, J. (1976) Purification of rat brain choline acetyltransferase. J. Neurochem. 26, 543-548.

Rotter, A., Birdsall, N.J.M., Burgen, A.S.V., Field, P.M., Hulme, E.C. and Raisman, G. (1979) Muscarinic receptors in the central nervous system of the rat. I. Technique for autoradiographic localization of the binding of [^{3}H] propylbenzilycholine mustard and its distribution in the forebrain. Brain Res. Rev. 1, 141-165.

Ryan, R.L. (1976) Choline acetyltransferase from rat and bovine brain: Purification and some properties, Ph.D. thesis, University of Illinois; (Ryan, R.L. & McClure, W.D. 1979, Biochemistry, 18, 5357-5365).

Slemmon, J.R., Salvaterra, P.M. Crawford, G.D. and Roberts, E. (1982) Purification of choline acetyltransferase from

Drosophila melanogaster. J. Biol. Chem. 257, 3847-3852.
Sternberger, L.A. (1979) Immunocytochemistry, Second Edition. John Wiley & Sons, New York.
Storm-Mathisen, J. (1977) Localization of transmitter candidates in the brain: the hippocampal formation as a model. Prog. Neurobiol. 8, 119-181.
Su, Y.Y.T., Wu, J.-Y and Lam, D.M.K. (1980) Purification and some properties of choline acetyltransferase from catfish brain. J. Neurochem. 34, 438-445.
Woodward, W.R. Blank, N.K. and Seil, F.J. (1982) Choline acetyltransferase activity in mouse cerebellar culture. Brain Res. 241, 323-327.

NONENZYMATIC GALACTOSYLATION OF PROTEINS AND GALACTOSEMIA

J. A. Dain, J. C. Urbanowski, M. A. Cohenford, D. C. Shepard and J. B. Hitz
Department of Biochemistry and Biophysics
University of Rhode Island
Kingston, Rhode Island USA

Abstract

Hereditary galactosemia due to galactose-1-phosphate uridyl transferase deficiency produces severe illness often leading to death in the neonate. Surviving infants, if dietary galactose is not restricted, develop cataracts, jaundice and irreversible mental retardation. Presently, in patients with galactosemia, the relationship between elevated tissue concentrations of galactose and pathological complications has remained obscure. We suggested that galactosemic pathophysiology may be related to the nonenzymatic galactosylation of proteins *in vivo*.

Incubation of human serum albumin, human IgG and bovine myelin basic protein with D-(1-^{14}C)galactose (5 mM) or D-(1-^{14}C)glucose (5 mM) *in vitro* for 7 days under physiological conditions resulted in a time and concentration dependent accumulation of radioactivity into TCA precipitable material. The extent of nonenzymatic galactosylation of the proteins was more than 3-fold greater than the extent of nonenzymatic glucosylation. Galactosylated albumin was also isolated from the serum of a galactosemic infant, but not from the sera of three normal infants. Key Words: Nonenzymatic galactosylation-Galactosemia.

Address correspondence and reprint requests to Dr. Joel A. Dain, Department of Biochemistry and Biophysics, University of Rhode Island, Kingston, RI 02881, USA.

Several autosomal recessive genetic disorders which cause enzyme deficiencies in the galactose metabolic pathway can be identified in man (Levy and Hammerson, 1978).

$$\text{gal} + \text{ATP} \xrightarrow{\text{GK}} \text{gal-1-P} + \text{ADP}$$

$$\text{gal-1-P} + \text{UDP-glc} \overset{\text{GPUT}}{\rightleftharpoons} \text{glc-1-P} + \text{UDP-gal}$$

$$\text{UDP-gal} \overset{\text{E}}{\rightleftharpoons} \text{UDP-glc}$$

Fig. 1 The main galactose metabolic pathway gal = galactose, glc = glucose, GK = galactokinase, GPUT = galactose-1-phosphate uridyltransferase, E = UDP-gal 4'-epimerase.

Galactosemia, the most frequently detected of these disorders is characterized by a deficiency of GPUT (Fig. 1). This results in a biochemical phenotype that includes elevated concentrations of galactose, galactose-1-P and galactitol in tissue and body fluids (Segal, 1978). After appearing normal at birth, the clinical sequelae in neonates include vomiting, jaundice and often death due to *Escherichia coli* sepsis (Levy et al., 1977). With surviving infants, mental retardation, liver cirrhosis and cataracts occur as a chronic complication unless milk is withdrawn and the patient maintained on a galactose-free diet (Levy and Hammerson, 1978; Segal, 1978; Fishler et al., 1980). Recent long-term studies have shown that many treated patients have some type of intellectual neurological damage such as a delay in mental development, visual perceptual limitations and behavioral changes (Fishler et al., 1980). The occurrence of these milder, but significant, forms of neurological impairment should not be unexpected, because of the normal release of free galactose from the turnover of endogenous cellular and extracellular glycolipids and glycoproteins. This metabolic release of free galactose is, of course, unaffected by a strict galactose free diet.

The galactokinase defect (Fig. 1) which has not been as well studied as galactosemia is also associated with high tissue galactose concentrations, but unlike galactosemia the galactose-1-phosphate tissue concentrations are normal. However, if galactose is not restricted from the diet, early childhood cataracts are formed as well as some cases of mental retardation (Segal et al., 1979).

The epimerase defect is found only in the red blood cell where galactose and galactose-1-phosphate accumulate. It appears to be benign, apparently because galactose and galactose-1-phosphate can be adequately metabolized by other tissues.

Presently, in patients with galactosemia and also those with the galactokinase deficiency, the relationship between elevated tissue concentration of galactose and the observed pathologic complications has remained obscure. In diabetes mellitus, the relationship between high blood glucose and some of its pathophysiologic manifestations has been suggested to be directly attributable to the nonenzymatic incorporation of glucose into proteins in vivo (Cerami et al., 1979). In fact, levels of hemoglobin A_{1c}, a nonenzymatically glucosylated form of hemoglobin A, closely parallel blood glucose concentrations in patients with diabetes and its level is used to monitor long term dietary compliance (Bunn, 1981). This report will discuss evidence demonstrating that galactose is much more active than glucose in nonenzymatically glycosylating proteins under physiological conditions (Urbanowski et al., 1982 a,b,c) and proposes the hypothesis that the observed clinical symptoms in galactosemia could be caused by the altering of biological activity of functional proteins in vivo by nonenzymatic galactosylation.

Fig. 2 diagrams a proposed mechanism for the nonenzymatic galactosylation of a protein. The open chain configuration of D-galactose forms a Schiff's base with the primary amino and/or the ϵ-lysyl amino groups of the protein chain. This Schiff base can then undergo an Amadori rearrangement to form a stable ketoamine linkage.

MATERIALS AND METHODS

Materials and methods can be found in the articles by Urbanowski et al., 1982 a,b.

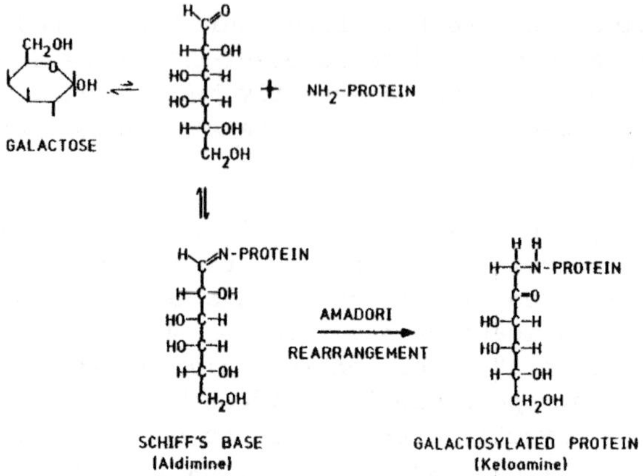

Fig. 2 Scheme for the nonenzymatic galactosylation of a protein.

RESULTS AND DISCUSSION

Nonenzymatic incorporation of D-galactose and D-glucose into protein

A time study comparing the nonenzymatic incorporation of 5 mM D-(1-^{14}C)galactose or D-(1-^{14}C)glucose under physiological conditions of temperature and pH into TCA precipitable serum proteins demonstrated both the initial velocity and the total incorporation of galactose into proteins was approximately 3-fold greater than that of glucose (data not shown). The (^{14}C)glycosyl linkages formed were not only stable to TCA precipitation, but, in addition, prolonged (48 hr) dialysis of the modified proteins against a large excess of the corresponding unlabeled monosaccharide failed to alter the amount of protein bound carbohydrate radiolabel. The galactosylated serum proteins were then separated on a Sephadex G-200 column. Most of the glycosylated proteins comigrated with the albumin (95%) and IgG (4%) fractions (Cohenford et al., 1982).

In the next study purified human serum albumin was incubated for different times (Fig. 3) under physiological conditions with D-glucose, a normal serum constituent, as well as D-galactose and D-galactose-1-phosphate, the carbohydrates abnormally present in the serum of untreated galactosemic infants.

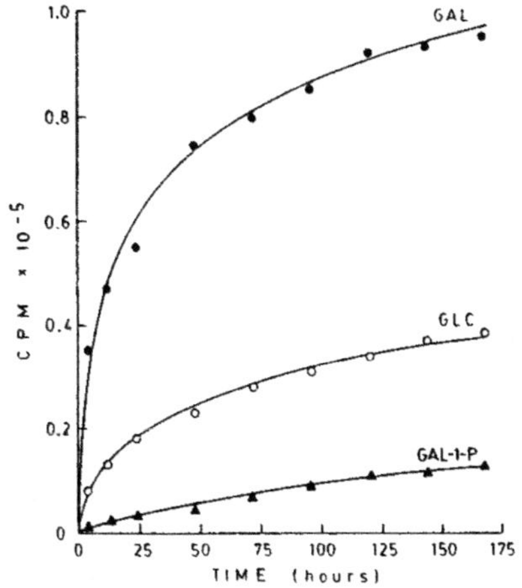

Fig. 3 In vitro incorporation of D-galactose, D-galactose-1-phosphate or D-glucose into human serum albumin. Albumin solutions (10 mg/ml) were prepared in 1 ml of Dulbecco's phosphate-buffered saline containing 1 μCi of either D-(1-^{14}C)galactose, D-(1-^{14}C)galactose-1-phosphate, or D-(1-^{14}C) glucose and 5 mM of either D-galactose, D-galactose-1-phosphate, or D-glucose, respectively. Incubations proceeded aseptically at 37°C in a shaking water bath. Aliquots were removed at indicated times, and radioactivity precipitable by trichloroacetic acid was determined as described by Urbanowski et al., 1982a. D-Galactose and D-glucose curves from Urbanowski et al., (1982a).

Again, as in the above experiment with the total serum proteins, galactose was found to nonenzymatically glycosylate human serum albumin both faster and more extensively than D-glucose under equivalent incubation conditions (Urbanowski et al., 1982). The reason for this, as has been suggested (Urbanowski et al., 1982a), may be due to the approximately 2-fold greater instability of the preferred conformation of D-galactose compared to the preferred chair conformation of D-glucose as a result of the axial orientation of the hydroxyl group on the fourth carbon of galactose. This steric hindrance would generate a higher percentage of straight chain form of the sugar which is the only form in which it can react with the protein (Fig. 2). D-galactose-1-phosphate was not effective in glycosylating albumin since it cannot form an open chain because it is locked into the ring structure by the phosphate ester at the 1 position. The minimal amount of galactosylation that is observed can be attributed to the formation of free galactose due to autolysis of the phosphate ester bond.

Recent studies in our laboratory have demonstrated that calf, rabbit and guinea pig myelin basic protein (Dain et al., 1981), as well as human IgG (Cohenford et al., 1982) are as readily nonenzymatically glycosylated by D-galactose and D-glucose as the human serum albumin. D-galactose was again the more reactive although the degree of increased reactivity varied with the particular protein. In addition, we have also found that L-fucose, sialic acid, D-mannose, N-acetylglucosamine and N-acetylgalactosamine, also products of brain glycoprotein and glycolipid degradation, will nonenzymatically glycosylate myelin basic proteins and albumin (Dain et al., 1981; McKinney et al., 1982).

The _in vitro_ albumin galactosylation studies were confirmed _in vivo_ (Urbanowski et al., 1982b) by obtaining serum from a 13-day-old boy who had been initially breast-fed without problems, but subsequently began to vomit after feedings and became jaundiced. His blood galactose was 3 mM and he had no detectable GPUT activity. His condition was diagnosed as galactosemia. Controls were serum obtained from three normal infants of the same age. The albumin from each sample was isolated and its sugar released by heating with 4N acetic acid for 5 hr at 95°. The galactosemic patient and the galactosylated control albumin released approximately 3 times as much 5-hydroxyfurfuraldehyde, derived from the ketoamine under these conditions as the control (Table 1). The enzymatic procedures, which specifically measured each sugar released, revealed the presence of galactose only from the albumins of the galactosylated control and the galactosemic child.

Table 1. Analyses of Carbohydrate Released from Albumin*

Source of Albumin	Thiobarbituric Acid Assay**	Enzymatic Assay+			
		Gal	Glc	Fru	Man
	pmol of HMF/ mg of albumin	pmol/mg of albumin			
Controls‡	510±76	0	289±33	0	98±15
Patient	1406	873	135	0	43

* From Urbanowski et al. (1982b) with permission of New Eng. J. Med.

** The color yield in the thiobarbituric acid assay, derived from the presumptive ketoamine formed from galactose, is not known. HMF denotes 5-hydroxymethylfurfuraldehyde.

\+ Gal denotes galactose, glc glucose, fru fructose, man mannose.

‡ Means ± S.E.M. in three subjects.

GLC analysis (Table 2) was used to confirm that galactose was the main carbohydrate released from both the galactosylated control and the galactosemic albumin and also to identify for the first time the presence of talose, the 2-hydroxy epimer of galactose as a hydrolysis product. The ratio of galactose to talose was approximately 3:1. A similar ratio of glucose and its 2-hydroxy epimer mannose was found in all samples confirming the finding of Bunn et al. (1975).

Our results suggest that the measurement of nonenzymatically galactosylated albumin could be a valuable sensitive clinical tool in monitoring long-term dietary compliance in either galactokinase deficient or galactosemic patients because this reaction is both linear with time and concentration dependent. The recent observation (Howard et al., 1981) that elevated levels of the glycosylated form of hemoglobin (HbA_1) in galactosemia could be reduced by dietary management, further supports the future usefulness of measurement of glycosylated proteins in the management of galactosemia and galactokinase deficiency. However, caution should be exercised because there is still "the need to establish both the time

Table 2. Identification of Carbohydrate Released from Albumin by Gas-Liquid Chromatography*

Source of Albumin	Gal	Glu	Fru	Man	Tal
		percent of total carbohydrate			
Controls‡	1.0±0.0	74±3.4	0	21±1.0	0
Patient	63	10	0	3	24

* From Urbanowski et al. (1982b) with permission of N. Eng. J. Med.

+ Gal denotes galactose, glc glucose, fru fructose, man mannose, and tal talose.

‡ Means ± S.E.M. in three subjects.

relationship between dietary indiscretion and elevation of HbA_1" (Howard et al., 1981), or we might add glycosylated albumin. It is also obvious (Urbanowski et al., 1982c), that the usefulness of this method would be greatly enhanced if a convenient method for differentiating between glucosylated and galactosylated protein can be developed.

In order to construct a biochemical relationship between high tissue concentrations of galactose and the clinical symptoms observed in galactosemia one can draw an analogy to a number of reports on the nonenzymatic incorporation of the glucose into proteins both *in vivo* and *in vitro* and its suggested contribution of this phenomena to the pathophysiology of diabetes mellitus (Urbanowski et al., 1982a,b,c). It should be assumed that in galactosemics elevated tissue concentration of galactose which is approximately three times as active as glucose would more quickly and extensively glycosylate protein and for this reason be responsible for the more devastating effects of this disease when compared to diabetes. More specifically, our results are compatible with the speculation that the nonenzymatic galactosylation of albumin may affect one of the functions of albumin which is to transfer metabolites such as bilirubin to the liver and may be the reason galactosemics are jaundiced. The nonenzymatic galactosylation of IgG may alter the immunogenicity of the IgG molecule and affect the galactosemic's ability to resist infection by *E. coli*. Lastly, galactosylation of brain proteins could affect their orientation in the nervous tissue and thus affect critical early orderly development of brain with the resultant effects on intellectual ability.

Acknowledgment: This work was supported in part by grant NS 05104 from the National Institutes of Health.

REFERENCES

Bunn, H. F. (1981) Evaluation of glycosylated hemoglobin in diabetic patients. Diabetes 30, 613-617.

Bunn, H. F., Haney, D. N., Gabbay, K. H., Gallop, P. M. (1975) Further identification of the nature and linkage of the carbohydrate in hemoglobin A_{1c}. Biochem. Biophys. Res. Commun. 67:103-109.

Cerami, A., Stevens, V. J., Monnier, V. M. (1979) Role of nonenzymatic glycosylation in the development of the sequelae of diabetes mellitus. Metabolism 28:431-437.

Cohenford, M. A., Urbanowski, J. C. and Dain, J. A. (1982) Nonenzymatic glycosylation of Human IgG. Fed. Proc. 41, 511.

Dain, J. A., Cohenford, M. A., Hitz, J. B., McKinney, R. A., Shepard, D. C., Urbanowski, J. C. (1981) Nonenzymatic glycosylation of myelin basic protein by D-galactose, N-acetylmannosamine, 2-deoxy-D-glucose, L-fucose, D-mannose, D-glucose and N-acetylneuraminic acid. Trans. Int. Soc. Neurochem. 8:151.

Fishler, K., Koch, R., Donnell, G. N., Wenz, E. (1980) Developmental aspects of galactosemia from infancy to childhood. Clin. Pediatr. 19:38-44.

Howard, N. J., Monaghan, H., Martin, J. M. (1981) Hemoglobin A_1 in galactosemia, a possible role in monitoring dietary compliance. Acta Paediatr. Scand. 70:695-711.

Levy, H. L., Hammersen, G. (1978) Newborn screening for galactosemia and other galactose metabolic defects. J. Pediatr. 92:871-877.

Levy, H. L., Sepe, S. J., Shih, V. E., Vawter, G. F., Klein, J. O. (1977) Sepsis due to *Escherichia coli* in neonates with galactosemia. N. Engl. J. Med. 297:823-825.

McKinney, R. A., Urbanowski, J. C. and Dain, J. A. (1982) Nonenzymatic glycosylation of albumin and fetuin by sialic acid. Biochem. Int. 4:127-130.

Segal, S. (1978) Disorders of galactose metabolism. In: Stanbury, J. B., Wyngaarden, J. B., Fredrickson, D. S., eds. The metabolic basis of inherited disease. New York: McGraw-Hill, pp 160-181.

Segal, S., Rutman, J. Y. and Frimpter, G. W. (1979) Galactokinase deficiency and mental retardation. J. Pediatr. 95, 750-752.

Urbanowski, J. C., Cohenford, M. A., Dain, J. A. (1982a) Nonenzymatic galactosylation of human serum albumin *in vitro* preparation. J. Biol. Chem. 257:111-115.

Urbanowski, J. C., Cohenford, M. A., Levy, H. L., Crawford, J. D., Dain, J. A. (1982b) Nonenzymatically galactosylated serum albumin in a galactosemic infant. N. Engl. J. Med. 306:84-86.

Urbanowski, J. C., Cohenford, M. A., Dain, J. A., Levy, H. L., Crawford, J. D. (1982c) Galactosylated Albumin in Galactosemia. N. Eng. J. Med. 306, 1428-1429.

CHEMICAL STUDIES ON THE PORCINE MYELIN BASIC PROTEIN

Chi-Ling Wang and Wen-Chang Chang

Institute of Biological Chemistry, Academia Sinica and Institute of Biochemical Sciences National Taiwan University, Taipei, Taiwan, R.O.C.

SUMMARY

An encephalitogenic myelin basic protein was isolated from fresh frozen pig brains by acid-acetone extraction and purified to homogeneity by CM-cellulose column chromatography and gel-filtration on a Sephacryl S-200 column. In addition to its ability to induce experimental encephalomyelitis in guinea pigs, this protein exhibited the characteristic properties of mammalian myelin basic proteins: high content of glycine and basic amino acid residues; presence of methylated arginine residues and absence of cystine or cysteine; absence of α-helix as revealed by circular dichroic studies in aqueous solution and high isoelectric point (pI>10). The molecular weight as determined by SDS-polyacrylamide gel electrophoresis was around 20,000, in contrast to 36,000 obtained by gel-filtration on a Sephadex G-100 column. The latter value suggests that the protein might exist as dimers in aqueous solution.

INTRODUCTION

The myelin basic proteins (MBP) from the central nervous systems (CNS) of bovine (Eylar et al., 1971), human (Carnegie, 1971) and rat (Martenson et al., 1972; Dunkley and Carnegie, 1974) species have been isolated and sequenced. Many of these proteins have been isolated and characterized. However, early chemical studies on the porcine basic protein have left some critical problems unanswered. Thus, in a series of studies Tomasi and Kornguth (1967; 1968) purified a basic protein from pig brain and concluded that the basic protein was a tissue-specific nucleohistone by immunohistological studies, although the chemical properties resembled those of MBP in the mammalian CNS. Later, Uyemura et al. (1970) purified a basic protein from pig spinal cord, but its chemical properties remain to be characterized. This chapter describes the isolation and purification procedures, and characterization of the porcine MBP.

MATERIAL AND METHODS

Fresh pig brains were obtained from a local slaughterhouse. CM-cellulose was from E. Merck, West Germany. Sephacryl S-200 was the product of Pharmacia, Sweden. Complete Freund's adjuvant was purchased from Difco Laboratories, U.S.A.

Extraction of the MBP

Acid acetone powder (AAP) was prepared from frozen whole brains by a procedure similar to that of Swann and Li (1979), but the precipitation step with NaCl was omitted. Briefly, the whole brain tissue, after rinsing with ice-cold water, was homogenized and added to 2 liters of acetone: hydrochloric acid:water (40:1:5, by vol) at 4°. The mixture was stirred for 1 hr, after which the insoluble material was removed by centrifugation and re-extracted with 1 liter of 80% acetone. The combined extract was poured into 10 vol of cold acetone with stirring. The mixture was kept at 4° for 2 days. The acetone was then siphoned off, the residue filtered and washed twice with acetone, and was dried under vacuum over NaOH pellets. The resultant AAP was purified as described below.

CM-cellulose column chromatography

The AAP was subjected to cation-exchange chromatography on a CM-cellulose column (3.2 x 40 cm) according to Li et al. (1975). The column was equilibrated with 0.01 M ammonium acetate (NH_4OAc) (pH 4.5) and then charged with the AAP dissolved in the starting buffer. After the unretarded fraction came off, the column was eluted with four gradient buffers produced by a mixing chamber connected to a reservoir containing successively, 0.1 M, 0.2 M, 0.5 M and 1.0 M NH_4OAc. Protein content in the eluate was monitored by measuring the absorbance at 280 nm. The protein fractions were pooled and lyophilized.

Gel-filtration and homogeneity test

The basic protein eluted from CM-cellulose column chromatography was further purified on a Sephacryl S-200 column equilibrated with 0.05 M acetic acid. The homogeneity of the basic protein was tested by disc-gel electrophoresis (Reisfeld et al., 1962) at pH 4.3 on 15% polyacrylamide gels, or on 12.5% polyacrylamide gel slabs in the presence of 0.1% sodium dodecyl sulfate (SDS) (Neville, 1971).

Assay for encephalitogenic activity

The basic protein dissolved in normal saline was emulsified with an equal volume of Freund's complete adjuvant and injected subcutaneously (0.2 ml per animal) into the hind foot pad of male guinea pigs of about 300 g body weight at a dose of 50 ug per animal. The treated animals were observed daily for development of clinical signs such as paralysis of the hind legs and death. The brain and cervical regions of the spinal cord of the control and treated animals were further examined for histological and morphological changes after fixation in 10% formalin, embedding in paraffin, sectioning and staining with hematoxylin-eosin.

Amino acid analysis

Amino acid analyses were performed according to the accelerated procedure of Spackman et al. (1958) in a Yanagimoto LC-5A amino acid analyzer. Samples were hydrolyzed with constant boiling HCl for 24 hr at 107° in evacuated sealed tubes. The hydrolysates were dried under reduced pressure before analysis. The tryptophan content was determined by hydrolysis with 4 N methanesulfonic acid at 107° for 24 hr. The methylated arginine content was determined under the same conditions against authentic standards: N^G,N^G-dimethyl-L-arginine, N^G,N'^G-dimethyl-L-arginine and N^G-monomethyl-L-arginine (Calbiochem-Behring Corp., U.S.A.).

Molecular weight determination

The molecular weight of the basic protein was determined by SDS-polyacrylamide gel electrophoresis using 12.5% polyacrylamide gel slabs (Neville, 1971). The calibration curve was established by using a LMW Calibration Kit from Pharmacia Fine Chemicals, Sweden. To detect the aggregated forms in the absence of detergent, gel-filtration on a Sephadex G-100 column (equilibrated with 0.05 M Tris, 0.5 M NaCl, 0.02% $NaNO_3$, pH 7.5) was also used for the estimation of the apparent molecular weight.

Circular dichroic spectroscopy

The circular dichroic spectra were measured in a JASCO J-20 spectropolarimeter under constant nitrogen flush. The data were expressed as mean residue ellipticity. The mean residue weight was calculated from the amino acid composition.

Isoelectric focusing

Isoelectric focusing in the pH range of 9 to 11 was carried out in a 110 ml apparatus (LKB) with synthetic carrier ampholyte, Ampholine. The bottom electrode was chosen as the anode and filled with the following electrode solution: 0.2 ml of conc. H_2SO_4 and 12 g of sucrose in 14 ml of H_2O. The column was thermostatted at 4° throughout the experiment. A 0-50% linear sucrose gradient which contained 30 mg of purified basic protein and pH 9-11 Ampholine (final concentration 1%), was then applied by means of a gradient mixer and a peristaltic pump. The remaining space at the upper electrode was filled with cathodic solution (0.1 g NaOH in 10 ml H_2O). Electrofocusing was carried out with increasing voltage (from 600 to 1200 V) until it reached a constant minimal power, which usually took about 60 hr. The content was then eluted in 2 ml fractions and analyzed for protein content by the absorbance at 280 nm. The pH values were determined with an Orion pH meter (model 601 A).

RESULTS AND DISCUSSION

Preparation and homogeneity of the porcine MBP

Typical fractionation patterns of the MBP are shown in Figs. 1 and 2. The purified material was homogeneous as tested by disc electrophoresis (Fig. 3) and SDS-polyacrylamide gel electrophoresis. Table 1 shows the yields of material at various stages of the purification. Swann and Li (1979) were the first to use the acid-acetone extraction method to isolate MBP from bovine and turkey brains. As compared to the classical procedures (see, for example, Eylar et al., 1974), it is much simpler and equally efficient. The present study again demonstrates that acid-acetone extraction can be successfully applied to obtain a crude MBP from porcine brains with good yield.

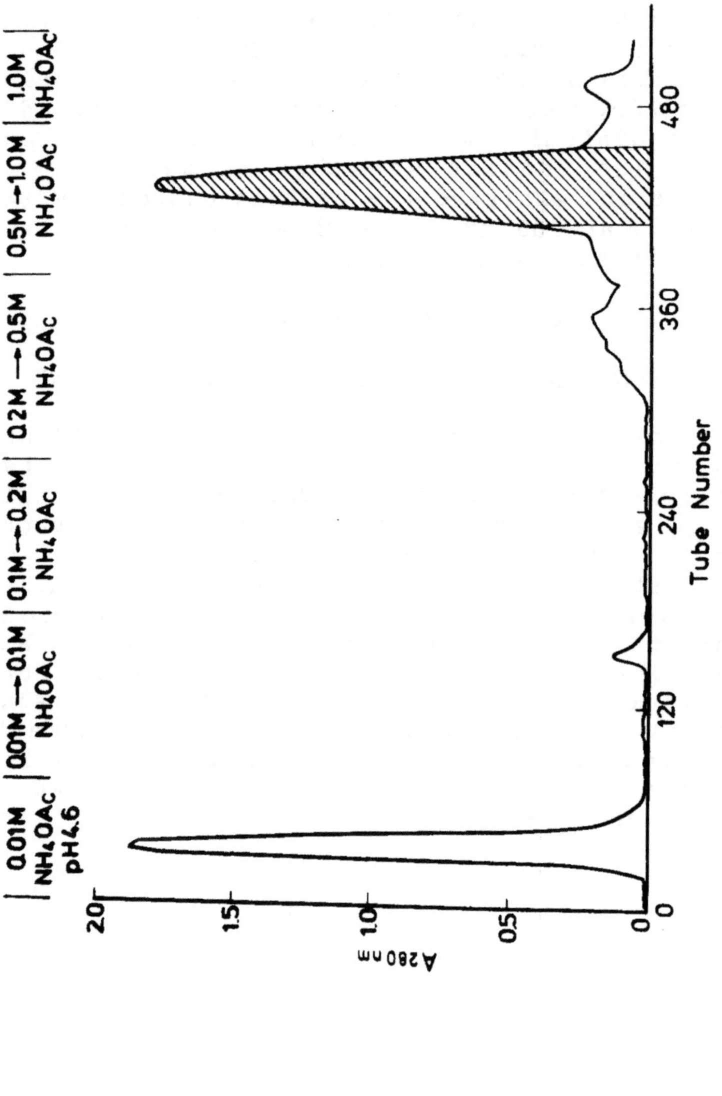

Fig. 1. CM-cellulose column chromatography of porcine AAP (2.1 g). The column (3.2 x 40 cm) was eluted as indicated at a flow rate of 4.2 ml/tube/6 min. The shaded peak was pooled and lyophilized as crude MBP.

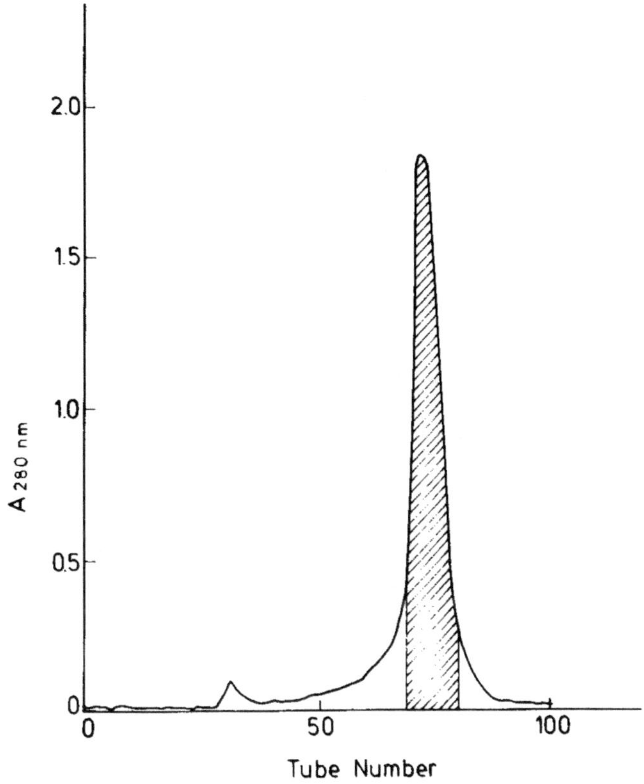

Fig. 2. Gel-filtration of the crude porcine MBP (173 mg) on a Sephacryl S-200 column (2.1 x 90 cm) equilibrated with 0.05 M acetic acid. The flow rate was 4.2 ml/tube/10 min. The shaded peak was recovered as pure MBP.

Fig. 3. Polyacrylamide gel electrophoresis of purified porcine MBP (50 ug) on 15% gel at pH 4.3. The anode was on top of the gel.

Table 1. Yields obtained at different steps in the purification of porcine MBP from 1000 g of fresh frozen brains.

Steps	Yield, g
AAP	28
CM-Cellulose	4.0
Sephacryl-S-200	3.7

Encephalitogenic activity

The purified MBP at a dose of 50 ug per 300 g body weight was able to induce typical experimental allergic encephalomyelitis in all 5 guinea pigs tested. The clinical symptoms, including paresis and paralysis of hind legs, developed 7-8 days after challenge with the MBP. All the animals died around day 11. Histological examinations of the brains and spinal cords revealed perivascular lymphocytic cuffing in all the tissues examined. Both clinical and histological examinations clearly established the encephalitogenic activity of the porcine MBP.

Amino acid composition and methylated arginine

The amino acid composition of the porcine and bovine proteins have very similar amino acid compositions (Table 2). There is a marked difference in the serine content which might be erroneous because the data for the porcine protein were not corrected for destruction. There are also gross similarities in amino acid compositions among the porcine MBP, the porcine spinal basic protein (Uyemura et al., 1970) and the porcine brain-specific histone (Tomasi and Kornguth, 1967; 1968), suggesting that they may be related to some degree (Table 2). Of particular interest is the possible relationship between the porcine MBP and the brain-specific histone because both were isolated from porcine brains, and they showed very similar amino acid compositions. However, it is not clear whether the latter is active in inducing experimental allergic encephalomyelitis.

The symmetrical dimethylarginine, N^G,N'^G-dimethyl-L-arginine, was found to be present in the porcine MBP at a concentration of 1.3% of the arginine content. This value was within the range reported for the MBP from other species (Deibler and Martenson, 1973). However, the monomethylated arginine which is present in most of the MBP so far studied (Deibler and Martenson, 1973), co-eluted with arginine on our amino acid analyzer. Therefore, its content in the porcine MBP could not be determined.

Table 2. Amino acid composition of the porcine MBP and comparison with some related proteins.

amino acid	porcine MBP[a]	porcine spinal basic protein[b]	porcine brain-specific histone[c]	bovine MBP[d]
Lys	12.8	7.40	8.1	13
His	9.3	5.41	5.9	10
Arg	16.5	9.92	9.9	18
Asp	11.6	6.62	6.2	11
Thr	7.9	4.98	4.3	7
Ser	12.6	8.78	8.3	18
Glu	9.65	7.22	6.6	10
Pro	13.9	7.84	7.8	12
Gly	27.0	13.40	15.1	25
Ala	14.1	8.28	9.2	14
1/2CySSCy	0	0	0	0
Val	2.3	2.19	1.7	3
Met	1.9	1.69	1.3	2
Ile	2.78	2.19	1.9	3
Leu	9.42	6.22	5.8	10
Tyr	3.95	2.33	2.9	4
Phe	8.0	4.47	4.9	8
Trp	0.50	0.77	not determined	1
N^G,N'^G-dimethyl-arginine	0.22	not determined	not determined	

[a]Present study. The value for each residue indicates relative molar ratio.
[b]Uyemura et al. (1970). Values are expressed as moles of amino acids per 100 moles of total amino acids recovered.
[c]Tomasi and Kornguth (1967). Values are expressed in the same way as b.
[d]Based on sequence data (Eylar et al., 1971; Brostoff et al., 1974).

Molecular weight

The molecular weight of the porcine MBP was determined as 20,000 by SDS polyacrylamide gel electrophoresis. This value is close to that of the myelin basic proteins from most other vertebrate species (for a summary of of sequence data, see Martenson, 1981). However, abnormally high apparent molecular weight of 36,000 was obtained for this porcine protein by gel-filtration on a column of Sephadex G-100 in aqueous solution, indicating that this protein may exist as dimers in the absence of detergent.

Circular dichroism and isoelectric point

The circular dichroic spectrum for porcine MBP is shown in Fig 4. Data clearly indicate that the protein is present in an unordered conformation in aqueous solution at neutral pH, in agreement with all the MBP reported so far (Eylar et al., 1974). When the protein was subjected to isoelectric focusing, the isoelectric point was found to be above 10 (Fig. 5). This is expected for a protein with such a high content of basic amino acid residues.

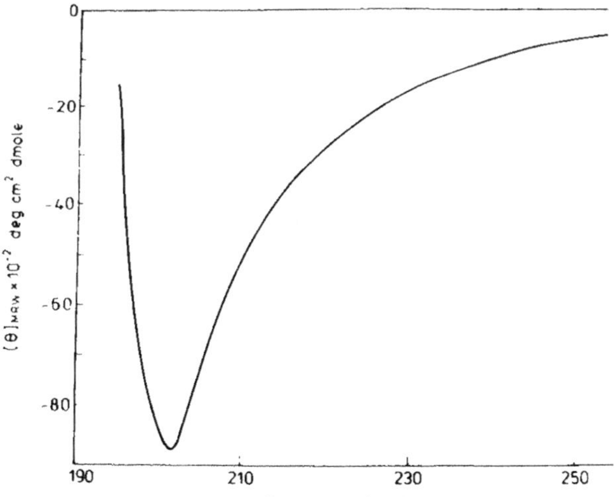

Fig. 4. The circular dichroism of porcine MBP in 0.05 M Tris buffer, pH 7.4.

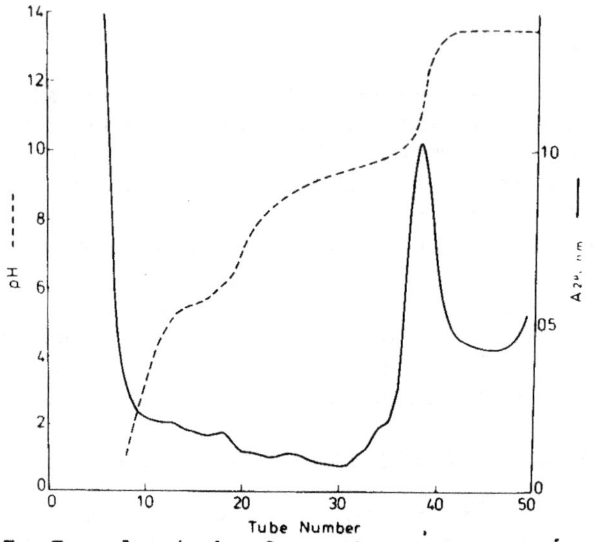

Fig. 5. Isoelectric focusing of porcine MBP (30 mg) in a pH gradient of 9-11 using a LKB column (110 ml) and Ampholine as carrier. The focusing was carried out at 4° for 72 hr.

ACKNOWLEDGMENT. We thank Mr. Ching-Ho Wang of the Provincial Research Institute for Animal Health for technical assistance in the histological studies.

REFERENCES

Brostoff, S.W., Reuter, W., Hichens, M. and Eylar, E.H. (1974) Specific cleavage of the A_1 protein from myelin with cathepsin D. J. Biol. Chem. 249, 559-567.

Carnegie, P.R. (1971) Amino acid sequence of the encephalitogenic basic protein from human myelin. Biochem. J. 123, 57-67.

Deibler, G.E. and Martenson, R.E. (1973) Determination of methylated basic amino acids with the amino acid analyzer. Application to total acid hydrolyzates of myelin basic proteins. J. Biol. Chem. 248, 2387-2391.

Dunkley, P.R. and Carnegie, P.R. (1974) Amino acid sequence of the smaller basic protein from rat brain myelin. Biochem. J. 141, 243-255.

Eylar, E.H., Brostoff, S., Hashim, G., Caccam, J. and Burnett, P. (1971) Basic A_1 protein of the myelin membrane. The complete amino acid sequence. J. Biol. Chem. 246, 5770-5784.

Eylar, E.H., Kniskern, P.J. and Jackson, J.J. (1974) Myelin basic proteins. In: Methods in Enzymology (Fleischer, S. and Packer, L., eds.) Vol. 32, pp. 323-341, Academic Press, New York.

Li, C.H., Danho, W.O., Chung, D. and Rao, A.J. (1975) Isolation, characterization, and amino acid sequence of melanotropins from camel pituitary glands. Biochemistry 14, 947-952.

Martenson, R.E., Deibler, G.E., Kies, M.W., McKneally, S.S., Shapira, R. and Kibler, R.F. (1972) Differences between the two myelin basic proteins of the rat central nervous system. A deletion in the smaller protein. Biochim. Biophys. Acta 263, 193-203.

Martenson, R.E. (1981) Prediction of the secondary structure of myelin basic protein. J. Neurochem. 36, 1543-1560.

Neville, Jr., D.M. (1971) Molecular weight determination of protein-dodecyl sulfate complexes by gel electrophoresis in a discontinuous buffer system. J. Biol. Chem. 246, 6328-6334.

Reisfeld, R.A., Lewis, U.J. and Williams, D.E. (1962) Disk electrophoresis of basic proteins and peptides on polyacrylamide gels. Nature 195, 281-283.

Spackman, D.H., Stein, W.H. and Moore, S. (1958) Automatic recording apparatus for use in the chromatography of amino acids. Anal. Chem. 30, 1190-1206.

Swann, R.W. and Li, C.H. (1979) Characterization of turkey myelin basic protein isolated by a simple procedure. Int. J. Peptide Protein Res. 14, 495-503.

Tomasi, L.G. and Kornguth, S.E. (1967) Purification and partial characterization of a basic protein from pig brain. J. Biol. Chem. 242, 4933-4938.

Tomasi, L.G. and Kornguth, S.E. (1968) Characterization and immunochemical localization of a basic protein from pig brain. J. Biol. Chem. 243, 2507-2513.

Uyemura, K., Tobari, C. and Hirano, S. (1970) Purification and properties of basic proteins in pig spinal cord and peripheral nerve. Biochim. Biophys. Acta 214, 190-197.

THE EXCITATION PROPERTIES OF NERVE MEMBRANES

Chun Chiang

Institute of Physics, Academia Sinica,

Nankang, Taipei, Taiwan, R.O.C.

Introduction

The Hodgkin-Huxley theory (1952a,b,c,d) was highly successful in the early days in explaining the change of sodium and potassium permeability of the neural membrane and enjoyed a wide acceptance. The basic ideas of the Hodgkin-Huxley theory are:

1. The sodium or potassium current depends on the driving force ($E - E_{Na}$) or ($E - E_K$), where E is the applied electric field, and E_{Na} or E_K is the equilibrium potential for sodium or potassium ions.

2. Na^+ and K^+ conductance are independent and have different voltage and time dependency.

3. Activation (m) and inactivation (h) of I_{Na} are parallel processes, and they follow an exponential time course.

4. The dependence of g_{Na} and g_K on voltage may be due to charged molecules, or the dipoles opening or closing of a pathway. Putting the above ideas into a mathematical expression, they are:

$$g_{Na} = \bar{g}_{Na}\, m^3 h,$$

$$g_K = \bar{g}_K\, n^4$$

$$\frac{dn}{dt} = (\underline{n} - n)/\tau_n \qquad (1)$$

$$\frac{dm}{dt} = (\underline{m} - m)/\tau_m,$$

$$\frac{dh}{dt} = (\underline{h} - h)/\tau_h$$

$$I = g_K\, n^4(V - V_K) + g_{Na}\, m^3 h(V - V_{Na})$$

$$+ g_L(V - V_L) \qquad (2)$$

where τ_n, τ_m, τ_h, $\underline{n}$, $\underline{m}$, $\underline{h}$ are voltage dependent empirical parameters, and g_{Na}, g_K and g_L are constants. The main characteristics for the equation are sodium activation, sodium inactivation, and potassium activation with delay. Note also that since τ_n, τ_m, τ_h, $\underline{n}$, $\underline{m}$, $\underline{h}$ are voltage dependent empirical parameters, an infinite number of parameters are probably present in the equation.

A convenient interpretation of this equation is that there are three activation particles and one inactivation particle for the sodium channel gate, and there are four activation particles for the potassium channel gate with some time delay. This formulation may satisfactorily explain the voltage clamp data (Cole, 1949; 1968) and the nerve action potential (Hodgkin-Huxley, 1952d).

Recently, it was found that some experimental data do not conform with the Hodgkin-Huxley formulation. Some of these may be listed as follows:

1. Under certain conditions, the m^3h and n^4 kinetics cannot fit the data, unless the n is raised to the 25th power (Cole and Moore, 1960). This means that there are 25 gating particles involved. It is likely that in one case there are 4 gating particles involved, whereas in another case there are 25 gating particles involved.

2. Some data indicate that the activation and inactivation are not independent as Hodgkin-Huxley originally formulated (for review, see Goldman, 1976).

3. Keynes (1975a,b) has estimated the charges of inactivation gating particles and concluded that if the inactivation gating particles exist, they should be detectable. However, separate inactivation gating current has not been observed. Thus, multi-state processes different from Hodgkin-Huxley formulation should be called for.

4. According to Hodgkin-Huxley formulation, the <u>h</u> is zero at high voltage and inactivation should have been completed. However, Chandler and Meves (1970) have shown that inactivation is not complete, and they proposed that during small depolarization, the channel is closed and during high depolarization, the channel is opened again. However, this proposal only describes the phenomenon without offering too much explanation regarding the mechanism of opening or closing.

5. Na permeability of the resting axonal membrane is found to be roughly equal to K permeability (Tasaki, 1963).

To cope with these new data, new formulations which are different from the classical one proposed by Hodgkin-Huxley are required. Although several multi-state kinetic models have been proposed, most are confronted only with restricted success (for review, see Goldman, 1976). The present paper offers a new model at the molecular level to explain some of the excitation properties of the neural membrane.

A New Model

Voltage-Clamp Currents

To formulate a new excitation model for neural membrane, we present the basic concepts as follows:

1. There are dipoles in the membrane. Hodgkin and Huxley (1952d) have suggested the possibility of dipoles in the membranes. Recently, Wei (1969a,b; 1971a,b; 1972) has also endorsed the dipole model. Howarth et al. (1968) found that heat is produced in the early stage of nerve impulse and reabsorbed during a later stage. Fraser and Frey (1968) reported electromagnetic emission at micron wavelength from active nerves. These phenomena were regarded by Wei as the evidence for the existence of dipoles in nerve excitation. The dipoles can be arranged in orientation I with energy E_1 (arrow to the right in Fig 1) or in orientation II (arrow to the left) with energy E_2 ($E_2 > E_1$). The ratio of dipoles in orientation I to orientation II in the membrane is determined by the voltage, and the dipole force is proportional to the difference between the number of dipoles in the two orientations.

2. It is further assumed that dipoles can interact with phonons (a vibration state) in the membrane matrix (Chiang, 1978; 1979; 1980). When the dipoles flip-flop from orientation II to orientation I, photons are

emitted and these photons are absorbed to be phonons in the membrane matrix. The agitation of phonons can, however, also disturb the orientation of dipoles:

$$\text{Dipoles} \xrightleftharpoons{\text{Emit photons}} \text{Phonons}$$

3. The forces acting on sodium ions or potassium ions are the externally applied voltage, the diffusional force due to concentration difference between the outside and inside of the membrane, the dipole barrier force (Wei, 1971b), and the friction force due to friction caused by the movement of ions in the membrane matrix. Note that in the Hodgkin-Huxley formulation, only applied voltage and diffusional force are utilized.

4. Newton's second law is used to describe the ion transport; namely, the change of the velocity of the ions dv/dt, or the change of the current dI/dt (since current is proportional to velocity), is proportional to the summation of the above mentioned four forces.

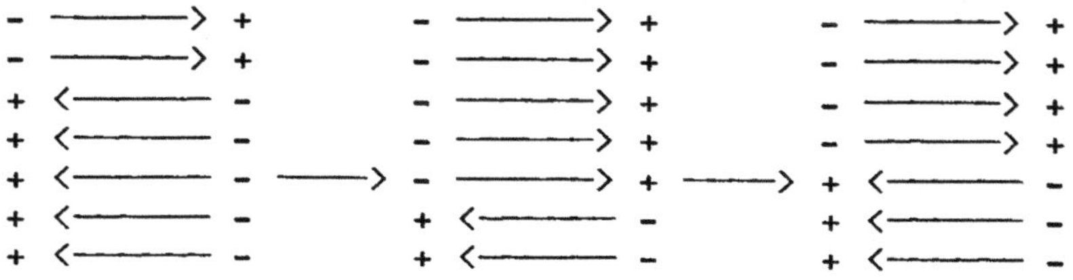

a) Normal Dipole Orientation (Resting State)

b) Over Shoot Orientation (Excitation)

c) New Equilibrium Orientation

Fig 1. A schematic diagram showing the average dipole orientation in membrane. (a) Resting state. (b) During excitation process. Too many dipoles from high energy orientation (arrow to the right) flip-flop to the low energy orientation (arrow to the left). (c) New equilibrium orientation. Some dipoles flip-flop back.

With the above concepts and using some mathematical formulation, it is possible to manipulate and derive the voltage clamp equation, the nerve action potential equation and equations for related phenomena such as time constants, gating currents, etc. Conceptually, we may understand the process as follows: During depolarization, the dipoles would flip-flop to a low energy orientation (Fig 1b). The flip-flop of dipoles from high energy to low energy would emit photons, which are partially absorbed by the membrane matrix to become phonons. These phonons in turn can disturb the dipole population to allow more dipoles to flip-flop to the low energy orientation. This is an over-shoot process; namely, more dipoles are in the low energy orientation due to the disturbance of phonons. As the phonons decay, some dipoles at the low energy orientation can flip-flop back and reach a new equilibrium population (Fig 1c). A transition number of dipoles may also be detected as in Fig 2. Since the dipole force is proportional to the number of dipoles, this diagram also shows a transient appearance of the dipole force. As for sodium ions, suppose the dashed line represents that the net force for Na^+ movement is zero (Fig 2a). Then the solid line represents the driving force for sodium ions and this approximately represents the sodium current. As for the potassium ions, since they move in an opposite direction to that for sodium ions, the dipole force for potassium ions appears as shown in Fig 2b, which is the reverse of that in Fig 2a. Suppose the dashed line represents the reference line and that the net force is zero for potassium transport. Then the solid line represents the driving force for potassium ions, and also approximately represents the potassium current. Since the response of the potassium ions is relatively slow due to its mass and friction, the initial current may appear as the circle line. Thus, potassium has a relatively large time delay. This may explain the Hodgkin-Huxley conductance curve (1952b). From this argument, the sodium activation, sodium inactivation, potassium activation, and potassium delay can all be explained coherently with no separate assumptions needed. A closer look at the same data (for example, Chandler and Meves, 1970; Goldman and Schauf, 1973; Hahin and Goldman, 1978) shows that the initial current may not start from the zero point and have an initial drop similar to Fig 2b. Usually, this initial current is regarded as the capacitance current. However, if it is capacitance current, it should be dependent on initial voltage and it should decay very fast. This seems not to be the case. Thus, there should be an initial drop of potassium current, and the friction and the mass of the ions are not great enougn to mask the initial response (drop) of the potassium current. Hodgkin-Huxley's calculation of action potential (1952d), which shows that the initial drop of voltage is smaller than observed, may also be due to their lack of consideration of the initial drop of potassium current in their equation.

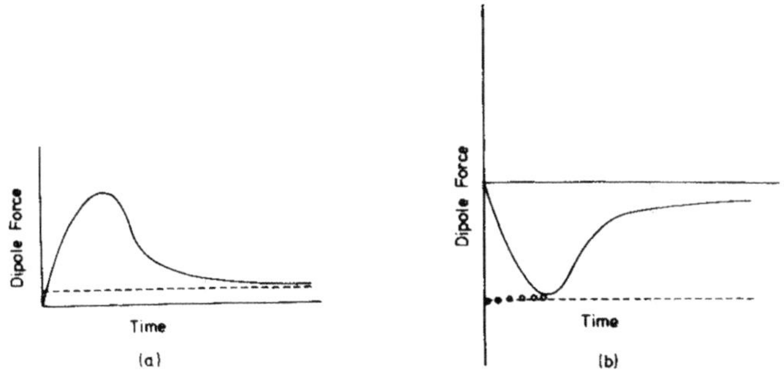

Fig 2. Change of dipole orientation or dipole force with time. (a) Transient of dipole force for Na. (b) Transient of dipole force for K. Na and K move in opposite directions. Thus, the dipole force for each ion is also opposite.

The voltage-clamp equations for sodium current and potassium current may be calculated as follows (Chiang, 1978):

$$I_{Na} = C_0 e^{-\lambda_0 t} + C_1 e^{-\lambda_1 t} + C_2 e^{-\lambda_2 t} + C_3 + C_4 V + C_5 \tanh\left(\frac{U - \frac{p}{w} V}{kT}\right) \qquad (2)$$

$$I_K = C_0' e^{-\lambda'_0 t} + C_1' e^{-\lambda_1 t} + C_2' e^{-\lambda_2 t} + C_3' + C_4' V + C_5' \tanh\left(\frac{U - \frac{p}{w} V}{kT}\right) \qquad (3)$$

where $U = (E_2 - E_1)/2$, P is dipole moment, w is width, k is Boltzmann constant, T is temperature, t is time, and the rest of the parameters are constant at a given condition. If we let $C_0 = 0.49$, $C_0' = -4.9$, $C_1' = 1.6$, $C_2 = 4.3$, $C_2' = -3.2$, $C_3 + C_4 + C_5 \tanh[(U - \frac{p}{w} V)/kT] = 0$, $C_3' + C_4' V + C_5' \tanh[(U - \frac{p}{w} V)/kT] = 1.3$, $\lambda = 4.5$, $\lambda_0' = 0.2$, $\lambda_1 = 3.0$, $\lambda_2 = 1.5$, I_{Na} and I_K can be calculated and plotted as shown in Fig 3.

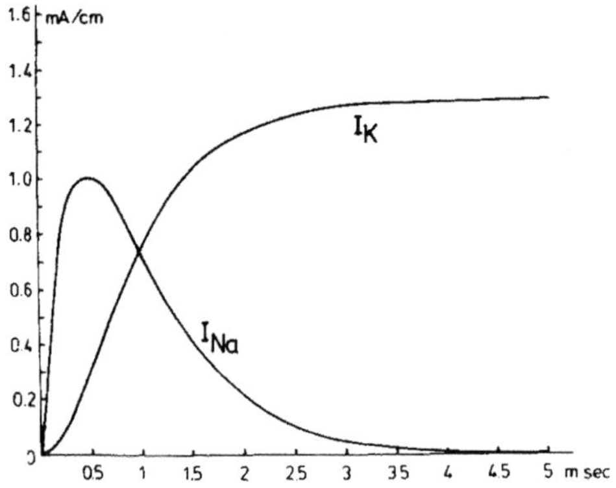

Fig 3. Na current and K current calculated according to eqs. (2) and (3).

If t is very large, then

$$I_{Na} = C_3 + C_4 V + C_5 \tanh \left(\frac{U - \frac{p}{w} V}{kT} \right) \tag{4}$$

This equation indicates that the sodium current depends on voltage. If we let U = 12 meV, kT = 23.5 meV, $\frac{p}{w}$ = −0.35 e, C_3 = 1.755 mA/cm^2, C_4 = 0.053 Ω^{-1}/cm^2, C_5 = −3.9 mA/cm^2, equation (4) is plotted in Fig 4 and shown to be in agreement with the data of Chandler and Meves (1970), whereas no assumption such as "small depolarization closes the gate and large depolarization opens the gate again" made by them is needed. In Fig 4, the lower part is dominated by dipole force, and the upper part is dominated by externally applied force.

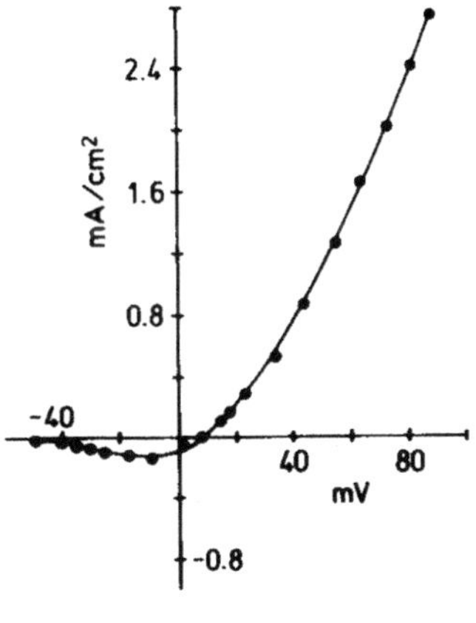

Fig 4. Voltage vs. current at equilibrium.

Nerve Action Potential

Nerve action potential is a very important phenomenon in neural physiology (Frankenhaeuser and Huxley, 1964). When the applied voltage across the membrane reaches a certain threshold value, the nerve is excited and becomes uncontrolled. The basic reason for this action potential according to this model is due to the unbalanced flow of inward sodium current and outward potassium current. The early inward flow of sodium current increases the voltage inside of membrane, whereas the delayed flow of outward potassium current tends to decrease the voltage. When the inward sodium current equals to the outward potassium current, the voltage reaches the maximum value as shown in Fig 5. Note that while the unequal flow of sodium and potassium current determines the voltage, the voltage also determines the flow of sodium and potassium current according to the principles given in the section on voltage-clamp current; i.e, acceleration of ion movements is determined by the four forces, including the voltage. The only difference is that now the voltage varies instead of being held constant.

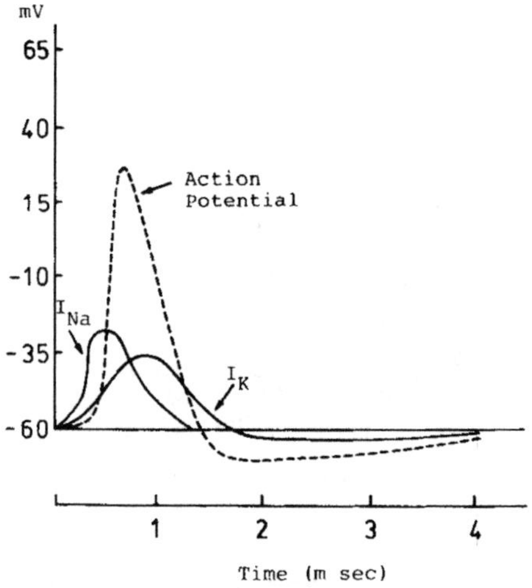

Fig 5. Schematic diagram showing the Na and K current and voltage during nerve action potential. Note that when $I_{Na} = I_K$, the voltage is maximum.

The above ideas may be put into a mathematical form, and a nerve action potential equation may be derived (Chiang, 1978):

$$V = a_0 + a_1 e^{-\alpha_1 t} + a_2 e^{-\alpha_2 t} + a_3 e^{-\alpha_3 t} + a_4 e^{-\lambda_1 ' t} + a_5 e^{-\lambda_2 ' t} \quad (5)$$

Let $a_0 = -60$, $a_1 = -250$, $a_2 = 7$, $a_3 = -7$, $a_4 = -500$, $a_5 = 750$, $\alpha_1 = 1.3$, $\alpha_2 = 0.3$, $\alpha_3 = 0.02$, $\lambda_1' = 4$, $\lambda_2' = 2$.

This nerve action potential equation is plotted in Fig 6.

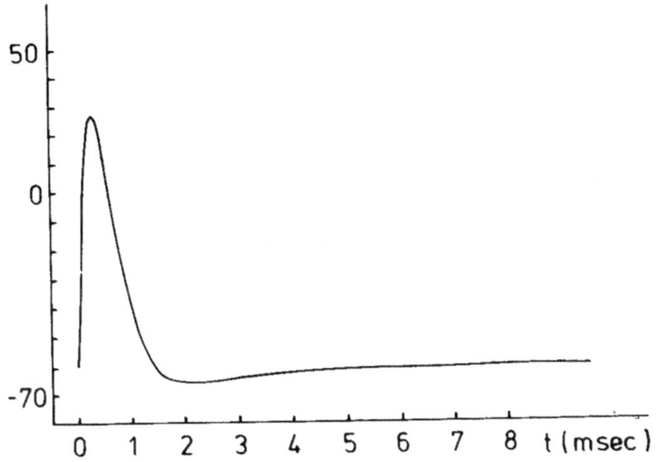

Fig 6. Nerve action potential calculated according to eq. 5.

After-Potential Oscillation

Under certain conditions, it is observed that the voltage has some small oscillations after the initial voltage impulse (Fig 7). This phenomenon may be explained as due to the "overshoot" within the construction of the present model. In approaching the resting state of the membrane after the nerve impulse, if the sodium and potassium current is decaying in a "proper" way, the voltage would behave "properly" with no oscillation, as shown in Fig 6. However, if too much potassium current is moving outward, then the nerve is hyperpolarized; if too much sodium current is moving inward, then the nerve is depolarized. This unbalanced "overshoot" and interplay of hyperpolarization and depolarization can give rise to the after-oscillation phenomenon. The mathematical condition for this oscillation phenomenon is that ${}_2$ ${}_3$ are conjugate roots of a cubic equation (Chiang, 1978). In this case, the equation for nerve action potential with after-potential oscillation is

$$V = a_0 + a_1 e^{-\alpha_1 t} + a_4 e^{-\lambda_1 ' t} + a_5 e^{-\lambda_2 ' t} + a_6 e^{-\alpha_2 ' t} \sin(\omega t + \theta) \qquad (6)$$

If we let $a_6 = 5$, $\alpha_2{}' = 0.2$, $\omega = \frac{2}{5}\pi$, $\theta = 0$, and the rest of the parameters as given previously, the equation is plotted as shown in Fig 7.

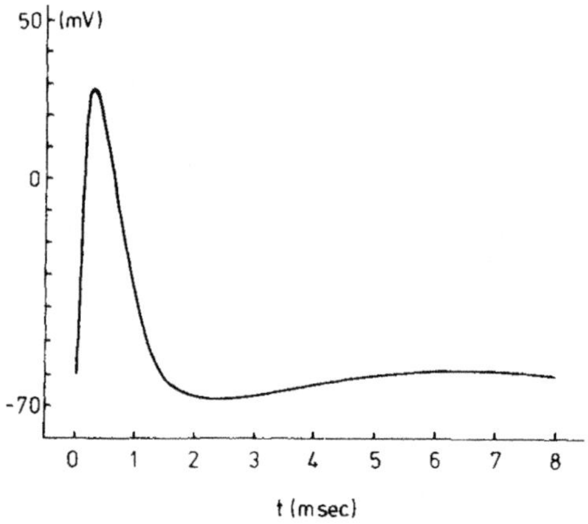

Fig 7. Nerve action potential with after-potential oscillation calculated according to eq. 6.

<u>Voltage Dependence of Time Constants</u>

The time constants $1/\lambda_{1,2}$ may be calculated to be (Chiang, 1979):

$$\frac{1}{\lambda_{1,2}} = \frac{1}{2b'}\{1 + \tau(b\beta + b')\operatorname{sech}(\frac{U - \frac{p}{w}V}{KT}) \mp$$

$$([1 + \tau(b\beta + b')\operatorname{sech}(\frac{U - \frac{p}{w}V}{KT})]^2 -$$

$$4b'\tau\operatorname{sech}(\frac{U - \frac{p}{w}V}{KT}))^{\frac{1}{2}}\} \qquad (7)$$

Let $\tau = 0.7$ msec, $\beta = 0.3$, $b' = 6$, $b = 3$, $w = 70$ Å, $U = 11$ meV, $p = -50$eÅ, $T = 273$. This equation is plotted in Fig 8. The circles are experimental data from Meves (1974). It can be seen that the agreement is good. Keynes and Rojas (1974) also developed a model for this time constant. Their equation predicts that the time constant will decay to zero at high voltage, and thus would not agree with the experiments within this range. The present model, however, has no discrepancy in this respect.

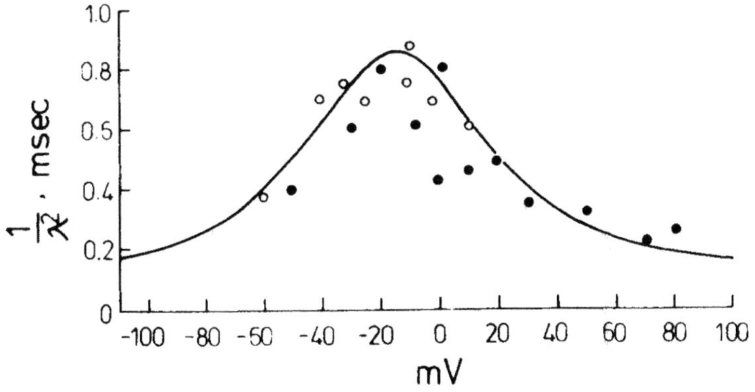

Fig 8. Time constant vs. voltage calculated according to eq. 7.

Gating Current

Hodgkin and Huxley (1952a-d) have expressed the ideas that membrane permeability change is probably due to the orientation change of the charged molecules. This orientation change should give rise to the gating current. However, the gating currents are very small and thus are very difficult to observe. Only recently, with some chemical and electrical techniques, Armstrong and Bezanilla (1973, 1974, 1975, 1977), Keynes and Rojas (1974, 1976), Keynes (1975a,b), Meves (1974) and Nonner et al. (1975) were able to observe these currents.

According to the present model, the movement of dipoles under the influence of externally applied voltage should give rise to the gating currents, and the gating current I_g should be (Chiang, 1979):

$$I_g = g_1 e^{-\lambda_1 t} + g_2 e^{-\lambda_2 t} \qquad (7)$$

This equation has two exponential components. Earlier, the observed gating currents were interpreted as having only one exponential component; however, recent experiments reveal that the gating currents can be more accurately described as the sum of more than one exponential (Goldman and Hahin, 1978; Nonner et al., 1978), in agreement with the prediction of this model.

Excitation Processes in Block Diagram

The entire excitation process involving the interplay of voltage, dipoles, phonons, Na^+,K^+, friction and diffusions may also be conveniently expressed in a block diagram (Fig 9). The arrow indicates the direction of influence, and the circles indicates the summation of all forces. Thus, the voltage can influence the dipole barrier; dipole barrier and phonons can influence each other; the value of dI_{Na}/dt or dI_K/dt will influence the instantaneous value of I_{Na} or I_K. Also the combination of voltage, the dipole barrier force, the diffusional force E_{Na} or E_K, the friction force I_{Na} or I_K will determine the value of dI_{Na}/dt of dI_K/dt. For the case of the voltage-clamp experiment, the externally applied voltage is fixed and the dI/dt or I is the observed data. For the case of action potential, the voltage is not fixed and is determined by the difference between I_{Na} and I_K. In this case, the whole process forms a closed loop and becomes autonomous. Thus, an impulse may be generated and small ripples of oscillations may also be formed under proper conditions.

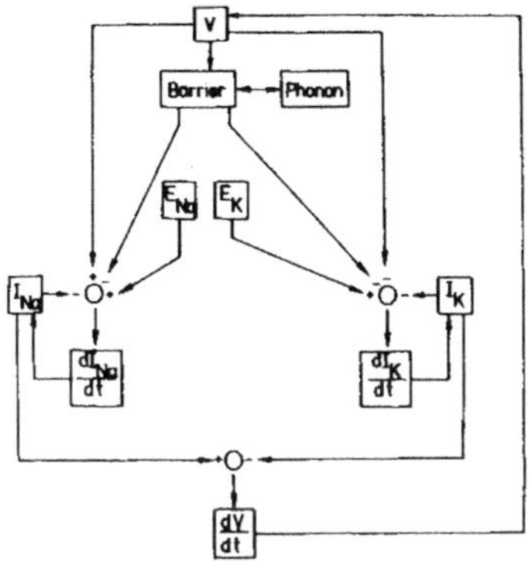

Fig 9. Block diagram showing the interplay of various factors.

In conclusion, with the assumption of coupling between the membrane dipoles and the phonons, it is possible to explain sodium activation, sodium inactivation, potassium activation with time delay, the nerve action potential, after-potential oscillation and gating currents in a coherent way. While it is possible that some polishing and modification may be needed in the future, it seems at present that this theory is able to explain a large amount of data satisfactorily.

REFERENCES

Armstrong, C.M. and Bezanilla, F. (1973) Currents related to the movement of the gating particles of the sodium channels. Nature (Lond.) 242, 459-461.

Armstrong, C.M. and Bezanilla, F. (1974) Charge movement associated with the opening and closing of the activation gates of the Na channels. J. Gen. Physiol. 63, 533-552.

Armstrong, C.M. and Bezanilla, F. (1975) Currents associated with the ionic gating structures in nerve membrane. Ann. N.Y. Acad. Sci. 264, 265-277.

Armstrong, C.M. and Bezanilla, F. (1977) Inactivation of the sodium channel. II. Gating current experiments. J. Gen. Physiol. 70, 567-590.

Chandler, W.K. and Meves, H. (1970) Sodium and potassium current in squid axons perfused with fluoride solutions. J. Physiol. 211, 623.

Chiang, C. (1978) On the nerve impulse equation: the dynamic responses of nerve. Bull. Math. Biol. 40, 247-255.

Chiang, C. (1979) On the gating current and the dipole transition. Bull. Math. Biol. 41, 841-848.

Chiang, C. (1980) A proposed dynamic process of nerve action potential. Physiol. Chem. Phys. 12, 265-269.

Cole, K.S. (1949) Dynamic electrical characteristics of the squid axon membrane. Arch. Sci. Physiol. 3, 253-258.

Cole, K.S. (1968) In: Membranes, Ions and Impulses. pp. 569, Berkeley: University of California Press.

Cole, K. and Moore, J. (1960) Potassium ion current in the squid giant axon: dynamic characteristic. Biophys. J. 1-14.

Frankenhaeuser, B. and Huxley, A.F. (1964) The action potential in the myelinated nerve fibre of *Xenopus laevis* as computed on the basis of voltage clamp data. J. Physiol. (Lond.) 171, 302-315.

Fraser, A. and Frey, A.H. (1968) Electromagnetic emission at micro wavelengths from active nerves. Biophys. J. 8, 731-734.

Goldman, L. (1976) Kinetics of channel gating in excitable membranes. Q. Rev. Biophys. 9, 491-526.

Goldman, L. and Hahin, R. (1978) Sodium tail currents in Myxicola. Biophys. J. 21, 41a.

Goldman, L. and Schauf, C.L. (1973) Quantitative description of sodium and potassium currents and computed action potentials in Myxicola giant axons. J. Gen. Physiol. 61, 361-384.

Hahin, R. and Goldman, L. (1978) Initial conditions and the kinetics of the sodium conductance in Myxicola giant axons. J. Gen. Physiol. 72, 863-877.

Hodgkin, A.L. and Huxley, A.F. (1952a) Currents carried by sodium and potassium ions through the membrane of the giant axon of Loligo. J. Physiol. 116, 449-472.

Hodgkin, A.L. and Huxley, A.F. (1952b) The components of membrane conductance in the giant axon of Loligo. J. Physiol. 116, 473-496.

Hodgkin, A.L. and Huxley, A.F. (1952c) The dual effect of membrane potential on sodium conductance in the giant axon of Loligo. J. Physiol. 116, 497-506.

Hodgkin, A.L. and Huxley, A.F. (1952d) A quantitative description of membrane current and its application to conduction and excitation in nerve. J. Physiol. 117, 500-544.

Howarth, J.V., Keynes, R.D. and Ritchie, J.M. (1968) The origin of the initial heat associated with a single impulse in mammalian non-myelinated nerve fibres. J. Physiol. (Lond.) 194, 745-793.

Keynes, R.D. (1975a) The ionic channels in excitable membranes. In: Energy Transformation in Biological Systems. Ciba Foundation Symposium 31, 191-203, Amsterdam.

Keynes, R.D. (1975b) Organization of the ionic channels in nerve membranes. In: The Nervous System. Vol. 1: Basic Neurosciences (Tower, D.B., ed.) Raven Press, New York.

Keynes, R.D. and Rojas, E. (1974) Kinetics and steady-state properties of the charged system controlling sodium conductance in the squid giant axon. J. Physiol. 239, 393-434.
Keynes, R.D. and Rojas, E. (1976) The temporal and steady-state relationships between activation of the sodium conductance and movement of the gating particles in the squid giant axon. J. Physiol. 255, 157-189.
Meves, H. (1974) The effect of holding potential on the asymmetry currents in squid axons. J. Physiol. 243, 847-867.
Nonner, W., Rojas, E. and Stampfli, R. (1975) Displacement currents in the node of ranvier. Pflugers Arch. 354, 1-18.
Nonner, W., Rojas, E. and Stampfli, R. (1978) Asymmetrical displacement currents in the membrane of frog myelinated nerve: early time course and effects of membrane potential. Pflugers Arch. 375, 75-85.
Tasaki, I. (1963) Permeability of squid axon membrane to various ions. J. Gen. Physiol. 51, 755-772.
Wei, L.Y. (1969a) Role of surface dipoles on axon membrane. Science, N.Y. 163, 280-282.
Wei, L.Y. (1969b) Molecular mechanisms of nerve excitation and conduction. Bull. Math. Biophys. 31, 39-58.
Wei, L.Y. (1971a) Possible origin of action potential and birefringence change in nerve axon. Bull. Math. Biophys. 33, 521-537.
Wei, L.Y. (1971b) Quantum theory of nerve excitation. Bull. Math. Biophys. 33, 187-194.
Wei, L.Y. (1972) Dipole theory of heat production and absorption in nerve axon. Biophys. J. 12, 1159-1170.

Part IV: Gangliosides

GANGLIOSIDES OF THE NEURONAL MEMBRANE AND THE NEURITOGENIC PHENOMENON.

Robert W. Ledeen, Mary C. Byrne, Fred J. Roisen
Glee Yorke and Joseph R. Sclafani

The Departments of Neurology and Biochemistry, Albert Einstein College of Medicine, Bronx, NY 10461 and the Department of Anatomy, UMDNJ-Rutgers Medical School, Piscataway, NJ 08854

INTRODUCTION

Gangliosides, since their discovery over four decades ago, have been proposed to subserve a variety of functions commensurate with their broad distribution among vertebrate cells and their remarkable structural diversity (Macher and Sweeley, 1978; Brunngraber, 1979; Svennerholm et al, 1980; Rapport and Gorio, 1981; Hakomori, 1981; Ledeen and Yu, 1982; Wiegandt, 1982; Makita et al, 1982; Ledeen, 1983). Although it is no longer possible to view these substances as neuronal markers, special interest still attaches to their putative role(s) in neurons owing to their unusually high concentration in the central nervous system. Thus, their concentration in gray matter exceeds that in liver by approximately 14-fold and that of muscle by approximately 60-fold (Svennerholm ėt al, 1972; Seyfried et al, 1978; Ledeen and Yu, 1982). At the same time it is recognized that great diversity exists among neurons, those of the peripheral nervous system, for example, possessing considerably lower concentrations than those of brain. Molecular patterns are also found to vary among different neuronal types.

The fact that gangliosides contribute importantly to the carbohydrate network on the surface of the neuron has suggested to many a possible role in neuronal recognition processes attendant to development and differentiation.

Reinforcing that concept have been reports from many laboratories documenting pronounced changes in ganglioside concentrations and composition of brain and retina during development. This has encompassed a variety of species including chick (Garrigan and Chargaff, 1963; Engel et al, 1979; Rosner, 1975; Dreyfus et al, 1975; Irwin and Irwin, 1979; Panzetta et al, 1980), Rana pipiens (Yiamouyiannis and Dain, 1968), rat (Pritchard and Cantin, 1962; James and Fotherby, 1963; Suzuki, 1965; Kishimoto et al, 1965; Rosenberg and Stern, 1966; Spence and Wolfe, 1967; Wells and Dittmer, 1967; Vanier et al, 1971; Tettamanti, 1971; Merat and Dickerson, 1973; Berra et al, 1978; Yusuf and Dickerson, 1978; Irwin and Irwin, 1979; Horowitz and Schanberg, 1979; Vrbaski, 1980), mouse (Kostic et al, 1969; Bosch et al, 1973; Lees et al, 1977; Baumann et al, 1979; Willinger and Schachner, 1980), rabbit (Sonnino et al, 1981), pig (Berra and Galeone-Bertona, 1966; Merat and Dickerson, 1973) and human (Svennerholm, 1964; Suzuki, 1965; Vanier et al, 1971; 1973; Yusuf et al, 1977; Martinez and Ballabriga, 1978; Cestaro et al, 1979). Comparison of four vertebrate classes in regard to altered synthetic patterns during development led to the conclusion that ganglioside biosynthesis follows Haeckel's principle of ontogeny recapitulating phylogeny (Hilbig et al, 1981).

The use of primary cultures has also proved elucidating (Schengrund and Repman, 1977; Dreyfus et al, 1980). Thus, neurons from 8-day-old chick embryo hemispheres were found to show two discrete periods of change: a first phase corresponding to cell division in which ganglioside content increased slightly, and a second period corresponding to cell maturation in which all the major CNS gangliosides accumulated (Dreyfus et al, 1980). The period of synapse formation was marked by decrease of GD3 and rapid increase of GD1a. Similar use was made of aggregating cell cultures from chick optic tectum, in which transient appearance of four new gangliosides was observed early in development (Engel et al, 1979).

Cultured cells have been employed to study neuronal differentiation in at least two additional ways: (1) observation of ganglioside changes in cells reacting to differentiation inducers (e.g., butyrate, nerve-growth factor), and (2) neurite growth induced by exogenous gangliosides. The latter phenomenon has been studied with a neuroblastoma cell line (Roisen et al, 1981a; 1981b;

Dimpfel et al, 1981; Leon et al, 1982), as well as with primary cultures (Roisen et al, 1981a; 1981b; Hauw et al, 1981; Seifert, 1981). Thus, addition of bovine brain ganglioside mixture to Neuro-2A cells in culture produced a dramatic increase in the length and number of cell processes. Other changes included an increase in cyclic AMP content and ornithine decarboxylase, a decrease in thymidine incorporation into cellular DNA and significant elongation of G_1 phase (Roisen et al, 1981a; 1981b; Dimpfel et al, 1981; Leon et al, 1982).

Such results may be contrasted with most other reports in the literature in which neurotrophic activity has been associated with peptides (Varon and Adler, 1980; 1981; Harper and Thoenen, 1980). It thus seemed imperative to determine whether the reported effect was in fact due to the gangliosides themselves or peptide contaminants. An instructive example was the recent observation of growth-promoting activity toward B104 cultures which was found to be caused by peptide contaminants in the ganglioside preparations (Seifert, 1981). This aspect deserves emphasis in view of growing evidence that a number of brain peptides are solubilized in chloroform-methanol (Brunngraber et al, 1976; Wu et al, 1980) in addition to the well-known examples of myelin proteins. Furthermore, some of these peptides are now known to copurify with gangliosides. The present study was undertaken to develop methods for removing such contaminants and to ascertain the biological activity of the resulting purified gangliosides. The fact that such preparations retained full activity reinforces the assignment of neuritogenic activity to these glycolipids.

MATERIALS AND METHODS

Materials

Various types of Sephadex (DEAE A-25, LH-60, LH-20) were purchased from Pharmacia Fine Chemicals (Piscataway, NJ). Iatrobeads silicic acid was purchased from Iatron Laboratories, Inc. (Tokyo). Florisil, 100-200 mesh, was obtained from Fisher Scientific Co. (Fairlawn, NJ). Enzymes purchased from Sigma Chemical Co. (St. Louis, MO) included: neuraminidase type VI from *C. perfringens*; pronase E, type XIV, from *Streptomyces griseus*; trypsin, type III, from bovine pancreas. Trypsin inhibitor (soybean) was obtained from GIBCO (Grand Island, NY).

Ganglioside Isolation

Gray or white matter from fresh bovine brain was extracted with chloroform (C)-methanol (M)* 1:1 and the gangliosides isolated by DEAE-Sephadex chromatography, base treatment, dialysis and silicic acid (Iatrobeads) chromatography as previously described (Ledeen et al, 1973; Ledeen and Yu, 1982). In some runs an additional step was introduced to ensure removal of sulfatides: the sample in methanol containing 0.1N HCl was warmed to 37° for 2 hrs, neutralized, evaporated to dryness at room temperature and the resulting cerebrosides separated from the gangliosides on an Iatrobeads column.

Ganglioside Purification

The gangliosides thus obtained were found to contain considerable though variable amounts of protein, and a small amount of phospholipid as well. Further purification was achieved by the following chromatographic sequence: Sephadex LH-60, Sephadex LH-20 and Iatrobeads. The LH-60 procedure was patterned after that of Bizzozero et al. (1982). Four to 6 mg ganglioside mixture in 3-5 ml of C-M 1:1 containing 5% 0.1N HCl was applied to a column of Sephadex LH-60 (19 mm O.D. x 1000 mm) prepared in the above solvent. Elution was carried out with the same solvent, 3 ml fractions being collected at approximately 0.1 ml per min. Fractions were assayed with resorcinol and fluorescamine, and the resorcinol-positive fractions were pooled. Some of the trailing fractions were rejected when fluorescamine assay was high. The pooled fractions were neutralized with dilute aq. NaOH and evaporated to dryness. The residue was dissolved in M containing 5% water (W) and applied to an 18 g column of Sephadex LH-20, previously equilibrated with the same solvent (column dimensions: 16 mm O.D. x 490 mm). Flow rate was approximately one ml per min. The pooled resorcinol-positive fractions were evaporated to dryness, dissolved in C-M-W 85:15:1 and applied to a 4 g column of Iatrobeads prepared in the same solvent. Following elution with 20 ml of the above solvent, gangliosides were eluted with 250 ml of C-M 60:40. Evaporation of solvent gave the purified gangliosides.

*Abbreviations: C, chloroform; M, methanol; W, water; BBG, bovine brain gangliosides; N-2A, neuro-2A murine neuroblastoma.

The latter chromatography was patterned after that of Matsumoto et al (1964).

Ganglioside Fractionation

The above mixture of purified gangliosides was fractionated into mono-, di-, tri- and polysialo species on DEAE-Sephadex. A step-wise gradient employing ammonium acetate in methanol was used following approximately the procedure of Skrivanek and Livermore (1981). Salt was removed by dialysis.

Neuraminidase

A disialoganglioside fraction prepared as above (ca. 300 μg sialic acid) was dissolved in 200 μl of 0.02 M acetate buffer (pH 5) and one unit of neuraminidase in 200 μl of the same buffer added in 20 μl portions over 5 hrs. The solution was overlayered with benzene from the beginning and incubated at 37°; after the last addition the reaction was permitted to continue overnight. Addition of C-M terminated the reaction, and the resulting residue was removed by centrifugation. Ganglioside in the supernatant was purified on an Iatrobeads column. It was then applied to a DEAE-Sephadex column as above and the product was eluted as a monosialoganglioside. Identity of GM1 was confirmed by TLC.

Trypsin and Pronase Treatment

A sample of bovine brain ganglioside (BBG) was dried down and dissolved in one ml of phosphate buffer, pH 7.5. Forty units of trypsin were added and the mixture incubated at 37° for 40 min. This was treated with 70 units of trypsin inhibitor followed by 40 units of pronase. The mixture was incubated at 37° for 1 hr and the reaction stopped with C-M 1:1. After centrifugation the product in the supernatant was purified as above by DEAE-Sephadex, Sephadex LH-20 and Iatrobeads. One sample was treated with pronase alone followed by acetylation-deacylation (see below).

Acetylation-Deacylation

A BBG sample purified as above and subjected to pronase digestion was further purified as follows: it was solubilized in 0.25 ml acetic anhydride + 0.25 ml pyridine, allowed to stand overnight at room temperature and then dried by evaporation with addition of toluene. The residue was dissolved in C, applied to a 5 g column of Florisil (pre-washed) and impurities eluted with the same solvent. Gangliosides were eluted with C-M 90:10. The acetylated product was saponified in 0.1 N KOH/CH_3OH and the gangliosides isolated on a Sephadex LH-20 column as above. This procedure was a modification of that described by Saito and Hakomori (1971).

Assays

Ganglioside sialic acid was assayed by the resorcinol method (Svennerholm, 1957). Protein assays were attempted by the modified Lowry procedure according to Lees and Paxman (1972). It was also determined by addition of individual amino acids following acid hydrolysis and analysis of the latter. Serine was excluded from the calculation. Fluorescamine was employed to detect substances with free amino groups (Bohlen et al, 1973).

Cell Culture

Neuro-2A murine neuroblastoma cells were grown as previously described (Roisen et al, 1981a; 1981b). Cells for assay were plated on glass cover slips in medium with or without gangliosides. The concentration of the latter was 40-50 µg sialic acid per ml. A sprouting index was calculated as the product of average process length times number of processes per cell.

RESULTS

Ganglioside Purification

Despite the extensive isolation procedure involving DEAE-Sephadex, base treatment, dialysis and silicic acid (Iatrobeads) chromatography, the resulting gangliosides contained variable amounts of protein and phospholipid contaminants. The proteins proved difficult to quantify as even use of the Lowry modification for proteolipids (Lees and Paxman, 1972) failed to give values in agreement with amino acid analysis; the latter alone was thus taken as the reliable estimate.

TABLE 1

REDUCTION IN PROTEIN CONTAMINATION DURING PURIFICATION OF BOVINE BRAIN GANGLIOSIDES (BBG).

Stage	Protein Content wt. ratio: protein/ganglioside x 100
original BBG	6.5
Sephadex LH-60	1.7
Sephadex LH-20	0.32
Iatrobeads	negligible

Protein content was quantified by summation of individual amino acids following acid hydrolysis and analysis. The summation excluded tyrosine and phenylalanine (obscured by hexosamine) as well as serine, cystine and tryptophan. Based on comparison of various lipophilic proteins, these 5 amino acids were assumed to comprise 19% of protein weight and the summation was corrected for this amount.

Table 1 indicates the reduction in protein content which occurred during a typical purification sequence. The level of protein contamination of the starting BBG was quite variable, but substantial reductions were consistently observed at each stage. Because of the difficulty in running the LH-60 columns, we found it convenient to substitute an LH-20 column, two of the latter thus being employed in a purification cycle. Both procedures were found to remove the bulk of the phospholipid contaminant, which eluted from LH-60 and LH-20 immediately after the ganglioside.

Test for Neuritogenesis

Neuro-2A cells grown for 24 hrs in the presence of purified bovine brain ganglioside mixture showed significantly enhanced neurite formation (Fig. 1). Scanning electron microscopic examination of cells treated in this manner previously revealed rapid production of microvilli which resembled dendritic spines (Roisen et al, 1981a; 1981b). These projections were not limited to the neurite processes and were found heavily concentrated on the perikaryal surface.

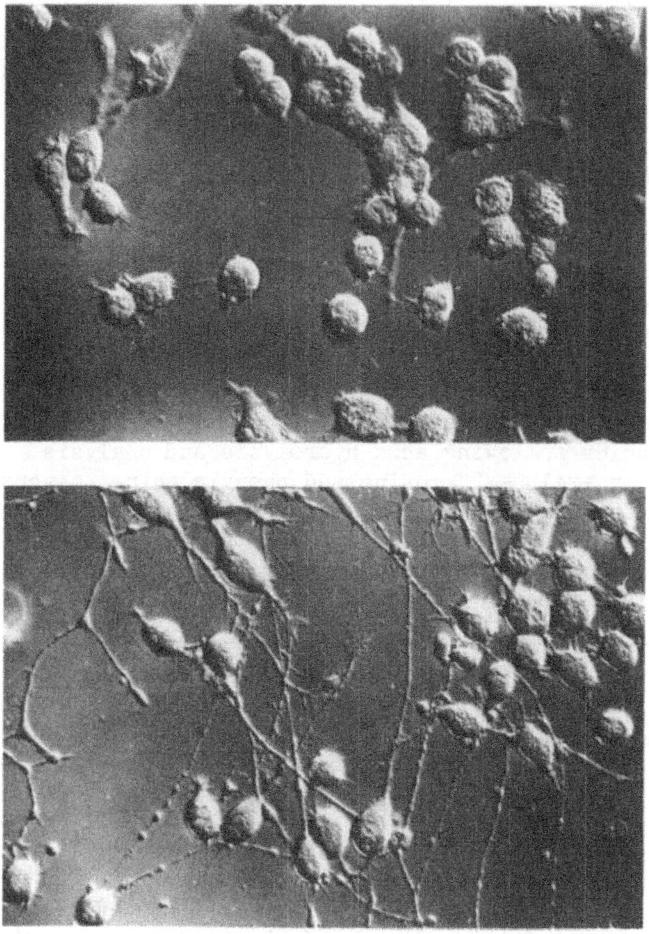

FIGURE 1. NEURITE INDUCTION BY GANGLIOSIDES IN N-2A CELLS.

Upper: Photomicrograph of cells grown for 24 hours on a glass surface in standard medium: Minimum Essential Medium (Eagle) with Hanks' balanced salt solution (90%), heat-activated fetal calf serum (10%), non-essential amino acids, and gentamicin (10 mg %). Nomarski optics, 350X.

Lower: Cells grown under the same conditions as above except that the standard medium was supplemented with 150 µg/ml bovine brain gangliosides mixture. Nomarski optics, 350X.

TABLE 2

SPROUTING INDEX OF N-2A CELLS TREATED WITH BOVINE BRAIN GANGLIOSIDES NOT SUBJECTED TO FURTHER PURIFICATION.

Source	Sprouting Index		
Gray Matter	Sample	Standard	Control
1	36	40	10
2	46	25	4
3	36	40	10
4	38	36	15
White Matter			
1	toxic	40	10
2	5	40	5
3	25	40	5
4	40	35	15

N-2A cells were plated on a glass surface, treated with standard medium alone (control) or supplemented with 150 μg ganglioside mixture. Standard consisted of a purified BBG sample from Fidia Pharmaceutical Co. (Abano Terme, Italy).

Gangliosides isolated from bovine gray matter, without extensive additional purification, gave generally consistent enhancement of the sprouting index (Table 2). When white matter was the source, however, results varied considerably ranging from strong enhancement to no sprouting or even toxic destruction. We attributed this to inhibitory factors present in the white matter which neutralized the effect of gangliosides and which copurified with the latter to a variable degree. Contaminants from laboratory reagents might also be a factor. In a few instances even gray matter gangliosides were found to be ineffective, further illustrating the necessity to exercise care in excluding harmful artifacts during isolation and purification.

Following extensive purification of BBG through the sequence Sephadex LH-60, Sephadex LH-20 and Iatrobeads, neurite inducing activity of BBG was not diminished (Table 3). When such gangliosides were further subjected

TABLE 3

SPROUTING INDEX OF N-2A CELLS TREATED WITH PURIFIED BOVINE BRAIN GANGLIOSIDES.

	Sprouting Index		
	Sample	Standard	Control
purification (3 stage)	35	38	5
trypsin + pronase	38	40	10
pronase + acetylation	40	35	15
neuraminidase + DEAE-Sephadex	22	38	9

N-2A cells were plated on a glass surface, treated with standard medium alone (control) or supplemented with 150 µg ganglioside mixture. Standard consisted of a purified BBG sample from Fidia Pharmaceutical Co. (Abano Terme, Italy).

to both trypsin and pronase, followed by purification, activity was retained. An alternate procedure, involving pronase treatment followed by acetylation, Florisil chromatography and deacetylation, also gave a product with undiminished activity. Finally, a disialoganglioside fraction isolated on DEAE-Sephadex was converted with neuraminidase to a product which was then isolated as a monosialoganglioside on DEAE-Sephadex, and the latter showed a substantial sprouting index.

DISCUSSION

The results of this study confirmed the neuritogenic effect of gangliosides on Neuro-2A cells and all but eliminated the possibility that protein contaminants which co-isolate with gangliosides were responsible for the activity. They also highlighted the difficulty in removing lipophilic proteins of this type, three chromatographic steps being required following initial isolation of BBG. This phenomenon brings to mind the report of "gangliophilic proteins" in erythrocytes which showed similar solubility properties and bound tenaciously to gangliosides (Watanabe et al, 1980).

Our results initially suggested a difference between gangliosides from bovine gray and white matter since the latter appeared, at least in first trials, to have no neurite-promoting effect. However, other white matter preparations were then tried and found to be as effective as gray matter gangliosides. We attribute this to removal of inhibitory substances. Since the cells are quite sensitive to a variety of noxious materials, we have found that care is necessary to guard against artifacts of this type; these can arise from tissue constituents co-purifying with the gangliosides or extraneous toxins introduced during workup. It was very likely this type of phenomenon which previously misled us into believing that gangliosides from normal cat brain were inactive, compared to those from feline GM1-gangliosidosis which were active (Ledeen et al, 1982); highly purified gangliosides from both types of brain are now known to be active (unpublished observations).

The mechanism by which exogenous gangliosides induce neuronal-like differentiation of neuroblastoma cells as well as primary culture of neurons is unknown. It is possible the process requires insertion into the plasma membrane, and in that regard it has been demonstrated that a portion of gangliosides in the medium is able to enter the membrane and assume true orientation within the bilayer (Callies et al, 1977; Leon et al, 1981; Radsak et al, 1982). These appear to function as native gangliosides (Moss et al, 1976). It has been pointed out that when N-2A and other cells are caused to differentiate by serum withdrawal or dibutyryl cyclic AMP, the ganglioside content of the cells increases along with a change in pattern (Rebel et al, 1980). In similar fashion, the striking morphological differentiation some epithelioid cells undergo when treated with butyrate is accompanied by elevation of GM3 and the sialosyltransferase responsible for its synthesis (Simmons et al, 1975; Taki et al, 1979). It is well to emphasize that some neuronal cell types do not show neurite stimulation with exogenous gangliosides (Roisen et al, 1981a; 1981b).

A prospective role for gangliosides in neurite growth within the developing brain was earlier suggested by the developmental studies referred to above. Another highly relevant finding was the discovery of aberrant secondary neurites in pyramidal cells of brain afflicted with ganglioside storage disease (Purpura and Baker, 1978; Walkley et al, 1981). Studies employing lectin labeling of sprouting neurons suggested that gangliosides may be the predominant glycoconjugate in the growing neurite (Pfenninger and Maylie-Pfenninger, 1978). It is of some interest that in vivo studies have suggested facilitation of nerve regeneration by exogenous gangliosides (Ceccarelli et al, 1976; Caccia et al, 1979; Gorio et al, 1980; Norido et al, 1981; Norido et al, 1982). More work is required to determine whether a connection exists between ganglioside-induced neuritogenesis at the cellular level, as described in this report, and nerve growth facilitation by these substances at a much higher level of orientation.

ACKNOWLEDGEMENTS

This study was supported by PHS grants NS 03356, NS 04834 and NS 11299.

REFERENCES

Baumann N., Sonnino S., Ghidoni R., Harpin M.L., and Tettamanti G. (1979) Developmental profile of a new ganglioside, carrying 9-O-acetyl, N-acetyl neuraminic acid, in the brain of normal mice and dysmyelinating quaking mutants, in A Multi-disciplinary Approach to Brain Development (Di Benedetta C., Balazs R., Gombos G., and Porcellati G., eds.), pp. 51-56. Elsevier/ North Holland, Amsterdam.

Berra B., Cestaro B., Omodeo-Sale' F., Venerando B., Beltrame D., and Cantone A. (1978) Gangliosides and neuraminidases in foetal rat brain. Bull. Mol. Biol. Med. 3, 86-97.

Berra B. and Galeone-Bertona L. (1966) I gangliosidi del cervello di maiale durante lo sviluppo. Bioch. Biol. Sper. 5, 466-470.

Bizzozero O., Besio-Moreno M., Pasquini J.M., Soto E.F. and Gómez C.J. (1982) Rapid purification of proteolipids from rat brain subcellular fractions by chromatography on a lipophilic dextran gel. J. Chromatography 227, 33-44.

Bosch M.A., Rebel G., Sarlieve L.L., and Neskovic N.M. (1973) Brain gangliosides of myelin synthesis-deficient mice. Lipids 8, 436-438.

Brunngraber E.G. (1979) Neurochemistry of Aminosugars, Charles C. Thomas, Illinois.

Brunngraber E.G., Tettamanti G., and Berra B. (1976) Extraction and analysis of materials containing lipid-bound sialic acid, in Glycolipid Methodology (Witting L.A., ed), pp. 159-186. American Oil Chemists' Society, Illinois.

Caccia M.R., Meola G., Cerri C., Frattola L., Scarlato G. and Aporti F. (1979) Treatment of denervated muscle by gangliosides. Muscle & Nerve 2, 382-389.

Callies R., Schwarzmann G., Radsak K., Siegert R. and Wiegandt H. (1977) Characterization of the cellular binding of exogenous gangliosides. Eur. J. Biochem. 80, 425-432.

Ceccarelli B., Aporti F., and Finesso M. (1976) Effects of brain gangliosides on functional recovery in experimental regeneration and reinnervation, in Ganglioside Function - Biochemical and Pharmacological Implications (Porcellati G., Ceccarelli B., and Tettamanti G., eds.) pp. 275-293, Plenum Press, New York and London.

Cestaro B., Cervato G., Lindi C., Venerando B., and Berra B. (1979) Developmental changes of particulate neuraminidase and ganglioside content in human brain. Bull. Mol. Biol. Med. 4, 205-213.

Dimpfel W., Möller W., Mengs U. (1981) Ganglioside-induced neurite formation in cultured neuroblastoma cells, in Gangliosides in Neurological and Neuromuscular Function, Development, and Repair (Rapport M.M. and Gorio A., eds.) pp. 119-134, Raven Press, New York.

Dreyfus H., Lonis J.C., Harth S., Mandel P. (1980) Gangliosides in cultured neurons. Neuroscience 5, 1647-1655.

Dreyfus H., Urban P.F., Edel-Harth S., and Mandel P. (1975) Developmental patterns of gangliosides and of phospholipids in chick retina and brain. J. Neurochem. 25, 245-250.

Engel E.L., Wood J.G. and Byrd F.I. (1979) Ganglioside patterns and cholera toxin-peroxidase labeling of aggregating cells from the chick optic tectum. J. Neurobiol. 10, 429-440.

Garrigan O.W. and Chargaff E. (1963) Studies on the fucolipids and the cerebrosides of chicken brain during embryonic development. Biochim. Biophys. Acta 70, 452-464.

Gorio A., Carmignoto G., Facci L. and Finesso M. (1980) Motor nerve sprouting induced by ganglioside treatment. Possible implications for gangliosides on neuronal growth. Brain Res. 197, 236-241.

Hakomori S.-I. (1981) Glycosphingolipids in cellular interaction, differentiation, and oncogenesis, in Annual Review of Biochemistry, Vol. 50 (Snell E.E., Boyer P.D., Meister A. and Richardson C.C., eds.), pp. 733-764. Annual Reviews, California.

Harper G.P. and Thoenen H. (1980) Nerve growth factor: Biological significance, measurement, and distribution. J. Neurochem. 34, 5-16.

Hauw J.J., Fenelon S., Boutry J.-M., Nagai Y., and Escourolle R. (1981) Effects of brain gangliosides on neurite growth in guinea pig spinal ganglia tissue cultures and on fibroblast cell cultures, in Gangliosides in Neurological and Neuromuscular Function, Development, and Repair (Rapport M.M. and Gorio A., eds.) pp. 171-175, Raven Press, New York.

Hilbig R., Rösner H. and Rahmann H. (1981) Phylogenetic recapitulation of brain ganglioside composition during ontogenetic development. Comp. Biochem. Physiol. 68B, 301-305.

Horowitz A.J. and Schanberg S.M. (1979) Hormonal effects on the development of rat brain gangliosides - I. Cortisol. Biochem. Pharmacol. 28, 881-895.

Irwin L.N. and Irwin C.C. (1979) Developmental changes in ganglioside composition of hippocampus, retina, and optic tectum. Develop. Neurosci. 2, 129-138.

James F. and Fotherby K. (1963) Distribution in brain of lipid-bound sialic acid and factors affecting its concentration. J. Neurochem. 10, 587-592.

Kishimoto Y., Davies W.E., and Radin N.S. (1965) Turnover of the fatty acids of rat brain gangliosides, glycerophosphatides, cerebrosides, and sulfatides as a function of age. J. Lipid Res. 6, 525-531.

Kostic D., Nussbaum J.L., and Mandel P. (1969) A study of brain gangliosides in "Jimpy" mutant mice. Life Sci. 8, 1135-1143.

Ledeen R. (1983) Gangliosides, in Handbook of Neurochemistry, 2nd ed, Vol. 3 (Lajtha A., ed.) in press. Plenum Press, New York.

Ledeen R.W., Roisen F.J., Byrne M.C., Yorke G., and Sclafani J.R. (1982) Enhancement of neurite development by gangliosides in vitro, in Transactions of the American Society for Neurochemistry, Vol. 13, pp. 130, Abstract.

Ledeen R.W. and Yu R.K. (1982) Gangliosides: structure, isolation, and analysis, in Methods in Enzymology, Vol. 83 (Ginsburg V., ed.), pp 139-191. Academic Press, New York.

Ledeen R.W., Yu R.K., and Eng L.F. (1973) Gangliosides of human myelin: sialosylgalactosylceramide (G_7) as a major component. J. Neurochem. 21, 829-839.

Lees M., and Paxman S. (1972) Modification of the Lowry procedure for the analysis of proteolipid protein. Anal. Biochem. 47, 184-192.

Lees A., Sarlieve L.L., Meskovic N.M., Wintzerith M. and Mandel P. (1977) Changes in brain components during the development of mice homozygous for the locus "DWARF" (dw). Neurochem. Res. 2, 11-25.

Leon A., Facci L., Toffano G., Sonnino S. and Tettamanti G. (1981) Activation of (Na^+,K^+)-ATPase by nanomolar concentrations of G_{M1} ganglioside. J. Neurochem. 37, 350-357.

Leon A., Facci L., Benvegnù D. and Toffano G. (1982) Morphological and biochemical effects of gangliosides in neuroblastoma cells. Develop. Neurosci. 5, 108-114.

Macher B.A. and Sweeley C.C. (1978) Glycosphingolipids: structure, biological source, and properties, in Methods in Enzymology, Vol. L (Ginsburg V., ed.), pp. 236-250. Academic Press, New York.

Makita A., Handa S., Taketomi T., and Nagai Y., eds. (1982) Advances in Experimental Medicine and Biology, Vol. 152 New Vistas in Glycolipid Research. Raven Press, New York.

Martinez M. and Ballabriga A. (1978) A chemical study on the development of the human forebrain and cerebellum during the brain "growth spurt" period. I. Gangliosides and plasmalogens. Brain Res. 159, 351-362.

Matsumoto M., Matsumoto R. and Folch-Pi J. (1964) The chromatographic fractionation of brain white matter proteolipids. J. Neurochem. 11, 829-838.

Merat A. and Dickerson J.W.T. (1973) The effect of development on the gangliosides of rat and pig brain. J. Neurochem. 20, 873-880.

Moss J., Fishman P.H., Manganiello V.C., Vaughan M. and Brady R.O. (1976) Functional incorporation of ganglioside into intact cells: Induction of choleragen responsiveness. Proc. Nat. Acad. Sci. USA 73, 1034-1037.

Norido F., Canella R. and Aporti F. (1981) Acceleration of nerve regeneration by gangliosides estimated by the somatosensory evoked potentials (SEP). Experientia 37, 301-302.

Norido F., Canella R. and Gorio A. (1982) Ganglioside treatment of neuropathy in diabetic mice. Muscle & Nerve 5, 107-110.

Panzetta P., Maccioni H.J.F., and Caputto R. (1980) Synthesis of retinal gangliosides during chick embryonic development. J. Neurochem. 35, 100-108.

Pfenninger K.H. and Maylie-Pfenninger M.-F. (1978) Characterization, distribution, and appearance of surface carbohydrates on growing neurites, in Neuronal Information Transfer (Karlin A., Vogen H.J. and Tennyson, V.M., eds.), pp. 373-386. Academic Press, New York.

Pritchard E.T. and Cantin P.L. (1962) Gangliosides in maturing rat brain. Nature 193, 580-581.

Purpura D.P. and Baker H.J. (1978) Meganeurites and other aberrant processes of neurons in feline G_{M1}-gangliosidosis: A Golgi study. Brain Res. 143, 13-26.

Radsak K., Schwarzmann G. and Wiegandt H. (1982) Studies on the cell association of exogenously added sialoglycolipids. Hoppe-Seyler's Z. Physiol. Chem. 363, 263-272.

Rapport M.M. and Gorio A., eds. (1981) Gangliosides in Neurological and Neuromuscular Function, Development, and Repair, Raven Press, New York.

Rebel G., Robert J. and Mandel P. (1980) Glycolipids and cell differentiation, in Structure and Function of Gangliosides (Svennerholm L., Mandel P., Dreyfus H. and Urban P.-F., eds.) pp. 159-166, Plenum Press, New York and London.

Roisen F.J., Bartfeld H., Nagele R. and Yorke G. (1981a) Ganglioside stimulation of axonal sprouting in vitro. Science 214, 577-578.

Roisen F.J., Bartfeld H., Rapport M.M. (1981b) Ganglioside mediation of in vitro neuronal maturation, in Gangliosides in Neurological and Neuromuscular Function, Development, and Repair (Rapport M.M. and Gorio A., eds.) pp. 135-150, Raven Press, New York.

Rosenberg A. and Stern N. (1966) Changes in sphingosine and fatty acid components of gangliosides in developing rat and human brain. J. Lipid Res. 7, 122-131.

Rösner H. (1975) Changes in the contents of gangliosides and glycoproteins and in the ganglioside pattern of the chicken brain. J. Neurochem. 24, 215-216.

Schengrund C.-L. and Repman M.A. (1977) Cell culture of sixteen-day-old rat embryo cerebra and associated changes in ganglioside pattern. J. Neurochem. 29, 923-927.

Seifert W. (1981) Gangliosides in nerve cell cultures, in Gangliosides in Neurological and Neuromuscular Function, Development, and Repair (Rapport M.M. and Gorio A., eds.) pp. 99-117, Raven Press, New York.

Seyfried T.N., Ando S. and Yu R.K. (1978) Isolation and characterization of human liver hematoside. J. Lipid Res. 19, 538-543.

Simmons J.L., Fishman P.H., Freese E. and Brady R.O. (1975) Morphological alterations and ganglioside sialyltransferase activity induced by small fatty acids in HeLa cells. J. Cell Biol. 66, 414-424.

Skrivanek J.A. and Livermore G.H. (1981) Separation of major ganglioside species using DEAE-Sephadex, in Transactions of the American Society for Neurochemistry, Vol. 12, p. 236. Abstract.

Sonnino S., Ghidoni R., Masserini M., Aporti F. and Tettamanti G. (1981) Changes in rabbit brain cytosolic and membrane-bound gangliosides during prenatal life. J. Neurochem. 36, 227-232.

Spence M.W. and Wolfe L.S. (1967) Gangliosides in developing rat brain. Can. J. Biochem. 45, 671-688.

Suzuki K. (1965) The pattern of mammalian brain gangliosides. III. Regional and developmental differences. J. Neurochem. 12, 969-979.

Svennerholm L. (1957) Quantitative estimation of sialic acids. II. Colorimetric resorcinol-hydrochloric acid method. Biochim. Biophys. Acta 24, 604-611.

Svennerholm L. (1964) The distribution of lipids in the human nervous system. I. Analytical procedure: Lipids of foetal and newborn brain. J. Neurochem. 11, 839-853.

Svennerholm L., Bruce A., Mansson J.-E., Rynmark B.-M. and Vanier M.-T. (1972) Sphingolipids of human skeletal muscle. Biochim. Biophys. Acta 280, 626-636.

Svennerholm L., Mandel P., Dreyfus H. and Urban P.-F., eds. (1980) Advances in Experimental Medicine and Biology, Vol. 125, Structure and Function of Gangliosides. Plenum Press, New York.

Taki T., Hirabayashi Y., Kondo R., Matsumoto M. and Kojima K. (1979) Effect of butyrate on glycolipid metabolism of two cell types of rat ascites hepatomas with different ganglioside biosynthesis. J. Biochem. (Tokyo) 86, 1395-1402.

Tettamanti G. (1971) Brain gangliosides in development, in Chemistry and Brain Development (Paoletti R. and Davison A.N., eds.), pp. 75-89, Plenum Press, New York.

Vanier M.T., Holm M., Öhman R. and Svennerholm L. (1971) Developmental profiles of gangliosides in human and rat brain. J. Neurochem. 18, 581-592.

Vanier M.T., Holm M., Mansson J.E. and Svennerholm L. (1973) The distribution of lipids in the human nervous system. V. Gangliosides and allied neutral glycolipids of infant brain. J. Neurochem. 21, 1375-1384.

Varon S. and Adler R. (1980) New growth factors and control of nerve growth. Curr. Top. Dev. Biol. 16, 207-251.

Varon S. and Adler R. (1981) Trophic and specifying factors directed to neuronal cells in Advances in Cellular Neurobiology, Vol. 2 (Fedorff S., ed.), pp. 115-163, Academic Press, New York.

Vrbaški S.R. (1980) Cerebellum gangliosides in postnatal rats. J. Neurochem. 34, 1032-1033.

Walkley S.U., Wurzelmann S. and Purpura D.P. (1981) Ultrastructure of neurites and meganeurites of cortical pyramidal neurons in feline gangliosidosis as revealed by the combined Golgi-EM technique. Brain Res. 211, 393-398.

Watanabe K., Hakomori S., Powell M.E. and Yokota M. (1980) The amphipathic membrane proteins associated with gangliosides: The Paul-Bunnell antigen is one of the gangliophilic proteins. Biochem. Biophys. Res. Commun. 92, 638-646.

Wells M.A. and Dittmer J.C. (1967) A comprehensive study of the postnatal changes on the concentration of the lipids of developing rat brain. Biochemistry 6, 3169-3195.

Wiegandt H. (1982) The gangliosides, in Advances in Neurochemistry, Vol. 4, (Agranoff B.W., and Aprison M.H., eds.), pp. 149-231. Plenum Press, New York.

Willinger M. and Schachner M. (1980) GM1 ganglioside as a marker for neuronal differentiation in mouse cerebellum. Devel. Biol. 74, 101-117.

Wu P.-S., Ledeen R.W., Udem S. and Isaacson Y.A. (1980) Nature of the sendai virus receptor: Glycoprotein versus ganglioside. J. Virology 33, 304-310.

Yiamouyiannis J.A. and Dain J.A. (1968) The appearance of ganglioside during embryological development of the frog. J. Neurochem. 15, 673-676.

Yusuf H.K.M. and Dickerson J.W.T. (1978) Disialoganglioside G_{D1a} of rat brain subcellular particles during development. Biochem. J. 174, 655-657.

Yusuf H.K.M., Merat A. and Dickerson J.W.T. (1977) Effect of development on the gangliosides of human brain. J. Neurochem. 28, 1299-1304.

NEURONAL DEVELOPMENT AND REGENERATION: A ROLE FOR GANGLI-OSIDES

G. Toffano, D. Benvegnù, R. Dal Toso, L. Facci and A. Leon
Fidia Research Laboratories, Dept. of Biochemistry
Via Ponte della Fabbrica 3/A, 35031 Abano Terme, Italy

INTRODUCTION

Until recently, the nervous system was thought of as a highly static and immutable system which could be "wired" only during development and, once formed, the only changes in its circuitry were due to neuronal death. Such a concept however posed serious limitations for comprehension of phenomena such as behavioural plasticity of the mature nervous system and of functional recovery following lesions. Currently, it has been well demonstrated that neuronal circuitry is highly adaptable, i.e. capable of remodeling or establishing new connections, not only development, but also in the mature central nervous system in response to perturbations such as lesions (Bjorklund and Stenevi, 1979; Cotman el al., 1981). Indeed basic neurobiologic research has increasingly focused its attention on the study of factors which determine neuronal development and regeneration. Studies with neuronal tissue culture techniques, which allow controlled manipulation of neuronal microenvironment, have indeed clarified that normal neuronal ontogeny (including neuroblast migration to its specific localization, acquisition of a permanent postmitotic status, transmitter choice and activity, neurite outgrowth, elongation, selection and functional target innervation) is highly dependent on the occurrence of an orderly sequence of extrinsic neurotrophic, neurite-promoting, guidance and inhibitory signals (Varon et al., 1982). These signals may derive directly from the humoral environment or be associated with the surrounding and target cell surfaces or extracellular matrices and apparently operate within a determined period of time with remarkable precision. Similarly, CNS regeneration is or is not successful depending on the availability or absence of adequate environmental influences (Benfey and Aguayo, 1982). Starting from these premises, a great importance has been put on identification and isolation of the proposed extrinsic factors. The comprehension of the cellular and molecular mechanisms through which the nervous system develops and regulates its own response to normal and pathological situations is likely to progress parallelly with the success of such a study.

Factors Affecting Neuronal Development and Regeneration

At present, the only well-characterized macromolecular factor which controls neuronal performances is nerve growth factor (NGF). In this context, it has been demonstrated that both normal developing and adult deafferented irisés are capable of synthesizing NGF (Ebendal et al., 1980). Similar environmental cues may be operative both during neuronal development and regeneration. Hence denervation presumably reveals an extreme expression of inherent latent properties of neuronal connections (Cotman et al., 1981).

However, functional responsiveness of neurons to neurotrophic factors ultimately depends on their intrinsic capabilities of regulating recognition and transduction of the signal. Thus the plasma membranes and their components play an active fundamental role in regulating neuritogenesis, repair, and presumably maintainance of correct intercellular relationships. Consequently, in order to comprehend the mechanism regulating the above processes, it is not only necessary to characterize the existing extrinsic cues biochemically, but also to identify and understand the functional role of intrinsic neuronal components in the regulation of cell surface responses.

Gangliosides in Neuronal Development and Regeneration

It has in recent years become increasingly evident

that gangliosides contribute significantly to the regulation of surface-related cellular activities of both developing and mature neurons. Such molecules are normal membrane constituents preferentially localized in the outer leaflet of the plasma membrane and particularly abundant in the nervous system (Ledeen, 1978). They are suggested as the predominant form of glycoconjugate in the growing neurite (Pfenninger and Maylie-Pfenninger, 1978), while their content and complexity changes in various neuronal developmental stages (Dreyfus et al., 1981; Willinger and Schachner, 1980) and in neuropathological conditions in the adult (Purpura et al., 1977; Bass, 1981). The crucial problem is understanding how the membrane processes inducing the participation of ganglioside components are associated with acquisition and maintenance of proper neuronal activity. A role for gangliosides in regulating membrane processes is most probably correlated with the physico-chemical characteristics of the molecule per se and its asymmetric localization in the outer membrane. Although their exact organization into the membrane is still unknown - a large proportion of the gangliosides are now known to occur within cryptic areas along the cell surface - their ceramide moiety presumably anchors the molecule into the outer leaflet of the lipid bilayer and as such may serve to regulate the structural rigidity of the surface membrane (Hakomori, 1981). In contrast, the oligosaccharide portion, characterized by the presence of one or more sialic acid residues, protrudes towards the outer environment thereby permitting their interaction with exogenous ligands, defining also their extrinsic mobility and contributes to the cell surface carbohydrate content and overall charge density. Indeed, the presence of gangliosides in neuronal membrane preparations has been shown to affect adenylate cyclase (Partington and Daly, 1979), phosphodiesterase (Davis and Daly, 1979) and Na^+,K^+ ATPase activity (Leon et al., 1981), and to modify choline uptake (Schulze and Rommelspacher, 1978) and dopamine release (Cumar et al., 1978). Gangliosides have been implicated as physiologic receptors for exotoxins (Yamakawa and Nagai, 1978), viruses (Holmgren et al., 1980), interferon (Ankel et al., 1980), wheat germ agglutinin (Redwood and Palefka, 1976), serotonin (Ochoa and Bangham, 1976) and glycoprotein hormones (Kohn, 1977). However, excluding the specific binding of cholera toxin with monosialoganglioside (Critchley et al., 1982), the observed interactions of gangliosides with various biofactors have also been suggested to be purely ionic and non-specific. If this would be the case, gangliosides may not be the receptor but rather the modulator of receptor-ligand interactions (Hakomori, 1981). Gangliosides have furthermore been suggested to operate as cell surface markers, antigens and to be involved in cell-cell recognition (Hakomori, 1981).

In view of the above-mentioned functional notions and certainly in an oversimplified manner, one may envisage hypothetical roles for gangliosides in regulating: a) intrinsic plasma membranes properties; b) cell-ligand recognition; c) cell-cell interactions. More specifically:

a) as intrinsic mobile cryptic surface components, gangliosides may influence membrane fluidity and dynamic organization of cell surface components thereby affecting enzyme activities, ionic fluxes, release and uptake processes, and turnover of membrane components;
b) as potential cell surface recognition molecules, gangliosides may modulate cell surface recognition mechanisms thereby mediating neural responses to extrinsic trophic and specific factors;
c) as cell surface antigens, gangliosides may play a major role in determining specific cell-cell surface interactions thereby affecting directional guidance of growing neurites and functional communications between neurons and neuro-glial elements.

Clonal and primary cell cultures are appropriate model systems for the study of the above molecular mechanisms. It is well recognized that such in vitro systems provide unique opportunities for the study of the fine interplay occurring between extrinsic and intrinsic cellular influences. These systems permit manipulation not only of the environmental signals but also of the ganglioside content on the neuronal surface. The latter effect may be achieved by addition of exogenous gangliosides to culture medium.

Neuroblastoma Cells

Neuroblastoma cell cultures have extensively been utilized as a model system for the study of the molecular mechanisms of potential ganglioside involvement in the regulation of neuritogenesis. Under suitable experimental conditions neuroblastoma cells may be induced to express various functions characteristic of mature neurons (Denis-Donini and Augusti-Tocco, 1980). When grown in conditions favouring cell proliferation, a low and less complex ganglioside pattern, as compared to normal mature neuronal cells, is observed (Dawson, 1979). However, when the morphological differentiation of neuroblastoma cells is stimulated, a significant increase in the amount of membrane ganglioside occurs (Dawson, 1979; Rebel et al., 1980). These data may be tentatively interpreted as suggesting a role for membrane gangliosides in regulation of neuronal surface phenomena associated with neuronal growth and differentiation. Indeed dramatic changes in glycolipid composition, surface organization and metabolism are known to be associated with oncogenic transformation (Hakomori, 1981). As occurs with SV-transformed and 3T3-untransformed cells (Hakomori, 1981), the addition of gangliosides to culture medium of neuroblastoma cells greatly affects cell morphology and growth (Fig. 1). Neurite fiber formation increases dramatically within 20-24 hours (Leon et al., 1982; Morgan and Seifert, 1979; Dimpfel et al., 1981; Roisen et al., 1981). The morphological effect is correlated with an increase in intracellular cAMP content and a decrease in the rate of ^{3}H-thymidine incorporation into the cellular DNA. These modifications presumably reflect ganglioside-induced elongation of G_1 phase of cell cycle (Leon et al., 1982).

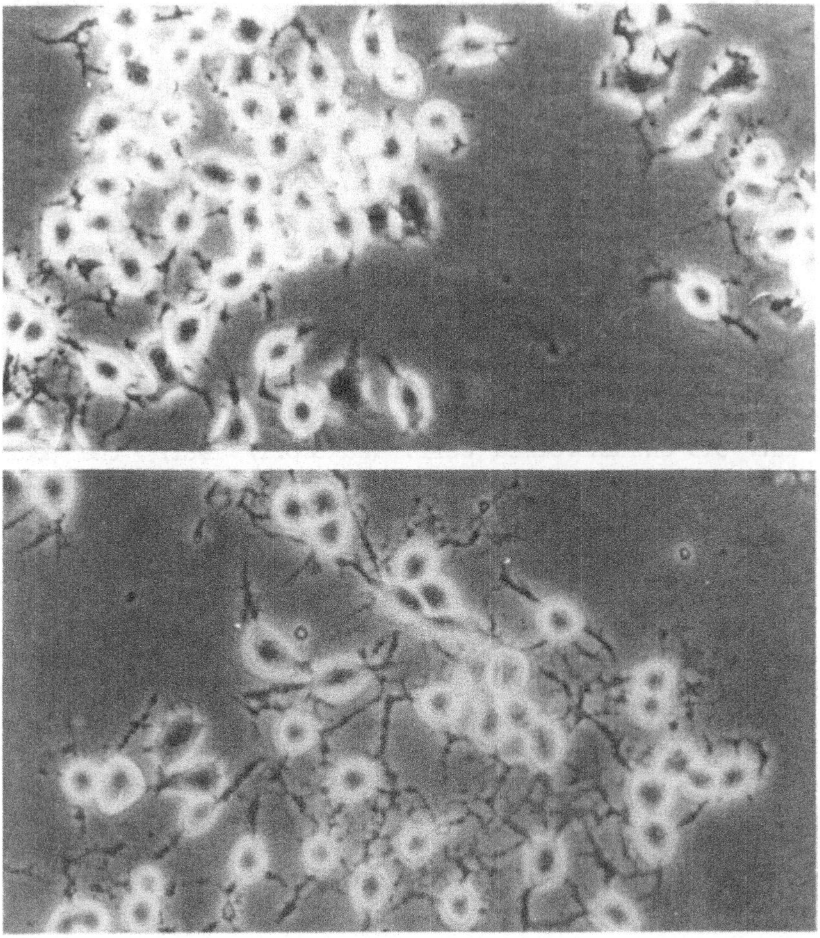

a) (top) Control cells were treated for 2 hrs with DMEM followed by removal of medium and replacement with DMEM - supplemented 10% fcs;
b) (bottom) Cells were preincubated for 2 hrs in DMEM containing monosialoganglioside (10^{-4}M). Medium was then replaced with DMEM-supplemented 10% fcs.

Fig. 1: Photomicrographs of N_2A neuroblastoma cells after 24 hours of growth in Dulbecco's Modified Eagle Medium (DMEM)-supplemented 10% fetal calf serum (fcs). Cells were plated at a density of 2-4 x 10^5 cells/100 mm Falcon dishes in DMEM + 10% fcs. 24 hours later medium was removed and cells were treated as indicated above.

The cause-effect relationship of cell surface ganglioside modification together with the molecular mechanisms underlying potential ganglioside control of neuronal growth and neurite formation are still obscure. The technique of artificial enrichment of cell surface gangliosides by addition of exogenous ganglioside to culture medium theoretically allows controlled manipulation of cell surface ganglioside content and composition, thereby permitting detailed analysis of ganglioside effects. However, although gangliosides added to culture medium have been reported to accumulate at the plasma membrane (Callies et al., 1978) and to cause acquisition of new biologically active receptors (Moss et al., 1976), there is considerable controversy as to the mechanism of ganglioside insertion into the membrane. Exogenously supplied gangliosides have been reported to be more labile to sialidase action than the native membrane gangliosides (Keenan et al., 1975). Both labile and stable interactions of gangliosides with cell surface have been reported (Radsak et al., 1981). Indeed as shown in Table 1, incorporation studies done utilizing ^{3}H-GM_1 labelled on the terminal galactose moiety (Toffano et al., 1980) indicate that the exogenous ganglioside apparently bound to neuroblastoma cells can be subdivided into a) a component removed by washings with phosphate buffer saline, b) a component removed by washing with serum containing media, c) a component released by incubation with trypsin, and d) a component stable to trypsin treatment.

Wiegandt et al. (1982), utilizing techniques such as electron spin resonance and measurement of lateral diffusion of fluorescent labelled gangliosides, recently presented evidence indicating that the trypsin stable portion of the exogenously supplied gangliosides may well be bound to the cells by insertion into the lipid phase of the plasma membranes. Although the biological implications of such incorporation remain to be elucidated, its occurrence favours the possibility that exogenous gangliosides added to culture medium of cells can reach a position in the lipid phase of the outer membrane comparable to those of endogenous glycolipids. Such an experimental strategy may prove to be of fundamental value not only in establishing the function of membrane gangliosides in regeneration of neuronal development but also in comprehending the biochemical mechanisms underlying the pharmacological effects of gangliosides in peripheral and central nervous system regeneration (Ceccarelli et al.,

1976; Gorio et al., 1980; Wojcik et al., 1982).

Table 1: Different forms of ^3H-GM_1 association to N_2A cells. Quantitative aspects.

Form of association	Associated ^3H-GM_1 nmoles/mg cell protein	% of total associated	% of total added
Total associated (loosely + stably associated)	9.85 ± 0.4	100 ± 4	1.98 ± 0.1
Loosely associated	3.88 ± 0.2	39.4 ± 1	0.77 ± 0.04
Total stably associated	5.92 ± 0.3	60.1 ± 3	1.18 ± 0.04
- serum labile form	4.00 ± 0.4	39.8 ± 4	0.80 ± 0.06
- trypsin labile form	0.97 ± 0.1	9.7 ± 1	0.19 ± 0.02
- trypsin stable form	1.05 ± 0.1	10.3 ± 1.1	0.21 ± 0.02

Neuroblastoma cells (2-4 x 10^5 cells/100 mm Falcon dishes) were incubated for 2 hours with 5 ml DMEM containing ^3H-GM_1 (10^{-4} M, 1 μCi/ml). Medium was removed and cells either immediately detatched with PBS + 0.5 mM EDTA pH 7.4 (total associated) or submitted, prior to detatchment, to washings with PBS (total stable associated) followed by washings with serum-containing medium (serum stable and labile form) and subsequently with 1% trypsin/PBS solution (trypsin labile and stable form). Bound total radioactivity was measured following cell centrifugation and solubilization of pellet. TLC identification of radioactivity was performed following cell centrifugation and ganglioside extraction of pellet according to Tettamanti et al. (1973). In all cases recovery of ^3H-GM_1 exceeded 95% of total radioactivity.

Primary Neuronal Cell Cultures

Primary neuronal cell cultures, even if heterogeneous, offer the means to study a larger spectrum of behaviours of " normal" neural cells including neural-glial interactions and formation of functional synapses with specific target cells. By careful manipulation of the culture system, those aspects of cell development which are intrinsically programmed can be readily distinguished from the aspects - including neuronal cell survival - which are dependent on extrinsic factors. Indeed such model systems have permitted identification and distinction among various macromolecular factors directed to developing neurons (Ebendal et al., 1980; Varon and Adler, 1981). Furthermore they have also proven useful for the identification in vivo of similar factors in the immediate environment of the regenerating neurons (Lundborg et al., 1982).

The membrane ganglioside molecule, as already mentioned, can on one side interact with biological ligands and on the other modulate its recognition by the cell surface. Both possibilities may be investigated by studying their effects on the morphological and biochemical response of dissociated fetal chick dorsal root ganglionic cells to NGF. These cells, although initially thought of as requiring NGF for survival and acquisition of differentiated properties only during a limited period of embryonic development, are now known to require NGF even after birth if they are deprived of their ganglionic non-neural partners (Varon and Adler, 1981). As shown in Fig. 2, the addition to culture medium containing NGF of low concentrations of monosialoganglioside - concentrations which result solely in a trypsin-stable type of association (Facci, Leon, Toffano and Tettamanti, manuscript in preparation) - significantly increases the complexity of neurite formation. Similar morphological results have been reported elsewhere (Roisen et al., 1981; Hauw et al., 1981). However, before speculating on the underlying molecular mechanisms, biochemical data are necessary to support the apparent, ganglioside-induced, increased cell differentiation. Gangliosides have also been shown to affect PC12 morphological response to NGF. Such an effect does not involve a modification of the ligand-cell interaction but presumably reflects a modulation by gangliosides of the NGF-induced metabolic shift (Gorio et al., 1981; Ferrari et al., submitted).

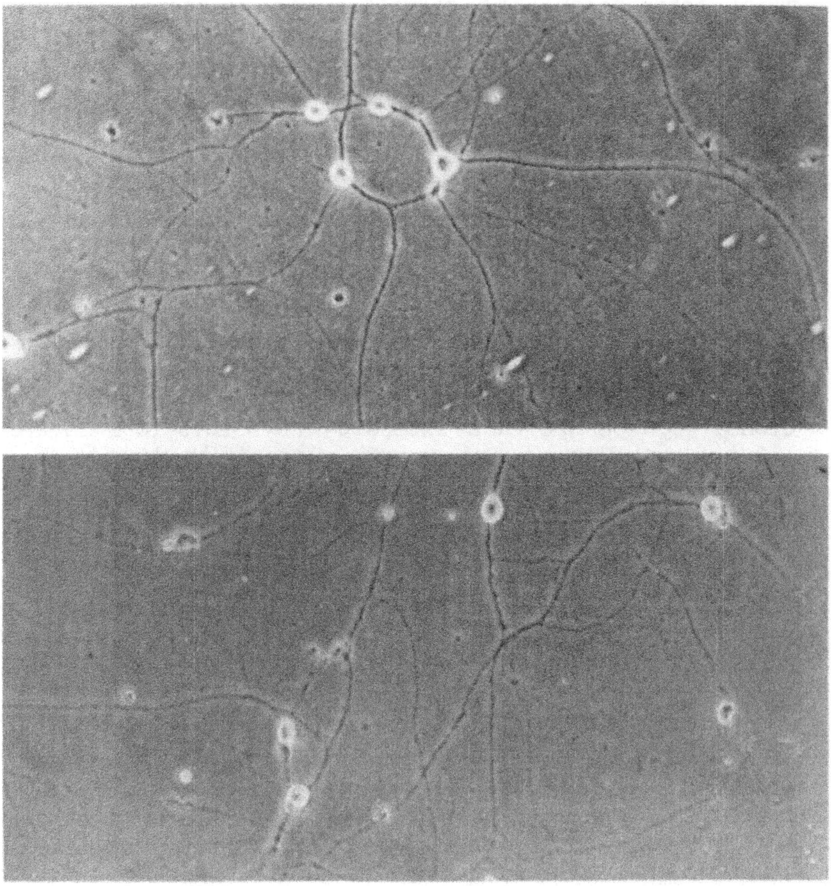

a) (top) Control cells;
b) (bottom) Ganglioside-treated cells. Note increased neurite branching and complexity.

Fig. 2: Photomicrographs of fetal chick dissociated dorsal root ganglionic cells after 10 days in culture. Dorsal root ganglia were dissected from 8 day-old embryos. Enriched dissociated neuronal cell cultures were prepared according to Barde et al. (1980), and plated at a density of 5 ganglia/30 mm Falcon dishes on highly adhesive collagen substratum (Adler et al., 1979).
The culture media consisted of DMEM + medium 199 (3 : 1 v/v) containing 5% fetal calf serum, 5% horse serum and 10 ng/ml NGF (2.5S subunit). Monosialoganglioside was added (on plating day) at a concentration of 10^{-7}M.

During both development and regeneration, the neurite processes emanating from the neuron must seek out and form specific functional connections with appropriate target cells. Such a process has been shown to be highly dependent on specific cellular interactions and membrane recognition processes (Moscona and Hausman, 1977). The nigro-striatal system offers an appropriate experimental model for exploring a possible ganglioside involvement. Fetal mesencephalic dopaminergic neurons, isolated and dissociated before the establishment of any contact with the striatum, have been shown to differentiate in culture. In co-culture with dissociated mouse embryo striatal cells their maturation is significantly enhanced (Prochiantz et al., 1981). An unknown striatal-bound dopaminergic factor, which most probably provides highly specific recognition sites favouring the functional contact of dopaminergic fibers, has been hypothesized. Similar mechanisms have also been suggested to occur during regeneration of the nigro-striatal system. Monosialoganglioside addition, namely GM_1 (10^{-7} to 10^{-8}M), to fetal striatal mesencephalic co-cultures affects neurite outgrowth. Increased neurite complexity may be observed (Fig. 3). Preliminary experiments indicate the occurrence of a parallel increase in ^{3}H-dopamine uptake. An increased ganglioside-induced rate of dopaminergic neurite outgrowth or facilitation of dopaminergic-striatal cell recognition processes may perhaps occur. Nevertheless further experiments are needed to characterize the effect. It is pertinent at this point to mention that ganglioside administration *in vivo* accelerates recovery of lesioned striatal-dopaminergic pathway (Toffano et al., submitted).

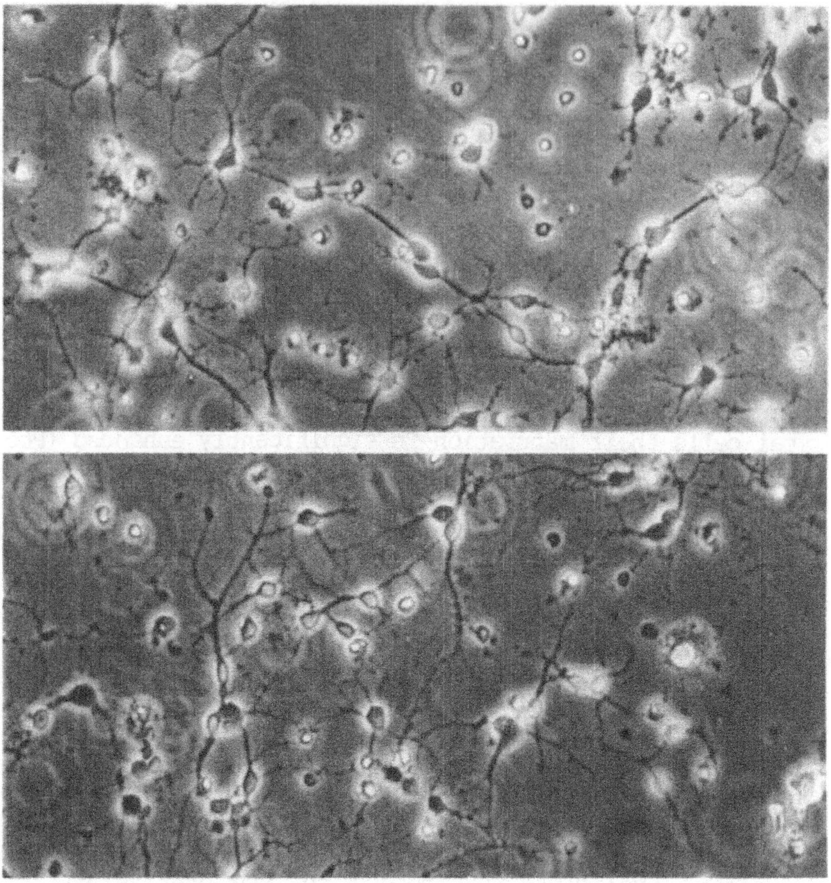

a) (top) Control cells;

b) (bottom) Ganglioside-treated. Note also in this model system the occurrence ot a greater neurite branching.

Fig. 3: Photomicrographs of fetal mouse mesencephalic striatal co-cultures after 4 days in culture. Cells were dissected from 13 (mesencephalic) and 15 (striatal) day-old embryos, mechanically dissociated and plated, at a density of 5 x 10^5 striatal cells + 5 x 10^5 mesencephalic cells/30 mm Falcon dishes, on hydrated collagen substratum. Media consisted of Basal Eagle Medium (BME) : F12 (1 : 1 v/v) buffered with 10 mM Hepes supplemented with insulin (25 μg/ml), transferrin (100 μg/ml), progesterone (20 nM), putrescine (60 μM), selenium (30 nM) and 0.6% glucose. Monosialoganglioside was added at a concentration of 10^{-7}M on plating day.

CONCLUSION

The aim of this chapter was to provide perspectives supporting evidence and model systems which may be utilized for studying the role of gangliosides in the regulation of neuronal development and regeneration.

Clonal and primary cell cultures are appropriate model systems for the understanding of those molecular mechanisms underlying neuronal development and regeneration. These in vitro systems give unique opportunities for the study of the fine interplay between extrinsic and intrinsic cellular influences. They permit manipulation not only of environmental factors but also of membrane constituents which can play a role in the regulation of cell surface responses. The association of such studies should lead to definition of pharmacological and clinical strategies aimed to improve neural regeneration, especially of CNS. This could be attained either by minimizing the effect of deleterious environmental situations or by increasing the availability and/or the potency of existing growth-promoting factors. The latter condition may also be achieved indirectly by enhancement of neuronal intrinsic regenerative capacities and sensitivity towards the factors. Indeed, recent pharmacological attempts utilizing gangliosides, typical neuronal cell surface components, have been successful in improving nerve development and regeneration both in vitro and in vivo. It has been shown that gangliosides stimulate morphological and biochemical differentiation of clonal and primary cell cultures, in agreement with previous reports (Leon et al., 1982; Gorio et al., 1981; Roisen et al., 1981, Dimpfel et al., 1981; Morgan and Seifert, 1979). Most probably the comprehension of the mechanisms underlying such effects will provide the rational basis for the clarification of the effects of gangliosides on nerve regeneration in vivo (Ceccarelli et al., 1976; Gorio et al., 1980; Wojcik et al., 1982; Toffano et al., submitted).

REFERENCES

Adler R., Manthorpe M., and Varon S. (1979) Separation of neuronal and nonneuronal cells in monolayer cultures from chick embryo optic lobe. Develop. Biol. 69, 424-435.

Ankel H., Krishnamurti C., Besancon F., Stefanos S., and Falcoff E. (1980) Mouse fibroblast (type I) and immune (type II) interferons: pronounced differences in affinity for gangliosides and in antiviral and antigrowth effect in mouse leukemia L-1210R cells. Proc. Natl. Acad. Sci. USA 77, 2528-2532.

Barde Y.-A., Edgar D., and Thoenen H. (1980) Sensory neurons in culture: changing requirements for survival factors during embryonic development. Proc. Natl. Acad. Sci. USA 77, 1199-1203.

Bass H. (1981) Ganglioside sialic acid as a quantitative neurochemical index of the integrity of synaptic function in cognitive disorders of development and aging, in Gangliosides in Neurological and Neuromuscular Function, Development and Repair (Rapport M.M., and Gorio A., eds), pp. 29-43. Raven Press, New York.

Benfey M., and Aguayo A.J. (1982) Extensive elongation of axons from rat brain into peripheral nerve grafts. Nature 296, 150-152.

Bjorklund A., and Stenevi U. (1979) Regeneration of monoaminergic and cholinergic neurons in the mammalian central nervous system. Physiol. Rev. 59, 62-100.

Callies R., Schwarzmann G., Radsak K., Siegert R., and Wiegandt H. (1977) Characterization of the cellular binding of exogenous gangliosides. Eur. J. Biochem. 80, 425-432.

Ceccarelli B., Aporti F., and Finesso M. (1976) Effects of brain gangliosides on functional recovery in experimental regeneration and reinnervation. Adv. Exp. Med. Biol. 71, 275-293.

Cotman C.W., Nieto-Sampedro M., and Harris E.W. (1981) Synapse replacement in the nervous system of adult vertebrates. Physiol. Rev. 61, 684-784.

Critchley D.R., Streuli C.H., Kellie S., Ansell S., and Patel B. (1982) Characterization of the cholera toxin receptor on Balb/c 3T3 cells as a ganglioside similar to, or identical with, ganglioside GM_1. Biochem. J. 204, 209-219.

Cumar F.A., Maggio B., and Caputto R. (1978) Dopamine release from nerve endings induced by polysialogangliosides. Biochem. Biophys. Res. Commun. 84, 65-69.

Davis C.W., and Daly J.W. (1979) Activation of rat cerebral cortical 3',5'-cyclic nucleotide phosphodiesterase activity by gangliosides. Mol. Pharmacol. 17, 206-211.

Dawson G. (1979) Complex carbohydrates of cultured neuronal and glial cell lines, in Complex Carbohydrates of Nervous Tissue (Margolis R.U., and Margolis, R.K., eds), pp. 291-325. Plenum Press, New York.

Denis-Donini D., and Augusti-Tocco G. (1980) Molecular and lectin probe analyses of neuronal differentiation, in Neuronal Development in Model Systems. Neuronal Development, part II (Moscona A.A., and Monroy A., eds), pp. 323-348. Academic Press, New York.

Dimpfel W., Moller W., and Mengs U. (1981) Ganglioside-induced formation in cultured neuroblastoma cells, in Gangliosides in Neurological and Neuromuscular Function, Development and Repair (Rapport M.M., and Gorio A., eds), pp 119-133. Raven Press, New York.

Dreyfus H., Harth S., Massarelli R., and Louis J.C. (1981) Mechanisms of differentiation in cultured neurons, involvement of gangliosides, in Gangliosides in Neurological and Neuromuscular Function, Development and Repair (Rapport M.M., and Gorio A., eds) pp. 151-170. Raven Press, New York.

Ebendal T., Olson L., Seiger A., and Hedlund K.O. (1980) Nerve growth factors in the rat iris. Nature 286, 25-28.

Gorio A., Carmignoto G., Facci L., and Finesso M. (1980) Motor nerve sprouting induced by ganglioside treatment. Possible implication for gangliosides neuronal growth. Brain Res. 197, 236-241.

Gorio A., Aporti F. and Norido F., with the technical collaboration of Canella R. (1981) Ganglioside treatment in experimental diabetic neuropathy, in Gangliosides in Neurological and Neuromuscular Function, Development and Repair (Rapport M.M., and Gorio A., eds), pp. 259-266. Raven Press, New York.

Hakomori S. (1981) Glycosphingolipids in cellular interaction, differentiation and oncogenesis. Ann. Rev. Biochem. 50, 733-764.

Hauw J.J., Fenelon S., Boutry J.M., Nagai Y., and Escourolle R. (1981) Effects of brain gangliosides on neurite growth in guinea pig spinal ganglia tissue cultures and on fibroblast cell cultures, in Gangliosides in Neurological and Neuromuscular Function, Development and Repair (Rapport M.M., and Gorio A., eds), pp. 171-175. Raven Press, New York.

Holmgren J., Svennerholm L., Elwing H., Fredman P., and Strannegard O. (1980) Sendai virus receptor: proposed recognition structure based on binding to plastic-adsorbed gangliosides. Proc. Natl. Acad. Sci. USA 77, 1947-1950.

Keenan T.W., Schmid E., Franke W.W., and Wiegandt H. (1975) Exogenous glycosphingolipids suppress growth rate of transfomed and untransformed 3T3 mouse cells. Exp. Cell. Res. 92, 259-270.

Kohn L.D. (1977) Relationships in the structure and function of cell surface receptors for glycoprotein hormones, bacterial toxins, and interferon. Ann. Rep. Med. Chem. 112, 361-372.

Ledeen R.W. (1978) Ganglioside structures and distribution: are they localized at the nerve ending? J. Supramol. Struct. 8, 1-17.

Leon A., Facci L., Toffano G., Sonnino S., and Tettamanti G. (1981) Activation of (Na^+, K^+)ATPase by nanomolar concentrations of GM_1 ganglioside. J. Neurochem. 37, 350-357.

Leon A., Facci L., Benvegnù D., and Toffano G. (1982) Morphological and biochemical effects of gangliosides in neuroblastoma cells. Develop. Neurosci. 5, 108-144.

Lundborg G., Longo F.M., and Varon S. (1982) Nerve regeneration model and trophic factors in vivo. Brain Res. 232, 157-161.

Morgan J.I., and Seifert W. (1979) Growth factors and gangliosides: a possible new perspective in neuronal growth control. J. Supramol. Struct. 10, 111-124.

Moscona A.A., and Hausman R.E. (1977) Biological and biochemical studies on embryonic cell recognition, in Cell and Tissue Interactions (Lash, J.W., and Burger, M., eds), pp. 173-186. Raven Press, New York.

Moss J., Fishman P.H., Manganiello V.C., Vaughan M., and Brady R.O. (1976) Functional incorporation of ganglioside into intact cell: induction of choleragen responsiveness. Proc. Natl. Acad. Sci. USA 73, 3480-3483.

Ochoa E.L.M., and Bangham A.D. (1976) N-acetylneuraminic acid molecules as possible serotonin binding sites. J. Neurochem. 26, 1193-1198.

Partington C.R., and Daly J.W. (1979) Effect of gangliosides on adenylate cyclase activity in rat cerebral cortical membranes. Mol. Pharmacol. 15, 484-491.

Pfenninger K.H., and Maylie-Pfenninger M.-F. (1978) Characterization, distribution and appearance of surface carbohydrates on growing neurites, in Neuronal Information Transfer (Karlin A., Vogen, H.J., and Tennyson, V.M., eds) pp. 373-386. Academic Press, New York.

Prochiantz A., Daguet M.C., Herbet A., and Glowinski J. (1981) Specific stimulation of in vitro maturation of mesencephalic dopaminergic neurones by striatal membranes. Nature 293, 570-572.

Purpura D.P., Pappas G.D., and Baker H.J. (1977) Meganeurites and aberrant processes of neurones on GM_1-gangliosidosis. A Golgi study. Brain Res. 145, 13-26.

Radsak K., Schwarzmann G., and Wiegandt H. (1981) Studies on the cell association of exogenously added sialoglycolipids. Z. Physiol. Chem. 363, 263-272.

Rebel G., Robert J., and Mandel P. (1980) Glycolipids and cell differentiation, in Structure and Function of Gangliosides (Svennerholm L., Mandel P., Dreyfus H., and Urban P.-F., eds), pp. 159-166. Plenum Press, New York.

Redwood W., and Palefka T. (1976) Interaction of wheat germ agglutinin with phosphatidylcholine liposomes containing incorporated monosialogangliosides. Biochim. Biophys. Acta 455, 631-643.

Roisen F.J., Bartfeld H., Nagele L., and Yorke G. (1981) Gangliosides stimulation of axonal sprouting in vitro. Science 214, 577-578.

Schulze G., and Rommelspacher H. (1978) The influence of gangliosides on high affinity uptake of choline in vitro. Abstract 11th Collegium International Neuro-Psychopharmacologicum Congress, Vienna July 9-14, pp 427.

Tettamanti G., Bonali F., Marchesini S., and Zambotti V. (1973) A new procedure for the extraction, purification and fractionation of brain gangliosides. Biochim. Biophys. Acta 296, 160-170.

Toffano G., Benvegnù D., Bonetti A.C., Facci L., Leon A., Orlando P., Ghidoni R., and Tettamanti G. (1980) Interactions of GM_1 ganglioside with crude rat brain neuronal membranes.J. Neurochem. 35, 861-866.

Varon S., and Adler R. (1981) Trophic and specifying factors directed to neuronal cells, in Advances in Cellular Biology, Vol. 2, pp. 115-163. Academic Press, New York.

Wiegandt H., Kanda S., Inoue K., Utsumi H., Nojima S., Elson E.L., and Reidler J.A. (1982) Studies on the membrane association of glycosphingolipids. Abstract Special FEBS Meeting on Cell Function and Differentiation, Athens, April 25-29, pp. S-11.

Willinger M., and Schachner M. (1980) GM_1 ganglioside as a marker for neuronal differentiation in mouse cerebellum. Develop. Biol. 74, 101-107.

Wojcik M., Ulas J., and Oderfeld-Nowak B. (1982) The stimulating effect of ganglioside injections on the recovery of choline acetyltransferase and acetylcholinesterase activities in the hippocampus of the rat after septal lesions. Neurosci. 7, 495-499.

Yamakawa T., and Nagai Y. (1978) Glycolipids at the cell surface and their biological functions. Trends Biochem. Sci. 3, 128-131.

HETEROSIS FOR MYELIN CONTENT IN MICE:

A NEW MODEL FOR STUDYING MYELINATION

R.K. Yu, H. Ebato, and T.N. Seyfried

Dept. of Neurology, Yale University

New Haven, CT 06510, U.S.A.

Introduction

The myelin sheath is a lipid-laden neural membrane whose chief function is to insulate nerve fibers and facilitate saltatory conduction of nerve impulses along the axons. The formation of myelin is one of the most spectacular events in nervous system development. In man, myelination commences in late fetal life or soon after birth (Adams and Leibowitz, 1972; Fishman et al. 1975; Meier, 1976); in rodents, it occurs most rapidly between 14 and 30 postnatal days of age and then progresses at a much slower rate throughout adult life (Norton and Poduslo, 1973; Matthieu et al. 1973). Studies from a number of laboratories have documented that myelinogenesis, which involves the synthesis and assembly of myelin constituents, proceeds in a regulated and ordered fashion (for review, see Benjamins and Smith, 1977; Braun et al. 1980; Norton, 1981; Waehneldt and Linington, 1980). Such a complex developmental event is necessarily under genetic control as has already been demonstrated in several mouse mutants which exhibit a retardation or disruption of myelinogenesis (Hogan, 1977; Baumann, 1980). In this paper, we describe a unique murine model which exhibits a more heavily myelinated central nervous system (CNS).

Heterosis for Brain Myelin Content in Mice

Heterosis or hybrid vigor is a well-known genetic

phenomenon where the mean value of a phenotypic trait is greater in F_1 hybrids than in their inbred parents (Shull, 1952). The underlying genetic mechanism for the heterotic effect is not known, but may be attributable to the accumulation of favorable dominant alleles contributed by both parents, overdominance at single loci (heterozygosity), the interaction of non-allelic genes (epistasis), or any combination of these mechanisms (Strickberger, 1978). These types of gene action generally result in a physiologically advantageous state for the F_1 hybrids.

During the course of studying audiogenic seizure (AS) susceptibility in mice, we found that the concentrations of GM1 ganglioside were higher in the brains of AS susceptible DBA/2J (D2) mice than in the brains of AS resistant C57BL/6J (B6), LG/J, C3H/HeJ, and BALB/CJ mouse strains at 21 days postnatal (Seyfried et al. 1978; 1979). Since GM1 ganglioside is known to be highly enriched in CNS myelin (Suzuki et al. 1967, 1968; Ledeen et al. 1973; Yu and Yen, 1975; Yu and Iqbal, 1979; Ueno et al. 1978; Cochran et al. 1982), our results suggest that D2 mouse brains are more heavily myelinated than those of other mouse strains. This finding prompted us to examine the biochemical genetic relationship between brain myelin content and AS susceptibility. We measured the concentration of GM1 ganglioside and cerebroside, two glycolipids that are abundant in myelin, in the brains of D2, B6 and (B6 x D2) F_1 hybrid mice at 16 $\pm$ 1 days postnatal. Instead of finding an intermediate or low level of these glycolipids in the F_1 hybrids, as might be expected, since these mice have an intermediate AS susceptibility (Seyfried, 1982), we found that the concentrations of these glycolipids were actually higher in the F_1 hybrids than in either parental strain (Fig. 1). In addition, the water content, which is known to be inversely proportional to the myelin content, was lower in the F_1 hybrids (Fig. 1). These observations provided the first evidence that the F_1 hybrids possess a more heavily myelinated CNS than either parental strain. This phenomenon thus represents a classical example of heterosis.

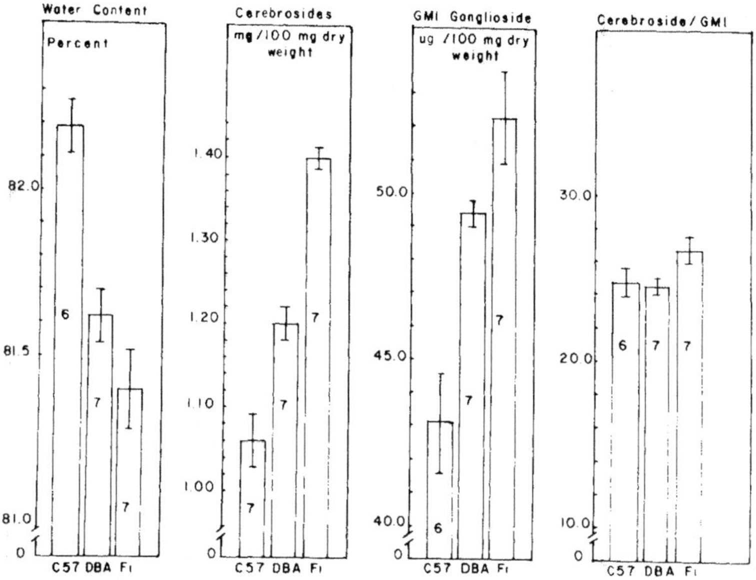

Fig. 1 Glycolipid and water content in brains of B6, D2 and F_1 hybrid mice at 17 days of age. Cerebroside content is based on the weight of cerebroside, and GM1 ganglioside content is based on lipid-bound sialic acid. Values represent mean ± SEM. Numbers in the bar represent number of samples analyzed.

To further determine the specificity of this phenomenon, we analyzed several other myelin-associated biochemical parameters in the brains of B6, D2 mice and their F_1 hybrids (Ebato et al. 1983). We isolated myelin from fresh brains employing the two cycle-sucrose density gradient procedure of Norton and Poduslo (1973). The results are shown in Table 1.

Table 1. Composition of 17-day-old mouse brains

	Brain wet wt (mg)	Total protein (mg/g wet wt brain)	Myelin yield (mg/brain)	Myelin protein (mg/brain)	Myelin protein (μg/mg dry wt myelin)
B6	399.2 ± 6.1 (N = 7)	97.8 ± 4.2 (7)	2.0 ± 0.1 (4)	0.40 ± 0.06 (4)	200.4 ± 2.1 (4)
D2	334.9 ± 4.8 [xx] (N = 7)	112.0 ± 5.4 (7)	2.1 ± 0.1 (3)	0.56 ± 0.08 [xx] (3)	256.5 ± 3.6 [x] (3)
F_1	392.7 ± 3.1 (N = 9)	111.2 ± 4.3 (9)	2.6 ± 0.1 [x] (4)	0.74 ± 0.05 [xx] (4)	283.6 ± 4.2 [xx] (4)

The total amount of myelin was highest in the F_1 hybrids, followed by D2 and B6 mice. The difference between the amounts of myelin in the F_1 hybrid and D2 and B6 mice was statistically significant. These results provide direct evidence of heterosis for total brain myelin content. Interestingly, the amount of myelin protein was also highest in the F_1 hybrid mouse brains, followed by that in D2 and B6 brains. No heterotic effect was found for whole brain weight or total brain protein concentration (Table 1).

The heterotic effect was also observed for the "myelin-associated" enzymes: 2′,3′-cyclic nucleotide 3′-phosphohydrolase (CNPase), 5′-nucleotidase, and carbonic anhydrase. Among the three enzymes assayed, the heterotic effect was greatest for CNPase activities. The activity of CNPase is known to be closely associated with myelin (Kurihara and Tsukada, 1967, 1968; Waehneldt, 1975; Sprinkle et al. 1978, 1980 a, b; Sims and Carnegie, 1978; Drummond and Dean, 1980; Wells and Sprinkle, 1981), and the enzyme may actually correspond to the Wolfgram proteins (Drummond and Dean, 1980; Sprinkle et al. 1980a, b). Evidence has been provided that the enzyme is also present in the perikarya and plasma membranes of

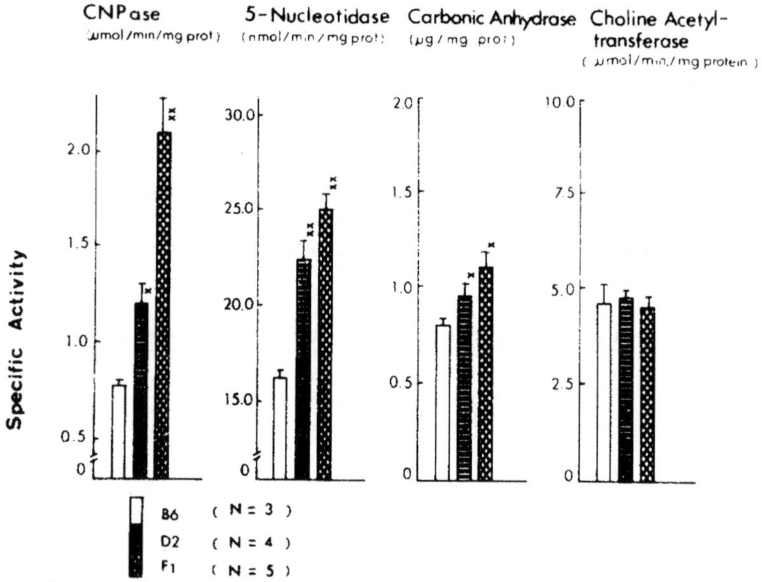

Fig. 2. Specific enzyme activities of 17-day-old mouse brains.

oligodendroglia (Poduslo and Norton, 1972; Poduslo, 1975). It is not surprising, therefore, that the magnitude of the heterotic effect was greater for CNPase than for 5′-nucleotidase and carbonic anhydrase. The latter two enzymes, although present in myelin (Kreutzberg et al. 1978; Cammer et al. 1976, 1977, 1979; Sapirstein and Lees, 1978), are also present in glial cell membranes (Kreutzberg et al. 1978; Roussel et al. 1979).

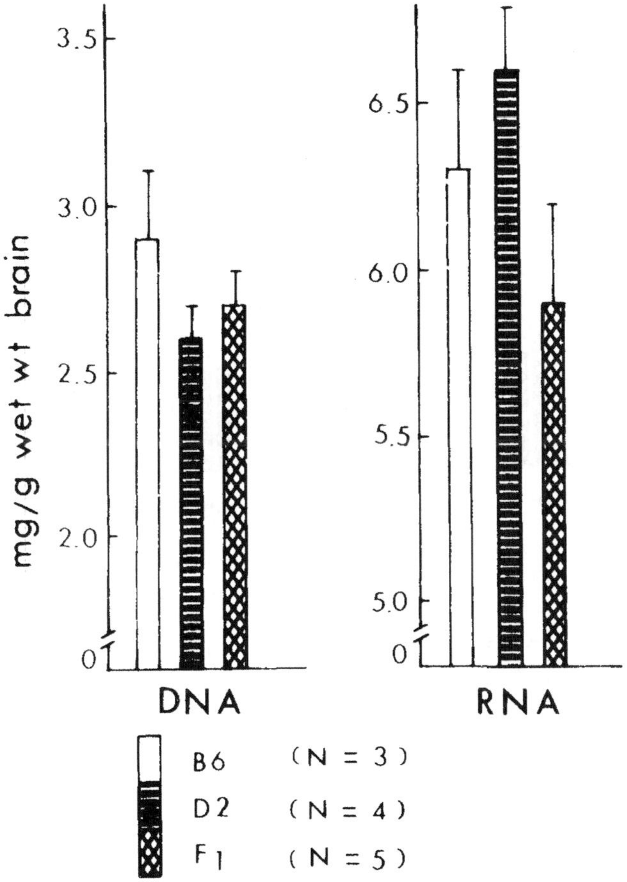

Fig. 3 Nucleic acid content in 17-day-old mouse brains. Values represent mean ± SEM. N = number of samples analyzed.

The heterotic effect was not observed for such nonmyelin components as choline acetyltransferase (Fig. 2), and the concentration of DNA, and RNA (Fig. 3), and gangliosides G_{D3}, G_{D1a}, G_{T1a}, G_{D1b}, G_{T1b} and G_{Q1b} (Fig. 4). In fact, the concentration of G_{D1a}, G_{T1a}, and G_{T1b} was lower in the brains of F_1 hybrids than in B6 brains. Taken together, our data indicate that the heterotic effect is specific for myelin content.

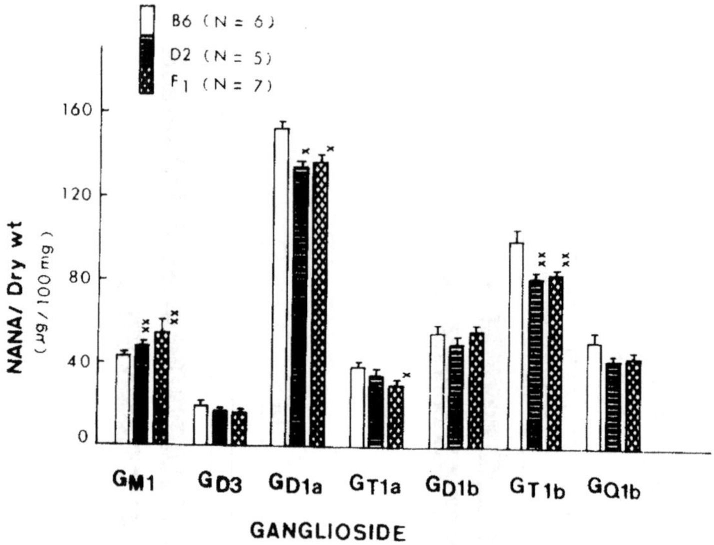

Fig. 4. Brain ganglioside content in 17-day-old mouse brains. Values represent mean ± SEM. N = number of samples analyzed.

Since the heterotic effect was observed for the concentration of protein in the isolated myelin, we compared the myelin protein composition of the three genotypes using sodium dodecyl sulfate-polyacrylamide gel electrophoresis (SDS-PAGE), as described previously (Yu et al. 1982). The results are shown in Fig. 5.

All the characteristic myelin proteins were observed, specifically, Wolfgram proteins (WP), proteolipid protein (PLP), DM-20, large and small myelin basic proteins (LBP and SBP) and a 17 K protein which corresponded to the pre-small myelin basic protein reported by Barbarese et al. (1978). To further separate the various basic proteins, we also carried out an SDS-PAGE in the high urea (8μM) gel

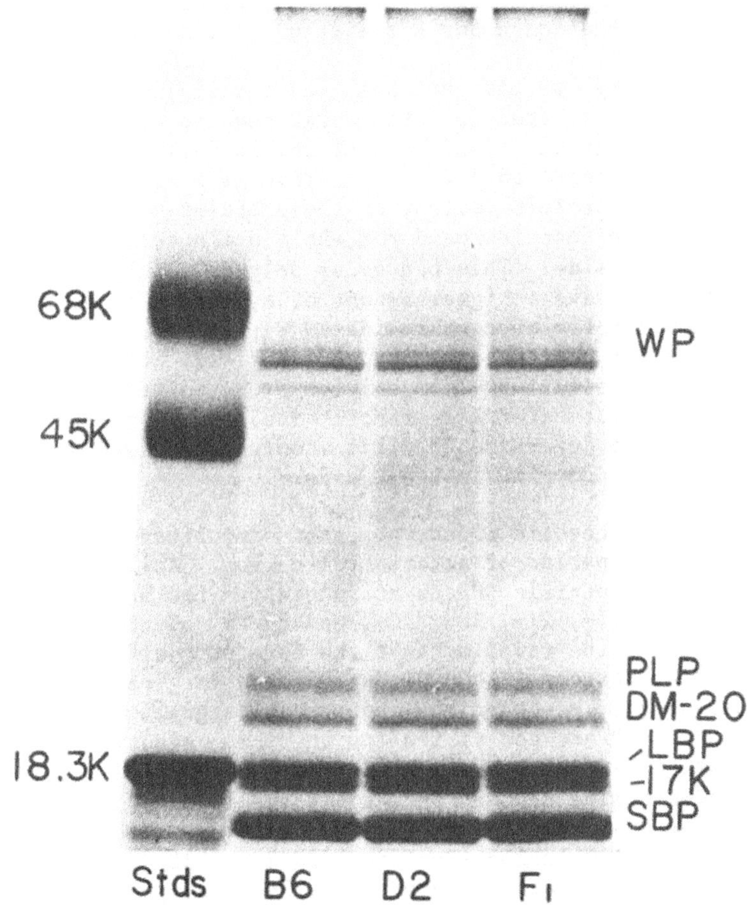

Fig. 5 SDS-PAGE of myelin proteins. The gel system contained 12% (wt/wt) acrylamide. After electrophoresis, the protein bands in the gel were stained with Coomassie Blue R250. SBP and LBP, small and large myelin basic proteins; WP, Wolfgram proteins; PLP, proteolipid protein. Standards were bovine myelin basic protein (18.3 kilodaltons), ovalbumin (45 kilodaltons), and bovine serum albumin (68 kilodaltons).

system of Swank and Munkres (1971) (data not shown). In both gel systems, we did not find any discernible differences in the protein compositions of the three genotypes, including the four myelin basic proteins (pre-large, 21.5K; large, 18.5K; pre-small, 17K; and small, 14K) previously reported for mouse brain myelin (Barbarese et al. 1977; 1978). Our results thus indicate that the heterotic effect is specific for the total myelin protein content rather than for a specific protein or proteins.

Although we have not analyzed the lipid content of the myelin isolated from the three genotypes, it is apparent that the lipid content should follow the following trend: B6 > D2 > F_1. This is a reversal of the myelin protein content of these brains. However, a heterotic effect is found for whole brain cerebroside and G_{M1} ganglioside. This can occur only when the brains of F_1 hybrids have a higher amount of myelin. In addition, since the ratio of cerebroside to G_{M1} in whole brain is similar for the three genotypes, (Fig. 1), it is possible that the myelin lipid composition is not altered as a result of the heterotic effect. Studies are now in progress to determine the lipid content and composition of the isolated myelin in these mice.

The heterotic effect appears to manifest itself only during the period of active myelination. Analysis of the total and specific CNPase activities in the brains of 10-, 12-, 14-, 17-, 21-, 30-, 40-, and 80-day-old B6, D2 and F_1 hybrid mice revealed that the heterotic effect was greatest between 17 and 30 days of age (Fig. 6). At 80 days-of-age, we could not observe any significant differences in CNPase activities for the three genotypes. The greatest rate of accretion of CNPase activities for the F_1 hybrid mice occured between 14 and 17 days of age. A similar developmental profile was also observed for cerebroside accumulation (Seyfried and Yu, 1980). Our data thus suggest that the heterotic effect may involve an accelerated or enhanced synthesis of myelin during a critical period of development.

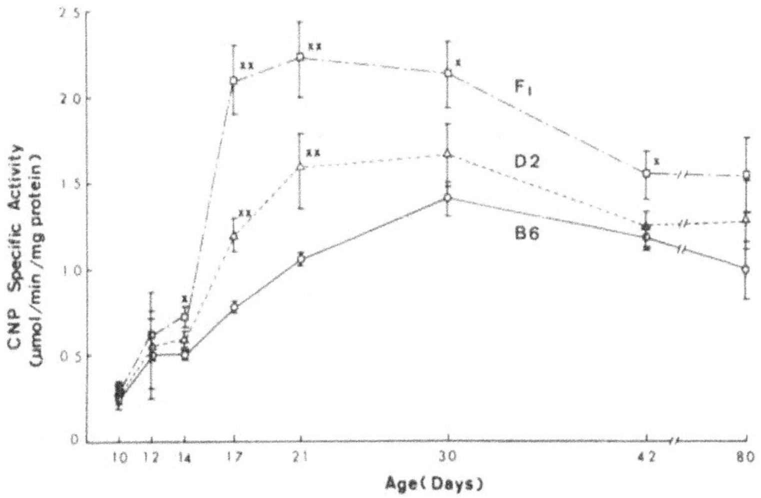

Fig. 6 Developmental profiles of specific CNPase activities in B6, D2 and F_1 mouse brains. The number of independent determinantions for each genotypes is at least 3. Bars represent SEM. Significance levels, shown only for F_1 vs B6 and D2 vs B6, are: *, $P > 0.05$, **$P > 0.01$.

Molecular basis of the heterotic effect for myelin content.

In order to better understand the molecular basis of the heterotic effect for myelin content, we carried out an in vitro myelin basic protein (MBP) synthesis experiment. We focused on the cell-free synthesis of MBPs because these proteins are representative of myelin (Barbarses et al. 1978) and are the most well-studied (Carey and Campagnoni, 1979; Campagnoni et al. 1980; Matthees and Campagnoni, 1980; Barbarese and Pfeiffer, 1981; Fan, 1982; Yu and Campagnoni, 1982). Polysomal mRNA from the whole brains of 16-day-old B6, D2 and F_1 mice was isolated according to the procedure of Palmiter (1974). Poly (A+)-containing mRNA was then isolated from the polysomal mRNA preparations by chromatography on oligo (dT)-cellulose (T-3 grade) using the procedure of Aviv and Leder (1972), as modified by Hirsh and Penman (1974). The final mRNA pellet was dissolved in sterile distilled water and stored at -70° until use.

Translation of poly (A^+) mRNA was then carried out according to the procedure of Pelham and Jackson (1976) employing a rabbit reticulocyte-lysate translational system containing L-[2,4,5-^{3}H]-leucine. Following translation, the radioactivity incorporated into total protein was determined by cold trichloroacetic acid precipitation followed by liquid scintillation counting. The translation products were further characterized by immunoprecipitation of MBPs (Rhoads et al. 1971) followed by SDS-PAGE (Swank and Munkres, 1971). The gels were sliced into individual bands and examined for radioactivity according to the procedure of Carey and Campagnoni (1979).

We could not find any differences among the three genotypes for the yield of total polysomes and the amount of poly (A^+) mRNA. Under optimal conditions for protein translation (0.5 μg total mRNA added per incubation mixture and 1 hr.), no difference was observed for the radioactivity incorporated into the total translated proteins for the three genotypes. However, the radioactivity incorporated into the MBP fraction of the F_1 hybrid brains was greater than in D2 (65% increase) and in B6 (74% increase). The ratios of radioactivities in the four MBPs were not different for the three genotypes. Since the translation was carried out under identical _in vitro_ conditions, the results suggest that there must be more MBP-mRNA present in the brains of 16 day-old F_1 hybrid mice than in the brains of B6 and D2. Although we cannot completely rule out an effect of translational regulation of myelin basic protein synthesis, our data nonetheless strongly suggest that the heterotic effect for myelin content is regulated at the transcriptional level for all four MBP-mRNAs.

Concluding Remarks

Unlike several single gene mouse mutants (jimpy, quaking, etc.) in which there is a retardation or disruption of normal myelinogenesis, the heterotic effect for myelin content displays an accelerated or enhanced myelinogenesis during a critical period of development. Therefore it may serve as a useful animal model for studying the various genetic and biochemical factors that regulate myelination. Further knowledge in these areas should be invaluable in our understanding of myelination under normal and abnormal conditions.

Acknowledgments

This work was supported by USPHS grants NS11853 (RKY), IK04NS0059, and NS-17704 (TNS), and by grant RG-1289-B-3 of the National Multiple Sclerosis Society (RKY). We are grateful to Dr. John Carson of the University of Connecticut, Farmington, CT, for his advice on the *in vitro* myelin basic protein synthesis experiment.

References

Adams, C.W.M., & Leibowitz, S. (1972) Morphological aspects of myelin and demyelination. in "*Research on Multiple Sclerosis*" Charles Thomas, Springfield, Ill. p. 26.

Aviv, H., & Leder, P. (1972) Purification of biologically active globin messenger RNA by chromatography on oligothymidylic acid-cellulose. *Proc. Natl. Acad. Sci.*, U.S.A. 69:1408-1412.

Barbarese, E. & Pfeiffer, S.E. (1981) Developmental regulation of myelin basic protein in dispersed cultures. *Proc. Natl. Acad. Sci.* U.S.A. 78:1953-1957.

Barbarese, E., Braun, P.E., & Carson, J.H. (1977) Identification of prelarge and presmall basic proteins in mouse myelin and their structural relationship to large and small basic proteins. *Proc. Natl. Acad. Sci.* U.S.A. 74:3360-3364.

Barbarese, E., Carson, J.H., & Braun, P.E. (1978) Accumulation of the four myelin basic proteins in mouse brain during development. *J. Neurochem.* 31:799-782.

Baumann, N. (1980) *Neurological Mutations Affecting Myelination*. Elsevier, Amsterdam.

Benjamins, J.A. & Smith, M.E. (1977) Metabolism of myelin in *Myelin*. (Morell, P. Ed.), Plenum Press, NY, Chap. 7, pp. 233-270.

Braun, P.E., Pereyra, P.M., & Greenfield, S. (1980) *Prog. Clin. Biol. Res.* 49:1-17.

Cammer, W. (1979) Carbonic anhydrase activities in myelin from sciatic nerves of adult and young rats: quantitation and inhibitor sensitivity, *J. Neurochem.* 32:651-654.

Cammer, W., Bieler, L., Fredman, T., & Norton, W.T. (1977) Quantitation of myelin carbonic anhydrase: Development and subfractionation of rat brain myelin and comparison with myelin from other species. Brain Res. 138, 17-28.

Cammer, W., Fredman, T., Rose, A.R., and Norton, W.T. (1976) Brain carbonic anhydrase: Activity in isolated myelin and the effect of hexachlorophene. J. Neurochem. 27, 165-171.

Campagnoni, A.T., Carey, G.D., & Yu, Y.T. (1980) In vitro synthesis of the myelin basic proteins: subcellular site of synthesis. J. Neurochem. 34:677-686.

Carey, G.D. & Campagnoni, A.T. (1979) In vitro synthesis of myelin basic proteins in the developing mouse brains: Properties of a homogenate system. J. Neurochem. 33:81-87.

Cochran, F.B., Yu, R.K., and Ledeen, R.W. (1982) Myelin gangliosides in vertebrates. J. Neurochem. 39:773-779.

Drummond, R.J. and Dean G. (1980) Comparison of 2′, 3′-cyclic nucleotide 3′-phosphodiesterase and the major component of Wolfgram protein W1. J. Neurochem. 35: 1155-1165.

Ebato, H., Seyfried, T.N., and Yu, R.K. (1983) Biochemical study of heterosis for brain myelin content in mice. J. Neurochem. 40:440-446.

Fan, K. (1982) mRNA-directed cell-free myelin basic protein synthesis: No evidence of neuronal control. Neurosci. Lett. 29:89-94.

Fishman, M.A., Agrawal, H.C., Alexander, A., Golterman, J., Martenson, R.E., & Mitchell, R.F. (1975) Biochemical Maturation of Human Central Nervous system myelin. J. Neurochem. 24:689-694.

Hirsh, M. & Penman, S. (1974) Post-transcriptional addition of polyadenylic acid to mitochondrial RNA by cordycepin-insensitive process. J. Mol. Biol. 83:131-142.

Hogan, E.L. (1977) Animal models of genetic disorders of myelin. in "Myelin" (Morell P. ed) pp. 489-520. Plenum, NY.

Kreutzberg, G., Barron, K.J., & Shubert, P. (1978) Cytochemical localization of 5′-nucleotidase in glial plasma membranes. Brain Res. 158:247-257.

Kurihara, T. & Tsukada, Y. (1967) The regional and subcellular distribution of 2′, 3′-cyclic nucleotide 3′-phosphohydrolase in the central nervous system. J. Neurochem. 14:1167-1174.

Kurihara, T. & Tsukada, Y. (1968) 2′, 3′-cyclic nucleotide 3′-phosphohydrolase in the developing chick brain and spinal cord. J. Neurochem. 15:827-832.

Ledeen, R.W., Yu, R.K., & Eng, L.F. (1973) Gangliosides of human myelin: Sialosylgalactosylceramide (G_7) as a major component. J. Neurochem. 21:829-839.

Matthees, J. & Campagnoni, A.T. (1980) Cell-free synthesis of the myelin basic proteins in a wheat germ system programmed with brain messenger RNA. J. Neurochem. 35:867-872.

Matthieu, J.M., Widmer, S., & Herschkowitz, N. (1973) Biochemical changes in mouse brain composition during myelination. Brain Res. 55:391-402.

Meier, C. (1976) Some observations on early myelination in the human spinal cord. Light and electron microscope study. Brain Res. 104:21-32.

Norton, W.T. (1981) Biochemistry of myelin. Adv. Neurol. 31:93-121.

Norton, W.T. & Poduslo, S.E. (1973) Myelination in rat brain: Method of myelin isolation. J. Neurochem. 21:749-758.

Palmiter, R.D. (1974) Magnesium precipitations of ribonucleoprotein complexes. Expedient techniques for the isolation of undegraded polysomes and messenger rigonucleic acid. Biochem. 13:3606-3615.

Pelham, R.B. & Jackson, R.J. (1976) An efficient m-RNA-dependent translation system from reticulocyte lysate. Eur. J. Biochem. 67:247-257.

Poduslo, S.E. (1975) The isolation and characterization of a plasma membrane and a myelin fraction derived from oligodendroglia of calf brain. J. Neurochem. 24:647-654.

Poduslo, S.E. & Norton, W.T. (1972) Isolation and some chemical properties of oligodendroglia from calf brain. J. Neurochem. 19:727-736.

Rhoads, R.E., McKnight, G.S., & Schimke, R.T. (1973) Quantitative measurement of ovalbumin messenger ribonucleic acid activity. J. Biol. Chem. 248:2031-2039.

Roussel, G., Delaunoy, J.-P., Nussbaum, J.L., & Mandel, P. (1979) Demonstration of a specific localization of carbonic anhydrase C in the glial cells of rat CNS by an immunohistochemical method. Brain Res. 160:47-55.

Seyfried, T.N. (1982) Developmental genetics of audiogenic seizure susceptibility in mice. In: Genetic Basis of the Epilepsies (Anderson, V.E., Hauser, W.A., Perry, J.K., & Sing, L.F., eds.) Raven Press, NY (In press).

Seyfried, T.N. & Yu, R.K. (1980) Heterosis for brain myelin content in mice. Biochem. Genetics. 18:1229-1238.

Seyfried, T.N., Glaser, G.H., & Yu, R.K. (1978) Cerebral, cerebellar and brain stem gangliosides of audiogenic seizure susceptible mice. J. Neurochem. 31:21-27.

Seyfried, T.N., Glaser, G.H., & Yu, R.K., (1979) Genetic variability for regional brain gangliosides in five strains of young mice. Biochem. Genetics 17:43-55.

Shull, G.H. (1952) Beginnings of the heterosis concept. In Heterosis (Gowen, J.W., Ed.) Hafner, New York pp.14-48.

Sims, N.R. & Carnegie, P.R. (1978) 2′, 3′-cyclic nucleotide 3′-phosphodiesterase. Adv. Neurochem. 3: 1-41.

Sprinkle, T.J., Grimes, M.J., & Eller, kA.J. (1980a) Isolation of 2′, 3′-cyclic nucleotide 3′-phosphodiesterase from human brain. J. Neurochem. 34:880-887.

Sprinkle, T.J., Wells, M.R., Garver, F.A., & Smith, D.B. (1980b) Studies on the Wolfgram high molecular weight CNS myelin proteins: Relationship to 2′, 3′-cyclic nucleotide 3′-phosphodiesterase. J. Neurochem. 35:1200-1208.

Sprinkle, T.J., Zaruba, M.E., & McKhann, G. (1978) Activity of 2′, 3′-cyclic nucleotide 3′-phosphhodiesterase in regions of rat brain during development: Quantitative relationship to myelin basic protein. J. Neurochem. 30:309-314.

Strickberger, M.W. (1976) Genetics McMillan, NY pp. 794-798.

Suzuki, K., Poduslo, S.E., & Norton, W.T. (1967) Gangliosides in myelin fraction of developing rats. Biochim. Biophys. Acta. 144:375-381.

Suzuki, K., Poduslo, S.E., & Poduslo, J.F. (1978) Further evidence for a specific ganglioside fraction closely associated with myelin. Biochim. Biophys. Acta 152:576-586.

Swank, R.T. & Munkres, K.D. (1971) Molecular weight analysis of oligopeptides by electrophoresis in polyacrylamide gel with sodium dodecyl sulfate. Analyt. Biochem. 39:462-477.

Taylor, J.M. & Schimke, R.T. (1973) Synthesis of rat liver albumin in a rabbit reticulocyte cell-free protein synthesizing system. J. Biol. Chem. 248:7661-7668.

Ueno, K., Ando, S., & Yu, R.K. (1978) Gangliosides of human, cat, and rabbit spinal cords and cord myelin. J. Lipid Res. 19:863-871.

Wells, M.R. & Sprinkle, T.J. (1981) Purification of rat 2′, 3′-cyclic nucleotide 3′-phosphodiesterase. J. Neurochem. 36:633-639.

Waehneldt, T.V. (1975) Ontogenetic study of a myelin-derived fraction with 2′, 3′-cyclic nucleotide 3′-phosphohydrolase activity higher than that of myelin. Biochem. J. 151:435-437.

Waehneldt, T.V. & Linington, C. (1980) Organization and assembly of the myelin membrane, in "Neurological Mutations Affecting Myelination." (Baumann, N. Ed.) Elsevier, Amsterdam, pp. 389-412.

Yu, Y.-T. & Campagnoni, A.T. (1982) In vitro synthesis of the four mouse myelin basic proteins: Evidence for the lack of a metabolic relationship. J. Neurochem. 39:1559-1568.

Yu, R.K. & Iqbal, K. (1979) Sialosylgalactosylceramide as a specific marker for human myelin and oligodendroglia: Gangliosides of human myelin, oligodendroglia and neurons. J. Neurochem. 32:293-300.

Yu, R.K. & Yen, S.I. (1975) Gangliosides in developing mouse brain myelin. J. Neurochem. 25:229-232.

Yu, R.K., Ueno, K., Glaser, G.H., & Tourtellotte, W.W. (1982) Lipid and protein alterations of spinal cord and cord myelin of multiple sclerosis. J. Neurochem. 39:464-477.

GANGLIOSIDE-MEDIATED INTERNALIZATION OF CHOLERAGEN

Ingming Jeng

Missouri Institute of Psychiatry
University of Missouri-Columbia
5400 Arsenal Street, St. Louis, MO
63139

INTRODUCTION

Systems using proteins as receptors for extracellular ligands have been extensively characterized in the last decade (Kaplan, 1981). Only recently has attention been given to the receptor function of gangliosides (Holmgren et al., 1980). Gangliosides serve as receptors for a variety of ligands including interferon, viruses, toxins and glycoprotein hormones. Ganglioside G_{M1}-choleragen distinguishes itself as one of the most promising systems for biochemical studies. Ganglioside G_{M1} exists ubiquitously in all eukaryotic cells. In addition, the central role of the metabolism of cyclic AMP, which is activated by choleragen, certifies the importance of the G_{M1}-choleragen system (Lin and Taniuchi, 1980).

The initial step of choleragen-cell interaction is the high affinity, specific binding of choleragen to ganglioside G_{M1}. This association triggers the penetration of choleragen so that it can modify intracellular adenylate cyclase. Although the initial and final steps of choleragen action have been exhaustively characterized, the intervening steps have received little study.

In this article, current knowledge of the ganglioside G_{M1}-dependent internalization of choleragen will be reviewed. Since the molecular events in the receptor-mediated endocytosis of choleragen are scarcely known, an attempt is made to fill the gaps of our current understanding of this important process with some logical hypotheses. It is hoped that they will provide useful guidelines for future endeavors.

Although no two receptor systems are completely identical, approach and strategy can often be successfully transplanted from one system to another. For this reason, relevant information about other protein receptor systems is also presented.

Receptor-mediated internalization of ligand is most likely a continuous dynamic process. To simplify discussion, the following stages of choleragen internalization are separately discussed: a) Association between ligand and receptor, b) lateral movement of ligand-receptor complex in the plane of membrane, c) transmembrane movement of ligand-receptor complex, d) lysosomal degradation of ligand, and e) receptor processing.

ASSOCIATION BETWEEN LIGAND AND RECEPTOR

Choleragen (MW 84,000 daltons) is composed of three dissimilar peptides (Moss and Vaughan, 1979): The A-1 subunit (one each, MW 21,000 daltons), the A-2 subunit (one each, MW 7,000 daltons), and the B subunit (five each, MW 11,600 daltons). The A-1 and A-2 subunits are held together by disulfide bonds. The A subunits are apparently linked to the B subunit through non-covalent interactions.

The initial steps in the choleragen action is the high affinity, specific association between its B subunit and the oligosaccharide moiety of ganglioside G_{M1} in the plasma membrane (Fishman et al., 1980). The evidence for this conclusion is overwhelming. The sole function of the binding of the B subunit of choleragen to membrane is to inject the choleragen molecule intracellularly so that the A-1 subunit can gain access to and activate adenylate cyclase. This phenomenon can be best illustrated by the observation that the activation of adenylate cyclase by the A-1 subunit in broken cells no longer requires the B subunit (Gill and King, 1975).

The activity of the receptors can be controlled either by affinity for ligands or by the number of receptors. There are wide ranges of G_{M1} content in different cell types. Furthermore, there are at least three different ways to manipulate the number of G_{M1} in the cultured cells. a) Exogenous gangliosides can be inserted into membrane (Jeng et al., 1981; Fishman et al., 1980). b) Certain chemical agents, such as sodium butyrate, can increase markedly the G_{M1} content of cells. c) Finally, neuraminidase treatment of cells can convert G_{D1a}, G_{D1b}, G_{T1a}, and G_{T1b} to G_{M1}. This treatment converts about 90% of brain gangliosides into ganglioside G_{M1} (neuraminidase does not remove the sialic acid of G_{M1}). All of the above treatments are known to endow cells with new functional receptors for choleragen. Responsiveness of cells to choleragen is drastically increased by the above treatments.

The ease with which the numbers of receptors can be tailored to experimental needs has provided investigators with an unmatched flexibility in studying many otherwise unsolvable problems. Similar techniques for receptors which are proteins rather than glycolipids have been achieved only recently. The spontaneous insertions of exogenous protein receptors into membrane are accomplished in the presence of added fusogenic agents (Schramm, 1979; Doyle et al., 1979) or even in the absence of fusogenic agents.

Although receptor densities (G_{M1} contents) vary greatly with cells, the affinities of the interactions of different cells are quite similar. The affinity of interaction between gangliosides and choleragen at 4° is stronger than that at 37°. This reflects the slower dissociation rate of ligands from receptors at a lower temperature. The internalization rate is highly temperature-dependent and is at least one order of magnitude faster than the dissociation rate. For this reason, the interaction between ligand and the receptor is best studied at 4°, at which internalization is eliminated completely (Van Sande et al., 1979).

LATERAL MOVEMENT OF LIGAND RECEPTOR COMPLEX

One of the unique properties of ganglioside as a receptor is its size. The molecular weight of epidermal growth factor receptor is 170,000 daltons. Due to the smaller size of ganglioside (MW 1,545), neuroblastoma cells can accomodate as many as ten million copies of G_{M1} per cell. (As a basis of reference, there are about 30,000 to 50,000 copies of low density lipoproteins (LDL) receptor molecules per fibroblast.) The incredibly high receptor density has a unique and unusual consequence in the lateral movement of ligand receptor complex.

Unoccupied receptors are freely mobile and are distributed randomly and diffusely in the membrane in the absence of ligand. (Some receptors, for example LDL receptors, are sequestered in a specialized structure called coated pits, which occur even in the absence of LDL.) G_{M1} is normally spread evenly over the entire membrane. Upon choleragen binding to G_{M1}, patches form in the membrane. According to current concept, intracellular membrane-associated actin or myosin may be accumulated into patches that locate directly under the occupied receptor. In the case of choleragen, it was found that cytochalasin B and colchicine blocked the redistribution of the ligand-receptor complex, suggesting that microfilaments and microtubulers are involved in the cap formation. The initial clustering of choleragen ganglioside complexes into small patches is an energy-independent process. At this stage, ligand-receptor complexes become immobilized and further movement in the plane of membrane requires metabolic energy and is temperature-dependent.

The redistribution of choleragen-ganglioside complexes in the plane of membrane has been demonstrated by three different laboratories. Fluorescein-conjugated antibodies against choleragen (Revesz and Greaves, 1975), fluorescein-labeled choleragen (Craig and Cuatrecasas, 1975; Bennett et al., 1976) and fluorescence-labeled sialoglycolipid (Sedlacek et al., 1976) were used to demonstrate cap formation after the binding of choleragen to ganglioside in the cell surface. It was reported that the cap formation induced by choleragen was markedly reduced by the pretreatment of cells with proteases, suggesting that protein may be involved in the cap formation.

Ganglioside G_{M1} tends to polymerize (Formisano et al., 1979), and the choleragen B subunit exists as a ring-shape pentamer (Gill, 1976). The possible role of multiple attachment of choleragen to ganglioside is intriguing. The effects of receptor density and multivalent binding on the activation of adenylate cyclase by choleragen in whole cells have been elucidated (Fishman, 1980; Fishman and Atikkan, 1980). The lateral mobility of toxin-receptor complexes is required to achieve multivalent binding in cells with few receptors.

In contrast, cells with a high receptor density can achieve multiple attachment without lateral mobility. Neuroblastoma cells in which lateral mobility is not required for the internalization of choleragen have exceedingly high receptor density. This incredibly high receptor density allows multivalent interaction between receptor and ligand without the "usual" reorganization of complexes in the membrane.

In summary, the lateral movement of choleragen-receptor complexes is required to achieve multivalent binding in cells with a low receptor density. The cap formation is dependent on metabolic energy, time and temperature. The lag period, the time required for choleragen to penetrate into cells and to release its active fragment, is dictated by G_{M1} content of the cell. Choleragen-ganglioside complex has to form a cap to be internalized, as is the case in any receptor system where protein serves as a receptor. We do have an unusual situation, in that extraordinary high receptor density in some cells makes the lateral movement of ligand receptor complex unnecessary for the cap formation of ligand receptor polymers.

TRANSMEMBRANE MOVEMENT OF LIGAND RECEPTOR COMPLEX

Ganglioside G_{M1} was found, by immunoelectron microscopic methods, to be exclusively on the external side of membrane. Ganglioside is too short (about 30 A^{o}) to span the membrane. It is almost inevitable that proteins (or other membrane components) participate in the internalization of ganglioside-choleragen complexes.

It was found that dansylcadaverine, a potent transglutaminase inhibitor, blocked the internalization of choleragen in a variety of cells (Van Sande et al., 1979; Jeng, unpublished observation). Pastan and his coworkers (Davies et al., 1980; Levitzki et al., 1980) advanced a proposal that the clustering of ligand-receptor complexes is reversible, but the subsequent cross-linking of ligand-receptor complexes by transglutaminase leads to the progressive accumulation of coated vesicles. This hypothesis has been confirmed repeatedly in many systems. The observation that dansylcadaverine inhibits the internalization of choleragen suggests that transglutaminase might also participate in the ganglioside-mediated penetration of choleragen.

In summary, ganglioside as a receptor has properties that are very similar, if not

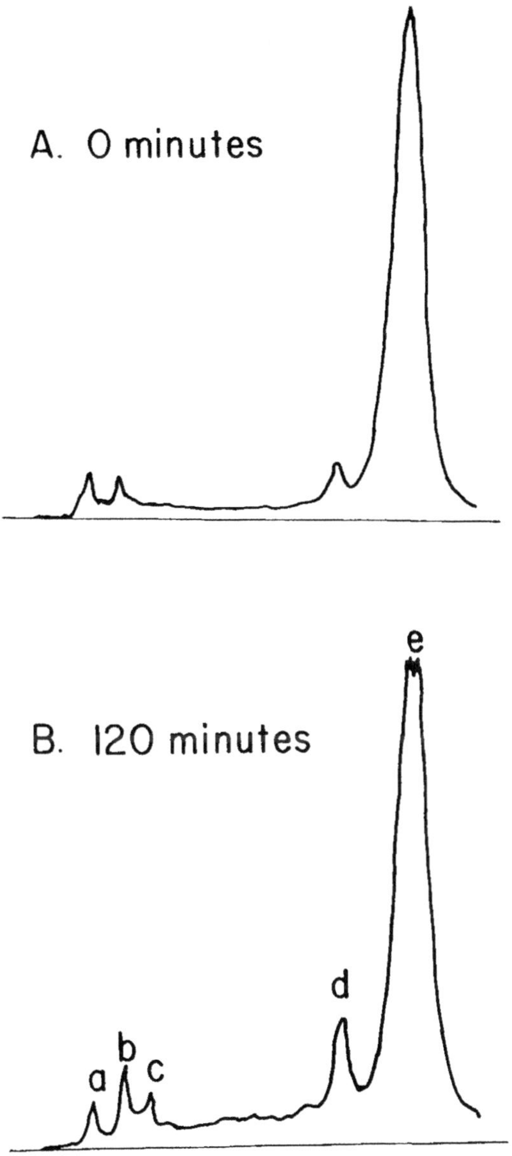

Figure 1: Cross-linking of ^{125}I-choleragen B subunit to membrane proteins.

identical, to those mechanisms of internalization found in other systems in which proteins serve as receptors.

The nature of transglutaminase requires its substrates to be protein. However, the receptor in the choleragen-G_{M1} system is a glycolipid which can not be recognized and modified by transglutaminase. Therefore, a protein intimately connected with ganglioside G_{M1} must provide this essential link between transglutaminase and ganglioside G_{M1}. One possible hypothesis is that G_{M1} serves as the recognition site for choleragen, and this putative ganglioside binding protein serves as a substrate for transglutaminase. G_{M1} forms a tight complex with membrane-bound protein. This complex should be considered as the unit of receptor in receptor-dependent endocytosis.

An attempt was made to identify the protein adjacent to ganglioside in brain membrane. Radioactive choleragen B subunit was allowed to associate with ganglioside G_{M1} noncovalently. Choleragen-membrane complex was isolated for a cross-linking study. A bifunctional reagent, dimethylsuberimidate, was added so that choleragen could be linked irreversibly to protein adjacent to ganglioside G_{M1} in the membrane. We found (Fig 1) that the formation, a band of 160,000 daltons (band C), was time and membrane-dependent. Whether or not this protein plays a role in ganglioside activity remains to be determined.

The B subunit plays a direct role in the entry process for the A subunit. How the B subunit helps the internalization of A subunit is not clear at the present time. It is known that binding of G_{M1} induces conformational changes in the B subunit. The conformational changes in the B subunit may result in the exposure of the hidden hydrophobic region which is capable of associating with membrane hydrophobic protein during the internalization process.

LYSOSOMAL DEGRADATION

Endocytosis leads to the deposition of ligand receptor into the lysosome where the ligand is destroyed. Receptor is usually spared from destruction and acts as a "ligand shuttle," a repeated transfer of ligand from the cell surface to the site of destruction.

In an elegant discussion of membrane economics in endocytotic systems, Duncan and Pratten (1977) pointed out that when two spherical particles fuse, some membrane materials have to be removed to preserve the spherical shape of the final particle. It is possible that some membrane materials, including receptor, are removed during the fusion of endocytotic vesicles with lysosomes. Recently, compartmental analysis and computer modeling revealed that the dissociation of ligand and receptor precedes ligand degradation; i.e., ligand remains associated with receptor after internalization, but the complex becomes dissociated only prior to degradation by lysosome (Bridges et al., 1982).

The involvement of lysosomal processing in the choleragen activation of adenylate cyclase by choleragen was established. It was found that the lysosomotropic agents, chloroquine and ammonium chloride, blocked the activation of adenylate cyclase by choleragen, indicating that the lysosme participates in releasing active fragments from choleragen (Lin and Taniuchi,1980).

We studied the sensitivity of the degradation of ^{125}I-choleragen to lysosomotropic agents and transglutaminase inhibitor. We measured the degradation of choleragen by trichloroacetic acid-soluble counts released in the medium. It was found that choleragen degradation is time and temperature-dependent. Furthermore, chloroquine inhibited the degradation of choleragen drastically. This indicated the participation of lysosome in the degradation of choleragen.

The effects of various compounds on the fate of ^{125}I-holotoxin were also determined. It was found that both the lysosomotropic agents and transglutaminase inhibitors significantly increased cell-associated counts (Table 1), which represent both the total of cell surface-bound and internalized choleragen. Further analysis showed that the counts were dissociable from complex by nonradioactive choleragen and were not affected by chloroquine and danslycadaverine. This indicates that these two compounds did not interfere with the interaction between ligand and receptor. Methylamine, a known inhibitor of transglutaminase, increased the reversible bindings only slightly (20%). Irreversible bindings which represent internalized choleragen were increased by all three reagents.

The lysosomal degradation of choleragen deserves additional comments. Although the A-1 peptide, the active component, can be generated from intact holotoxin by a denaturating agent and a reducing agent in vitro, an active fragment generated in vivo has not yet been identified. Usual extensive lysosomal degradation of ligand may be inappropriate for the processing of choleragen since the processed product has to catalyze the ADP-ribosylation of GTP binding protein of adenylate cyclase.

PROCESSING OF RECEPTORS

Receptors, whose major function is to transfer ligands intracellularly, internalize a large number of ligands per unit time. It is essential that the receptor be reutilized. How the lysosome allows receptor to escape from destruction is not known. We have only a hint about the release of the receptor. Sutrina and Chen (1982) found that the lack of lysosomal degradation of the ceramide-containing LDL in Faber's disease fibroblasts (ceraminidase deficient) results in the failure of recycling of LDL receptor. Thus, the degradation of ligand by lysosome is intimately connected with the return of receptor to

TABLE 1: EFFECTS OF VARIOUS COMPOUNDS ON THE FATE OF ^{125}I-HOLOTOXIN

		Cell-Associated	Irreversible Binding	Reversible Binding
	mM	Radioactivity X 10^{-3}		
Control		72	48	24
Methylamine	2	88	59	30
	4	84	56	28
	6	89	60	29
Dansylcadaverine	.1	78	56	22
	.2	78	57	22
	.5	82	58	23
Chloroquine	.1	85	63	23
	.3	79	55	25

the exterior side of membrane. Recently, it was demonstrated that recycling of receptor may be dependent on extracellular environment, illustrating the complexity of the process.

Ganglioside receptor belongs to the class of receptors in which the main function of receptors is to internalize ligands to intracellular sites. The recycling of the receptor is a demonstrated necessity for this class of protein receptor. The unique situation of the smaller size in G_{M1} may make this an exception to the rule.

DISCUSSION

In terms of receptor-dependent endocytosis, the ganglioside-choleragen system apparently shares, at least in part, the cellular apparatus of other receptor systems in which proteins serve as receptors. The unexpected requirements of transglutaminase and lysosome in the expression of choleragen activity suggests that choleragen is internalized by a "normal" protein receptor route. This may argue in favor of the involvement of protein(s) in the choleragen internalization. The mechanism of choleragen internalization is far from established, but sufficient information is now available to warrant attempts to identify protein(s) in this important process.

REFERENCES

Bennett, V., Craig, S., Hollenberg, M.D., O'Keefe, E., Sahyoun, N. and Cuatrecasas, P. (1976) Structure and function of cholera toxin and hormone receptors. J. Supramol. Struc. 4, 99-120.

Bridges, K., Harford, J., Ashwell, G. and Klausner, R.D. (1982) Fate of receptor and ligand during endocytosis of asialoglycoproteins by isolated hepatocytes. Proc. Natl. Acad. Sci. USA 79, 350-354.

Craig, S.W. and Cuatrecasas, P. (1975) Mobility of cholera toxin receptors on rat lymphocyte membranes. Proc. Natl. Acad. Sci. USA 73, 3844-3848.

Davies, P.J.A., Davies, D.R., Levitzki, A., Maxfield, F.R., Milhaud, P., Willingham, M.C. and Pastan, I.H. (1980) Transglutaminase is essential in receptor-mediated endocytosis of 2-macroglobulin and polypeptide hormones. Nature 283, 162-167.

Doyle, D., Hou, E. and Warren, R. (1979) Transfer of the hepatocyte receptor for serum asialoglycoprotein to the plasma membrane of a fibroblast. J. Biol. Chem. 254, 6853-6856.

Duncan, R. and Pratten, M.K. (1977) Membrane economics in endocytic studies. J. Theor. Biol. 66, 727-735.

Fishman, P.H. (1980) Mechanism of action of cholera toxin: Studies on the lag period. J. Memb. Biol. 54, 61-72.

Fishman, P.H. and Atikkan, E.E. (1980) Mechanism of action of cholera toxin: Effect of receptor density and multivalent binding on activation of adenylate cyclase. J. Memb. Biol. 54, 51-60.

Fishman, P.H., Moss, J. and Osborne, J.C., Jr. (1978) Interaction of choleragen with the oligosaccharide of ganglioside G_{M1}: Evidence for multiple oligosaccharide binding sites. Biochem. 17, 711-716.

Fishman, P.H., Pacuszka, T., Hom, B. and Moss, T. (1980) Modification of ganglioside G_{M1}. J. Biol. Chem. 255, 7657-7664.

Formisano, S., Johnson, M.L., Lee, G., Aloj, S.M. and Eddhoch, H. (1979) Critical micelle concentration of gangliosides. Biochemistry 18, 1119-1125.

Gill, D.M. (1976) The arrangement of subunits in cholera toxin. Biochemistry 15, 1242-1248.

Gill, D.M. and King, C.A. (1975) Mechanism of action of cholera toxin in pigeon erythrocyte lysates. J. Biol. Chem. 250, 6424-6432.

Holmgren, J., Elwing, H., Fredman, P., Strannegard, O. and Svennerholm, L. (1980) Gangliosides as receptors for bacterial toxins and sendai virus. In: Structure and Function of Gangliosides (Svennerholm, L., ed.) Plenum Press, N.Y. 125, 453-470.

Jeng, I., Walz, M.A., Klemm, N., Harnacker, B. and Su, S.N. (1981) Insertion of ^{125}I-ganglioside into neuronal membrane. In: Glycoconjugate (Yamakawa, T., Osawa, T. and Handa, eds.) Japan Sci. Soc., Tokyo. p331.

Kaplan, J. (1981) Polypeptide-binding membrane receptors: Analysis and classification. Science 212, 14-20.

Levitzki, A., Willingham, M. and Pastan, I.H. (1980) Evidence for participation of transglutaminase in receptor-mediated endocytosis. Proc. Natl. Acad. Sci. USA 77, 2706-2710.

Lin, M.C. and Taniuchi, M. (1980) Inhibition of cholera toxin activation of the adenylate cyclase system in intact HeLa cells. J. Cyclic Nucleotide Res. 6, 359-367.

Moss, J. and Vaughan, M. (1979) Activation of adenylate cyclase by choleragen. Ann. Rev. Biochem. 48, 581-600.

Revesz, T. and Greaves, M. (1975) Ligand induced redistribution of lymphocyte membrane ganglioside G_{M1}. Nature 275, 103-106.

Schramm, M. (1979) Transfer of glucagon receptor from liver memoranes to a foreign adenylate cyclase by a membrane fusion procedure. Proc. Natl. Acad. Sci. USA 76, 1174-1178.

Sedlacek, H.H., Stark, J., Seiler, F.R., Ziegler, W. and Wiegandt, H. (1976) Cholera toxin induced redistribution of sialoglycolipid receptor at the lymphocyte membrane. FEBS Lett. 61, 272-276.

Sutrina, S.L. and Chen, W.W. (1982) Metabolism of ceramide-containing endocytotic vesicles in human diploid fibroblasts. J. Biol. Chem. 257, 3039-3044.

Van Sande, J., Pochet, R. and Dumont, J.E. (1979) Dissociation by cooling of hormone cholera toxin activation of adenylate cyclase in intact cells. Biochim. Biophys. Acta 585, 282-292.

BIOCHEMICAL STUDIES OF TAY-SACHS DISEASE

Yu-Teh Li and Su-Chen Li

Dept. of Biochemistry, Tulane University
Medical Center, New Orleans, Louisiana 70112
U.S.A., and Delta Regional Primate Research
Center, Covington, Louisiana 70433, U.S.A.

Tay-Sachs disease (GM2-gangliosidosis) is one of the most common ganglioside storage diseases characterized by the accumulation of a large amount of GM2 ganglioside (Tay-Sachs ganglioside) in the neural tissues. The catabolism of this ganglioside has been a subject of interest to many investigators. In 1968, Robinson and Stirling described the presence of two β-hexosaminidase isozymes (β-hexosaminidase A and B) in human spleen. Soon thereafter, Okada and O'Brien (1969), Hultberg (1969), and Sandhoff (1969) found that the tissues of classical Tay-Sachs patients were missing in β-hexosaminidase A isozyme. These two findings constitute the two most important events in elucidating the etiology of Tay-Sachs disease. So far, three biochemically distinct types of GM2-gangliosidosis have been described (O'Brien, 1978). Type-B (classical Tay-Sachs disease) has been established to be caused by the deficiency of β-hexosaminidase A, and Type-O (Sandhoff's disease), by the deficiency of both β-hexosaminidases A and B. Type-AB (late infantile or juvenile GM2-gangliosidosis) is characterized by the presence of a normal or elevated level of both β-hexosaminidases A and B. These three types were classified by analyzing the β-hexosaminidase isozymes in the tissues of patients using synthetic substrates such as p-nitrophenyl-β-N-acetylglucosaminide.

In order to understand the biochemical etiology of different GM2-gangliosidoses, we have undertaken a study of the normal catabolism of GM2 ganglioside *in vitro*. In 1973 we found that, in the absence of detergent, the crude β-hexosaminidase isolated from human liver could hydrolyze GM2 ganglioside better than the purified β-hexosaminidase A or B, even though there were no differences in their activities toward synthetic substrates such as p-nitrophenyl-β-N-acetylglucosaminidase. We further found that a highly purified β-N-acetylhexosaminidase A isolated from human liver or urine required a heat-stable activator to stimulate the hydrolysis of GM2 ganglioside (Li *et al.*, 1973). This activator was apparently removed from the enzyme during purification. We subsequently isolated this activator in highly purified form from human liver (Li *et al.*, 1981) and brain (Hirabayashi *et al.*, 1981). This presentation briefly reviews our work on the isolation and characterization of the protein activator which stimulates the enzymic hydrolysis of GM2 ganglioside, and the studies of this activator in three types of GM2-gangliosidosis.

Isolation of the Protein Activator which Stimulates the Enzymic Hydrolysis of GM2 Ganglioside

We have purified a protein activator (GM2-activator) which is specific for the hydrolysis of GM2 ganglioside from human liver (Li *et al.*, 1981). The purification procedure includes adjustment of the pH of the liver extract to pH 4.3, followed by ammonium sulfate precipitation, Sephadex G-200 filtration, and column chromatographies on DEAE-Sephadex A-50, Matrex Gel Blue A, and Octyl-Sepharose 4B. The GM2-activator is moderately heat-stable up to 50°C in crude states, but unstable in the purified form even at 37°C. The molecular weight of this activator is about 23,500 as determined by Sephadex G-75 gel filtration and pI is about pH 4.8. The activator moved as one major protein band when examined by a flat bed agarose gel isoelectrofocusing. Since the storage of GM2 ganglioside in Tay-Sachs patients occurs mainly in the central nervous system, we extended our study to GM2-activator in the brain. By using the same procedure applied for the liver including ampholyte displacement chromatography (Leaback and Robinson, 1975), we have also isolated the same activator from the acetone dried powder of normal human brain (Hirabayashi et al., 1982). The level of GM2-activator in the brain was found to be three times higher than that in the liver. The properties and the specificity of the activator isolated from the brain were identical to that isolated from the liver. The antiserum against the brain activator also cross-reacted with the liver activator (Hirabayashi et al., 1982). The molar ratio of enzyme to the activator to obtain the maximal hydrolysis of GM2 ganglioside was about 1:1, while the molar ratio of GM2 ganglioside to the activator was about 300:1. These results may suggest that the activator interacts with β-hexosaminidase A rather than the lipid substrate. We further found that this activator is extremely species specific. Only the activators isolated from the brains of rhesus monkey and baboon, but not those from porcine and bovine brain could stimulate the hydrolysis of GM2 ganglioside catalyzed by human hepatic β-hexosaminidase A. This may indicate the existence of an intimate yet unexplained relationship between the activator and β-hexosaminidase A. This property is quite different from the activator which stimulates the enzymic hydrolysis of GM1 ganglioside (Li et al., 1979).

The Levels of GM2-Activator and β-Hexosaminidases in the Brains of Three Types of GM2-Gangliosidosis

Table I summarizes the in vitro analysis of GM2-activator in 10 g each of the brains from one case of Type-B, two cases of Type-O and two cases (Cases 1 and 2) of Type-AB GM2-gangliosidosis (Goldman et al., 1980). Two normal brains, ages 2.5 and 2.7 years, were used as controls. All pathological brains, except Case 2 Type-AB GM2-gangliosidosis, contained higher levels of GM2-activator than the two controls. The level of GM2-activator in the brain of Case 2 Type-AB GM2-gangliosidosis was found to be extremely low.

The level of total β-hexosaminidase activity in the three types of GM2-gangliosidosis varied. Two cases of Type-O were devoid of this enzyme as expected. The case of Type-B and Case 1 of Type-AB had only slightly higher

TABLE I

The Levels of GM2-Activator and Total β-Hexosaminidases in the Brains of Three Types of GM2-Gangliosidosis*

	$N_{2.7}$	$N_{2.5}$	Type-AB Case 1	Type-AB Case 2	Type-O Case 1	Type-O Case 2	Type-B
Tissue (g)	10 10	10 10	20.3	10.4	10	10	10
Total β-hexosaminidase (unit)	5.4 5.2	5.0 4.8	7.9	14.2	0	0	7.2
GM2-activator (x 10^{-5} unit)	0.89 0.65	0.82 0.54	2.94	0.04	2.26	3.20	3.80

*Analyses were carried out according to the procedure described by Li et al. (1981).

activity of this enzyme than that in the normals. It is noteworthy that the level of this enzyme in Case 2 of Type-AB was three times higher than that of the two controls.

Immunological Detection of GM1-Activator and GM2-Activator in the Pathological Brains

In order to critically establish the presence or absence of GM2-activator, the activator fractions from different brain samples were further examined by immunoelectrophoresis using the antiserum against the brain GM2-activator. Since normal tissues contain both GM1-activator and GM2-activator (Li *et al.*, 1979), these two activators were examined simultaneously. As shown in Fig. 1, all the brain samples, with the exception of Case 2 Type-AB GM2-gangliosidosis, contained both GM1-activator and GM2-activator. Case 2 Type-AB GM2-gangliosidosis, on the other hand, contained only GM1-activator, but was devoid of GM2-activator. These results agreed with the *in vitro* assay based on the stimulation of GM2 hydrolysis as presented in Table I.

Properties of β-Hexosaminidases from the Brains of Two Cases of Type-AB GM2-Gangliosidosis

The results of immunological analysis and the *in vitro* hydrolysis of GM2 ganglioside indicated that Case 1 and Case 2 Type-AB GM2-gangliosidosis were not the result of the same biochemical lesions. The accumulation of GM2 ganglioside in Case 2 could be explained by the absence of GM2-activator. However, Case 1 contained an elevated level of GM2-activator with no deficiency of β-hexosaminidase activity toward the hydrolysis of synthetic substrates (Table I). Thus, the accumulation of GM2 ganglioside in Case 1 could not be explained on the basis of the deficiencies of the activator or the enzyme. We subsequently examined the ability of β-hexosaminidases prepared from the brains of Case 1, Case 2, and a normal control to hydrolyze GM2 ganglioside, asialo GM2 and

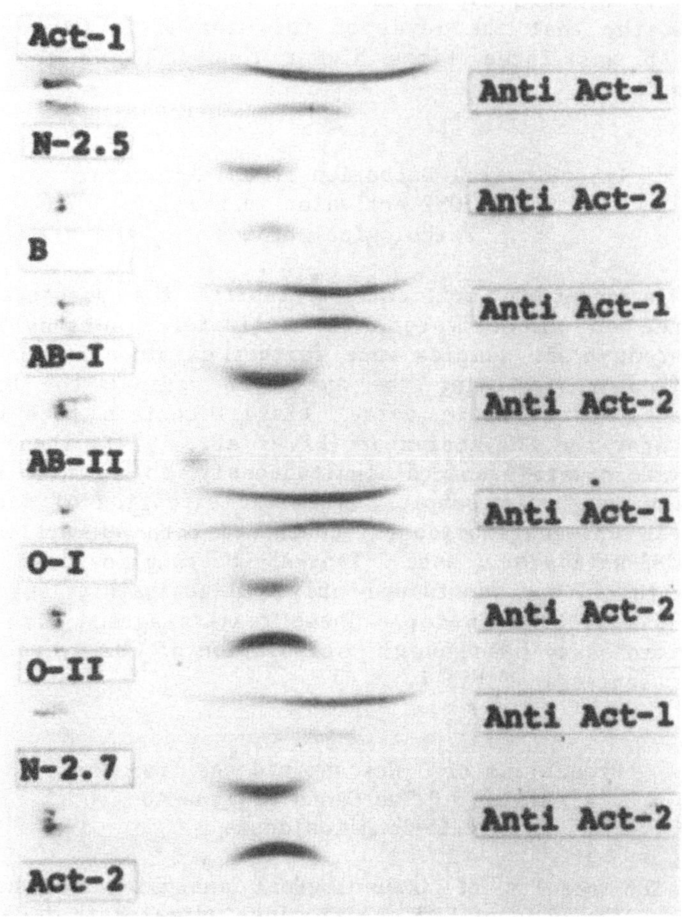

Fig. 1. Detection of GM1-activator and GM2-activator in the pathological brains by immunoelectrophoresis. The samples in the wells were: Act-1, pure GM1-activator; Act-2, pure GM2-activator; N-2.5 and N-2.7, activator fractions derived from the two normal controls of 2.5 and 2.7 years of age. The activator fractions obtained from three types of GM2-gangliosidosis are: B, Type-B; O-I, Case 1 of Type-O; O-II, Case 2 of Type-O; Anti-Act-1, rabbit antiserum against GM1-activator; Anti-Act 2, guinea pig antiserum against GM2-activator.

$GbOse_4Cer$. As shown in Table II, neither β-hexosaminidase A_1 nor A_2 (Li et al., 1981) isolated from the brain of Case 1 could hydrolyze any of the aforementioned glycosphingolipids in the presence of the purified GM2-activator. However, both were able to cleave asialo GM2 and $GbOse_4Cer$ in the presence of sodium taurodeoxycholate. The A_1 and A_2 isozymes from Case 1 exerted only very slight activity toward GM2 ganglioside even in the presence of sodium taurodeoxycholate. It is noteworthy that among β-hexosaminidases A_2 prepared from Case 1, Case 2, and the normal control, only the enzyme isolated from Case 1 showed no activity or very little activity toward the hydrolysis of GM2 ganglioside in the presence of either GM2-activator or sodium taurodeoxycholate. The β-hexosaminidases B from all three brains were active toward the hydrolysis of asialo GM2, and $GbOse_4Cer$. They exhibited, however, only a slight activity toward GM2 ganglioside in the presence of sodium taurodeoxycholate.

Discussion

Our biochemical studies on the catabolism of GM2-ganglioside led us to discover the presence of the protein activator for the hydrolysis of GM2 ganglioside catalyzed by β-hexosaminidase A. Through the studies of this activator in the three types of GM2-gangliosidoses, we have revealed for the first time the heterogenous nature of Type-AB GM2-gangliosidosis.

The accumulation of GM2 ganglioside in the brains of two cases of Type-AB GM2-gangliosidosis (Goldman et al., 1980) was clearly due to the two entirely different causes. Case 1 was apparently due to a mutation of the structural gene for β-hexosaminidase A in such a manner that the enzyme became nonresponsive to the stimulation of GM2-activator for the hydrolysis of GM2 ganglioside, asialo GM2 and $GbOse_4Cer$ (Table II), but still retained the activity toward synthetic substrates. The β-hexosaminidase A_1 from this case was found to cross-react with the antibody against the α-chain of normal β-hexosaminidase A (Hirabayashi et al., 1982). Thus, Case 1 may be the result of mis-sense mutation of β-hexosaminidase A. Case 2, on the other hand, was found to be deficient in GM2-activator as shown by an in vitro assay as well as by

TABLE II

Specificity of β-Hexosaminidases from Normal and Patient Brains*

Substrate	Addition of	Source of β-Hexosaminidase							
		Case 1 Type AB			Case 2 Type AB			Normal	
		A_1	A_2	B	A_1	A_2	B	A_2	B
GM2	Control (15 h)	0	0	0	3.0	7.0	0	6.0	0
	TDC (15 h)	3.3	1.9	14.6	10.3	57.9	5.8	63.6	13.8
	Act-2 (15 h)	0	0	0	27.1	54.0	0	49.5	0
Asialo GM2	Control (15 h)	0	0	0	0.1	0.2	0	0.2	0
	TDC (1 h)	86.9	77.2	80.0	62.1	82.2	72.3	72.2	83.5
	Act-2 (15 h)	0	0	0	24.7	30.2	0	28.7	0
$GbOse_4Cer$	Control (15 h)	0	0	0	0	0	0	0	0
	TDC (1 h)	95.1	99.1	99.2	84.2	92.4	79.6	96.9	94.2
	Act-2 (15 h)	0	0	0	4.6	9.9	0	19.1	0

*Analyses were carried out according to the procedure described by Li _et al_. (1981).

TABLE III

New Classification of Tay-Sachs Disease

In addition to the presence or absence of β-hexosaminidase A (Hex A) and GM2-activator, this classification also includes the ability of Hex A to cleave GM2 ganglioside in the presence of the activator.

Type	Hex A	Hex B	Activator	GM2-cleavage by Hex A
B	-	+	+	
0	-	-	+	
AB[1]	+	+	-	+
New[2]	+	+	+	-

[1] Proposed nomenclature for this variant: Type-AB-Activator deficient variant

[2] Proposed nomenclature for this variant: Type-AB-β-Hex A defective variant

an immunological assay. The gene mutation for GM2-activator, therefore, may represent a no-sense mutation. This is the first report of no-sense mutation of GM2-activator in Type-AB GM2-gangliosidosis. It is very interesting to note that these two cases were clinically different. As reported by Goldman *et al.* (1980), Case 2 suffered an earlier onset of symptoms and a more rapid clinical course than Case 1 whose clinical course was somewhat milder and more protracted. Also, the accumulation of GM2 ganglioside and asialo GM2 was more marked in Case 2. The result of our analysis on Case 2, the case of activator deficiency, is consonant with the findings in two other patients (Conzelmann and Sandhoff, 1978; Hechtman *et al.*, 1982) based on *in vitro* assay. The *in vitro* assay, however, cannot distinguish no-sense mutation from mis-sense mutation.

Table III summarizes the classification of various types of GM2-gangliosidoses revealed up to this date. In addition to the presence or absence of β-hexosaminidase A and GM2-activator, this classification also includes the ability of β-hexosaminidase A to cleave GM2 ganglioside in the presence of the activator.

Acknowledgment

This research was supported by United States Public Health Service Grants NS 09626 and RR 00164.

References

Conzelmann, E. and Sandhoff, K. (1978) AB variant of infantile GM2 gangliosidosis: Deficiency of a factor necessary for stimulation of hexosaminidase A - catalyzed degradation of ganglioside GM2 and glycolipid GA2. Proc. Natl. Acad. Sci. U.S.A. 75, 3979-3983.

Goldman, J. E., Yamanaka, T., Rapin, I., Adachi, M., Suzuki, K., and Suzuki, K. (1980) The AB-variant of GM2-gangliosidosis: Clinical, biochemical, and pathological studies of two patients. Acta Neuropathol. (Berl.) 52, 189-202.

Hechtman, P., Gordon, B. A., and Ng Ying Kin, N. M. K. (1982) Deficiency of the hexosaminidase A activator protein in a case of GM2 gangliosidosis; variant AB. Pediatr. Res. 16, 217-222.

Hirabayashi, Y., Li, Y.T., and Li, S.-C. (1982) The protein activator specific for the enzymic hydrolysis of GM2 ganglioside in normal human brain and brains of three types of GM2-gangliosidosis. J. Neurochem., in press.

Hultberg, B. (1969) N-Acetylhexosaminidase activities in Tay-Sachs disease. Lancet 1195.

Leaback, D. H. and Robinson, H. K. (1975) Ampholyte displacement chromatography - A new technique for separation of proteins illustrated by the resolution of β-N-acetyl-D-hexosaminidase isoenzymes unresolvable by isoelectric focusing or conventional ion-exchange chromatography. Biochem. Biophys. Res. Commun. 67, 248-254.

Li, S.-C., Hirabayashi, Y., and Li, Y.-T. (1981) A protein activator for the enzymic hydrolysis of GM2 ganglioside. J. Biol. Chem. 256, 6234-6240.

Li, S.-C., Nakamura, T., Ogamo, A., and Li, Y.-T. (1979) Evidence for the presence of two separate protein activators for the enzymic hydrolysis of GM1 and GM2 gangliosides. J. Biol. Chem. 254, 10592-10595.

Li, Y.-T., Mazzotta, M. Y., Wan, C. C., Orth, R., and Li, S.-C. (1973) Hydrolysis of Tay-Sachs ganglioside by β-hexosaminidase A of human liver and brain. J. Biol. Chem. 248, 7512-7515.

O'Brien, J. S. (1978) The gangliosidoses, in The Metabolic Basis of Inherited Disease (Stanbury, J. B., Wyngaarden, J. B., and Fredrickson, D. S., eds.), pp. 841-865. McGraw Hill, New York.

Okada, S. and O'Brien, J. S. (1969) Tay-Sachs disease: Generalized absence of a β-D-N-acetylhexosaminidase component. Science 165, 698-700.

Robinson, D. and Stirling, J. L. (1968) N-Acetyl-β-Glucosaminidases in Human Spleen. Biochem. J. 107, 321-327.

Sandhoff, K. (1969) Variation of β-N-acetylhexosaminidase-pattern in Tay-Sachs Disease. FEBS Lett. 4, 351-354.

GANGLIOSIDES IN PERIPHERAL NERVOUS SYSTEM DURING NORMAL CELL-DIFFERENTIATION AND ONCOGENESIS

F.B. Jungalwala, K.H. Chou and C.E. Nolan

E.K. Shriver Center for Mental Retardation
200 Trapelo Road, Waltham, MA 02254 USA

Gangliosides have been considered to function as cell-surface markers. They have been implicated to be involved in the molecular mechanisms of several biological phenomena that occur on the cell-surface, such as contact inhibition, intracellular adhesions, immunochemical tissue specificity, hormone receptor function and internalization of various external macromolecular substances (review by Hakomori, 1981 and Yogeeswaran, 1983). Gangliosides are present in relatively high concentrations in mammalian neural tissues. Although a considerable amount of information on the composition and metabolism of gangliosides is available for the central nervous system (review by Ledeen, 1979), only a few studies have been reported on the gangliosides of the peripheral nervous system. The major gangliosides of the PNS generally appear to be monosialic acid-containing species, GM_1 and GM_3 (Svennerholm et al., 1972; Fong et al., 1976). However, Yates et al. (1976) have reported only a small amount (<10%) of monosialic acid-containing gangliosides in rabbit sciatic nerve. In human peripheral nerves, a glucosamine-containing monosialoganglioside having a sialic acid residue on the terminal galactose, called sialosyl-lactoneotetrosylceramide (LM_1), has been reported as the major ganglioside (Li et al., 1973; Fong et al., 1976).

Peripheral nerve is considerably simpler than CNS. It consists mainly of Schwann cells and myelin with axon fibers. During development of the nerve, immature Schwann cells proliferate and differentiate into mature Schwann cells and produce myelin. It was of interest to determine the developmental profile of gangliosides in PNS in order to correlate with the morphological changes that occur in the nerves.

Trigeminal nerve is the largest cranial nerve in the rat. In the CFE strain of rats, trigeminal nerve was the most consistent site where Schwann cell tumors, neurinoma, were produced by a single transplacental injection of a chemical carcinogen ethylnitrosourea (Cravioto et al., 1973). Solid neurinomas were produced in the trigeminal nerve of more than 80% of the fetuses after about 200 days of exposure to the drug _in utero_. It was of interest to determine the changes in the composition of gangliosides and their metabolism that occur in the trigeminal nerve soon after the exposure to ethylnitrosourea in order to understand how specific alterations in the membrane gangliosides are brought about and regulated during normal and abnormal differentiation and oncogenesis. Here we first describe the ganglioside composition and metabolism in the normal trigeminal nerve during development and then describe the changes that occur as a result of cell-differentiation induced by ethylnitrosourea.

The length of the trigeminal nerve of a rat is doubled in 30 days after birth and tripled by 90 days to about 3 cm. The wet weight of the nerve is increased seven-fold by 30 days and 14-fold by 90 days to about 40 mg. The dry weight of the nerve is increased 15-fold by 30 days and 30-fold by 90 days to about 8 mg. As the Schwann cells of the nerve differentiate between 2-20 days after birth, the DNA content is sharply increased (Fig. 1); after 20 days the DNA content remained practically the same indicating little cell-proliferation. The deposition of protein and lipid in the nerve slightly lags the DNA deposition during early age. However, the lipid sialic acid content which represents total ganglioside in the nerve basically followed the rate of DNA deposition.

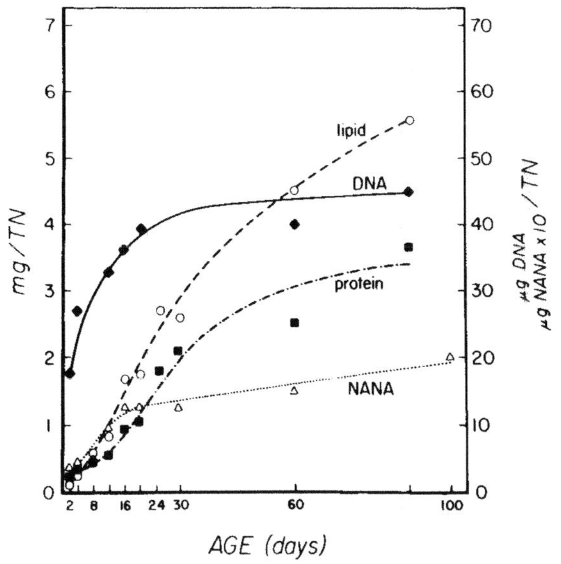

FIG. 1. Chemical constituents of trigeminal nerve with development. NANA, N-acetylneuraminic acid. Symbols ○ ,lipid; ◆ ,DNA; ■ ,protein; △ ,NANA.

The TLC pattern of gangliosides of rat trigeminal nerve during development is shown in Fig. 2. The ganglioside composition of the rat nerve is entirely different from that of rat brain. The nerve contained considerable amounts of GM_3, GD_3 and a monosialoganglioside that migrated slightly above rat brain GM_1. These gangliosides were present in only minimal amounts in rat brain. The latter of the three gangliosides in the nerve was characterized as sialosyl-lactoneotetrosylceramide (LM_1) by analysis of the products of acid hydrolysis and enzymatic degradation. As noted previously, LM_1 is also the major ganglioside of the human peripheral nerves (Li et al., 1973; Fong et al., 1976). With the development of the trigeminal nerve, the mole percentage composition of GM_3 gradually declined, whereas that of LM_1 increased (Figs. 2,3). In young animals GD_3 was resolved into two bands (Fig. 2).

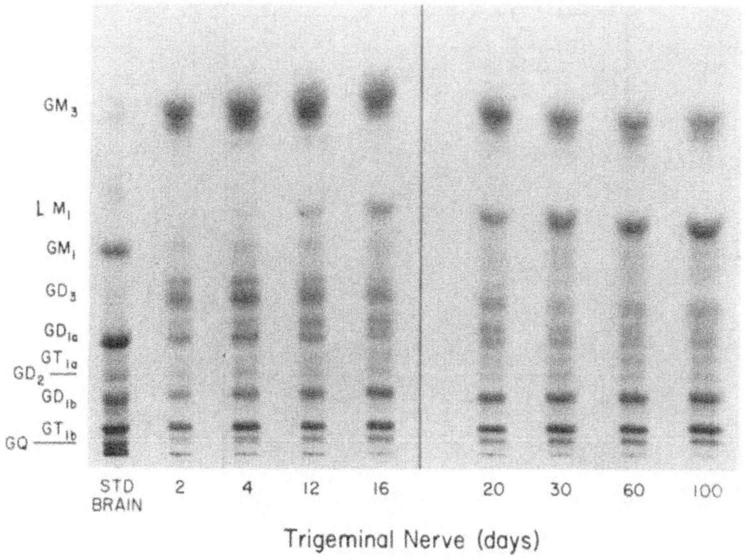

FIG. 2. Thin-layer chromatogram of ganglioside of trigeminal nerve during development. Std.: Ganglioside from rat brain. Numerals indicate the age of the rat. For each sample, gangliosides having approximately 10 µg of sialic acid were applied to HPTLC plates, which were developed with chloroform:methanol:water (55:45:10) containing 0.25% $CaCl_2$. Spots were visualized with resorcinol-HCl reagents.

The concentration of GD_3, particularly the upper band, was reduced with development (Fig. 2,3). GD_{1b}, GT_{1b} and GQ also increased with development (Fig. 2,3). Rat brain GM_1 was present only in minimal amounts in the nerve. Just after birth the major gangliosides of the trigeminal nerve were GM_3 (50 mole %) and GD_3 (18 mole %). However, in the adult nerve LM_1 and GM_3 were the major gangliosides, each representing about 25-30 mole %.

The rate of deposition of individual gangliosides in the rat trigeminal nerve during post-natal development was compared with that of myelin marker

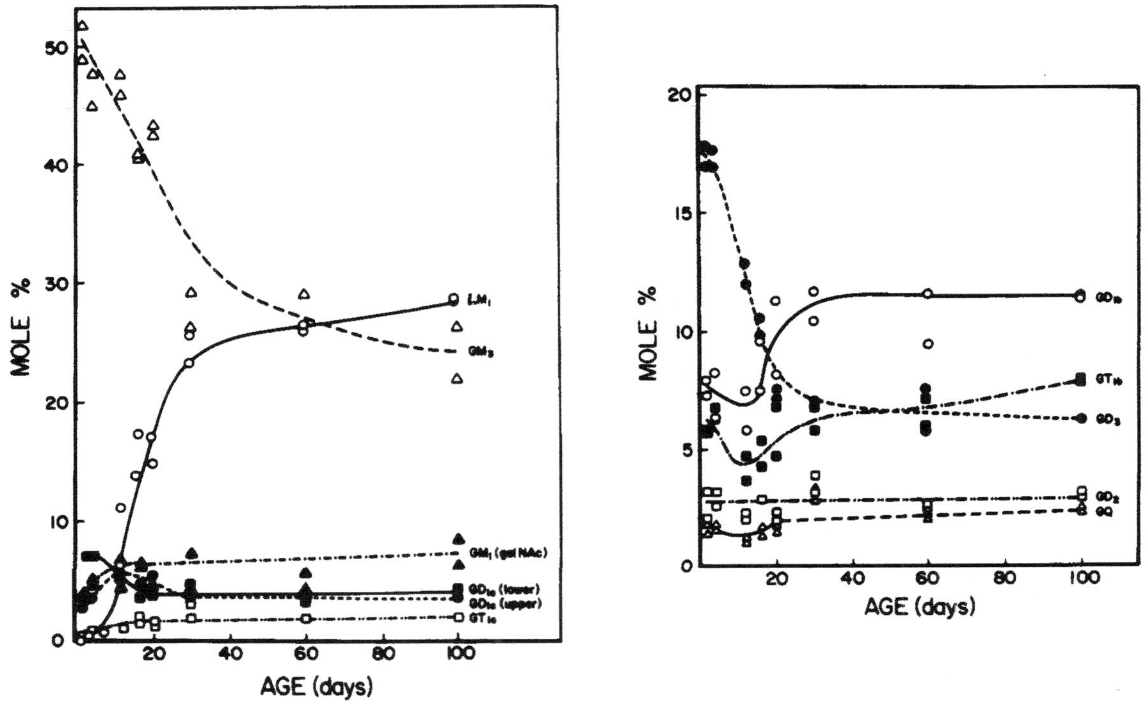

FIG. 3 Percentage compositions of trigeminal nerve gangliosides during development. A(left): ○, LM_1; △, GM_3; ▲, GM_1; ■, GD_{1a}(lower); , GD_{1a}(upper); ⊓, GT_{1a}. B(right): ⌒, GD_{1b}; ■, GT_{1b}; ●, GD_3; □, GD_2; ▲, GQ.

lipids cerebroside and sulfatide. The deposition of LM_1 practically followed the myelin marker lipids. However, the rate of deposition of other gangliosides was not identical to the myelin marker lipids. These results indicate that GM_3 and GD_3 are the major gangliosides in the immature and proliferating Schwann cells. This is quite in agreement with the previous work which showed that GM_3 and GD_3 are the major gangliosides of non-neuronal cells (Kuhn and Wiegandt, 1964; Yates and Thompson, 1978), fetal rat brain cells (Robert et al., 1977; Irwin et al., 1980) and transformed glial cultured cells (Stoolmiller et al., 1979). GD_3 and GM_3 have been shown to decline as the cells mature in the fetal rat and mouse brain (Irwin et al., 1980; Yavin and Yavin, 1979). Our results are consistent with these observations which show that as Schwann cells mature the mole percentage composition of GM_3 and GD_3 declines. Thus the two gangliosides could represent markers of immature glial cells and kinetic markers of differentiation and maturation of these (Schwann) cells. Irwin et al. (1980) also have concluded that the predominance of GM_3 and GD_3 may be a general feature of early neurogenesis in the CNS of mammals and birds.

Our developmental studies also show that as the nerve matures and deposits myelin, LM_1 accumulates in the nerve which would suggest that LM_1, like cerebroside, is mostly localized in rat PNS myelin. It should be pointed out here that in rat CNS myelin the major ganglioside is GM_1 (Suzuki et al., 1967).

The fatty acid composition of individual isolated ganglioside was determined (Table 1). GM_3 and GD_{1a}, which were resolved into two separate bands on TLC plates, were analyzed separately. In both, the upper band contained a greater proportion of long chain fatty acid (>20:0) than the lower band. The fatty acid composition of ganglioside LM_1 was entirely different from that of GM_1. LM_1 had more than 50% of total fatty acids as long chain fatty acid (>20:0). Significant amounts of long chain fatty acids were present only in GM_3, LM_1 and GD_{1a}; in this respect these gangliosides are different from rat CNS gangliosides in which mostly 18:0 is found. (Suzuki et al., 1968).

TABLE 1. Percentage composition of fatty acids of major gangliosides of adult rat trigeminal nerve.

Fatty acid	GM_3 Upper	GM_3 Lower	LM_1	GM_1	GD_3 Lower	GD_{1a} Upper	GD_{1a} Lower	GD_{1b}	GT1b	GQ
16:1	0.25	0.7	-	-	2.7	-	-	2.0	-	1.0
16:0	2.5	26.1	5.6	6.2	7.3	-	-	6.6	2.7	21.4
18:1	2.0	4.2	2.1	-	0.8	-	1.6	2.2	-	1.2
18:0	5.7	27.3	20.7	54.4	58.9	21.7	64.7	71.0	65.5	62.1
20:1	0.4	1.0	0.9	-	-	-	-	-	-	-
20:0	7.6	18.3	11.0	21.2	19.8	13.6	26.1	13.8	26.2	14.4
22:0	23.4	10.6	20.0	8.8	4.4	36.2	2.5	4.9	4.9	-
23:0	1.4	0.7	-	-	-	-	-	-	-	-
24:1	7.7	1.1	7.9	-	1.9	-	1.3	-	-	-
24:0	36.9	7.5	21.9	9.6	3.2	28.5	2.5	-	-	-
25:0	-	0.6	-	-	-	-	-	-	-	-
26:1	6.2	1.7	2.6	-	-	-	1.5	-	-	-
26:0	5.6	-	-	-	-	-	-	-	-	-
>20:0	81.2	22.2	52.4	18.4	9.5	64.7	7.8	4.9	4.9	-

The fatty acid analysis was done by glc. Average percentage composition of the two different samples of each ganglioside is given. (-), not detectable.

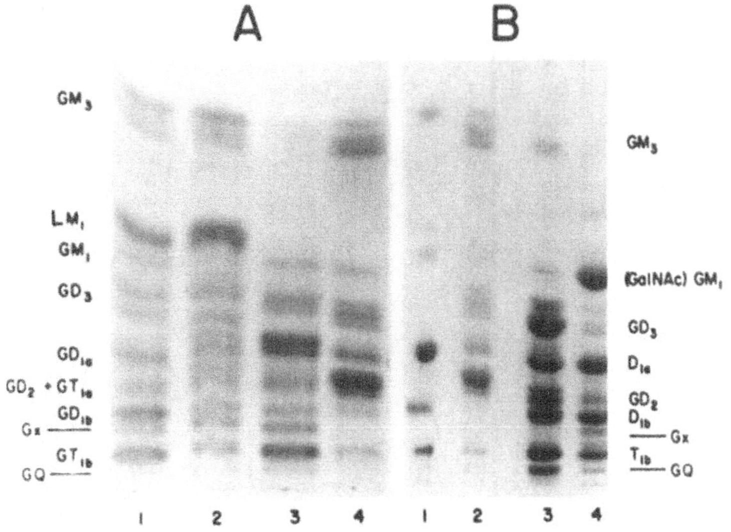

FIG. 4. Comparison of TLC pattern of gangliosides from various neural sources. A: lane 1, adult rat trigeminal nerve; lane 2, adult rat sciatic nerve; lane 3, adult rabbit sciatic nerve; lane 4, adult chicken sciatic nerve. B: lane 1, mixed standards, which included GM_3 (isolated from dog erythrocytes), GM_1, GD_{1a}, GD_{1b}, and GT_{1b} (all from Supelco); lane 2, adult chicken sciatic nerve; lane 3, 9-day-old chicken brain; lane 4, adult mouse brain. Chromatographic conditions were the same as in Fig. 2.

The TLC patterns of gangliosides isolated from sciatic nerve of rat, rabbit and chicken were compared with that of rat trigeminal nerve (Fig. 4). The ganglioside pattern of adult rat sciatic nerve was almost identical to that of adult rat trigeminal nerve, but it was entirely different from other species. LM_1 was not found in rabbit and chicken nerves. GD_{1a} and GT_{1b} were the major gangliosides of the rabbit sciatic nerve, whereas GT_{1a}, GD_3 and GM_3 were the major gangliosides of chicken sciatic nerve. Mouse sciatic nerve also contained some LM_1 (5.2%) but little GM_1 (1.6%). The other major gangliosides were GM_3 (59%) and GD_{1a} (21.1%) (data not shown).

Previous results suggested that LM_1 was primarily localized in myelin, whereas the rest of the gangliosides were in the other subcellular membranes. Myelin, microsomal and a 'membrane fraction' were isolated from fresh adult sciatic nerves by a modified procedure based upon density gradient centrifugations (Koul, et al., 1980). The distribution of marker enzymes in these sub-cellular fractions of sciatic nerve is given in Table 2. Myelin fraction from the nerve contained very little 2':3'-cyclic nucleotide 3' phosphohydrolase activity, which is considered to be a myelin marker enzyme in the CNS. However, it is now becoming evident that this enzyme is not a myelin marker enzyme in the PNS (Matthieu et al., 1980; Crammer and Zimmerman, 1982). The highest activity of this enzyme was, in fact, in the membrane fraction. 5'-nucleotidase is considered to be a plasma-membrane marker enzyme (Stefanovic et al., 1975). This enzyme activity was highest in the membrane fraction followed by that in the microsomal membranes (Table 2). CDP-choline: 1,2 diglyceride choline phosphotransferase and TPNH - cytochrome c reductase are considered to be microsomal membrane marker enzymes. These activities were found to be highest in the microsomal membranes with only negligible amounts in the myelin fractions, (Ruenwongsa et al., 1979). The membrane fraction, however, contained significant amounts of these enzyme activities.

These subcellular membranes were also characterized by their protein profiles on SDS polyacrylamide gel-electrophoresis. Myelin fractions contained only characteristic myelin proteins and did not contain typical high molecular weight proteins found in the microsomal and membrane fractions. Microsomal membranes contained several high molecular weight proteins but only negligible amounts of typical myelin proteins P_0, P_1 and P_2. The membrane fraction contained significant amounts of several high molecular weight proteins and the P_0 protein, but negligible amounts of P_1, P_2 and other basic proteins of myelin.

Table 2. Distribution of marker enzymes in different subcellular fractions of sciatic nerve of rat.

Enzymes	Light Myelin	Heavy Myelin	Membr. Fract.	Micro-somes
Protein (mg)	26.5	27.0	2.7	5.3
2':3' cyclic-nucleotide-3' phosphohydrolase (nmole/mg/min)	0.8	0.7	5.9	1.7
5' Nucelotidase (nmole/mg/min)	3.2	5.9	88.0	46.6
CDP:Choline trans. (nmole/mg/hr)	2.4	1.5	32.9	78.4
TPNH cyto. c red. (o.d/mg/min)	1.8	0.0	121.2	369.0
UDP-gal. trans (nmole/mg/hr)	0.2	0.2	2.5	7.9
Na^+,K^+ ATPase (μmole/mg/hr)	1.4	0.9	5.7	4.8
Mg^{2+},Ca^{2+},ATPase (μmole/mg/hr)	1.0	1.0	69.0	34.5
Acetylcholine esterase (μmole/mg/hr)	0.2	0.1	3.0	2.4

Subcellular fractionation was done with 8.0g of sciatic nerve by a modified procedure. Light myelin was prepared from 10,000 x g supernatant. Heavy myelin from 10,000 x g precipitate. Membrane fraction suspended in 0.85 M sucrose.

The subcellular fractions were also characterized by electron microscopy. Typical myelin-figures were seen for the myelin fraction. The membrane fraction was highly vesicular in nature and did not contain myelin-like figures, whereas the microsomal fraction was vesicular studded with ribosomal particles and contained polysomes.

The mole percent composition of gangliosides of different subcellular fractions of sciatic nerve is given in Table 3. Both myelin fractions contained mostly LM_1 and GM_3, about 65% and 22% of the total gangliosides, respectively. Other gangliosides were present in myelin but in negligible amounts (<5%). GM_1 and GM_3 were the major gangliosides of the microsomal membranes, about 30% each. LM_1 was also present in the microsomal membranes, about 15%. The major ganglioside of the membrane fraction was GM_3, about 40%. This fraction also contained a significant amount of LM_1, 24%. These results would suggest that LM_1 was probably synthesized in the microsomal membranes and was deposited in the myelin membranes via the membrane fraction. The fatty acid composition of GM_3 in the subcellular membranes was interesting. Myelin contained about 80% of long chain GM_3, whereas the microsomal and membrane fraction contained about 64% long chain GM_3.

GM_3 is considered to be a key intermediate in the metabolism of gangliosides. CMP- NeuNAc: ceramide dihexoside sialyltransferase catalyzes the synthesis of GM_3, whereas UDP-N-acetylgalactosamine: GM_3 galactosaminyl transferase converts GM_3 to GM_2, which is then converted to the higher gangliosides GM_1, GD_{1a} and GT_{1a}. The specific activity of both these enzymes was considerably higher in brain than in nerve during the entire developmental period (Fig. 5), reflecting higher amounts of gangliosides in the CNS than in PNS. The relative specific activity of galactosaminyltransferase in brain at 4 days was six times higher than sialyltransferase, suggesting that in brain the conversion of GM_3 to GM_2 is very active. This is consistent with the little amounts of GM_3 found in the brain. In the nerve, the relative specific activity of sialyltransferase was almost 3-4 times higher than galactosaminyltransferase, indicating that during early development of the nerve, more GM_3 is synthesized than

Table 3. Mole percent composition of gangliosides of different subcellular fractions of sciatic nerve.

Ganglioside	Light Myelin	Heavy Myelin	Membr. Fract.	Micro-somes
GM_3-Long chain	17.1	18.2	25.0	21.7
GM_3-Short chain	4.4	5.7	14.6	12.0
GX_2	-	-	1.4	1.0
LM_1 Upperband	-	-	6.9	2.6
LM_1 Lowerband	66.9	63.3	17.4	12.7
GM_1	-	-	8.6	30.4
GD_3	0.5	0.8	4.3	1.1
GD_{1a}	2.5	3.1	-	-
GT_{1a}	1.1	0.7	-	-
GD_2	-	-	1.8	1.5
GD_{1b}	1.9	1.7	6.7	6.6
GX_1	1.0	1.5	0.3	0.1
GT_{1b}	4.2	4.4	9.7	8.5
GQ	0.5	0.7	3.1	1.9
GM_3				
% Short chain	20.0	24.0	37.0	36.0
% Long chain	80.0	76.0	63.0	64.0

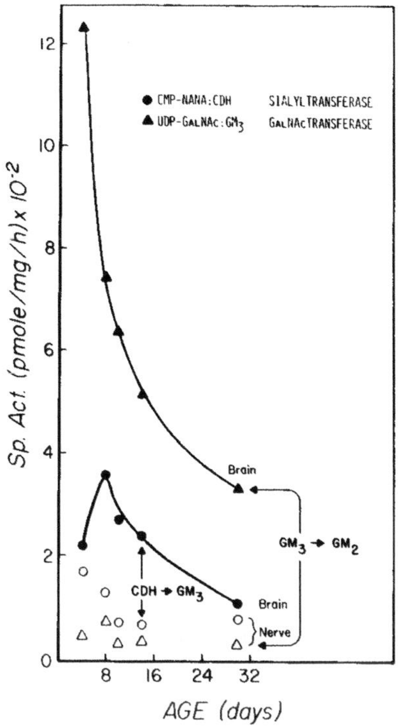

FIG. 5. Activity of CMP-N-acetylneuraminic acid: ceramide dihexoside sialyltransferase (● o) and UDP-N-acetylgalactosamine : GM_3-N-acetylgalactosaminyl-transferase (▲ △) in rat brain and trigeminal nerve during development.

is converted to higher gangliosides via GM_2. The decline in the activity of sialyltransferase with development correlates well with the decline in the amounts of GM_3 in the nerve.

Ethylnitrosourea is a potent carcinogen. A single dose (20-80 mg/kg body weight) administered intravenously to pregnant rats, on the 15th day post-coitus, produced neurogenic malignomas of the brain, spinal cord and cranial and peripheral nerves of the offspring. The sensitivity of the fetuses to the

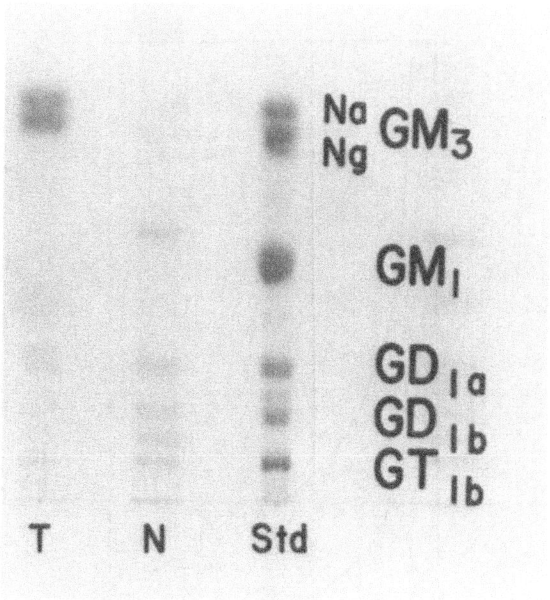

FIG. 6. Thin-layer chromatogram of gangliosides from trigeminal nerve of normal (N) and tumor (T) tissue. Std., mixed standard including Na- and Ng-GM_3 for NeuNAc isolated from dog erythrocyte and for NeuNGly isolated from horse erythrocyte respectively.

drug was highest in the last days before birth. The tumors of the cranial nerves were manifested at about 240 days after birth. The latency period and the location of the tumors could be varied with the dose of the drug and the strain of the rat. In the PNS, tumors mostly occurred in the trigeminal nerve (Chou et al. 1979). These tumors were histologically identified as Schwann cell tumors, called neurinomas. The ganglioside composition of these tumors was determined, and GM_3 accounted for almost 80-98% of the total gangliosides of the neurinomas from the trigeminal nerve (Fig. 6). LM_1 was not found in these tumors, whereas other higher gangliosides, although present, were in much smaller quantities compared to normal nerves.

Carcinogenesis is a multi-step process often involving the appearance of several new cell-populations between the initial target cells and ultimately well-differentiated tumors. In CFE strain of rats, a single exposure of about 25-50 mg of ethylnitrosourea/kg to a pregnant rat on 16-20th day of gestation consistently resulted in almost 80% yield of trigeminal nerve tumors in the offsprings (Swenberg et al., 1975). This consistent site of tumor location in the trigeminal nerve provides a unique opportunity to study the earliest preneoplastic and neoplastic cell-differentiation in the trigeminal nerve as a result of exposure to the chemical carcinogen. We have observed histopathologically distinct hyperplasia in the nerve at about 30 days after a single exposure of the drug ethylnitrosourea. The gangliosides were analyzed only in those trigeminal nerve areas where hyperplasia was noticed histologically at 30,45 and 90 days after the exposure to the drug. Controls included trigeminal nerves from animals of the same age but without the ethylnitrosourea exposure (Table 4). The ganglioside pattern showed that even at 30 days, the mole % of GM_3 was slightly increased in the exposed nerve compared to control, whereas other gangliosides GD_{1b}, GT_{1b} and GQ decreased. This trend and a significant decrease in the amount of LM_1 in the exposed nerve was most obvious at 45 and 90 days after the exposure. The increase in GM_3 of the drug-treated animal nerves was both in short chain and long chain fatty acid-containing GM_3, compared to the corresponding control nerves. The ratio of short chain:long chain fatty acid-containing GM_3 was slightly altered with normal development but considerably altered with the progression of hyperplasia (Table 4). In the normal animals the ratio was 1:4, which was progressively altered in the hyperplastic nerve of the treated animals and was 1:1 in the final tumors.

The synthesis of GM_3 by the enzyme CMP-NeuNAc: ceramide dihexoside sialyltransferase and of GM_2 by UDP-N-acetylgalactosamine: GM_3 galactosaminyl transferase was studied in the normal and ethylnitrosourea treated nerve homogenates (Fig. 7). In the normal nerve the sialyltransferase activity declined with development but in the ethylnitrosourea-exposed nerve the activity of the enzyme increased significantly and in final tumors this activity was about 6-fold

Table 4. Mole percent composition of ganglioside in normal and ENU exposed nerves

Ganglioside	Normal Nerve				ENU-exposed Nerve			
DAYS =	30	45	90	adult	30	45	90	Tumor
GM_3	48.3	46.1	30.3	24.0	55.4	56.9	66.1	98.0
LM_1	19.9	30.6	37.1	28.5	21.7	23.8	20.6	-
G_M1	4.0	1.0	2.7	7.5	1.0	2.1	1.2	-
GD_3	4.5	1.9	1.2	5.6	3.4	1.5	0.8	tr.
GT_{1a}	0.8	0.5	0.7	2.0	0.8	0.4	0.2	tr.
GD_2	1.4	1.3	1.2	3.1	1.2	0.8	0.6	-
GD_{1b}	11.6	9.4	13.1	11.5	9.3	7.0	5.7	2.0
GT_{1b}	7.2	6.7	9.8	8.0	5.5	5.9	3.9	tr.
GQ	2.6	2.4	3.9	2.5	1.8	1.6	0.9	tr.
GM_3								
% Short chain	25	21	22	21	25	28	37	50
% Long chain	75	79	78	79	75	72	63	50

Days represent age of the CFE strain of rats. One day prior to birth the fetuses were exposed transplacentally with a single dose of ENU 20 mg/kg body wt. of the mother. tr.-trace amounts.

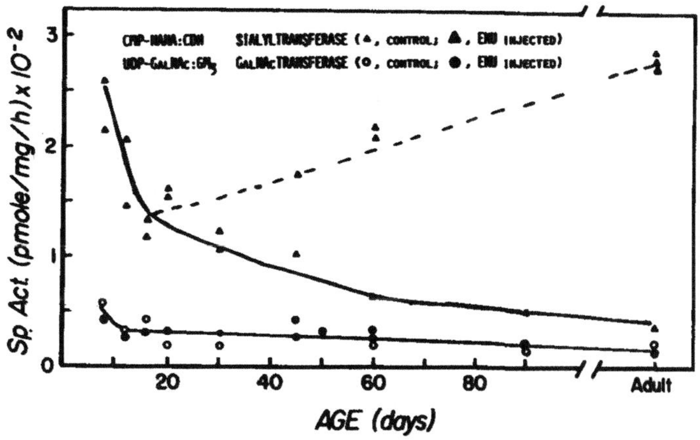

FIG. 7. Biosynthesis of gangliosides GM_3 and GM_2 in the normal and ENU-exposed trigeminal nerve during development.

higher than in the normal nerve. This high enzyme activity in the transformed nerve cells should account for the accumulation of GM_3 in these cells. The galactosaminyl transferase activity, however in the control and transformed nerve cells was similar, which would indicate that the excess of GM_3 in the nerve cells was not being converted to higher gangliosides at a preferred rate than that in the normal nerves.

Although considerable attention has been given to the carbohydrate structure of glycosphingolipids having a role in cellular interaction, differentiation and oncogenesis, the ceramide portion of the molecules has been ignored. Ceramide moiety has been simply regarded as the anchoring point to the membranes. It has been, however, observed that in various membranes the fatty acid and sphingosine moieties of glycosphingolipids vary a great deal, and this molecular heterogeneity may have important structural and dynamic roles in biological membranes. Recently Kannagi et al.(1982) have suggested that the ceramide portion of these lipids plays a role in selection of glycosylation and in defining the carbohydrate reactivity to its ligands. This suggestion is based on the observations that the ceramide composition shows a remarkable correlation with the terminal carbohydrate structure of "lacto series" glycosphingolipids isolated from human erythrocytes. A correlation between ceramide composition and carbohydrate structure among various glycosphingolipids in brain has been well-known. Altered glycosphingolipids in the transformed cells could result from altered fatty acyl composition of the key intermediates via ceramides and GM_3. A breakdown of normal relationship between ceramide and sugar synthesis could lead to pathological processes such as abnormal development, differentiation and oncogenesis (Kannagi et. al., 1982). Further studies on the specificity of the metabolic enzymes for the substrates with long and short chain fatty acid-containing ceramide, ceramide dihexoside and GM_3 should be of interest in understanding cell-differentiation and oncogenesis.

Acknowledgements. This work was supported by grants CA 16853, NS 10437 and HD 05515 from United States Public Health Services.

REFERENCES

Chou, K.H., Ambers, L.S.A. and Jungalwala, F.B. (1979) Ganglioside composition of chemically induced rat neural tumors and characterization of hematoside from neurinomas. J. Neurochem. 33, 863-873.

Crammer, W. and Zimmerman, T.R. (1982) 5'-Nucleotidase in sciatic nerves from adult and young rats: Recovery in myelin is lower than recoveries of 2',3'-cyclic nucleotide-3'-phosphohydrolase and carbonic anhydrase. Neurochem. Res. 7,229-234.

Cravioto, H., Weiss, J.F., Weiss, E., Goebel, H.H. and Ransohoff, J. (1973) Biological characteristics of peripheral nerve tumors induced with ethylnitrosourea. Acta Neuropathol. 23, 265-280.

Fong, J.W., Ledeen, R.W., Kundu, S.K. and Brostoff, S.W. (1976) Gangliosides of peripheral nerve myelin. J. Neurochem. 26, 157-162.

Hakomori, S. (1981) Glycosphingolipids in cellular interaction, differentiation and oncogenesis. Ann. Rev. Biochem. 50, 733-764.

Irwin, L.N., Michael, D.B. and Irwin, C.C. (1980) Ganglioside patterns of fetal rat and mouse brain. J. Neurochem. 34, 1527-1530.

Kannagi, R., Nudelman, E. and Hakomori, S. (1982) Possible role of ceramide in defining structure and function of membrane glycolipids. Proc. Natl. Acad. Sci. 79, 3470-3474.

Koul, O., Chou, K.H. and Jungalwala, F.B. (1980) UDP - galactose ceramide galactosyltransferase in rat brain myelin subfractions during development. Biochem. J. 186, 959-969.

Kuhn, L. and Wiegandt, H. (1964) Weitere ganglioside aus meuschenhirn. Z. Naturforsch. 196, 256.

Ledeen, R.W. (1979) Structure and distribution of gangliosides in complex carbohydrates of the nervous system. Ed. by R.U. Margolis and R.K. Margolis, Plenum Press, N.Y. p. 1-43.

Li, Y.T., Mansson, J.E., Vanier, M.T. and Svennerholm, L. (1973) Structure of the major glucosamine - containing ganglioside of human tissues. J. Biol. Chem. 248, 2634-2636.

Matthieu, J.M., Costantino-Ceccarini, E., Beny, M. and Reigner, J. (1980) Evidence for the association of 2',3'-cyclic-nucleotide 3'-phosphodiesterase with myelin related membranes in peripheral nervous system. J. Neurochem. 35, 1345-1350.

Robert, J., Rebel, G. and Mandel, P. (1977) Glycosphingolipids from cultured astroblasts. J. Lipid Res. 18, 517-522.

Ruenwongsa, P., Singh, H. and Jungalwala, F.B. (1979) Protein-catalyzed exchange of phosphatidylinositol between rat brain microsomes and myelin. J. Biol. Chem. 19, 9385-9393.

Stefanovic, V. Mandel, P. and Rosenberg, A. (1975) Concanavalin A inhibition of ecto - 5'-nucleotidase of intact cultured C_6 glioma cells. J. Biol. Chem. 250 7081-7083.

Stoolmiller, A.C., Dawson, G., Kemp, S.F. and Schachner, M. (1979) Synthesis of glycosphingolipids in mouse glial tumors. J. Neurochem. 32, 637-641.

Suzuki, K., Poduslo, S.E. and Morton, W.T. (1967) Gangliosides in the myelin fraction of developing rat. Biochim. Biophys. Acta 144, 375-381.

Suzuki, K., Poduslo, J.F. and Poduslo, S.E. (1968) Further evidence for a specific ganglioside fraction closely associated with myelin. Biochim. Biophys. Acta 152, 576-586.

Swenberg, J.A., Clendenon, N., Denlinger, R. and Gordon, W.A. (1975). A sequential development of ethylnitrosourea induced neurinomas: Morphology, biochemistry and transplantability. J. Natl. Cancer Inst. 55, 147-152.

Svennerholm, L., Bruce, A., Minsson, J., Rynmark, B. and Vanier, M. (1972) Sphingolipids of human skeletal muscle. Biochim. Biophys. Acta. 280, 626-636.

Yates, A.J., Bauchard, J.P. and Wherrett, J.K. (1976) Changes in the sciatic nerve of the rabbit and its tissue constituents during development. J. Neurochem. 23, 993-1003.

Yates, A.J. and Thompson, D.K. (1978) Ganglioside composition of peripheral nerve undergoing Wallerian degeneration. J. Neurochem. 30, 1649-1651.

Yavin, E. and Yavin, Z. (1979) Ganglioside profile during neural tissue development. Dev. Neurosci. 2, 25-37.

Yogeeswaran, G. (1983) Cell surface glycolipids and glycoproteins in malignant transformation. Adv. Cancer Res. 38, in press.

Part V: Abstracts

BENZODIAZEPINE-BINDING SITE AND GABA-BINDING SITE

Li-ren Chang and Eric A. Barnard

Department of Biochemistry, Imperial College, London, U.K. and Department of Pharmacology, National Defense Medical Center, Taipei, Taiwan R.O.C.

ABSTRACT

There is a body of experimental evidence to suggest that the benzodiazepine-binding site, the GABA-binding site and the chloride ionophore are functionally correlated. To elucidate this issue, the interactions among these components in both membrane-bound state and soluble state are examined. The results showed that 1% Triton X-100 extract of rat forebrain membranes possesses both ^{3}H-flunitrazepam binding activity and ^{3}H-muscimol binding activity, which co-migrate on sucrose gradients with a sedimentation coefficient of 9.5S, and both activities have the same isoelectric point of pI5.0. By electron irradiation inactivation of intact membranes, it was shown that both activities are lost at precisely the same rate, and that the complex has a molecular size in the membrane of about 220,000 daltons. The findings are compared with a number of related studies reported in the literature and a proposed model of supramolecular unit responsible for the pharmacodynamics of benzodiazepines in the mammalian central nervous system is discussed.

RESPONSES OF THE SINGLE SYMPATHETIC GANGLIONIC NEURONES TO ACETYLCHOLINE AND TO DOPAMINE

R.S. Hsu and B. Libet

Department of Physiology, School of Medicine,
University of California, San Francisco, CA
94143 USA.

ABSTRACT

The responses of the single superior cervical ganglion cells of the rabbits to acetylcholine (ACh) and to dopamine (DA) applied iontophoretically and in bulk were studied by the intracellular recording techniques. ACh elicited both nicotinic fast (f-) and muscarinic slow (s-) depolarization (DP) in about 30% of the principal ganglion cells perfused with d-tubocurarine. ACh elicited f-DP alone in about 20% of cells and s-DP in about 50%. ACh could elicit two forms of hyperpolarizing (HP) responses in the same cells that responded with f-DP and/or s-DP. Pretreatment with bethanechol (BCh) tended to reduce the short-latency HP response to ACh; pretreatment with bretylium eliminated the longer-latency HP response, while pretreatment with both BCh and bretylium essentially eliminated all HP responses to ACh. Neither f-DP nor longer-latency HP responses were observed in cells tested with BCh. It is suggested that the short-latency HP response is mediated by a muscarinic release of DA from SIF cell interneurones, and the longer-latency HP response is mediated by a non-muscarinic release of noradrenaline probably from dendrodendritic junctions between principal ganglion cells.

DA applied iontophoretically induced HP responses in many cells that exhibited no HP response to ACh and the

DA-induced HP responses could be observed regardless of the depth of cells lying in the ganglion. DA added in bulk to the perfusing medium could apparently elicit an HP response in principal ganglion cells located in the outer 30 µm of the ganglion, but not in deeper-lying cells. The lack of response to DA of the deeper-lying cells may be due to the inactivation of diffusing DA by COMT.

That the receptors on the principal ganglion cells responsible for each type of response to ACh are distributed in a differential spatial manner relative to the position of the ACh delivery tip is suggested to explain the variability of responses to ACh of the principal ganglion cells.

EFFECT OF CONVULSIVE AND ANTI-CONVULSIVE AGENTS ON BRAIN FREE FATTY ACIDS AND DIACYLGLYCEROLS.

M.Z. Huang and Grace Y. Sun

The Institute of Neurosciences, National Yang-Ming Medical College, Taipei, Taiwan, R.O.C. and Sinclair Research Farm, Univ. of Missouri, Columbia, MO 65201 USA

ABSTRACT

Brain free fatty acids (FFA) and diacylglycerols (DG) are maintained in a dynamic equilibrium with the membrane phospholipids. Many factors, including ischemia and drug induced convulsion, are known to disturb the equilibrium resulting in alteration of the FFA and DG level. The biochemical mechanism responsible for the disturbances is not yet fully understood. In this study, the FFA and DG in brain due to carbamylcholine and pentylenetetrazol induced convulsion and the suppressive action by scopolamine and diazepam was investigated. Carbamylcholine injected into rat brain intracerebrally gave rise to clonic and tonic seizures as well as exhibited some of the cholinergic symptoms. An increase in FFA and DG level was observed. The increase in FFA (but not DG) was blocked by pretreatment of scopolamine. Scopolamine injection (i.p.) alone could reduce the FFA level in brain but not DG. In all cases, subjecting the brain to 2 min post-decapitative ischemia resulted in a large increase in FFA and DG. Carbamylcholine stimulation potentiated the ischemia-induced increase in FFA and DG. Under this type of stimulation, arachidonic and stearic acids were preferentially released during stimulation and ischemia. Seizure was observed within two min after pentylenetetrazol administration (i.p.) to young rats. This drug also induced an increase in brain FFA but apparently was not effective in modulating the DG level. Pretreatment of rats with diazepam could effectively reduce the pentylenetetrazol-induced convulsion and FFA increase. A decrease in both FFA and DG was observed when rats were treated with diazepam (i.p.) alone. In general, FFA level in brain is increased during stimulation and decreased during depression. The differences in the FFA and DG changes as shown by convulsive and anti-convulsive agents further suggest that multiple biochemical mechanisms are present in mediating these changes.

LYSOPHOSPHOLIPASE IN BRAIN SYNAPTOSOMES

Steven F.L. Huang and Grace Y. Sun

Neuroscience Institute and Biochemistry Department, National Yang-Ming Medical College, Taipei, Taiwan, R.O.C. and Sinclair Research Farm, University of Missouri, Columbia, MO 65201 USA

ABSTRACT

Lysophospholipids are catalytic products of phospholipase A_1 and A_2. Their physiological significance in modulating membrane functions has been well recognized on account of their detergent properties. Lysophospholipids in brain are normally maintained at a very low level because two enzymic mechanisms are responsible for their degradation: (1) via reacylation back to the membrane phospholipids, and (2) further degradation by lysophospholipase. The lysophospholipase in brain is present in the membranous as well as cytosolic fractions. In brain synaptosomes, this enzyme exhibited a pH optimum of 8.0. It is sensitive to inhibition by both taurocholate and deoxycholate, but low level of Triton X-100 can stimulate its activity. The enzyme does not depend on calcium for activity; however, most divalent metal ions were potent inhibitors of the enzyme. The lysophospholipase did not show obvious specificity with respect to substrates with different polar head groups, but it preferred hydrolysis of saturated and monounsaturated acyl groups than the polyunsaturated acyl groups. Results here have indicated the role of lysophospholipase in protecting neural membranes from damage due to lysophospholipid accumulation. The coupled action of phospholipases A_1 and A_2 and lysophospholipase may be the underlying factor leading to the enormous amount of free fatty acid increase in membrane during brain stimulation and ischemia.

BUTYRATE-INDUCED REVERSE TRANSFORMATION OF CLONAL GLIOMA CELLS *IN VITRO*

Li-wen Ko

Institute of Neurosciences, National Yang-Ming Medical College, Taipei, Taiwan, R.O.C.

ABSTRACT

Clonal glioma (F98) cells were derived from an undifferentiated neurogenic tumor transplacentally induced by N-ethyl-N-nitrosourea in the CD Fischer rat. *In vitro*, these cells retain the characteristic autonomous proliferation and anaplastic phenotypic traits of the original neoplasm.

The unrestrained proliferative behavior of the glioma cells was arrested with morphologic features indicative of restoration of contact inhibition following butyrate treatment. Aquisition of additional morphologic and morphometric characteristics, such as appearance of intercellular junctional complexes, elaborate decoration of cytoskeletal networks, engorgement of dilated and interconnecting rough endoplasmic reticulum, fine dispersion of nuclear chromatin and marked reduction in nuclear-cytoplasmic ratio, is highly suggestive that the butyrate treatment has not only eliminated the malignant phenotypes, but also provoked the expression of normal biological traits of glia in these neoplastic cells. In addition, butyrate also prompted the *de novo* biosynthesis of macro-molecular constituents associated with glial differentiation, such as S-100 protein, glial fibrillary acidic protein and phospholipids with polyunsaturated fatty acids. It is of great interest that neuron-specific enolase, a glycolytic enzyme characteristic for neuronal maturation, was found to be expressed and activated in these glioma cells in the presence of butyrate. Furthermore, the behavioral and structural alterations correlate temporally with the induction of ornithine decarbosylase activity and of hyperacetylated nucleosomal histones following butyrate exposure.

The butyrate-induced reverse transformation of clonal glioma cells in vitro thus provides a very unique model system for studying cell proliferation and differentiation. The reversibility of butyrate-motivated phenomena further adds another attractive feature to the versatile model for the elucidation of regulatory mechanisms for genomic and phenotypic expressions of eukaryotes.

[1]L. Ko is a recipient of the Ching-Ling Basic Medical Research Chair Award (1981-1982).

[2]This investigation was supported by the grants from the National Science Council (NSC71-0412-B010-03) and from the Ching-Ling Neuroscience Foundation.

BUTYRATE-INDUCED EXPRESSION OF SELECTED GLIAL MARKERS BY CLONAL GLIOMA (F98) CELLS IN VITRO

Hon-reen Kuo and Li-wen Ko[1,2]

The Institute of Neurosciences, National Yang-Ming Medical College, Taipei, Taiwan, R.O.C.

ABSTRACT

The phenotypic expression of two glia-specific markers, S-100 protein and glial fibrillary acidic protein (GFAP), during exposure to sodium butyrate (SB) at 2.5 mM by clonal glioma cells was studied.

By immunocytochemical demonstration with the unlabeled antibody technique of peroxidase anti-peroxidase (PAP) reaction in combination with the phosphotungstic acid-hematoxylin (PTAH) staining method, specific for glial fibers, there was a significant increase in expression of both markers provoked in these anaplastic glioma cells by 48 hours following butyrate treatment. In contrast with the perinuclear localization of S-100 protein, GFAP was found primarily associated with the filamentous cytoskeletal network extending radially from the juxtanuclear region to the peripheral cytoplasm and protruding processes.

Experiments are now in progress to determine the significance of enhanced expression of these markers, normally associated with ontogenic differentiation of glial elements, by undifferentiated tumor cells in the presence of SB *in vitro*. It is the major objective of this investigation to provide experimental evidence for the *in vitro* morphological and biochemical differentiation accompanying reverse transformation of anaplastic glioma cells induced by butyrate. This should be of fundamental importance in designing prospective therapeutic regimens for gliomas with desired efficacy and specificity in the future.

[1] L. Ko is a recipient of the Ching-Ling Basic Medical Research Chair Award (1981-1982).

[2] This investigation was supported by grants from the National Science Council (NSC71-0412-B010-03) and from the Ching-Ling Neuroscience Foundation, the Republic of China.

ARACHIDONOYL-LABELED SYNAPTOSOMES: A MEMBRANE-SUBSTRATE FOR STUDYING PI-SPECIFIC PHOSPHOLIPASE C

Victor T.S. Leu and Grace Y. Sun

The Institute of Neurosciences and Biochemistry Department, National Yang-Ming Medical College, Taiwan and Sinclair Research Farm, Univ. of Missouri, Columbia, MO USA

ABSTRACT

Enhanced labeling of phosphatidylinositol (PI) in synaptosomes was achieved by incubating freshly isolated synaptosomes with [1-C^{14}] arachidonate in the presence of Mg^{++}-ATP, CoA and a small amount of lyso-glycerophosphoinositol (LPI). Under this incubation condition, labeled arachidonate was preferentially incorporated into PI. Washing the prelabeled synaptosomes with bovine serum albumin (BSA) and subsequently heating the membranes at 80°C for 5 min resulted in a membrane fraction enriched in labeled PI but devoid of enzymic activity for metabolizing the substrate. This membrane preparation was used to probe the PI-specific phospholipase C (PLC) activity in cells, body organs, and fluids. When prelabeled membranes were incubated with freshly isolated synaptosomes in the absence of Ca^{++}, practically no PLC activity was detected. PLC activity was low even in the presence of Ca^{++}; however, activity could be maximally activated by deoxycholate. The increase in labeled diacylglycerol in the presence of Ca^{++} and deoxycholate was attributed mainly to hydrolysis of PI. PLC in brain is present in both membranous and cytosolic fractions. Among the brain subcellular fractions, PLC activity is higher in plasma membranes than in microsomes and synaptic vesicles. Myelin membrane has practically no activity. Results show that PLC is ubiquitous in body organs and fluids but is mainly present in a dormant mode. The physiological significance of the dormant nature of PLC remains to be investigated.

DIFFERENTIAL EFFECTS OF DIVALENT CATIONS ON NERVE AND MUSCLE MEMBRANES

Shoei-Yn Lin-Shiau and Wei-Mei Fu

Pharmacological Institute, College of Medicine, National Taiwan University, Taipei, Taiwan, ROC.

ABSTRACT

Many divalent cations are known to influence biological activities. Among them, calcium is essential to many cellular functions. Elucidation of mechanism of action of these divalent cations may shed light on the molecular basis of cellular activities in the living organisms. In a serial study of effects of nine divalent cations (e.g., UO_2^{2+}, Cd^{2+}, Zn^{2+}, Mn^{2+}, Co^{2+}, Ni^{2+}, Cu^{2+}, Ba^{2+} and Sr^{2+}) on the neuromuscular transmission of the skeletal muscles, we found that these cations exerted different effects on the nerve and muscle transmission process. UO_2^{2+} was a unique cation capable of increasing acetylcholine release from nerve endings, while Cd^{2+} inhibited the release. Zn^{2+}, Mn^{2+}, Co^{2+} and Ni^{2+} not only inhibited transmitter release but also depressed muscle membrane functions. Both Cu^{2+} and Ba^{2+} induce muscle contractions prior to the blockade of neuromuscular transmission. We also examined the interaction of these cations with synaptosomes, liposomes and sarcoplasmic reticulum (SR) of the skeletal muscle. These cations inhibited $^{45}Ca^{2+}$ uptake and $(Na^{+}+K^{+})$-ATPase of synaptosomes, Ca^{2+}-ATPase of SR, and phospholipase A_2 and 5'-nucleotidase of <u>Naja naja</u> atra venom. In general, $(Na^{+}+K^{+})$-ATPase was the most and 5'-nucleotidase was the least sensitive to these cations. UO_2^{2+} and Cd^{2+} were most potent in inhibition of Ca^{2+}-uptake and $(Na^{+}+K^{+})$-ATPase of synaptosomes. Zn^{2+} markedly inhibited $(Na^{+}+K^{+})$-ATPase and phospholipase A_2 while Cu^{2+} was rather selective in inhibiting Ca^{2+}-ATPase of SR. Mn^{2+} and Ba^{2+} were only moderately or slightly active in depressing these activities. Interaction of these cations (except UO_2^{2+}) with synaptosomes, SR and liposomes enhanced the fluorescence intensity of 1-anilino-8-naphthalene sulfonate (ANS). Possible interaction of these cations with sulfhydryl or phosphoryl groups of membrane proteins and phospholipids was also explored. A proposed model for the mode of action of these cations on the cell membrane will be presented.

CHANGES IN PHOSPHOGLYCERIDES AND THEIR ACYL GROUP PROFILES OF GLIOMA CELLS DUE TO BUTYRATE TREATMENT

Synthia H. Sun, Grace Y. Sun and Li-wen Ko

The Institute of Neurosciences, National Yang-Ming Medical College, Taipei, Taiwan, R.O.C.

ABSTRACT

Butyrate treatment of F98 clonal glioma cell line is known to induce cell differentiation and result in expression of new phenotypes (Ko & Koestner, J. Natl. Cancer Inst. 65(5):1017-1027, 1980). In this experiment, the phospholipids and their acyl group profiles of cells grown in the presence and absence of butyrate (2.5 mM) were compared. The phospholipids in F98 glioma cells resembled those in tumors and contained an unusually high proportion of phosphatidylcholine in plasmalogen form (PCpl). Practically no cerebroside was observed. Butyrate treatment resulted in an increase in PC and phosphatidylinositols (PI) and a decrease in PCpl. PC in these glioma cells are comprised mainly of 16:0 and 18:1. Butyrate treatment resulted in a decrease in 16:0 and an increase in 18:1. PE of the glioma cells are enriched in 16:0, 18:0 and 18:1, and butyrate treatment resulted in a decrease in 16:1 and 20:4 but an increase in 18:0. The PCpl in glioma cells have an unusually high level of 16:0 and this fatty acid decreased upon butyrate treatment. The PI in glioma cells are practically devoid of 20:4 but upon butyrate treatment, there was a marked increase in this fatty acid. The PS in glioma cells are high in 18:0 and 18:1, and upon butyrate treatment, there was a decrease in 18:0 and an increase in 18:1. These results indicate that butyrate treatment has induced changes in phospholipids and their acyl group profiles towards a more differentiated state similar to those occuring during early brain development. (Supported in part by Chinese National Science Council and The Ching-Ling Foundation.)

THE ACTION OF THYROID HORMONES ON NEURAL MEMBRANES

Sue Yu and Albert Y. Sun

Institute of Neurosciences, National Yang-Ming Medical College, Taipei, Taiwan, R.O.C.

ABSTRACT

The action of thyroid hormones on heart resembles those of β-adrenergic stimulation. However, the negative feedback mechanism of thyroid hormones on thyroid stimulating hormone (TSH) secretion is still not clear. The purpose of this investigation is to test the hypothesis that the effect of thyroid hormones on central nervous system is related to norepinephrine (NE) stimulation. The effects of thyroxine (T_4) and tri-iodothyroxine (T_3 on NE-uptake was examined with synaptosomal particles and NE-bindings were studied with crude synaptic plasma membrane preparation from rat brain. Both T_3 and T_4 inhibited NE-uptake in a dose-response manner. However, T_3 and T_4 at low concentrations were found to enhance α- and β-adrenergic receptor binding. The result of this study demonstrated that thyroid hormones can regulate the action of norepinephrine, possibly through a concerted action involving pre- and post-synaptic mechanism. Furthermore, the inhibition of NE-uptake and the increase in receptor binding may be responsible for the increased neural sensitivity in the hyperthyroid state.

Index

E

F

G

R

S

T

V

GPSR Compliance

The European Union's (EU) General Product Safety Regulation (GPSR) is a set of rules that requires consumer products to be safe and our obligations to ensure this.

If you have any concerns about our products, you can contact us on ProductSafety@springernature.com

In case Publisher is established outside the EU, the EU authorized representative is:

Springer Nature Customer Service Center GmbH
Europaplatz 3
69115 Heidelberg, Germany

Batch number: 10755248

Printed by Printforce, the Netherlands